Physical Deposition Methods for Films and Coatings

Physical Deposition Methods for Films and Coatings

Lech Pawłowski
Faculty of Science and Technology
Institute of Research for Ceramics
University of Limoges
Limoges
France

WILEY

This edition first published 2025

Registered Office(s)
John Wiley & Sons, Inc., 111 River Street, Hoboken, NJ 07030, USA
John Wiley & Sons Ltd, New Era House, 8 Oldlands Way, Bognor Regis, West Sussex, PO22 9NQ, UK

For details of our global editorial offices, customer services, and more information about Wiley products visit us at www.wiley.com.

Wiley also publishes its books in a variety of electronic formats and by print-on-demand. Some content that appears in standard print versions of this book may not be available in other formats.

Library of Congress Cataloging-in-Publication Data applied for

Hardback ISBN 9781119713067

Cover Design: Wiley
Cover Image: SEM image of a cross-section of Cr3C2 – NiCr coating obtained by APS method
© Lech Pawłowski

Set in 9.5/12.5pt STIXTwoText by Straive, Pondicherry, India

This volume is dedicated to my wife Muryel and to our daughter Irene with all my gratitude for their support and patience.

Contents

Preface

Since the publication by Presse Polytechnique et Universitaire Romandes (PPUR) of the French edition of this book in 2003 under the title *Dépôts Physiques*, the techniques of physical methods of films and coatings deposition changed considerably. Many new countries started to develop and produce in this field, and the author hopes to reach these new readers. The economic growth of these methods has also resulted in new industrial applications of films and coatings. The new applications came from traditionally innovative fields such as microelectronics but also from the new domain grown rapidly in recent years such as biomaterials. Consequently, the content of the English edition covers these new fields and new applications.

Chapter 1 that describes the preparation of the substrates prior to films and coatings deposition. The chemical and physical properties of zones close to surfaces of substrates made of different types of materials were carefully described. The modern methods of cleaning and activation of surfaces using photons emitted by different types of industrial lasers were added to the methods using electric discharges. Also, methods describing the patterning of films and coating using different lithographic processes were described.

Some new atomistic films deposition techniques such as atomistic layer deposition (*ALD*) or plasma spray physical vapor deposition (*PS PVD*) were added to the second chapter. As the granular techniques using liquid feedstock, such as suspension plasma spraying (*SPS*) or solution precursor plasma spraying (*SPPS*), were intensively tested in recent years, they are also described in the present edition. An important activity, especially in research and development, concerning *additive manufacturing* being a part of bulk deposition methods was also considered in the chapter.

Chapter 3 includes the description of new knowledge concerning the simulation of atomistic layers' growth. Similarly, the phenomena of coating growth at the use of cold-gas spraying method (*CGSM*) and the splashing of micro- and nano-sized particles during liquid feedstock thermal spraying resulting in possible generation of columnar microstructure are part of the new knowledge added to this chapter. Finally, the description of bulk coatings' nucleation includes detailed description of the phenomena taking place at the deposition.

The methods of films and coating characterizations described in Chapter 4 include much more examples visualizing the usefulness of described methods. Some methods such as characterizing of the contact angle of a droplet on a solid surface or *in vitro* and in *vivo* testing of biomaterials were added.

Finally, Chapter 5 includes the description of properties of films and coatings obtained with new materials or new methods of deposition. Some new properties such as electron emission or results of biomaterials testing were also considered.

The author hopes that the readers from academia including students and professors as well as engineers and managers from industry will appreciate this edition of book in learning and keeping abreast with the development of films and coatings technologies.

June 2024

Lech Pawłowski
Professor Emeritus
University of Limoges
Faculty of Science and Technology
Institute of Research for Ceramics

Abbreviations and Symbols

Acronyms

1SLD One-step laser deposition
2SLD Two-step laser deposition
2D Two dimensions
3D Three dimensions
45S5 bioglass having composition in wt. % of 45 % SiO_2 + 24.5 % Na_2O + 24.5 % CaO + 6 % P_2O_5
ac alternative current
acac acetylocetonate, $C_5H_7O_2$
amu atomic mass unit defined as 1/12 of mass of carbon atom
AES Auger electrons spectroscopy
AFM Atomic force microscope
ALD Atomic layer deposition
ALE Atomic layer epitaxy
apatite Phosphates minerals such as hydroxyapatite (*HA*), fluorapatite, and chlorapatite
APS Atmospheric plasma spraying
ARE Activated reactive evaporation
AS Arc spraying
ASTM American Society for Testing and Materials (since 2001 *ASTM International*)
BCC Body-centered cubic
BARC Bottom organic antireflective coating
BEO Buffered oxide etch
BET theory of absorption of gas by the solid surface developed by S. Brunauer, P.H. Emmet, and E. Teller[1] being the base of determination for the specific surface and open porosity
Bronze alloy of Cu + (10–12) wt. % Zn

1 Published in a paper: S. Brunauer, P.H. Emmett, and E. Teller, 1938, Adsorption of gases in multimolecular layers, *J. Am. Chem. Soc.*, 60, 309.

CA	Contact angle
CBN	Cubic boron nitride, c-BN
CCD	Charge-coupled device
CD	Critical dimension of an element formed by lithography
CGSM	Cold-gas spray method
CIE	International Commision on Illumination
Colmonoy 5	alloy Ni + 12.6 wt. % Cr + 4.6 wt. % Fe + 4.0 wt. % Si + 2.5 B + 0.5 C
CST	Coefficient of surface tension, $CST = \frac{d\gamma}{dT}$
CSZ	Calcium oxide stabilized zirconia, e.g. *30CSZ* is ZrO_2 + 30 wt. % CaO
CVD	Chemical vapor deposition
CRT	Cathode ray tube
CZTS	Cu_2ZnSnS_4
cw	Continuous wave
dc	Direct current
DCB	Double cantilever beam
DLC	Diamond-like carbon
dmamb	1-dimethylamino-2-methyl-2-butoxy, $C_{14}H_{32}N_2O_2$
DoE	Design of Experiments
DP	Degree of polymerization
dpm	2,2,6,6-tetramethyl-3,5-heptane dione, $C_{11}H_{19}O_2$
DS	Detonation spraying
DSC	Differential scanning calorimetry
DT	Double torsion
DTA	Differential thermal analysis
DUV	Deep ultraviolet
EB CVD	Electron beam chemical vapor deposition
EB PVD	Electron beam physical vapor deposition
EBSD	Electron backscattered diffraction
e – beam	Electron beam
ECR	Electron cyclotron resonance
EDX	Energy-dispersive X-ray spectroscopy
EELS	Electron energy loss spectroscopy
EMPA	Electron microprobe analysis
em	Electromagnetic
FCC	Face-centered cubic
FGC	Functionally graded coatings
formalin	aqueous solution of formaldehyde, CH_2O
FS	Flame spraying
FWHM	Full width at half maximum: the parameter that describes the peaks
H13	Tool steel alloy containing molybdenum, chromium and vanadium
HA	Hydroxyapatite, $Ca_{10}(PO_4)_6(OH)_2$
HAZ	Heat-affected zone
HCP	Hexagonal close packed
HDPE	High-density polyethylene
hfac	hexafluoroacetylacetone, $C_5H_2F_6O_2$

HF CVD	Hot filament chemical vapor deposition
HRTEM	High-resolution transmission electron microscope
HB	Brinell hardness
HK	Knoop hardness
HM	Martens hardness
HR	Rockwell hardness
HV	Vickers hardness
HVOF	High-velocity oxy-fuel
GXRD	Grazing incidence and asymmetric X-ray diffraction
IBAD	Ion-beam-assisted deposition
i – beam	Ion beam
IBM	International Business Machines
IC	Integrated circuit
ICP	Inductively coupled plasma
ID	Internal diameter
Inconel 625	alloy Ni + 21.4 wt. % Cr + 9.1 wt. % Mo + 3.6 wt. % Nb
IP	Ion plating
IP PLD	Ion plating pulsed-laser deposition
IPS	Inert plasma spraying
ir	Infrared
ITO	mixture of In_2O_3 with 90 wt. % % of SnO_2
ITSC	International Thermal Spraying Conference
JCPDS	Joint Committee on Powder Diffraction Standards
LCVD	Laser chemical vapor deposition
LDPE	Low-density polyethylene
LP CVD	Low-pressure chemical vapor deposition
LPPS	Low-pressure plasma spraying
LSA	Laser surface alloying
LST	Laser surface texturing
MF	Mechanofusion
MIG	Metal inert gas
MIP	mercury intrusion porosimetry
MMC	Metal matrix composites
MO CVD	Metalorganic chemical vapor deposition
MOS	Metal oxide semiconductor
NA	Numerical aperture
Nd: YAG	Neodymium-doped yttrium aluminum garnet, $Nd{:}Y_3Al_5O_{12}$
Norem02	alloy Fe + 24.8 wt. % Cr + 4.6 wt. % Mn + 3.5 wt. % Ni + 3.0 wt. % Si + 2.1 wt. % Mo + 0.027 wt. % O
Novolak	polymer derived from phenol and formaldehyde used as photo-resist
OM	Optical microscopy
OM CVD	Organometallic chemical vapor deposition
PAA	Polyacrylic acid
PBTCA	2-Phosphonobutane-1,2,4-tricarboxylic acid
PC	Polycarbonate

PE	Polyethylene
PE ALD	Plasma-enhanced atomic layer deposition
PE CVD	Plasma-enhanced chemical vapor deposition
PEEK	Polyether ether ketone
PEI	Polyethyleneimine
PE PVD	Plasma-enhanced physical vapor deposition
PI	Polyimide
PLD	Pulsed laser deposition
PM	Photo-mask
PMMA	Poly (methyl methacrylate)
PP	Polypropylene
PR	Photo-resist
PS	Polystyrene
PS PVD	Plasma spray physical vapor deposition
PTFE	Polytetrafluoroethylene (trade name Teflon™)
PTA	Plasma-transferred arc
PTAW	Plasma transferred arc welding
PVD	*Physical vapor deposition*
RE PVD	Reactive atmosphere physical vapor deposition
PU	Polyurethane
PVC	Polyvinyl chloride
R&D	Research and development
RF	Radio frequency
RIE	Reactive ion etching
RMS	Root mean square
RS	Reactive sputtering or Raman spectroscopy
RT	Room temperature
SAD	Selected area diffraction
SAP	Selected area processing
SBF	Simulated body fluid
SDS	Sodium dodecyl sulfate
SE	spectroscopic ellipsometers
SEM	Scanning electron microscope
SHS	Self-propagating high-temperature synthesis
SOFC	Solid oxide fuel cells
SP	Spray pyrolysis
SPPS	Solution precursor plasma spraying
SPS	Suspension plasma spraying
Stellite 6	alloy of composition Co + (27–32 wt. %) Cr + (4–6 wt. %) W + (0.9–1.4 wt. %) C + Si, Fe, Ni, Mn, Mo
Stellite 21	alloy Co + 26.5 wt. % Cr + 5.6 wt. % Mo + 3.7 wt. % Ni + 1.2 wt. % Si + 1.4 wt. % Fe + 0.2 wt. % C
STM	Scanning tunneling microscope
STEM	Scanning transmission electron microscope
SZM	Structure zone model

TAT	Tensile adhesion test
TBC	Thermal barrier coating
TCP	Tricalcium phosphate, $Ca_3(PO_4)_2$
TCR	Temperature coefficient of resistance, Ohm/K
TEA	Transversely excited atmospheric
TEC	Thermal expansion coefficient, 1/K
TEM	Transverse electromagnetic mode, TEM_{00} , is the mode with Gaussian distribution of energy or transmission electron microscope
tetrahydrofuran	C_4H_8O
TGA	Thermo-gravimetric analysis
thd	2,2,6,6-tetramethylheptane-3,5-dionate
TIG	Tungsten inert gas
TOREXD	Total reflected X-ray diffraction
uv	ultraviolet
VLSI	Very large scale of integration
VPS	Vacuum plasma spraying
VSM	Vibrating sample magnetometer
WDX	Wavelength dispersive X-ray spectroscopy
WF	Work function, eV
WSP	Water-stabilized plasma
XPS	X-ray photoelectron spectroscopy
XRD	X-ray diffraction
XRF	X-ray fluorescence
YBCO	$YBa_2Cu_3O_{7-x}$ high-temperature superconductor
YIG	Yttrium iron garnet, $Y_3Fe_5O_{12}$
YSZ	Yttria-stabilized zirconia, e.g. *8YSZ* is $ZrO_2 + 8\,wt.\,\%\,Y_2O_3$

Symbols

A	surface area, m^2 , or absorptivity
a	thermal diffusivity, m^2/s, or absorption coefficient
$\vec{B}$	magnetic induction vector, T
C	concentration, kg/m^3 , or capacitance, F
C_D	drag coefficient, no dimensional (see Eqs. 2.28 and 2.29)
c	constant or light speed, $c = 299\,792\,458\,m/s$
c_p	specific heat, J/(kg.K)
D	diffusion coefficient, m^2/s
d	diameter or distance or resolution, m
d_{50}	average particles diameter, µm
d_c	size of crystal, m
d_{hkl}	distance between the crystal planes h, k, l
d_{vs}	diameter of a sphere having the same volume-to-surface ratio as the measured particle

E energy or internal free energy, J or eV ($1\,\text{eV} \approx 1.603 \times 10^{-19}$ J) or elastic modulus, GPa, or electric field, V/m

E_{bs} surface-binding energy can be estimated from heat of evaporation (see Eq. 2.9)

$\vec{E}$ electric field vector, V/m

e electron charge, $e = 1.602 \times 10^{-19}$ C or base of natural logarithm, $e = 2.718\,281\,828\,459\,793$

F force or load, N

$\vec{F}$ force or load vectors, N

f fluency being density of energy (J/m^2)

f_i mass fraction

G Gibbs energy, J/mol

G_{Ic} Critical value of energy density at mode I of fracture (forces acts in the direction of opening the crack), J/m^2

H magnetic field, A/m

$\vec{H}$ magnetic field, A/m

H_c coercive magnetic field (see Fig. 4.34), A/m

h Planck constant, $h = 6.626\,0755 \times 10^{-34}$ J.s, or heat transfer coefficient, W/(m^2.K), or depth, m

hkl Miller indices of crystal lattice plane *(hkl)*

I electric current, A, or radiation intensity, W/m^2

i imaginary unit, $i = \sqrt{-1}$

J flux of species, kg/(m^{2}s)

K shape factor in Eq. 4.7, $K \approx 0.9$

k Boltzmann constant, $k = 1.38064852 \times 10^{-23}$ kg/(s^2.K) or coefficient of extinction

Kn Knudsen number

K_{Ic} fracture toughness at mode I of fracture (forces act in the direction of opening the crack)

L absorption depth at which the radiation power density decreases *e* times or latent heat, J/kg

l distance, m

$\bar{l}$ mean free path

M metal or magnetization, A/m

$\vec{M}$ magnetization, A/m

m mass, kg, or integer number

N number of atoms or molecules or force acting normal to the surface, N

Nu Nusselt number

n index of refraction or integer number or molar concentration see (Eq. 2.16) in mol/m^3 or particles concentration, 1/m^3

P porosity, % or power, W

p pressure, Pa, or momentum being the product of mass and energy of a particle

p^* vapor pressure at given temperature, Pa

Pr Prandtl number

pH acidity scale basing onto [H$^+$] ions concentration, $pH = -\log_{10}[H^+]$

q electric charge, C, or density of power, W/m^2, or feed rate, g/min, or gas flow rate, slpm[2]

2 slpm = standard liter per minute.

R	reflectivity, dimensionless or universal gas constant $R = 8.314\,J/(mol.K)$ or flux of species arriving on the substrate, in $1/(m^2.s)$
R_a	average roughness defined as the arithmetical average of the absolute values of the amplitudes of the surface with regard to the mean line along the measured distance, μm
R_S	sheet resistance, Ohm/square
R_T	thermal contact resistance, $m^2.K/W$
R_t	roughness defined as the distance between the highest and the deepest point of the surface along the measured distance, μm
r	fraction or radius, m, or coefficient of reflection
r_D	radius of Debye, defined in Eq. 2.25
Re	Reynolds number, defined in Eq. 2.19
T	temperature, K, or force acting tangential to the surface, N, or transmissivity
T_c	temperature Curie, K, defined in Section 4.6.3
t	time, s, or coefficient of transmission
U	voltage, V
V	volume, m^3
v	velocity, m/s
$\vec{v}$	velocity vector, m/s
$\overline{v}$	mean velocity, m/s
w	width, m
Y	sputtering yield, atoms/incident ions
x, y, z	coordinates in *Cartesian* system, m

Indices

$\parallel$	parallel
$\perp$	perpendicular
0, 1, 2, 3, ...	following values
0.5	half of maximum (see Eq. 4.20)
A	acceleration or absorbed
a	anions
amb	ambient
bin	binding
c	cations or cylinder
cg	carrier gas
co	contact
cr	critical
D	drag
d	deposition or deposited layer
e	electron or evaporated atoms
ev	evaporation
f	formation or fracture
g	gap or *glass-transition* or gas

h	heavy species
I	intrinsic
lat	lateral
lg	interface liquid and gas
long	longitudinal
i	ion or an integer number 1, 2, 3,. . .
in	injector
ind	indenter
kin	kinetic
m	melting (molten) or molecule
max	maximal
n	neutral particles
ox	oxidation
p	proton or projectile or powder or particle
pl	plasma
pt	phase transformation
q	quenching
R	reflected
r	radius or reduced or relative or remanent
s	surface or solid
sg	interface, solid and gas
sl	interface, solid and liquid
sol	solidification
sp	splat
st	structural
sub	substrate
O	oxygen
t	total or target
T	thermal or transmitted
v	velocity or vapor

Exponents

−	negative charge
f	final
i	initial
*	excitation or precipitation or critical
+	positive charge

Greek Letters

α	correction factor in Eq. 2.9 varying between 0.2 and 0.4
β	field enhancement factor shown in Eq. 5.4

γ	surface tension or surface energy, N/m
δ	resolution, m, or thickness, m
ε	strain, % or emissivity or dielectric constant, F/m
ε_0	dielectric constant of vacuum, $\varepsilon_0 = 8{:}854 \times 10^{-12}$ F/m
η	dynamic viscosity, kg/(m.s)
θ	wetting angle, °, or angle of X-rays incidence in *Braggs' law*
λ	wavelength, m
μ	chemical potential or coefficient of friction or magnetic permeability
μ^0	chemical potential at normal pressure and *RT*
μ_0	magnetic permeability of vacuum, $\mu_0 = 4\pi \times 10^{-7}$ H/m
ν	frequency, Hz, or modulus of Poisson, 1/1
$\tilde{\nu}$	wavenumber, cm^{-1}
ξ	parameter defined in Eq. 3.5
ρ	density, kg/m^3, or resistivity, Ohm.m
σ	absorption cross-section, m^2, or stress, Pa
τ	duration, s
Φ	angle with regard to normal of light ray (see Fig. 4.35) or work function being a barrier height for electrons emitted from a solid body, eV (see Eq. 5.5)
Ψ	angle in Eq. 4.36, tan Ψ is a ratio of the magnitudes of the components of reflection coefficients
Δ	angle in Eq. 4.36 showing the phase shift between the components of incident and reflected wave or coverage factor in Eq. 5.8 defining the ration of volume of metal oxide to that of metal at high temperature
ΔE	difference of energy, J or eV
ΔG	total free energy, J/mol
ΔG°	standard Gibbs free energy of reaction at room temperature and normal pressure, J/mol
ΔG_v	volume free energy, J/mol
$\Delta_f G^\circ_{298}$	standard Gibbs energy of formation at room temperature and normal pressure, J/mol
ΔH	enthalpy required to the change of phase, J/kg
ΔT	difference of temperatures, K
ΔV	difference of volumes, m^3
$\Delta (2\theta)$	peak broadening at *FWHM* in a diffractogram, radians

Atomic Shells and Subshells

The atomic shells were introduced by ***Niels Bohr*** model of atom in which the electrons are on the shells around the nucleus. The letters describing the shells are, starting, from that being closest to nucleus: K, L, M, N, O, P. Each shell is composed of one or more subshells labeled: s, p, d, f, g. Each subshell holds at most the defined number of electrons, e.g. subshell s – not more than 2 electrons, subshell p – not more than 6 electrons, etc.

Introduction

A generic term "physical deposition methods of films and coatings" is often used but rarely defined in a rigorous way. In this work, this term is related to a deposition process, shown schematically in Fig. I.1.

This process is initiated by a source of the species forming the deposits (atoms, molecules, solid particles, etc.). These species are transported through a passive (vacuum or low pressure, inert gas, etc.) or active (plasma, reactive gas, etc.) medium to a substrate. The term "physical deposition" would relate to physical nature of generation of the constituent species from source. This mechanism of such generation can be:

- evaporation of solid sources in thermal evaporation;
- ejection of species by energetic ions in sputtering;
- laser ablation in the *PLD* technique;
- injection of reacting gaseous precursors in *CVD* techniques;
- injection of solid (powder particles) or liquid (suspension, solution) feedstock in thermal spraying;
- injection of powder particles in bulk methods.

In contrast, the chemical mechanism of generation of the species would be, for example, movement of ions toward an electrode being also a substrate in electrochemical processes. The passive environment does not modify species, unlike the active environment. For example, in a *CVD* process, a chemical reaction modifies initial chemical composition of the precursor gases. It should be noted that plasma in atomistic deposition processes is an active medium, because it can modify the chemical composition of the species (atoms, ions, molecules) constituting a deposit. This modification can be negligible in the granular- and bulk-coating processes in which the species are much larger (solid particles having more than 10^{12} atoms [1]), and the interaction time with the medium is short (in the range of milliseconds in thermal spraying) enough to avoid any important modification. Consequently, in these methods, the plasma is rather a passive medium.

The physical deposits described in this book are classified according to the size of the constituent species (see Table I.1) as: (i) atomistic; (ii) granular; and (iii) bulk deposition methods. In the atomistic techniques, atoms, ions, and molecules build up the films. The granular deposition methods use species such as particles typically about ten micrometers in size. These species may be formed by nucleation of liquids being precursors in some methods (e.g. solution thermal spraying). The solid particles are used as species in bulk

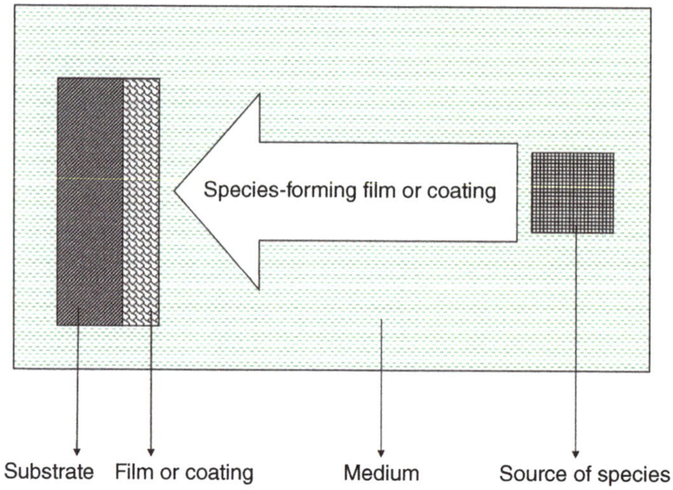

Fig. I.1 Diagram of a typical deposition process.

Table I.1 Physical deposits discussed in this work.

	Atomistic		**Granular**	**Bulk**	
Inert medium	Pulsed laser deposition	*Solid feedstock*	Flame spraying	*Arc melting of substrate*	*PTA*
	Evaporation		Arc spraying	*Laser melting of substrate (with or without pre-deposited coating)*	Cladding
Reactive medium	Evaporation in reactive atmosphere		*APS*		Laser surface alloying
	Chemical vapor deposition (*CVD*)		*HVOF*		Hard-phase dispersion
	Atomic layer deposition (*ALD*)				
Plasma	Sputtering		Detonation spraying		
	Ion plating		*VPS*		
	Plasma enhanced *CVD*	*Liquid feedstock*	Suspension plasma spraying		
			Solution precursor plasma spraying		
			Spray pyrolysis		

Source: Bunshah [2]/with permission of Elsevier.

Fig. I.2 A bracelet watch. *Source:* From https://www.dillards.com/p/movado-bold-evolution-gold-mesh-strap-bracelet-watch/509056845 [5]/Reproduced with permission from Dillard's/last accessed under April 05, 2023.

deposits. However, in bulk deposits the size of particles is generally greater than in granular deposits, and the particles are injected in molten substrate. Some explanation is necessary to make a difference between *film* and *coating*. The difference is thickness. A film is up to a few μm thick and a coating has a thickness greater than, say, more than 10 μm. This difference sticks also to the methods of manufacturing. Consequently, the methods with atomic species generate films and those having powder particles species generate coatings. As usual, there are many exceptions. For example, a thin-film technology of *EB PVD* enables manufacturing of thermal barriers being more than a few hundreds μm thick, and thermal spraying solution generates frequently a few μm thick film being frequently called coatings.

There are a lot of deposition techniques belonging to the classes defined above, and not all of them are described in this book. A selection had to be made to limit the description to the techniques most representative from a point of view of current or potential industrial applications. There are a number of books where an interested reader can find descriptions of other films and coatings techniques (e.g. [2–4]). The aim of this book is to present deposition techniques from the point of view of a user who is mainly interested in the properties of deposits leading to their practical application thanks to the value added to a product. Such a benefit can be, for example, a color of a watch bracelet.

In this case, it would be important that the film has a desired color and that its roughness is small. Film properties such as the modulus of elasticity or thickness do not play an important role in this application. Moreover, it seems reasonable that an atomistic deposition technique would be recommended to the film manufacturing. The coatings obtained by one granular deposits technique would be too heterogeneous (porosity, contact between the lamellae, etc.) and rough to fulfill the requested role.

Another example would be a deposit on hydraulic cylinders used in coal mines (Fig. I.3). This tool wears out quickly through abrasion and, therefore, the deposit must be well

Fig. I.3 Hydraulic cylinders in coal mines. *Source:* From https://www.laserline.com/de-int/laser-cladding/[6]/Reproduced with permission from Laserline GmbH/last accessed under April 05, 2023.

adherent, resistant to abrasive wear, and very thick in order to prolong its life. In this case, bulk deposits are much more advantageous.

The deposition techniques presented in this book should make possible to solve most of practical problems that can be solved with help of films and coatings in industry of today.

The book structure is linear and quite similar to a coating manufacturing procedure (see Fig. I.4) and describes:

- substrate preparation;
- methods of films and coatings deposition;
- treatment post-deposition;
- nucleation, growth and microstructure of films and coatings;
- methods of films and coatings characterization;
- properties of films and coatings.

Thanks to this structure, a reader could easily solve a problem related to a film or coating in service. Such a problem is somehow linked to one of the stages of development cycle. Consequently, all stages should be good enough to propose an appropriate solution. On the other hand, a good knowledge of the procedures included in the development cycle would make it easier to offer a deposit for any industrial need. Developing a repository for an industrial need, however, requires, in addition to knowledge of deposition techniques, a good knowledge of the useful materials that could be applied. Some of the materials are shortly discussed in some chapters of this book can serve as references to initiate a more in-depth study of specialized works such as, among many others, [7–9].

This book is addressed to the engineers in the films- and coatings-related industries who wish to deepen their knowledge of physical deposition processes and to the researchers starting or developing their activity on this field. The professors teaching the different

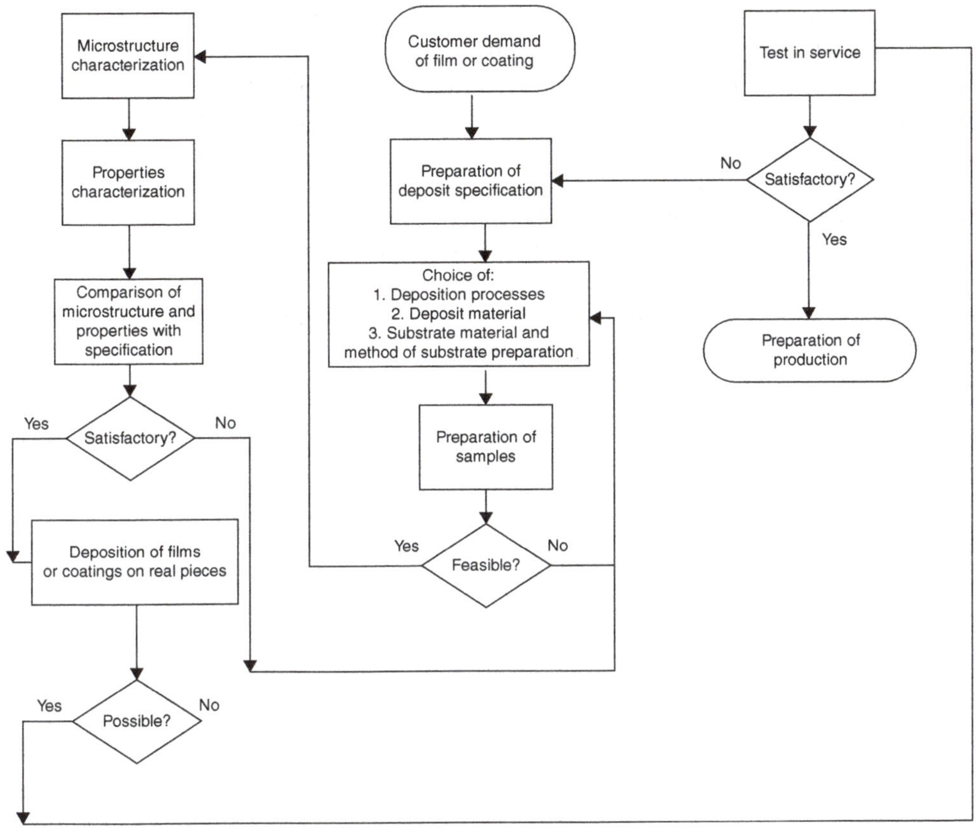

Fig. I.4 Cycle of development film or coating. *Source:* Pawłowski [7].

aspects of surface engineering and their students should find many useful information. For the purpose of this book, the author has used some information contained in his previous book on the field [11] and reviewed the literature from 2000 up to 2024 of the main scientific and technical journals in which appear the papers on physical methods of films and coatings deposition, of the presentations of scientific conferences, and of industrial exhibitions.

Bibliography

1 Pawłowski, L. (1999). Thin films technology vs. thermal spraying: towards peaceful co-existence. *J. Thermal Spray Technol.* 8 (2): 179–180.

2 Bunshah, R.F. (ed.) (1994). *Handbook of Deposition Technologies for Films and Coatings*. Park Ridge, NJ: Noyes Publications.

3 Martin, P.M. (2009). *Handbook of Deposition Technologies for Films and Coatings*. Amsterdam: Elsevier.

4 Pawłowski, L. (2008). *The Science and Engineering of Thermal Spray Coatings*, 2nde. Chichester: Wiley.

5 https://www.dillards.com/p/movado-bold-evolution-gold-mesh-strap-bracelet-watch/509056845 (accessed 18 March 2020).
6 https://www.laserline.com/de-int/laser-cladding/ (accessed 18 March 2020).
7 Pawłowski, L. (1997). Microstructure des dépôts physiques. *Galvano – Organo – Traitement de Surface* 97 (678): 633–637.
8 Callister, W.D. Jr. and Rethwisch, D.G. (2018). *Materials Science and Engineering: An Introduction*, 10the. New York: Wiley.
9 Richerson, D.W. and Lee, W.E. (2018). *Modern Ceramic Engineering. Properties, Processing and Use in Design*, 4the. Boca Raton, FL: CRC Press.
10 Inagaki, M. and Kang, F. (2014). *Materials Science and Engineering of Carbon: Fundamentals*, 2nde. Waltham, MA: Butterworth-Heinemann.
11 Pawłowski, L. (2003). *Dépôts Physiques*. Lausanne, Switzerland: PPUR.

1
Substrate Preparation

1.1 Introduction

Deposits must adhere well to the substrates in the *chosen places*. The adhesion of a deposit is achieved by several physico-chemical mechanisms in contact between the deposit material and that of the substrate (discussed in detail in Chapter 4). All these mechanisms work, however, under the condition that the surface of this contact is clean. A ***clean surface*** can be defined after Mattox [1] as "a surface which contains no significant amount of undesirable material." The degree of required property varies with the deposition technique.

Consequently, the films obtained with *atomistic deposition* methods require high level of cleanliness so that the atoms or molecules of a deposit that condense on the surface do not come into contact with a very fine of dust or another contamination of a small dimension. The thick coatings obtained by *granular deposition* techniques use generally molten particles to build up the deposit, and the required degree of purity is relatively smaller than that required for thin films. However, the adhesion of such coatings to the substrate must be improved by the surface activation. Similarly, the surfaces of the substrates coated by *bulk deposition* methods are molten to enable coating material particles to be injected what lowers considerably the purity requirement.

The impurities, which can be found on the surfaces of substrates and the methods of their cleaning, depend also on the material. The types of material determine the type bonding and the nature contaminants, which is different for metal and alloys, ceramics, and polymers. Consequently, the typical substrates materials used in different deposition methods are shortly described.

The purity of a surface of a substrate can be achieved by removing the contamination by following methods:

- chemical (chemical attack, dissolution of impurities, etc.);
- physical which is realized by:
  - high temperature treatment;
  - *mechanical cleaning* using particles abrasion (grit blasting) or water jet *ultrasonic cleaning*;
  - *photonic cleaning* using laser or by polychromatic *uv* radiation;
  - *plasma cleaning* using arc or ion bombardment.

Physical Deposition Methods for Films and Coatings, First Edition. Lech Pawłowski.

The films or coatings should be deposited in chosen places on the substrate. The places must be somehow defined. The chapter describes patterning using ***lithography*** and ***direct patterning***. The *lithography* is a classical masking technique, which has been used in semiconductor industry since many decades. It can be applied directly on the substrate if high resolution is desired. Alternatively, this technique can be applied to manufacture the *stencil masks* which, although less precise, can be used multiple times also in granular deposition methods. The resolution photolithographic masks depend on the *illumination sources* leading to the photo-polymerization (*uv* – radiation, X-rays, *e*-beam, *i*-beam, etc.). The *direct patterning* enables maskless protection of the substrates using different mechanisms such as oxidation of thin metal films or thermal decomposition of polymers [2]. This kind of patterning may be used in technology of atomistic films. Some granular coatings, such as cold-gas spraying method (*CGSM*), enable also direct patterning of sprayed coating.

The methods of cleaning methods are often empirical and are a part of a deposition procedure. The surface should be clean to obtain well-adhering film or coating. However, this condition is necessary but not sufficient to obtain a good adhesion of a deposit. Sometimes the surface has to be activated prior to deposition processes. In particular, the surface activation is essential to obtain adhesion of granular deposits. This results from the fact that one of principal adhesion mechanisms is the mechanical attachment of liquid particles to the top parts of surface irregularities. Consequently, it is necessary to roughen the surface by, e.g. grit-blasting. On the other hand, a surface of a polymer substrate must be activated in order to apply any deposit. Such activation can be done using plasma by, e.g. corona discharge. Finally, the application of a bond-coat can also be seen as activation of a substrate.

Surface cleaning and their activation are part of the pre-treatment of a substrate prior to film or coating deposition. Very often, the deposition does not directly follow the pre-treatment, and the substrate remains sometime in contact with the ambient atmosphere, which may result in a re-contamination. The appropriate storage of pre-treated substrates may avoid such recontamination.

1.2 Surfaces of Different Types of Substrates (Metals and Alloys, Ceramics, Polymers)

The substrates are made of different materials depending on the deposition method. Atomistic techniques are applied for the deposition of metals and alloys, ceramics (oxides, nitrides, carbides), carbon and polymers films on metals and alloys, ceramics (oxides and carbides), semiconductors, and polymers. The granular techniques are most frequently applied to the coatings of metals and alloys, ceramics (mainly oxides and carbides), and more seldom polymers and the composites (oxides – metals, polymers – oxides). The used substrates are mainly metals, alloys, and more seldom ceramics and polymers. Finally, the bulk deposition techniques are applied for coatings of metals and alloys, ceramics (oxides and carbides), and their composites (oxides – metals by the hard-phase dispersion technique) on the substrates of metals and alloys. Substrates used in bulk coatings techniques require less preparation than in others. Consequently, this chapter describes briefly metals/alloys, ceramics/glasses, and polymers and their physicochemical properties which play a role in films and coatings deposition.

Films and coatings are deposited onto surface of materials. Surfaces have different properties than bulk material. It results from the forces acting on the atoms/molecules (see Fig. 1.1). The surface is in contact with other phase (liquid or gas), and the forces of interaction with the atoms or molecules are much smaller than those in the solid. The difference between the forces can be described as ***surface energy***, being equal to the difference between the energy of all atoms/molecules in the surface and that they may have inside the solid [3].

The *surface energy* can be interpreted as a work of displacing the atoms/molecules (called particles thereafter) from the solid bulk to its surface. This energy decreases with the temperature[1] and is equal to zero at ***critical temperature*** and ***pressure*** when the phase solid and liquid/gas disappear and, consequently, the solid surface disappears too. The *surface energy* depends also on its morphology. The surfaces are seldom flat and have frequently peaks or corners shown in Fig. 1.2. The particles have lower number of neighbors, and the *surface energy* is greater. Consequently, the reactivity of the rough surface is greater than that of flat ones, and the corners and edges include more products of reaction than other places on the surface.

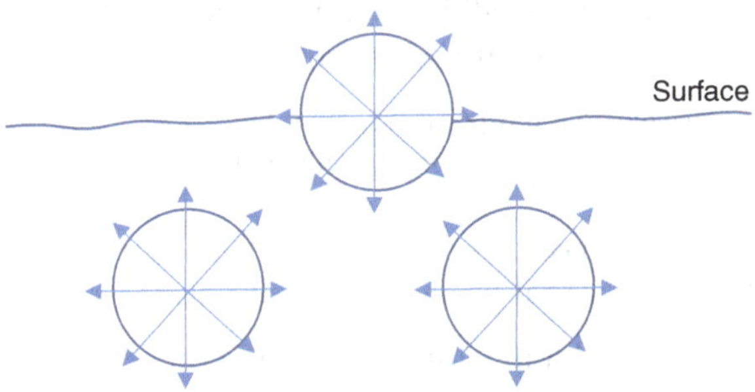

Figure 1.1 Forces acting on atoms or molecules on the surface and in the bulk of a solid. *Source:* Adapted from Burakowski and Wierzchoń [3].

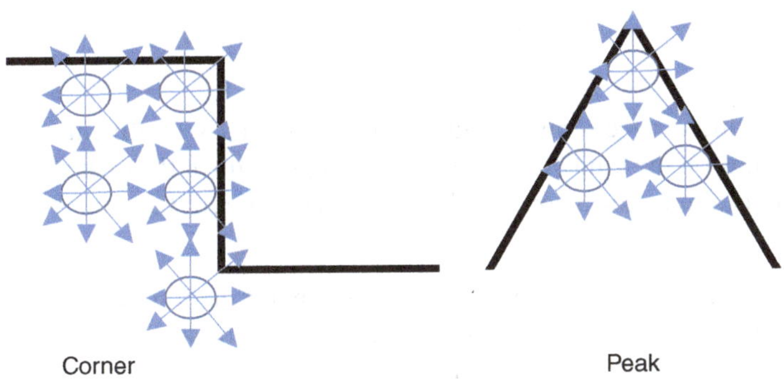

Figure 1.2 Forces acting on the surface including a corner and a peak.

1 For example, the *surface energy* of solid alumina at temperature of 2123 K is 0.905 N/m and of liquid alumina having temperature 2353 K is 0.7 N/m [4]

The particles on the solid surface are fixed and do not move. However, in many cases, the surface is heated and its temperature increases. At the temperatures close to the melting point, the particles can move and the number of irregularities decreases. It results from the tendency to decrease the surface area, as in the liquids. This tendency is an effect of interaction of the particles inside the body with the particles on its surface and is described by ***surface tension***, γ. This parameter can be described by the ratio of internal free energy, E, and the surface area, A, as:

$$\gamma = \frac{\partial E}{\partial A} \tag{1.1}$$

The *surface tension* is frequently used in describing the liquids. It reduces their surface area and makes the droplet spherical. It can be defined, for the solids after Kingery et al. [4], as the work done in creating new surfaces by adding new particles to this surface.

The real bodies are manufactured in different ways, and their surfaces result from the manufacturing process, which may be divided roughly as: (i) mechanical (grinding, polishing, etc.); (ii) thermal (casting, heating, etc.); and (iii) chemical (alloying, chemical reaction, etc.). Each of these processes is in practice associated with physical and chemical phenomena as diffusion, crystallization, melting, solidification, adsorption, and many others. The state of a surface may be roughly described using properties/data:

- thermodynamical (*surface energy, surface tension*, temperature, etc.…);
- mechanical (roughness, hardness, parameters describing elasticity, fracture, or stresses, etc.);
- chemical (chemical composition and crystal phases, etc.);
- physical (thermal or electrical conductivity, thermal expansion, optical emission, etc.).

Moreover, many of these properties change with the distance from the surface, and a couple of zones having different thickness (e.g. 3–5 zones models following [4]) can be distinguished. Finally, the manufacturing processes and the properties of surface depend strongly on the properties of the bulk material. That is why the useful information will be described for three types of materials: metals and alloys, ceramics, and polymers.

1.2.1 Metals and Alloys

The volume near to the metallic surface submitted to a thermomechanical treatment may include many zones depending on the kind of treatment, the temperature of the treatment, and its duration. To give an example, the 5-zone model, presented in Fig. 1.3, includes the following zones [3]:

- zone I being a *near-surface zone* in direct contact with the environment and includes some atoms and ions coming from the atmosphere around and from the surface of solids being in contact and the thickness of this zone is smaller than 1 nm;
- zone II called also *directed grains zone* includes the crystal grains deformed in a similar direction after the treatment;
- zone III known also as *thermal effects zone* which includes the grains influenced (size, phase composition, etc.) by temperature, and the thickness depends on temperature and on time of the treatment;

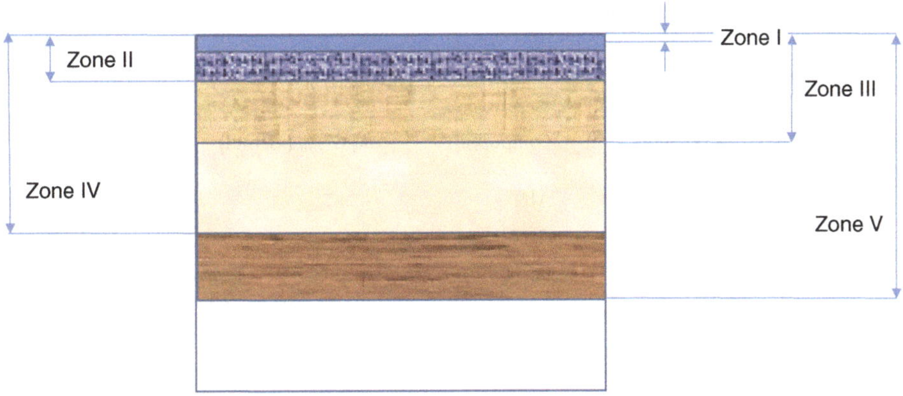

Figure 1.3 Five-zones of metallic surface after mechanical or thermomechanical treatment. Description of zones is in the text. *Source:* Adapted from Burakowski and Wierzchoń [3].

- zone IV is *textured grains zone* which includes the grains having similar crystallographic texture;
- zone V is known also as *crush zone* deformed plastically at thermomechanical treatment, and the thickness may reach a few to a hundred μm.

The metal or alloy surface in contact with atmosphere may adsorb species. The adsorbing surface is called ***adsorbent***, and the adsorbed species are called ***adsorbate***. The mechanism of adsorption may be [3, 5]: (i) physical, in which the species are attached by relatively weak ***van der Waals forces***; or, (ii) chemical, in which the *adsorbent* enters in reaction with *adsorbate*. Some authors add ***condensation adsorption*** in which the adsorbed gases transform in liquid. Such condensation may occur at the pressure, which is greater than a critical for given temperature [3]. Finally, ion-exchange adsorption is possible, in which the ions presented on the solid surface are replaced by the ions from the gas phase [5]. The possible adsorption mechanisms and different ***isotherm models*** enable predicting the adsorption amount on a solid surface at constant temperature.

Metals and alloys are generally highly reactive, and the contact with open atmosphere leads to a formation of oxides or nitrides. The oxidation of metallic surface in such atmosphere is more probable than its nitriding. This results from thermodynamic properties of oxides and nitrides. The Gibbs free energy of formation of oxides is much more negative as shown in Table 1.1. The comparison of the mass of formed oxides or nitrides depends however on the diffusion of oxygen and nitrogen across the oxide or nitride film. The dense oxide films as those of Al_2O_3 remain much thinner by forming an efficient barrier for oxygen diffusion [7].

The oxidation of a metal starts formation of metal ion and electron(s), which diffuse through the oxide film to meet and to enter in reaction with oxygen from air. The sketch of formation of oxide film oxidation in 1-D is shown in Fig. 1.4. The reaction occurring in the film is:

$$M + O \rightarrow MO \tag{1.2}$$

Table 1.1 Standard free energy of formation of some oxides and nitrides[a].

Oxides		Nitrides	
Compound	$\Delta_f G°_{298}$ (kJ/mol)	Compound	$\Delta_f G°_{298}$ (kJ/mol)
Al_2O_3	−1573	AlN	−287
CrO_3	−506.3	CrN	−92
Fe_2O_3	−744.8	Fe_2N	+12.6
CuO	−128.4	Cu_3N	+74.5
MoO_3	−669.4	Mo_2N	−50.2
TiO_2	−891.2	TiN	−309.2

[a] More negative Gibbs energy means that reaction is more probable. Positive energy of Gibbs means that the compound is stable. *Source:* Adapted from Elder et al. [6].

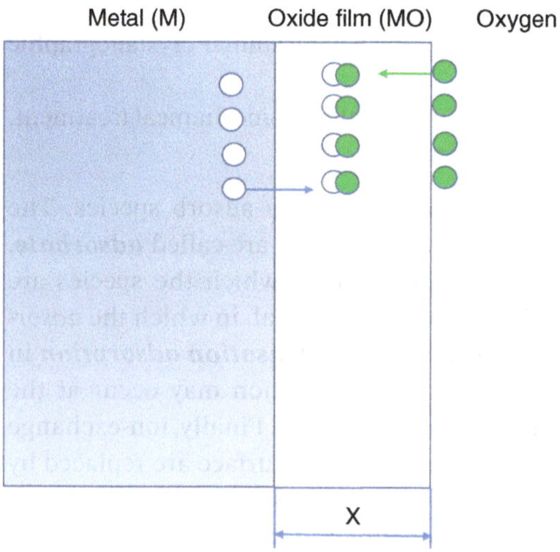

Figure 1.4 Mechanism of parabolic growth of oxide film.

The oxidation can be accompanied with three possible diffusions, namely those of [7]: (i) oxygen atoms (molecules) toward metal surface; (ii) metal atoms toward oxide film surface; (iii) electrons from metal substrate toward oxide film surface. The flux of each diffusing species results from its concentration gradient and can be described by ***Fick's first law***:

$$J = -D\frac{dC}{dx} \tag{1.3}$$

If we consider only oxygen, then the gradient of its concentration is equal simply to the ratio of its concentration to the film thickness, supposing that its concentration $C_{(O)} = 0$ at the metal surface:

$$\frac{dC_{(O)}}{dx} \approx \frac{C_{(O)}}{x} \tag{1.4}$$

Knowing that the velocity of oxide film growth, dx/dt is proportional to the flux of oxygen atoms, we may obtain the differential equation [6]:

$$\frac{dx}{dt} = -D\frac{C_{(o)}}{x} \tag{1.5}$$

The solution of the equation is $x^2 \sim t$ and shows that the square of oxide thickness is proportional to time of oxidation. This type of oxide growth is known as ***parabolic oxidation*** and describes the growth in continuous films. On the other hand, the growth of oxides films at high temperatures may lead to their fracturing because of the mismatch of thermal expansions coefficients of the metal and its oxide. This phenomenon may lead to the ***linear oxidation*** at which the oxide film thickness is proportional to oxidation time $x \sim t$. More details about oxidation mechanism can be found, e.g. in the textbook of Ashby and Jones [7].

The atoms well below the zone V (see Fig. 1.3) in the metals and alloys interact generally by metallic bonding (see Fig. 1.5). The valence electrons are not bound to any atom, and they may drift inside entire metal or alloy forming a "sea of electrons." The free electrons help in "gluing" of metal ions together and render metals and alloys good conductors of electricity and of heat. The bonding energy may reach 8.8 eV/atom for tungsten [8].

The atoms and ions in metals and alloys are arranged in the periodic network. Exceptionally, the rapid solidification may lead to the formation ***amorphous structure***, without any short- or long- distance order. The crystal structures can be described by geometry of ***unit cells***. There are, in total, seven crystal systems, and metals have frequently structures *FCC*, *BCC*, and *HCP* [8]. Most of crystalline solids are composed of small crystals rendering metals and alloys ***polycrystalline***. The crystal grains are in contact with each other by ***grain boundaries***. The boundaries weaken the bonding energy between atoms in neighboring grains. In addition, the diffusion through the boundaries is much easier than that through the crystals.

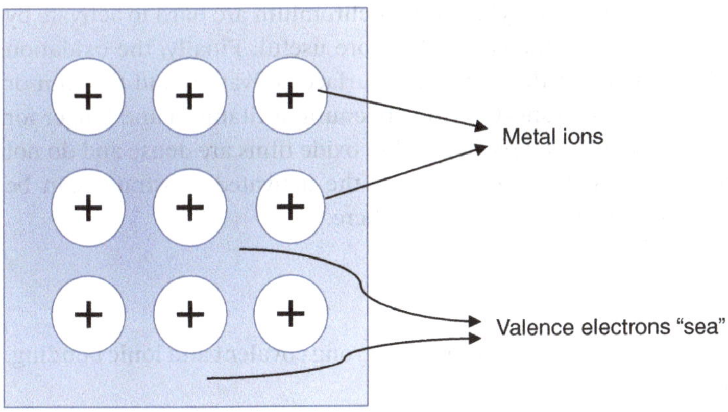

Figure 1.5 Metallic bonding showing metal ions and sea of valence electrons. *Source:* Adapted from Callister [8].

Table 1.2 Physicochemical and mechanical properties of some metals and alloys useful in surface preparation prior to films and coatings deposition [7, 9–12].

Metal or alloy	T_m, K	Elastic modulus, E, GPa	Time needed to obtain 1 mm thick oxide at temperature $0.7{*}T_m$, h
Al	933	69	Very long
Cr	2180	268	1600
Fe	1811	200	24
Cu	1358	117	25
Mo	2896	330	Very short
Ti	1941	114	<6
NiAl	1955	170	—
Stellite 6	1558–1683	237	—
Bronze	1223	96–120	—

The properties of metallic substrates, important for the preparation of films and coatings, include a few physicochemical and mechanical properties such as:

- melting point, helping in the choice the temperature of preparation treatment;
- oxidation resistance, some data concerning oxidation are discussed above (see Table 1.1 and Fig. 1.4);
- hardness, which renders the substrate activation by mechanical methods as sand-blasting is more or less difficult.

Table 1.2 is a collection of some useful data for typical metal and alloys. The melting point, T_m, is important to indicate the temperature, which can be used for the treatment prior to deposition of film and coatings. This temperature should be kept low for substrates of aluminum, copper, or *bronze*.

The modulus elasticity indicates the method that can be used to the substrate activation of substrate. Obviously, the substrate of molybdenum or chromium are hard to activate by mechanical methods, and chemical methods may be more useful. Finally, the oxidation activity determines time lapse between the substrate's surface activation and the film or coating deposition. This time should be short for molybdenum or titanium and longer for aluminum or chromium (which oxidizes rapidly, but thin oxide films are dense and do not grow up). This may also be useful to decide whether the activated substrates can be preserved for long periods in open or in an inert atmosphere.

1.2.2 Ceramics and Glasses

Ceramics and glasses can be defined as materials having strong covalent and ionic bonding. Examples of covalent ceramics are:

- different polymorphs of carbon, such as diamond, graphite, or fullerenes [8];
- compounds having structure similar to diamond, such as Si_3N_4 [13].

The ionic ceramic is composed of cations such as Al^{+3} or Ca^{+2} and anions as O^{2-}. The radius of anions, r_a, is generally greater than that of cations, r_c. The stability of ionic ceramic increases with the ratio of the radii r_c/r_a. If the ratio $r_c/r_a > 0.155$ then each cation is surrounded by, at least, three anions [8]. Stability concerns also an interaction between the ceramics surface and open atmosphere. As there are no chemical reactions of ceramics with oxygen or with nitrogen, this interaction is mainly physical and can be described by *surface energy*, *surface tension* (see Eq. (1.1)), adsorption, and diffusion.

The *surface energy* (see Table 1.3), which tends to have a possible low value, plays an important role in the distribution of impurities on the surface. The impurities arriving on the surface of ceramics will remain there more or less concentrated depending on their *surface tension*. Some values of *surface tensions* for liquids, which may be impurities, are collected in Table 1.4. The impurities having low *surface tension* would remain concentrated on the surface. Contrarily, those of high *surface tension* would tend to be less concentrated. The surface may lower its energy by adsorption of impurities. In the case of ionic crystals, their surfaces are filled with anions and cations and may attract ions, if there are any. The fractured ion crystal surface has high *surface energy* and attracts oxygen from air to lower this energy [4].

The surface of ceramics may be in contact with impurities being in gaseous and liquid phases. The contact with gaseous phase may be described by adsorption and diffusion. The contact with liquid phase needs an introduction of ***wetting***, which is generally analyzed for

Table 1.3 *Surface energy* of some ceramic materials and glasses [4, 14–16].

Ceramics	T_m, K	T, K	Surface energy, N/m
Al_2O_3	2327	1273	1.0
		2123	0.905
MgO	3073	298	1.0
SiO_2 + 20 wt.% Na_2O		1623	0.38
SiO_2 + 13 wt.% Na_2O + 13 wt. % CaO (liquid)			0.35
TiC	3430	1373	1.19
$CaCO_3$ (monocrystal)	1612	298	0.23

Table 1.4 *Surface tension* of some liquids having temperature of $T = 293$ K.

Liquid	γ, mN/m
Water	72.75
Ethanol	22.28
Pentane	15.7
Hexane	18.52

Source: Adapted from Snustad et al. [17].

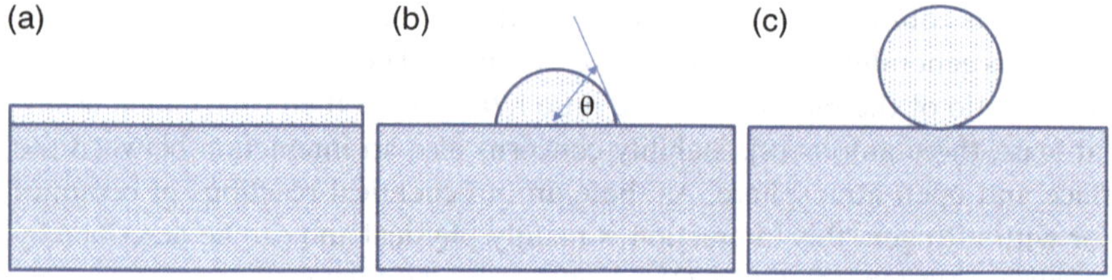

Figure 1.6 Behavior of liquid droplet on a solid surface: (a) spreading corresponding to $\theta = 0°$; (b) wetting corresponding to $\theta < 90°$; and (c) non-wetting corresponding to $\theta > 90°$ ($\theta = 180°$ in the Fig. 1.7c).

a liquid droplet on solid surface using angle θ as shown in Fig. 1.6. The droplet may spread entirely ($\theta = 0$). The wetting occurs for $\theta < 90°$ and non-wetting, for angle $\theta > 90°$. The *Young–Dupré* law enables predicting angle of wetting, θ, from *surfaces energies* of liquid and solid and from cohesion force of the liquid [18]. An important factor influencing wetting is surface roughness discussed, e.g. by Snustad et al. [17]

The crystal structure of ionic ceramics depends on the ratio of anions to cations and on number of anions. The anions, being much greater, can have a simple cubic structure (e.g. SiC) or *FCC* (e.g. MgO, $BaTiO_3$, or $MgAl_2O_4$). The silica and silicate structure is generally characterized by a tetrahedron in which every Si atom in the center is bonded by O being in the corners. Finally, covalent ceramics such as diamond, graphite, and fullerenes have the structure subsequently, *diamond cubic*, layers of hexagonally arranged carbon atoms and cluster of 60 carbon atoms, called C_{60} [8].

Ceramic materials are often manufactured from powders, which are sintered in temperatures lower than melting point. Consequently, such materials are polycrystalline and include some porosity. The glasses, being composed of silica and oxides of sodium, boron, or calcium among others, are generally amorphous. The electrical and mechanical properties of ceramics depend strongly on the porosity. In particular, the modulus of elasticity of porous ceramics modifies the modulus of no-porous material, E_0, in the following way [19]:

$$E = E_0\left(1 - c_1 P + c_2 P^2\right) \tag{1.6}$$

In the equation, the constants c_1 and c_2 depend on the shape of pores.

The ceramic materials may be conductors, semiconductors, and electrical isolators. This results from the electron energy gap between valence and conducting band, which can be at room temperature [4]:

- occupied, as in metals and in some oxides as CrO_2 or in TiO;
- partly occupied, as in semiconductors such as Si ($E_g = 1.1\,eV$) or GaAs ($E_g = 1.4\,eV$);
- empty, as in electric isolators as BN ($E_g = 5.9\,eV$) or Al_2O_3 ($E_g > 8\,eV$).

The elasticity modulus of some ceramic materials shows very high values for diamond and SiC (see Table 1.5). Another important property of these materials is their brittleness. The latter renders them hard to deform and to activate mechanically prior to the films and coatings deposition.

Table 1.5 Elasticity modulus for some ceramics [4, 20].

Material	Porosity, *P*, %	Elasticity modulus, *E*, GPa
Diamond		1164
SiC		469
TiC		310
Al_2O_3	5	365
ZrO_2 (stabilized)		151
BN		82.7
SiO_2 glass		72.5
Pyrex glass[a], SiO_2 + 12.6 wt. % B_2O_3 + 4.2 wt. % Na_2O + 2.2 wt. % Al_2O_3		70.0
Graphite	20	8.96

[a] Chemical composition corresponding to the equivalent glass *Corning 7740*. Diamond and pyrex glass are not porous

1.2.3 Polymers

Polymers are composed of hydrocarbons including ***macromolecules*** in which there are generally identical ***repeat units*** called also ***monomers***, such as ethylene, C_2H_4, or tetrafluoroethylene, C_2F_4. The *monomers* are interconnected by strong covalent bonds to form a chain becoming polyethylene (*PE*) and polytetrafluoroethylene (*PTFE*) [21, 22].

The molar mass of *macromolecules* may reach 10^3 or even 10^9 g/mol. The molecular mass of a polymer results from the degree of polymerization (*DP*) that represents the average number of *monomers* in a chain. In fact, each polymer has a lot of chains which are arranged in four possible ways corresponding to possible molecular structures of polymers. These structures, shown in Fig. 1.7, are generally called [21]:

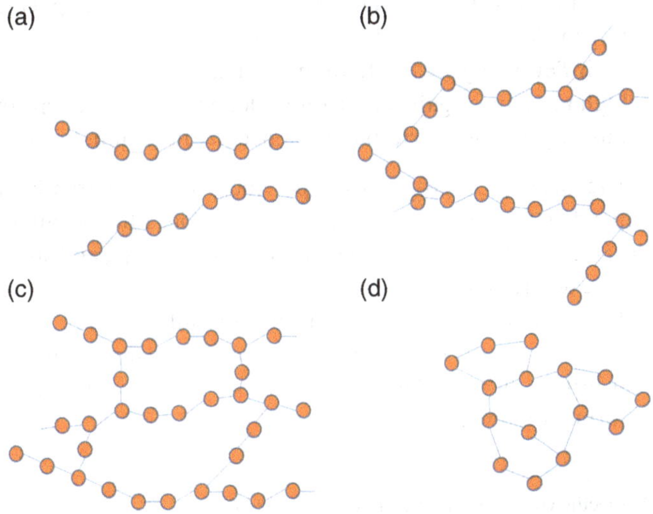

Figure 1.7 Possible molecular structure of polymers: (a) linear; (b) branched; (c) cross-linked; and (d) network or strongly cross-linked.

- ***Linear polymers,*** probably the most frequently used polymers, include the covalent bonded monomers in the chains, which are weakly bonded with other chains. The bonding between the chains is of hydrogen or *van der Waals type*. Examples of linear polymers *PE* are polystyrene (*PS*), poly (methyl methacrylate, *PMMA*), or high-density polyethylene (*HDPE*).
- ***Branched polymers*** include side branches lowering their density as in low-density polyethylene (*LDPE*).
- ***Cross-linked polymers*** have chains with additive molecules that are covalently bonded as in the rubbers.
- ***Network polymers*** or strongly cross-linked ones, which are composed of 3D covalently bonded monomers to which epoxy resin or polyurethane (*PU*) belong.

The classification of the polymers being most popular is related to their behavior in higher temperatures and under mechanical stress. Consequently, ***thermoplastics*** soften and deform easily in higher temperatures. The molecular structure of this type of polymers is linear or branched, and this group of polymers can be amorphous or crystalline (up to 90% of crystallinity). The opposite group of polymers is ***thermosets*** that do not soften with increasing temperature. *Thermosets* can have strongly cross-linked structures as epoxy resins. *Thermosets* with slightly cross-linked structure are called also ***elastomers***. Finally, there are also ***natural polymers*** such as [22]: (i) cellulose, which is main component of wood, cotton, and paper; (ii) keratin, present in wool fiber; and (iii) natural rubber.

The molecular structure of polymers determines their behavior in higher temperatures. The chains of molecules in *thermoplastics* and *elastomers* can easily move and change their position inside the body. Their specific volume[2] may considerably increase starting from the temperature called *glass transition*, T_g. The polymers have also melting point, T_m, but they do not have evaporation point, T_{ev}, as above the melting point, these materials decompose [22, 23].

The general properties of polymers can be synthesized after Seidel and Hahn [24]:

- density is in the range of $\rho = 0.8\text{–}2.2\,\text{g/cm}^3$;
- electrical resistivity is high leading to the frequent application of these materials as isolators in electric and electronic industry;
- high *TEC* being greater than in metals;
- high resistance against corrosion (depending on the type of polymers);
- rapid aging under external influence (*uv* radiation, thermal shocks, etc.).

The mechanical properties of polymers and, in particular, their stress vs. strain behavior depend on the type of polymers as seen in Fig. 1.8. The curve corresponding to *thermoplastics* (Fig. 1.8b) is similar to that of metals and alloys, and shows elastic behavior followed by plastic deformation.

The slope of the curve, i.e. Young modulus, *E*, for these materials depends however strongly on temperature and on the time of measurement. *Thermosets* (Fig. 1.8a) are harder and brittle without plastic deformation. Finally, *elastomers* (Fig. 1.8c) show only elastic deformation under very low stress.

2 Specific volume is inverse of density, i.e. $1/\rho$

The polymers used as substrates in films and coatings deposition are mainly *thermoplastics* and *thermosets*. Some properties of the most popular polymers are presented in Table 1.6.

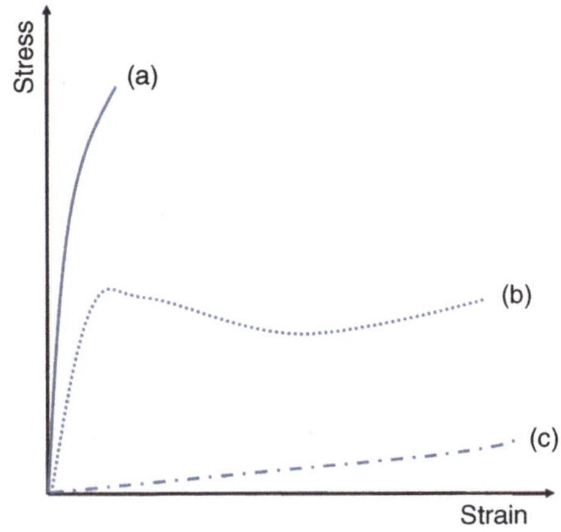

Figure 1.8 Typical strain–stress behavior of: (a) *thermosets*; (b) partly crystalline *thermoplastics*; and (c) *elastomers*. *Source:* Adapted from Callister [8].

Table 1.6 Some properties of frequently used polymers.

Type	Polymer	Glass transition temperature, T_g, K	Melting temperature, T_m, K	Density, ρ, kg/m^3	Modulus of elasticity at *RT*, *E*, GPa	References
Thermoplastics	Low-density polyethylene, *LDPE*	163	388	0.91–0.94	0.17–0.28	[8, 21, 23]
	High-density polyethylene, *HDPE*	183	410	0.95–0.97	1.08	
	Polyvinyl chloride, *PVC*	378	414	1.4	2.41–4.14	
	Polyamide, *PA6*	323	493	1.3	1.59–3.79	[21, 23, 25]
	Polycarbonate, *PC*	423	—	1.2	2.38	[21, 25]
Thermosets	Epoxy resin	380		1.2–1.4	2.1–5.5	[13]
	Phenol formaldehyde resin	—	—	1.27	8	
Natural polymer (wood)	Ebony	—	—	1.01	17.7	[23]
	Oak European	—	—	0.69	10.1	

1.3 Cleaning of Surface

The surface being prepared prior to a film or coating deposition has always some contaminants. These contaminants are hard to remove entirely but have to be reduced by cleaning to an acceptable level [26, 27]. The surface contaminations may result from:

- reaction with the environment (oxides, corrosion products, etc.);
- physical adsorption of the environment (water);
- traces of previous processes (polymer masks, oils, fingerprints, etc.);
- dust particles;
- degassing products generated during high-temperature action on the substrate (water, plasticizers, etc.).

The synthetic and naturals polymers are the class of materials used more and more frequently as the substrates for films, and interested readers may deepen the knowledge of chemistry and possible degradation of polymers in many references, such as [26].

The type of contaminants depend, to a degree, on the substrate material. Consequently, the surfaces of metal substrates are often covered with lubricating oils to protect them from corrosion during transport and storage. Ceramic substrates are frequently porous and can contain some water, which condensed from the vapors in the pores connected with the surface (called frequently *open pores*). Similarly, polymer substrates can also contain water and plasticizer added frequently to modify polymers' flexibility. Finally, all types of substrates can have dust or fingerprints on their surface.

The cleaning procedure depends also on the following film or coating deposition technique. Thin films obtained by atomistic methods need to be deposited onto surfaces and cleaned more carefully (having lower level of contaminants) than the coatings obtained by granular or bulk methods.

The cleaning methods are frequently composed of numerous steps. These steps can be roughly categorized as chemical and physical. An example of the TiNi-alloy surface preparation steps prior to Ti-film deposition by sputtering presented by Sonoda et al. [28] includes:

- polishing with emery paper;
- polishing with diamond paste and oil;
- washing with water and drying;
- ultrasonic cleaning with acetone.

It should be also stressed up that surface cleaning may be associated with the surface activation that is polishing with emery paper and diamond paste. In this chapter, the activation and chemical and physical methods cleaning are discussed separately.

1.3.1 Chemical Cleaning Methods

The chemical cleaning can be realized by [27, 29]:

- chemical etching, to be applied to remove some of the substrate making easier the removal of all type of contaminants;
- alkaline cleaning, converting organic fats into soaps easy to dissolve in water;

- chelating cleaning, to be applied for the phosphate contaminants;
- solvents cleaning, to be used to remove hydrocarbons as well as biofilm or protein contaminants;
- oxidation cleaning, to be used when oxides are volatile and is realized frequently using plasma discussed later in the section describing physical cleaning.

The chemical etching of metals and ceramics including composites as WC-Co is typically carried out using acids or bases. Table 1.7 shows some examples of chemical cleaning applied to treat different types of substrates. Polymer substrates have to be cleaned up with a detergent. Chemical cleaning can be also done with solvents. Polar contaminants should be cleaned up with water (used in solution with acids as shown in lines 2, 4, and 5 of Table 1.7), while non-polar contaminants, such as hydrocarbons, are to be cleaned with such solvents as ethanol, methanol, or trichloroethylene (see line 6 of Table 1.7).

Table 1.7 Examples of chemical cleaning.

No	Chemical cleaning using	Substrate	Film to be deposited	Deposition technique	Remarks	Reference
1	NaOH followed by HNO_3 or HNO_3/HF	Aluminum alloy	—		Second step is used to remove the residue of copper or silicon on the surface	[27]
2	Murakami solution: K_3 [Fe (CN) $_6$]: KOH: $H_2O = 1: 1: 10$	WC composite + 6 wt. % Co	*DLC*	*CVD*	Increase in Co/W ratio and of the roughness	[30]
3	NaOH and KOH heated solution				Removal of oxide product generated at plasma treatment	[31]
4	Solution of 2 vol. % HF in water, followed by rinsing in distilled water and washing with acetone	Si			Cleaning was followed by deposition of a bond-coat	[32]
5	$NH_4OH/H_20_2/H_20$ (1:1:5) at 348 K followed by rinsing in deionized water and by $HCl/H_20_2/H_20$ (1:1:6) at 348 K	Glass	Thin-film transistors	—	Preparation for liquid crystal displays	[33]
6	C_3H_6O (acetone) followed by blowing with N_2 and CH_3OH (methanol)		Ti	Magnetron sputtering	Preparation to laser-assisted micro-joining	[34]

The removal of proteins is more complicated. It was tested for the substrates used as biomaterials in medical applications such as Ti (99.3 wt. % pure), Au (86 wt. % pure), dental ceramics[3], and polymers as *PMMA* and *PTFE* with different solvents by Kratz et al. [29]. The authors made up to 50 rinsing cycles. They found out that the solvent known under tradename of *RIPA* Buffer was composed of[4]: $NH_2C(CH_2OH)_3 \cdot HCl$ (known as *Tris* HCL) + NaCl + nonylphenenoxypolyethoxylethanol (known as NP-40) + $CH_3(CH_2)_{11}SO_4Na$ (known as *SDS* or sodium dodecyl sulfate) was the best one compared to, e.g. ultrapure water, isopropanol, or only *SDS*.

Between the chemical cleaning steps, the substrate should be rinsed. The popular method is rinsing with deionized water. This water can be used a couple of times until the rinsing water reaches the prescribed electrical resistivity which gets lower with the dissolved impurities. Couillard et al. [33] started rinsing glass substrates with deionized water having resistivity greater than $\rho > 17\,M\Omega.cm$ and finished when this resistivity reached $\rho = 8\,M\Omega.cm$.

1.3.2 Physical Cleaning Methods

The physical cleaning leads to a removal of particles of different contaminants attached to the surface. The adhesion force depends on the surface's roughness and on the size and shape of the particles. The adhesive forces include after Cooper et al. [35]: (i) intermolecular *van der Waals forces*; (ii) various chemical bonds; and (iii) forces resulting from phenomena related to sintering such as diffusion. The authors determined experimentally and numerically the removal force of alumina particles having diameter of $d = 3.6\,\mu m$ to SiO_2 substrate having asperities up to about 2 nm to be about $F \approx 600\,nN$.

The physical cleaning methods are categorized in this section as: (i) thermal cleaning; (ii) particles cleaning which uses coarse (sand) and fine (atoms or ions) particles; and (iii) photons cleaning. The methods can be used in all films and coatings deposition methods depending on their specification resulting from the application.

1.3.2.1 Thermal Cleaning

Thermal cleaning is useful to evaporate such contaminants as water or hydrocarbons attached to metallic or ceramic substrates. Such cleaning may be associated with some negative effects such as: (i) diffusion of the contaminants in the substrate's interior; or (ii) oxidation of the substrate. To avoid these effects the heating can be applied after preliminary cleaning and in an inert atmosphere or under vacuum [27]. The heating may be realized in the conventional furnaces by convection or radiation and by using electromagnetic radiation at radio or microwave frequencies depending on substrate material.

The induction heating uses *RF* and can be applied for electrically conducting substrates. The heat is generated by *eddy currents*. Such heating can be generated very rapidly. Microwave heating can be applied for poor electrical conductors. The *dielectric heating* results from coupling of electromagnetic radiation with the dipoles in the substrate.

3 Product of VITA Zahnfabrik, Bad Säckingen Germany,
4 Proposed by, e.g. Merck, Darmstadt, Germany

1.3.2.2 Particles Cleaning

The particulate cleaning describes cleaning using small and large particles. Small particles such as ions, atoms, or molecules, which are used in cleaning by: (i) plasma; (ii) *i-beam* direct or being a part of sputtering; (iii) arc; and (iv) ultrasounds. The large particles, used for polishing, are in emery paper or diamond paste.

1.3.2.3 Plasma Cleaning

Plasma bombardment can be used with different types of plasma generators described by Thornton and Green [36]. This method is particularly effective *in situ* short before the deposition process. The *ex situ* method is often applied to clean up the surfaces of organic substrates. Cleaning with plasma is more efficient if the substrate is under negative potential to attract heavy positive ions. The electrically conductive substrates can compensate the arrival of such ion by a free electron and preserve a negative potential during all processing time. The dielectrics (oxides, polymers) do not have free electrons and would be charged with the positive ions and gradually repel them. This is why, the alternating current discharges should be used to generate plasma used to treat these types of materials. Moreover, the plasma for cleaning is typically generated with the use of an inert gas. The use of reactive gas to generate plasma may lead to plasma etching.

The plasma applied for cleaning is generated under low pressure being in the range of $p = 10^{-4} - 10\,\text{hPa}$. A few electrons (generated by field emission or ***photo – dissociation***) initiate their formation. The electrons are accelerated in the electric field and collide with atoms and molecules. If their energy is high enough ($E_e = 1\text{–}10\,\text{eV}$, see Table 2.8), collisions are inelastic and result in the ionization or dissociation of working gases. The electrons formed in the collisions are generated in cascade and fill the volume of gas by creating plasma. The plasma remains in equilibrium if the number of electrons generated is equal to the number of electrons lost by recombination with positive ions or with neutral species capable of forming negative ions. The plasma is electrically neutral on the macroscopic scale. In contrast, microscopically, there are significant variations in potential. The electric field of each charged species is compensated by another charged species. The compensation distance is called the ***Debye radius***. The temperature of the electrons in low-pressure plasmas is 10–100 times higher than the temperature of the ions (see Fig. 2.17).

The plasma is generated by electrical discharges. Cleaning is typically done with plasma generated by:

- *DC* discharge;
- *RF* discharge;
- microwave discharge;
- ***post-discharge***, that is, plasma downstream of a discharge.

The discharges of different types are shown schematically in Fig. 1.9.

Discharge in a *DC* diode system (Fig. 1.9a) begins with an electrical breakdown in the gas. The breakdown voltage depends on the gas and its pressure. The space between the electrodes becomes a conductive plasma being close to the potential of the anode. The voltage drops near the cathode. The main source of electrons maintaining the discharge is ion bombardment of the cathode (substrate). This bombardment results in the emission of secondary electrons. If the ions have low energy, secondary electrons are emitted by the

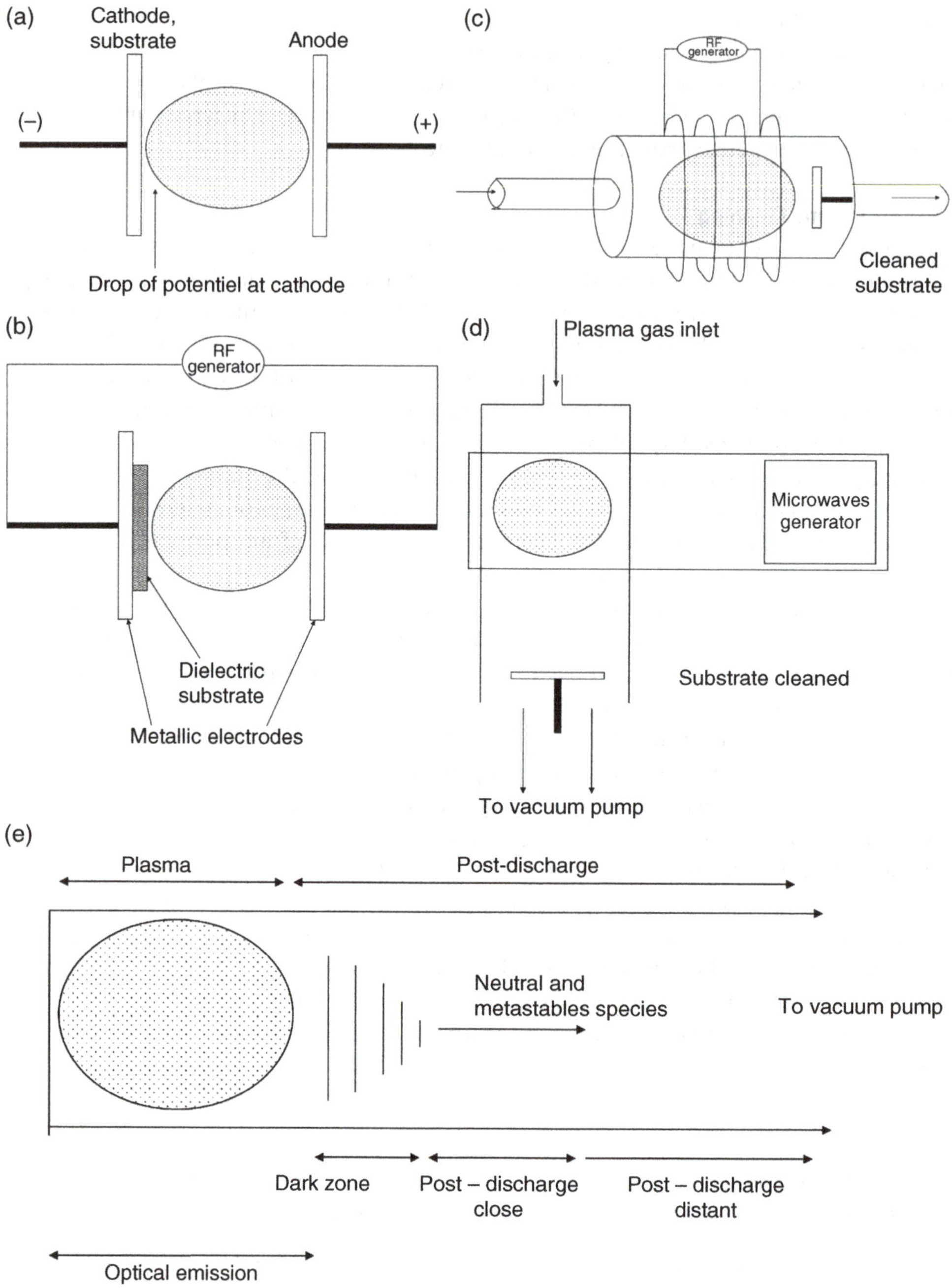

Figure 1.9 Types of plasmas generated by electrical discharges applied for cleaning or for etching surfaces of substrates: (a) *DC* diode discharge, (b) *AC* diode discharge, (c) *RF* discharge (*ICP*), (d) microwave discharge with a cleaned substrate downstream (in post-discharge), (e) characteristic areas of *post-discharge*.

Auger[5] mechanism, while the kinetic collision mechanism prevails for energy ions greater than 100 eV. The conductive substrate being put on the cathode potential is cleaned or etched by ion bombardment. The *DC* diode cleaning has the following advantages:

- plasma is uniformly distributed over large areas;
- system is simple;
- high-power densities are possible to reach.

A specific possibility of cleaning the oxides attached to a metallic substrate by an electric arc offers the technique of vacuum plasma spraying (*VPS*) belonging to the granular deposition techniques. This technique is sometimes used to spray superalloy coatings on turbine blades. After cleaning, the substrate is put on the negative potential and becomes a cathode, while the oxides contamination is the privileged place of attachment of electric arc. The arc heats up, melts, and removes the oxides.

Many problems are associated with the cleaning of dielectric substrates. The diode system, powered by alternating current at radio frequency, can be applied to clean such substrates (Fig. 1.9b). The diode system works usually at the frequency of $\nu = 13.56\,\text{MHz}$ and a peak-to-peak voltage greater than 1000 V [27]. The plasma is generated in this system by a capacitive coupling. The use of this system for conducting substrates would result in ion bombardment for half a cycle when the cathode is on negative potential, and the second electrode is bombarded during another half-cycle. Consequently, the substrate and the walls of an enclosure should be protected to prevent deposition of anode vapors. If the substrate is a dielectric plate fixed on a metal one, the system to be cleaned is then a capacitor with metal plate and plasma being conductors. The dielectric surface to be cleaned is periodically on a low positive potential and on a high negative potential because of inertia of the ions.[6] Cleaning by ion bombardment takes place when the surface of the dielectric is negative. *RF* plasma cleaning with capacitive coupling has the following drawbacks:

- *RF* energy losses in the system;
- lack of uniformity of the plasma on surfaces with complex geometry.

The inductively coupled plasma (*ICP*) system, shown in Fig. 1.9c, also works with the frequency of $\nu = 13.56\,\text{MHz}$. Its advantage is the lack of contact between the plasma and the electrodes. The gas is not contaminated and is therefore more useful to be used for plasma cleaning and etching the substrate.

The microwave discharge (Fig. 1.9d) uses frequencies from $\nu = 300\,\text{MHz}$ to 10 GHz. Microwave plasma is characterized by the higher degree of ionization compared to the plasmas described above. This plasma is often applied to clean or etch the substrates in the *post-discharge* configuration shown in Fig. 1.9e. Such treatment is recommended when

---

5 The mechanism of the excitation of an atom by an emission of electrons whose energy is weaker than the energy of incident ion. The emission of an Auger electron is associated by a photonic emission or another forms of an excitation

6 Ions are too inert to follow the electric field at a frequency greater than 100 kHz. Consequently, one electrode being initially more positive with respect to the earth potential collects electrons quickly, while the other electrode collects ions less quickly [37]. Finally, an electrode acquires a permanent negative potential and becomes a privileged target of ion bombardment.

there is a possibility that the surface of a substrate may be damaged by highly energetic species. These species are still present in the neighboring plasma area (***dark zone*** in Fig. 1.9e). However, their density decreases rapidly due to ***ambipolar diffusion***[7] and to the recombination of species on the walls of a reactor. Consequently, the emission of photons is less intense in this area than in the plasma. The next area downstream of the plasma is called the ***close post-discharge*** zone. It is characterized by the presence of excited molecules and of long-lived radicals. There are almost no ions, and the optical radiation is very weak. In the last ***distant post-discharge*** zone, there are only long-lived metastable species and free radicals. Mézerette et al. [38] reported that carbon contaminants are removed from an iron foil cleaned by $Ar\text{-}N_2$ microwave post-discharge at frequency of $\nu = 2.45\,\text{GHz}$ as CN^* radicals. In order to avoid nitrogen grafting to the substrate, the authors recommended heating up of substrate at post-discharge cleaning to $T = 373\,\text{K}$.

Other types of plasma applied to cleaning are, among others: (i) plasmas generated by electronic emission (e.g. by a hot wire); and (ii) plasmas confined in the magnetic field. These plasmas are generated in the installation for films deposition by *CVD* and *PVD* and are described in chapter 2.

1.3.2.4 Ions and Atoms Bombardment

As described above different types of plasma clean the substrates by heavy species bombardment. The bombardment may have many effects. These effects depend on plasma and on substrate and in particular on:

- mass, flow, and energy of bombarding species;
- nature and mass of the atoms of the substrate;
- nature and mass of the contamination atoms.

During the cleaning process only the first of these elements can be controlled. The energetic species interact during the bombardment with the ones being present on the surface (contamination, atoms of the substrate), and this interaction can result in the following effects [39]:

- desorption of contaminating species and setting them in motion;
- implantation of contaminating species in the substrate;
- creation of a structural defect caused by a cascade of collisions with the atoms of the substrate;
- implantation of a bombarding species in the atomic lattice of the substrate or in its structural defects;
- chemical reactions of the species present at processing;
- sputtering of the species present on the substrate's surface by the cascade of collisions which can be followed by back-scattering and their return back and re-deposition on the substrate;

  i.e.
- emission of secondary electrons;
- rebound of energetic species;

7 A movement of ions in one direction and electrons in the opposite direction

- transfer of the effects of the bombardment inside the substrate by the effect of a ***channeling***;
- heating of the substrate.

Among these effects, the most advantageous for cleaning are: (i) desorption of contaminating species; (ii) sputtering of the species present on the substrate's surface; and (iii) chemical reaction. These effects will be discussed in detail.

The choice of operational cleaning parameters should favor these effects. Desorption of contaminating species is desired because it leaves the surface clean of those species that are weakly attached to the surface by *van der Waals forces*. Contamination, which is better attached to the substrate's surface, can only be removed by sputtering.

Sputtering is, in fact, an effect used for one of the *PVD* deposition processes. The *PVD* deposition is done by using ions having energies sufficient to detach the atoms of a target and to move them toward a substrate to form a film. The cleaning and etching is made with the ions having energy ranging from 30 eV to 3 keV [40]. Consequently, Ochs et al. [41] used ions generated in the gases Ar, O_2, and Ar/O_2 having energy up to E_i = 0.35 keV to clean up the glass substrates by removal of carbon-containing contaminants. To generate them one can apply the plasma at high pressure or a post-discharge plasma with a weak polarization of the substrate. The application of ion -beams under low pressure enables avoiding the back-scattering and re-deposition discussed above. The *i*-beam can be directed under incident angle different to 0° (normal to the surface). To take an example, Goto et al. [42] used angle of 10–55° to clean up the ceramic substrate prior to the deposition of *YBCO* high-temperature superconductor film.

The chemical reaction with the plasma species does not depend very much on the energy of the ions bombarding the surface, which can remain very weak, but mainly on the chemical composition of the plasma. To clean the surface, it is preferable that the products of reaction are volatile. The process of surface cleaning by chemical reaction is called plasma etching and has been applied since the 1970s – in the electronics industry to treat semiconductor materials. The plasma gases used for plasma etching are typically chlorides or fluorides (Table 1.8).

Table 1.8 Substrates materials etched by different low-pressure plasmas.

Substrate	Plasma gas
Si	Cl_2, CCl_4, CF_4, NF_3, SF_6
SiO_2	CF_4/H_2, CHF_3, C_2F_6
Si_3N_4	CF_3Br, CF_4
WSi_2, $TaSi_2$	CF_4
Al	BCl_3, CCl_4, Cl_2
Ti, Mo	CF_4
Au	Cl_2
GaAs	Br_2, CCl_2F_2, Cl_2, CCl_4
InP	Cl_2, CCl_2F_2, CCl_4

Source: Curran [43]/IOP Publishing.

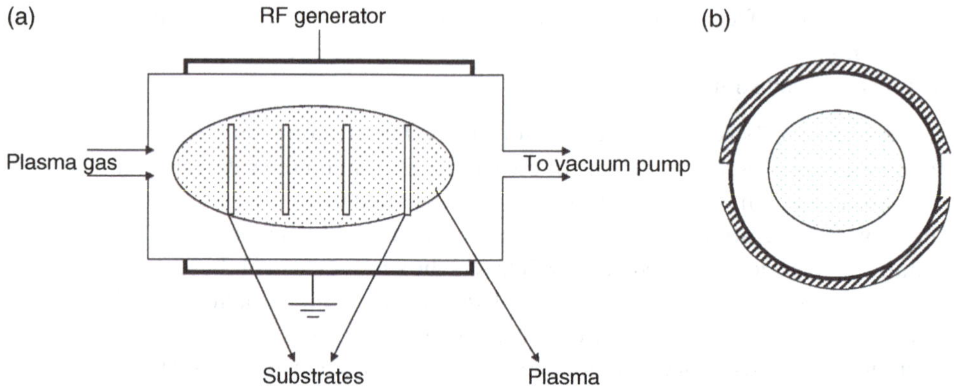

Figure 1.10 Reactor with capacitive coupling having cylindrical electrodes (*barrel etcher*): (a) longitudinal section including substrates; and (b) cross-section of the reactor.

The plasma etching is often carried out using RF plasma in cylindrical reactors using inductive (Fig. 1.9c) or capacitive coupling in a configuration with flat (Fig. 1.9b) or cylindrical electrodes as in ***barrel etcher*** shown in Fig. 1.10a,b. An example of plasma etching using such installation was GaAs substrate treated with plasma generated using the gas mixture of CCl_2F_2 and Ar. The etching rated was about 200 nm/min [43]. This process can be applied for removal of organic contaminants for example the remaining organic masks (photoresist) used in ***photolithography*** or oxidation of polymers with oxygen plasma.

1.3.2.5 Ultrasonic Cleaning

Ultrasonic cleaning uses ultrasounds having frequency ν = 20–40 kHz to agitate a liquid. The fluid is generally water, which can be used for ceramic or glass substrates, but other solvents can be used to clean metals and alloys. The process uses cavitation bubbles induced by the waves generated by ultrasound. The cleaning process is relatively short (few minutes), and the cleaning of blind holes in a substrate is possible. However, the action of the bubbles collapsing on a surface may lead to its erosion and fracturing [27]. To show an example, Jundale et al. [44] cleaned ultrasonically the amorphous glass substrates prior to the deposition of $CoFe_2O_4$ films by spray pyrolysis.

1.3.2.6 Photons Cleaning

The cleaning with the photons is realized nowadays mainly using lasers. The lasers used in industry are shown in Table 1.9. They emit radiation of different wavelengths, starting from *uv* (excimer laser) up to far *ir* (CO_2 lasers). The laser beam is focused on the surface of cleaned material. A part of laser power is reflected from the surface, while another part is absorbed. The reflectivity, R, and the depth of absorption, L, for some materials are shown in Table 1.10. It is possible to see that the major part of radiation at far *ir* ($\lambda \approx 10\,\mu m$) is well reflected by metals. The metals and alloys are to be treated rather by Nd: YAG laser. Consequently, the radiation of CO_2 laser matches better to treat ceramics. The optical properties result from the electronic structure of ceramics, which have completely filled valence bands, and no free electrons are available and the radiation is absorbed by the high-frequency

Table 1.9 Industrial lasers used in surface cleaning [15, 45, 46].

Parameters	CO_2 – laser	Nd: YAG – laser	Excimer laser	Diode laser	Fiber laser
Wavelength range, μm	10.6	1.06	0.2–0.3	0.4–3.3 depending on emitting diode	0.9–1.55
Continuous – C/ pulsed – P	C/P	P/C	P	P/C	P/C
Maximal average power, kW	25	2	0.4	4	20
Beam quality	Very high	Low	Low	High	Excellent

Table 1.10 Optical data for selected metals and oxides at the wavelengths about 1 and 10 μm.

Material of substrate	Wavelength, λ, μm	Reflectivity, R, dimensionless	Optical absorption depth, L, μm
Al	9.54	0.99	0.21
	0.83	0.87	0.022
Ni	9.54	0.98	0.14
	1.03	0.72	0.046
W	10	0.98	0.16
	1	0.58	0.068
SiO_2	10.6	0.2	40
	1.06	0.04	>106

Source: Pawłowski [47]/Springer Nature.

phonons. The detailed information concerning the interaction of lasers with different materials is presented in the chapter describing laser methods of films and coatings deposition.

The properties of the surfaces to be cleaned can be modified by covering them with a film of water. The water film evaporates explosively when intensively heated by laser [46]. The vapors expels the particles of contaminants. Contrary to such ***wet cleaning***, the ***dry cleaning*** by laser, working in pulsed mode, is associated generally with a laminar inert gas flow parallel to the surface which removes the detached particles. The contaminants particles may be detached by the following way [46, 48]:

- ***Laser-induced surface vibration***, which results from cyclic thermal expansion of substrate during incoming laser pulses. The heating and cooling cycles result in vibrations of cleaned surface transferred to the contaminating particles, which are ejected.
- ***Particle vibration*** results from direct absorption of laser energy by particles. The laser pulses induce a light pressure on the particles leading to their possible ejection from the cleaned surface.

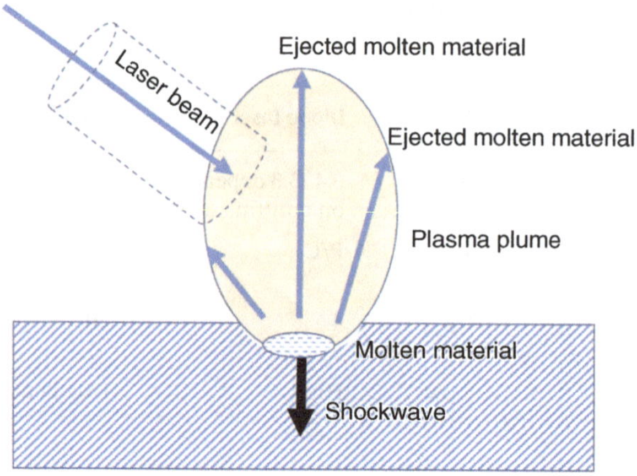

Figure 1.11 Sketch of laser ablation.

- ***Particle thermal expansion*** is induced by heating and cooling by laser pulses. The cyclic modification of particles size may lead to their detachment followed by ejection.
- ***Ablation***, shown schematically in Fig. 1.11, consist of formation of plasma plume by a laser pulse. The plume absorbs incoming energy and interacts with the cleaned surface by action of thermal shocks and rapid heating associated with evaporation of contaminants. Nd: YAG and excimer pulsed lasers are particularly well adapted to clean up by ablation the metallic substrates.

An important practical issue is time of cleaning. A choice of laser and its wavelength, the mode of action (continuous or pulsed), its density of energy or power are made in function of contaminant and of cleaned material (see Table 1.11). The surface is cleaned by the scans of a laser beam, and the velocity of the scanning is an important parameter to be optimized. It determines the time of treatment, which may be long especially for large surfaces to be treated. Laser cleaning is useful in many industrial applications, such as aeronautic, building, and maritime industries, as shown in Table 1.11. Razaba et al. [54] reviewed laser cleaning in automobile industry. Reitz [46] described laser cleaning of electronic components and artworks.

1.3.3 Monitoring of Surface

Monitoring can be roughly divided into two categories: (i) development of cleaning procedure; and (ii) routine monitoring of surfaces before films and coating deposition. The test at the development of the procedure includes mainly the microscopic observation of cleaned surfaces topography with the use of optical and scanning electron microscopes (*SEM*). Details having size in a nanometer range or even smaller can be observed with atomic force microscope (*AFM*). The surface chemistry, i.e. identification of elements, can be tested using X-rays photoelectron spectroscopy (*XPS*) or X-ray fluorescence (*XRF*). Finally, the crystal phases present on a surface can be identified using X-ray diffraction (*XRD*). All these methods are used also to characterize films and coatings, and are described in the chapter dealing with their characterization. The surface cleaning specifications may include other properties, such as hydrophobicity. In this case, it is useful to carry out the

Table 1.11 Examples of laser cleaning for different application.

Laser type → / Cleaning details ↓	CO_2	Nd: YAG	Excimer	Diode	Fiber
Wavelength, λ, μm	10.6	1.064	0.248	0.81	1.064
Mode pulsed/ continuous (P/C)	C	P	P	C	P
Laser power, kW	0.13	0.1–88	—	0.06	0.2–0.8
Pulses parameters		$\tau = 100\,ns$, $\nu = 1–30\,kHz$	$\tau = 25\,ns$, $\nu = 4\,Hz$, $E = 650\,mJ$		$\tau = 30\,ns$, $\nu = 20\,kHz$
Scan speed, mm/s	25	50	—	250	4000
Substrate	Ti alloy		Ag	Concrete	Al alloy
Contaminants	Oil film		Sulphur	Chlorinated rubber	Marine biofouling film
Industrial field of application	Aeronautic		—	Building	Maritime
Remarks	Ar with O_2 atmosphere	—	Laser KrF, fluence 200–400 mJ/cm^2	O_2, N_2, or Ar atmosphere	—
Reference	[49]	[50]	[51]	[52]	[53]

measurement of wetting angle. Tian et al. [53] made such measurements for different fiber laser fluencies by cleaning of biofouling from aluminum alloy surfaces. The authors found out that laser cleaning with the greatest fluency results in greatest wetting angle indicating surfaces with excellent hydrophobicity.

Mattox [27] proposed a couple of simple quality tests for glass surface cleaning. The *water-breaking test* consists of observations of water sheet, which breaks up on the surface impurities and remains continuous on the clean surface. Even simpler is a *water condensation test*, which shows the impurities of the surface covered with thin condensed water vapors. This test can be done each morning on the mirror of the bathroom after a shower.

1.4 Patterning of Films and Coatings

The films and coatings are frequently designed to have a particular shape, which results from the function to be fulfilled. The patterning is particularly often used in atomistic films. The desired planar shape of a film can be reached by different types of ***lithography***, which consists of masking the area where this film should not be deposited. An alternative technology is ***direct patterning*** which consists of deposition of the film in desired area on a substrate. Finally, the *mechanical* or *stencil masks* are used in thermal spray technology which is an example of particular coatings deposition. The patterning using ***mechanical masks*** is easier and cheaper than post-deposition grinding of coating. Such masking can be made using, e.g. temperature-resistant tapes, steel, aluminum foil, and wires. Dupuis et al. [55] applied such patterning to the manufacturing of a pyramidal array with the use of cold-spray technology.

An important issue is the size of the planar objects to be reached by patterning. This size decreases rapidly; the driving force behind this tendency comes from electronic and semiconductor industries manufacturing the integrated circuits (*IC*) using planar technologies on silicon substrates (so called *wafers*). The integration of these circuits increases steadily to reach a very large scale of integration (*VLSI*), and the size of planar elements composing a circuit stepped down to reach at present 10 nm [56]. The empirical ***Moore law***, valid in semiconductor industry since the beginning of 1970, predicts that the number of transistors in an *IC* grows exponentially (see Fig. 1.12).

Consequently, the patterning methods are different for preparation of micro-sized and nano-sized planar objects [57].

The *direct patterning* or *mask-less manufacturing* of planar objects can be made with use of different types of laser or with particles (electrons, ions) beams [2, 58]. In particular, laser patterning includes chemical transformation under action of photons. Finally, some granular techniques, such as cold spraying, enable the *direct patterning* of the planar objects of large sizes (cm). This technique is useful in a rapidly growing field of ***additive manufacturing***.

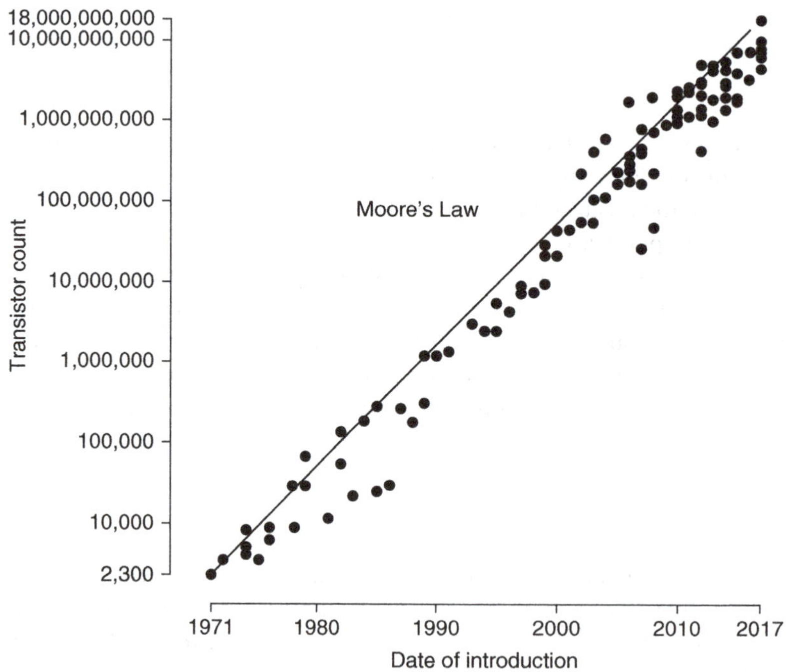

Figure 1.12 Illustration of empirical Moore's law valid in semiconductor industry [56]. *Source:* Reproduced with permission.

1.4.1 Lithography

Lithography means originally (ancient Greek) printing of an image drawn on a surface of stone on another smooth surface [59]. *Lithography* uses different ways of illumination such as: (i) *uv* light in *photolithography*; (ii) X-ray lithography; and (iii) particles (ions or electrons) beams lithography. The pattern is generated initially in a photo-mask, *PM* (to be used many times), and is shifted on a photo-resist, *PR*, being a *uv*-sensible polymer. The *PM* is made, typically, on a glass plate. The final pattern on a substrate is transferred by a development and removal by etching of *PR*. The sketch of photolithographic processes is shown in Fig. 1.13. The stages include:

- Cleaning of a substrate, which is followed, for *IC* preparation on a Si *wafer*, by oxidation to get a barrier layer and application of a photo-resist (*PR*) as seen in Fig. 1.13a. The *positive photo-resists* are easily soluble after *uv* illumination and negative ones are less soluble. Some authors recommend using bottom organic antireflective coating (*BARC*) to avoid the *uv* light reflection under *PR* and to improve its efficiency [60]. The *PR* is frequently applied by centrifugal spinning and baked (pre-baking).
- The *PM* is subsequently aligned above the *PR* (Fig. 1.13b). The mask can be in a direct contact (*contact printing*) or in a small distance of tenths of 15 μm to avoid contamination (*proximity printing*). Finally, the mask can be set at a distance of a few cm in so-called *projection printing*. The following step is the exposure to *uv* light illumination (Fig. 1.13c).
- The exposed part of positive *PR* is dissolved by a solution called frequently "developer" and removed from the substrate leaving the places on the substrate ready to be coated (Fig. 1.13d). The remaining *PR* should be hard-baked again to improve its resistance against deposited film. The etching process may be applied to clean up the part of the substrate to be deposited. The deposition of a film follows.
- The final stage of photolithography consists of removal of residual *PR* to leave a patterned film on the substrate (Fig. 1.13e). The removal can be made using oxygen-containing plasma, which oxidizes *PR* and detaches it from the substrate.

The processes are similar in all kinds of lithographic processes and will be analyzed more carefully in following sections. Some practical applications in different technologies of elements including films deposition will be also presented.

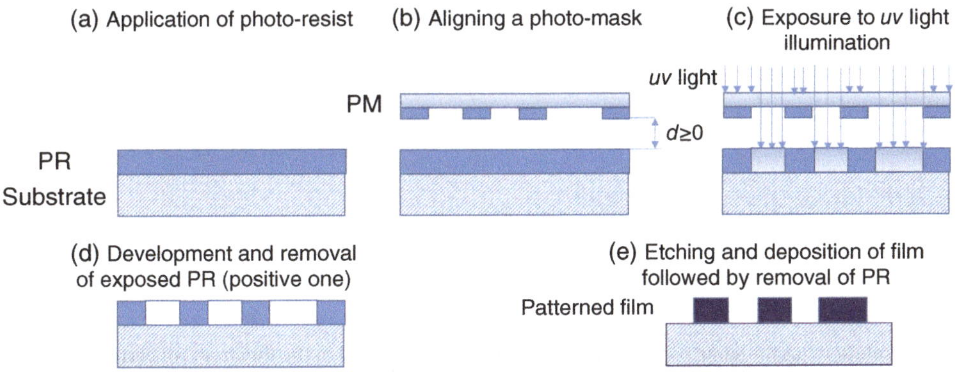

Figure 1.13 Sketch of patterning of a film with photolithography: (a) application of *PR*; (b) aligning of *PM*; (c) illumination with *uv* light; (d) development of *PR*; (e) removal of *PR*.

1.4.1.1 Illumination

Illumination is an important part of *photolithography* as well as *direct patterning*. It is realized using light or beams (X-ray, ions, or electrons). The description of *X-ray lithography* can be found elsewhere [59]. This section describes the particles' *e*- and *i*-beams. The simplified geometry of illumination at projection printing includes a lens and an aperture of a *PM* as shown in Fig. 1.14.

Light

An important parameter enabling obtaining a clear image of an aperture on the *PR* is resolution, which determines the minimum size of a pattern to be printed. The resolution[8] in the direction x or y, i.e. on the surface of *PR*, δ_x or δ_y, for light illumination is given, for the geometry shown in Fig. 1.14, in a medium having refraction index, n, by ***Abbe's equation*** [61]:

$$\delta_{x,y} = \frac{\lambda}{2n\sin(\theta)} \tag{1.7}$$

Knowing the value of refractory index for air $n = 1$, the value of product[9] $n \sin(\theta)$ can reach the value of 0.8 [62]. Consequently, it is possible to understand the reason of applying illumination with particularly short *uv*, being also called deep ultraviolet (*DUV*), such as emitted by excimer lasers having wavelength of $\lambda = 248$ nm or ArF with wavelength of $\lambda = 193$ nm. Similarly, the Nd: YAG laser may have shorter wavelength than $\lambda = 1062$ nm by using the different harmonics of laser frequency, namely $\lambda = 532$ nm or even $\lambda = 266$ nm [58]. Lithography with the wavelength of $\lambda = 248$ nm together with small thickness of *PR* was used to obtain the elements of semiconductor devices smaller than 50 nm [60]. However, to obtain the nanostructural size objects shorter-wavelength illumination sources are

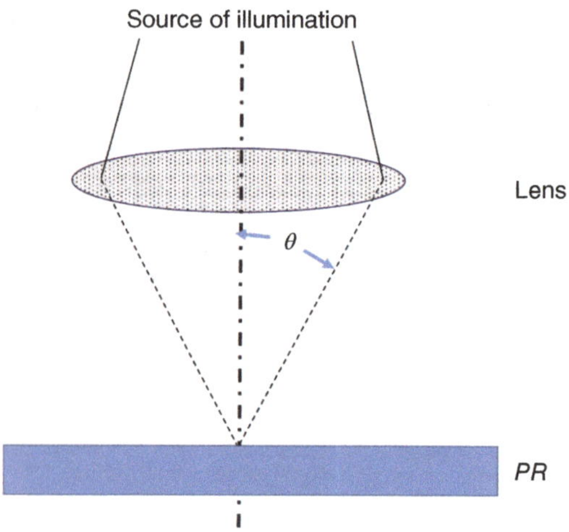

Figure 1.14 Simplified geometry of illumination at lithography.

8 The resolution has the same order of magnitude as, frequently used in the literature concerning lithography, critical dimension (*CD*) being a minimum size of a manufactured element

9 The product is also called numerical aperture, $NA = n \sin(\theta)$

recommended [57], such as *soft X-rays* having the wavelength of a few nanometers as well as particles beams described in the following section.

Particles Beams

Each particle can be associated with a wave following ***de Broglie hypothesis***. Its wavelength can be found by knowing particle's velocity and mass from the following equation:

$$\lambda = \frac{h}{mv} \tag{1.8}$$

The electrons and ions may be accelerated by electric field, and their wavelengths for different velocities are shown in Table 1.12. The application of the beams of electrically charged particles in lithographic process is however different from that of electrically neutral photons. The particles may interact with *PR* and induce some chemical and physical changes. Additionally, their penetration depths and profiles are different [57]. The *e-beam lithography* was used, e.g. for patterning of Mo/Si multilayer system to obtain structures having an accuracy of a few nanometers [63]. The *i-beam lithography* using heavy ions such as Ga^+ or of protons H^+ have an advantage, compared to *e*-beams, of greater penetration and giving straighter paths at patterning [59].

1.4.1.2 Masks

The masks used in lithographic processes can be made of different materials using different technologies as shown in Table 1.13.

The materials of masks include metals, ceramics, and polymers sometimes prepared as multi-films. The choice of materials is consistent with entire technology of planar objects manufacturing. The applications are mainly in microelectronic industry and the substrate is frequently silicon *wafer*. The patterning of the masks is realized by different methods including photolithography or "natural" patterning by the holes formed at growth of anodized alumina films. Finally, the geometry of films masking is discussed in the next section.

Table 1.12 Wavelengths corresponding to particles having different energies and velocities[a].

Particles	Velocities (km/s)	Kinetic energies, eV	Wavelengths
Electrons	100	0.0284	7.27 nm
	500	0.711	1.45 nm
	1000	2.84	0.727 nm
Protons, H^+	100	5.22	3.96 pm
	500	1300	0.793 pm
	1000	5220	0.396 pm

[a] Mass of an electron is $m_e \approx 9.11 \times 10^{-31}$ kg and mass of a proton is, $m_p \approx 1.67 \times 10^{-27}$ kg.

Table 1.13 Different masks used in lithographic processes of film patterning.

Mask material	Kind of patterning	Thickness, nm	Substrate	Geometry of film masking	Final application	Reference
Cr	Photolithography	4, 14, and 114	Si	Proximity printing	Capacitive microphone	[64]
Ni	*EBPVD*	15	Sapphire	Contact printing	Light-emitting diode	[65]
Al	Diamond tip of *AFM*		Si		Development of *AFM* lithography	[66]
PI/ Si/ *PMMA*	*e-beam* lithography	*PI* – 600, Si – 40, and *PMMA* – 200			Bragg-mirrors	[63]
Al_2O_3	Holes formed at anodizing	700	Al_2O_3		Nanodot arrays	[67]
SiN	—	500		Proximity printing	3D patterns	[68]
DLC	*ICP*	—	Si	—	Development of hard mask	[69]

1.4.1.3 Geometry of Lithography

The geometry of lithography depends mainly on the distance between the photo-resist (*PR*) and photo-mask (*PM*) during exposure to light or beams. The possible geometries, being also the ways of printing the form of the mask, are shown in Fig. 1.15. The methods include following ways of printing [59]: (i) contact; (ii) proximity; and (iii) projection.

The contact printing has an advantage of the highest possible resolution, but the mask gets used in contact with the *PR*. The damage of the mask is reduced at proximity and entirely eliminated at projection printing.

1.4.1.4 Etching and Removal Processes

Etching and removal occur many times during lithography. They are applied to the photo-resists (*PR*) exposed to illumination and are used to remove the residual one (see Fig. 1.13d, e).

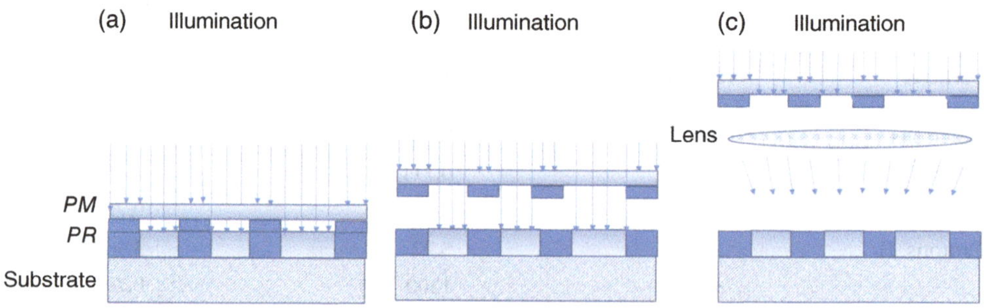

Figure 1.15 Possible geometries of *PM* and *PR* at following lithographic printing: (a) contact; (b) proximity; and (c) projection. *Source:* Adapted from Sebastian et al. [59].

Some planar technologies are realized with the *PM* being in direct contact with *PR* (see Fig. 1.15a). This requires also etching and removal of the masks and deposited films made of different materials. The etching can be roughly categorized as: (i) wet, realized with the liquids; and (ii) dry, done mainly using plasma gazes including reactive ions in a process called *reactive ion etching* (*RIE*).

Wet etching with the use of the alkaline aqueous solution (having $pH > 7$) is recommended for polymer *PR* after exposition to illumination [59]. Liu et al. [70] described a technology including etching of double-layer mask including Mo and SiO_2. The Mo layer was etched by a solution of HNO_3, H_3PO_4, CH_4COOH, and H_2O. SiO_2 was etched with a solution of HF, NH_4F with H_2O called buffered oxide etch (*BOE*).

Dry etching is frequently made using inductively coupled plasma (*ICP*) shown in Fig. 1.9c with an industrial frequency of $\nu = 13.56$ MHz. The *RIE* processing of a multilayer including Mo and Si was made with plasma gases including CF_4, SF_6, CHF_3, O_2, and Ar [63]. Nitride layers of GaN and InGaN were etched using plasma working gases Cl_2 with Ar [64]. Finally, the polymer *PM* was reportedly etched using O_2 with N_2 plasma working gases having optimized composition [60].

The final removal of polymer *PR* is also made using plasma containing O_2, which oxidizes it in a process called *plasma ashing* in semiconductors technology.

1.4.2 Direct Patterning

The very fine dimensions of photons or particles beams and the possibility of precise control of their movement by the robots controlled by computers enable the direct patterning. The process does not use masks and is called by some authors, such as Mayer and Allen [58], selected area processing (*SAP*). The patterning can be used to manufacture the unique products as well as many identical ones. The beams can be used to deposit directly the desired film or to create a pattern with a compound being as a gas or solid close to substrate's surface. The beams actions may have physical characters such as heating, scattering, or diffusion, or chemical ones such as photochemical synthesis or decomposition. These fundamental aspects are well described in the references [2, 58], and the sections below focus rather on the beams application.

1.4.2.1 Photons Patterning

The laser beams focused on a substrate may lead to local growth of films on the substrate by photon-activated reactions of the decomposition of gas- and solid-phase compounds and the reaction of oxidation. The gaseous compounds should be in the neighborhood of substrate surface, and the solid ones should be deposited on it. An example of reaction in gas phase is the decomposition of $(CH_3)_2Cd$ and subsequent nucleation and growth of Cd film on SiO_2 substrate [58]. The authors of [2] described solid-phase decomposition of $Cu(HCOO)_2$, at temperature of 620 K by action of the CO_2 laser working in continuous mode to form Cu film on SiO_2 substrate. Another possible reaction assisted by laser is local oxidation of Cr film deposited onto glass surface to form Cr_2O_3. This kind of patterning may be used to manufacture wear-resistant photo-masks.

Lasers may be also useful to etch the surface locally. The mechanisms of etching involve mainly [71]: (i) ablation; (ii) photo-chemical reactions in gas- and liquid phases;

and (iii) photo-thermal processes. The processes may occur simultaneously. For example, photo-thermal and photochemical processes enabled local etching and reduction of graphene oxide to graphene using a femtosecond laser having wavelength of 800 nm [72]. The examples of etching for all types of materials including metals, semiconductors, ceramics and polymers using different lasers are described in [2].

1.4.2.2 Particles Patterning

The *e*-beams have been used in mask-less patterning since 1970. The main advantage is high resolution of such patterning resulting from the wavelength associated with the particles being smaller than 10 nm as shown in Table 1.12. The patterning may be carried out on photo-resists such as *PMMA* or *Novolak* and is used frequently to fabricate the photo-masks. The process is however slower than *photolithography* [59]. The *i*-beams applied in ***mask-less lithography*** use light ions such as H^+ or heavy ones such as Ga^+. The interaction of heavy ions with the substrate surface may lead to the formation of defect sites. The sites are useful as nucleation centers on nonreactive surfaces [58]. These authors give the example of the beams applications. Consequently, the *e*-beam having energy of 3 keV was used for the nucleation of Fe film on Si by decomposition of gaseous-phase $Fe(CO)_5$ at temperature of 423 K. On the other hand, the Ga^+ beam having energy of 50 keV enabled growth of Fe and Al films on SiO_2 substrate. It is possible to use the heavy ions to get implanted as atoms in the substrate. The resolution of *i*-beams is very high, and the lines of GaAs having width of only 2–3 nm were manufactured with such beam [57].

1.5 Activation of Substrates Surface

The surface of a substrate is frequently activated to obtain or to improve the adhesion of films and coatings. The adhesion is related to the properties of the deposit and substrate, which determine the phenomenon occurring at their deposition and can be called ***intrinsic adhesion***. Other phenomena influencing adhesion occur during their service.

The *intrinsic adhesion* depends mainly on [73]:

- surface of the substrate including its roughness and *surface energy*;
- chemical affinity and, in particular, that of similarity of crystal lattices between a deposit and a substrate and on impurities sand structural defects in the film/coating;
- size, shape, and energy of the species building up the deposit resulting in the deposits microstructure;
- difference between the coefficients of thermal expansion, *TEC*, of the film/coating and the substrate in the case of a deposition process in the temperature much higher than the ambient one; and
- internal stresses in deposit and on their thickness.

The species forming film in atomistic methods of deposition are atoms and molecules. Their sizes are in the nanometers range, being much smaller than that used at granular deposition methods. The latter used the species having size in the range of tenths to hundreds of micrometers. Consequently, the adhesion mechanisms and the factors influencing it are totally different. In atomistic methods, these factors are related to nucleation of the islands forming film and may concern the atomic defects of substrate or small impurities attached to them. The species used at granular deposition as thermal spraying are 10^3–10^4 greater and arrive being molten on the substrate surface. The adhesion of coating depends in this case on the substrate roughness and on their chemical affinity. The latter may enable the rapid diffusion of the atoms of the impacting particle to the substrate. The rapidity of the diffusion results from the short time of solidification of particle (in the range of microseconds) and of its cooling (seconds to minutes) [74]. The methods of surface activation are consequently very different and are discussed separately for atomistic and granular methods of films and coatings deposition. The process of bulk coatings generation is associated with melting of substrate's surface. Consequently, the surface preparation is less important than an interaction between solid reinforcement and liquid metal [75].

The ***adhesion in service*** is related to the phenomena occurring at films and coatings service. For example high temperature at service of coatings being part of thermal barriers or mechanical stresses occur in the films on the cutting tools, and may depend on the nature and level of thermal stresses (e.g. thermal shocks) and mechanical stresses. The mechanical properties, especially of elasticity modulus, E, of the coating and of the substrate play also an important role.

The difference of these two types of adhesion becomes evident when a film/coating adheres to the substrate after deposition procedure but detaches off the substrate at its service. Another example is a *DLC* film, which has excellent mechanical properties but cannot withstand the stresses generated at application on the surface of cutting tools.

Intrinsic adhesion can be improved by the various pre- and post-deposition treatments, while *adhesion in service* can also be improved by applying a bond-coat. The bond-coat for metallic substrates is typically a metal or an alloy, whose properties (*TEC*, E) are frequently intermediate between the substrate and the deposit.

1.5.1 Activation of Substrates at Atomistic Deposition

The atomistic deposits starts to grow by formation of clusters of atoms or molecules followed, generally, by formation of islands. This growth, described in detail in the appropriate chapter, depends on many parameters including substrate and its contact with growing films. The analysis of evaporated Al and Ti films nucleation and growth on different substrates did show that the stresses generated at the films' growth might modify the type of growth from island to layer-by-layer growth [76–78]. The stresses resulted from the presence of gases incorporated in the substrate.

Careful substrate cleaning is necessary to reach good adhesion of deposits. The bond-coats may also improve the *intrinsic adhesion* of films very different from the substrate. Especially, the hard and wear-resistant diamond-like films require intermediate films to improve their adhesion to metallic substrates. Consequently, Ti bond-coat was applied to improve adhesion of *DLC* films to steel substrates [79]. The adhesion of such films to Al substrate was improved by adding a small quantity of aluminum atoms to the graphite target used for deposition. These atoms sputtered together with carbon ones forming a very thin intermediate layer resulting in stronger adhesion [80].

The methods of surface activation depend on the activated substrate material and are realized using liquids, photons, and large (sand) or small particles (molecules, atoms, and ions).

1.5.1.1 Liquids Activation

Liquids are frequently applied for substrate cleaning by chemical etching[10] in semiconductor industry. The GaAs wafers were cleaned up with some acids before the $(NH_4)_2S$ solution was applied. This solution enabled sulfidation of surface and avoidance of oxidation prior to the application of metallic layers [81]. The chemical treatment using HF solution was one of the methods applied to prepare Si substrate before deposition of Al_2O_3 thin film by atomic film deposition (*ALD*) method [82]. The same acid was a part of WC-Co substrate surface activation prior to deposition of *DLC* film by *CVD* method [83].

1.5.1.2 Photons Activation

The laser may activate the surface by evaporation of impurities and modification of its morphology. This activation is called laser surface texturing (*LST*). An example of such texturing including formation of dimples, crossed grooves, and concentric rings obtained on an aluminum alloy substrate using Nd: YVO_4 picosecond laser is shown respectively in Fig. 1.16a–d.

Table 1.14 shows a few examples of surfaces activation using *LST*. Such texturing can be made with *cw* laser, which heats up the surface. The heating may evaporate a lower melting point part of composite such as cobalt in WC-Co [85].

Another possible action is evaporation of small holes in thermoplastic polymer by short laser pulses. Such texturing enabled increase in the adhesion strength of evaporated aluminum coatings to *PEEK* from 0.7 to 5.5 MPa [86]. Finally, an appropriate pulsed laser texturing enables reaching the density of diamond clusters nucleation equal to about 5.3×10^7 clusters/cm^2 [87].

1.5.1.3 Particles Activation

Activation using coarse particles such as sand is a method used frequently as a pretreatment prior to granular coatings. Some authors used it however to prepare substrate before deposition of thin films. In particular, Zhang and Zhou [83] used sand-blasting to treat hard WC+ 6 wt. % Co substrate prior of *CVD* deposition of a *DLC* film and Grubova et al. [88] applied sand to blast on Ti-substrate prior to the deposition of Ag-doped *HA* by sputtering. It must be added that the sand-blasting was followed in cited studies by acid etching to complete the substrate activation.

The atomic-size species, i.e. ions, atoms, and molecules, are used frequently to activate surface. The species are part of plasma, and their energy of ions at treatment depends mainly on gas pressure and acceleration voltage. Some examples of atomic-size species treatment are shown in Table 1.15.

---

10 See also Section 1.3.1 of this chapter

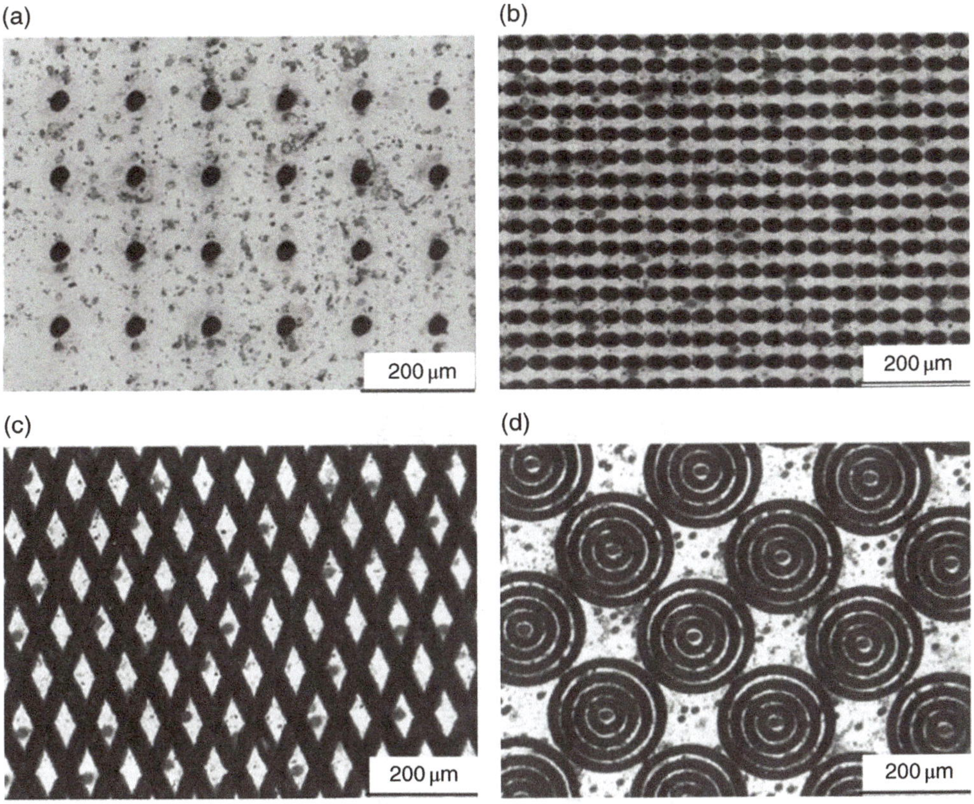

Figure 1.16 Optical micrographs of laser-treated surface of aluminum alloy (Alloy AA2024-T3 of Al + (3.8–4.9) wt. % Cu + (1.2–1.8) wt. % Mg) textured using Nd: YVO_4 picosecond laser (λ = 1064 nm) showing different densities of dimples (a, b), crossed grooves (c), and concentric rings (d). *Source:* From Min et al. [84]/Reproduced with permission from ELSEVIER.

The *i-beam* pretreatment renders possible the chemical reaction in the interface between film and substrate. Gheriani et al. [89] reported about formation of carbides such as Cr_7C_3 between steel substrate and Cr film. This reaction occurred after high-temperature annealing of coated samples rendering possible diffusion of small C atoms in in Cr film. The choice of ions and their energy used to the treatment must be adapted to the substrate material [90, 91]. Consequently, the application of heavier Xe^+ ions activated better the surface of polymers than Ar^+ and, on the other hand, the highest bias voltage of $U = -600\,V$ was most favorable for adhesion of *DLC* films to steel substrate.

Plasma treatment is useful for activation of polymers' surface. Corona discharge is frequently used in this process. The discharge, done in open atmosphere, breaks the polymer chains and creates the radicals [93]. The treatment using arc plasma torch also improved the adhesion of Al films to *PE* and *PEEK* polymers [86]. Finally, a plasma generated under vacuum by a high-voltage pulse (up to $U = -15\,kV$) in presence of different gases such as C_2H_4, N_2, or O_2 enabled to activate Cu substrate surface and render possible the deposition of *DLC* films [92].

Table 1.14 Examples of different substrates activated prior to atomistic film deposition using laser surface texturing.

Laser	Wavelength, nm	Substrate material	Film material	Method of film deposition	Final application	Remarks	Reference
Diode laser *cw*	940	WC + 6 wt. % Co	Polycrystalline diamond	*HFCVD*	Wear-resistant films	Co-binder was removed	[85]
Nanosecond laser[a]	355	Thermoplastic polymers: *PP*, *PEEK*, and *PE*	Al	Evaporation	Automotive, aerospace, and medical applications	Square grid pattern was textured	[86]
Nd: YAG nanosecond	1064	Nb	Boron-doped diamond	*HFCVD*	Waste water treatment electrodes	Laser spot has diameter of about 30 μm	[87]

[a] Laser type was not precised in the paper. The lasers Nd: YAG, diode, or fiber emit this wavelength.

Table 1.15 Examples of surface activation prior to atomistic film deposition using species having atomic sizes.

Method of treatment	Pressure at treatment, Pa	Substrate material	Film material	Method of film deposition	Final application	Remarks	Reference
Activation by ions	10	Steel 100C6	Cr	Sputtering	Wear-resistant films	Ar^+ ion treatment improved films adhesion and their microhardness	[89]
	—	Polymers as *PC*	HfO_2 or ZrO_2		Optical antireflection coatings	Ions Ar^+ and Xe^+	[90]
	11	Stainless steel 304	*DLC*	*PECVD*	Wear-resistant films	Ions Ar^+ accelerated by voltage up to 600 V	[91]
Plasma	0.7	Cu		Ion implantation		Preimplantation using N_2 and O_2, Ar, CO_2, and C_2H_4	[92]
	10^5	Thermoplastic polymers: *PP*, *PEEK*, and *PE*	Al	Evaporation	Automotive, aerospace, and medical applications	Arc-generated air plasma	[86]

1.5.2 Activation of the Substrates Before Granular Deposits

The granular deposits grow up starting from molten particles, which impact the substrate's surface. The size of particles varies from sub-micrometers occurring in the methods of spraying using liquid feedstock such as solution precursor plasma spraying (*SPPS*), suspension plasma spraying (*SPS*) or spray pyrolysis (*SP*) up to many tenths of micrometers in the methods using solid feedstock such as atmospheric plasma spraying (*APS*), cold-gas spraying method (*CGSM*), or arc spraying (*AS*). The liquid droplets are transformed into splats. The size and shape of solidified splats depend on their velocity at impact and on the topography and temperature of surface [74, 94]. The topography of the surface and, in particular, its roughness are essential in the mechanism of adhesion called ***mechanical anchorage***. The methods of activation modify this topography. The description of this modification is generally related to the roughness. However, the laser surface texturing (*LST*), which is at present intensively researched as activation method, enables more precise description of the holes array generated by laser [95]. The final result of activation is the improvement of coatings adhesion. The adhesion is generally characterized using standard tensile strength test ASTM C633–01[11].

1.5.2.1 Particles Activation

The substrates for granular deposits are activated mainly by grit-blasting. The method consists of accelerating the particles of hard material (corundum, hard steel, silica, silicon carbide) in a jet of compressed air. The ceramic particles have generally angular shape, and the metallic ones sometimes are spherical. The particles bombard the surface of a substrate making it rough. The industrial installations are equipped in several simultaneously acting nozzles being controlled numerically controlled automata. The main blasting parameters are shown in Table 1.16.

The grit particle size used in blasting depends on the thickness of the workpiece (fine grit is recommended for thin deposits) and the desired surface roughness of final deposit (fine grit blasting renders the surface smoother). Roseberry and Boulger [96] made the recommendations of grit size for the plasma-sprayed coatings of different thicknesses. Consequently, silica grit is often used for thin coatings applied without polishing, while alumina is used as a coarse one.

The grit's size determines the roughness of the treated surface. Mellali et al. [97, 98] analyzed substrates of different elasticity and found that the roughness after grit-blasting is lower for the substrate having greater elasticity modulus. Similarly, Wigren [99] after having studied different metals and alloys blasted with alumina grit found that the R_a parameter decreases linearly from 4.5 to 2.5 μm with the substrate's modulus of elasticity increasing from 40 GPa (Mg) to 240 GPa (Co-based Haynes 188 alloy composed of Co + 22 wt.% Ni + 22 wt. % Cr + 13 wt. % W + 3 wt. % Fe). Finally, grit blasting generates compression stresses in the substrate, and thin substrates can deform after blasting.

---

11 Standard Test Method for Adhesion and Cohesion of Thermal Sprayed Coatings of 10 March 2001

Table 1.16 Principal grit-blasting parameters.

Process element	Parameters
Grit	Material, particle size, hardness
Activated surface	*Young's* modulus, geometric dimension (size and thickness), hardness
Grit-blasting medium	Compressed air
Blasting set-up	Blasting in a booth or in open atmosphere
Technique of blasting	Processing time, air pressure, blasting distance

The research on the influence of grit-blasting parameters and on the surface roughness R_a or on the *intrinsic adhesion* of deposits to the substrate may be summarized in the following way:

- roughness, characterized by R_a:
  - reaches a maximum after some blasting time;
  - increases with the blasting pressure;
  - is not strongly influenced by blasting distance between nozzle and substrate;
- the grit blasting angle[12] may influence the roughness and the adhesion:

It is important to mention that the grit particles may get embedded in the treated surface. This embedment may decrease the adhesion of the coating. Subsequently, embedment is a major disadvantage of activation by grit blasting.

The *dry-ice* (solid CO_2) blasting enables avoiding embedment. The solid CO_2 sublimes without forming liquid phase and was applied *in situ* to activate Ti6Al4V (Ti + 6 wt. % Al + 4 wt. % V) alloy during spraying hydroxyapatite (*HA*) [100]. The roughness, R_a, of sprayed deposit was slightly smaller at *dry-ice* treatment (R_a = 8 μm) than after pre-treatment using alumina sand blasting (R_a = 12 μm). The bond strength of as-sprayed coatings increased however up to 55 MPa comparing to 22 MPa in samples prepared using sand-blasting.

1.5.2.2 Photons Activation

The laser surface texturing (*LST*) activation of substrates prior to granular coating deposition has been intensively researched in recent years. The authors drilled holes using different lasers in the substrate in the way shown in Fig. 1.16a and defined the geometry-treated surface by giving following parameters:

- distance between the holes, being generally equal in *x* and *y* directions;
- diameter and depth of the holes;
- angle of the hole axis (being angle of laser treatment) with regard to the normal to surface.

12 The blasting angle of 0° corresponds to the direction normal to the surface

The diameters of the holes are comparable to the estimated size of liquid droplets arriving on the substrate. Otherwise, the droplets cannot penetrate into the holes. Similarly, the angle of laser treatment increases the hole diameter as found by Kromer et al. [95] for the Al substrate drilled using pulsed Nd: YAG laser.

The description of different coating and different deposition methods obtained by *LST* pre-treatment are mentioned in Table 1.17. All authors used pulsed fiber laser and optimized the texturing geometry. The adhesion of coating deposited onto substrates textured using optimization improved considerably compared to other pretreatment methods. Kromer et al. [95] mentioned that for some *LST* conditions the adhesion strength was greater than coating cohesion. Texturing may modify the mechanical properties of substrate. The modification of microhardness was observed close to surface of aluminum alloy Al2017 submitted to texturing using pulsed fiber laser [103].

The lasers may be used *in situ*, i.e. during coating deposition. Two different Nd: YAG pulsed laser were used simultaneously to clean up and to heat up the substrate immediately before the plasma torch arrived at the treated area. However, the adhesion NiAl coatings to Al-alloy and Ti-alloy substrate decrease compared to grit-blasting pre-treatment, due to the decrease of surface roughness [104]. Finally, the shape of the laser-drilled holes at *LST* can be modified by 2-step processing. The first step enables holes drilling and the second one smoothens the surface and lowers its roughness. Such pre-treatment was made using nanosecond pulsed fiber laser, and the different laser fluences and scans velocities enabled 2 steps to be realized [105].

1.5.2.3 Water jet Activation

The treatment with a water jet does not have this drawback, but it is not yet used often in the industry. The research studies of Taylor and Knapp [106, 107], related to water jet treatment of Inconel alloys, enabled to find the essential parameters of this treatment (see Table 1.18).

The results of research show that the surface of alloy Inconel 718 starts to be rough at water pressure of $p = 200$ MPa. The interface between the plasma-sprayed MCrAlY coating and the substrate was perfectly clean compared to grit-blasted sample surface. Finally, the adhesion strength was between 70 and 80 MPa for coatings on substrates treated with water jet compared to the values of 46–78 MPa for coatings on the grit-blasted substrates.

1.5.2.4 Plasma Activation

Plasma treatment such as cleaning and activation can be realized under vacuum or under inert gas atmosphere. The realization of such pre-treatment in air may lead to the metal surface oxidation. Takeda and Takeuchi [108] studied pre-treatment using arc under vacuum. The authors focused on the cleaning of oxides from stainless steel substrate, which may be useful prior to vacuum plasma spraying (*VPS*). More recently, Lukauskaite et al. [109] used cathodic treatment prior to *APS*. The treatment was realized under protecting Ar in open atmosphere to activate and clean up aluminum alloy plates to be coated with NiAl. The final coatings' adhesion strength improved from 23 MPa for grit-blasted samples up to 32 MPa for cathodically treated ones.

Table 1.17 Examples of different substrates activated prior to granular coatings deposition using laser surface texturing.

Laser	Wavelength and pulse duration	Substrate	Geometry of textured array[a]	Initial powder and its size, μm	Method of coatings deposition	Final application	Remarks	Reference
Fiber laser	1064 nm 100 ns	Al 2017 alloy, Al + 4 wt. % Cu + Si, Mn, Fe	$D = 60\,\mu m$, $x = 100$, 150, 200 and 300, $L = 80\,\mu m$, $\alpha = 0$ and 30°	Ni + 5 wt. % Al, $d_{50} = 67\,\mu m$	*APS*	Bond-coat	Adhesion strength 34–52 MPa	[95]
		Steel ASTM 1045	$D = 40–120\,\mu m$, $x = 30–110\,\mu m$, $L = 50\,\mu m$	Ni60 alloy, Ni + 16 wt. % Cr + 9 wt. % Fe + 4 wt. % Si			Adhesion strength up to 50 MPa for $x = 30\,\mu m$	[101]
		Al alloy 7000, Mg alloy RZ6, Al + 20 vol. % SiC	$D = 50\,\mu m$, $x = 80$ and 150 μm, $L = 25\,\mu m$	Al alloy 6000, Mg alloy, and Al + 20 vol. % SiC, $d_{50} = 40\,\mu m$	*CGSM*	Light metal coating	Adhesion up to 50 MPa for Al-alloy	[102]
		Thermoset carbon fiber composite	$D = 50\,\mu m$, $x = 200$ and 300 μm, $L = 25\,\mu m$	Cu_05T wire	*AS*	Metallization	Adhesion up to 13 MPa	
		AM1 single crystal, Ni + Cr,Co, Ta, W, Al, Mo, Ti	$D = 60$ and 80 μm, $x = 100$ and 150 μm, $L = 40\,\mu m$	ZrO_2 + 7 wt. % Y_2O_3	*APS*	*TBC*	Adhesion up to 33 MPa	

[a] D is diameter of hole, x is the distance between holes' centers, L is their depth, and α is angle of between hole axis and normal to the surface.

Table 1.18 Typical parameters of surface activation using a water jet.

Parameter	Value
Water pressure, MPa	200–350
Water flow rate, L/min	3.5–4.5
Nozzle *ID*, mm	0.4
Distance between nozzle and substrate, cm	8
Scan speed of water jet, cm/min	30–130
Width of the trace of treatment in the substrate, mm	1.5
Shift of following traces, mm	0.8

Source: Adapted from Taylor [106].

References

1 Mattox, D.M. (1982). Adhesion and surface preparation. In: *Deposition Technologies for Films and Coatings* (ed. R.F. Bunshah), 63–82. Park Ridge, NJ: Noyes Publications.

2 Metev, S.M. and Veiko, V.P. (1998). *Laser-Assisted Microtechnology*, 2e. Berlin: Springer.

3 Burakowski, T. and Wierzchoń, T. (1995). *Surface Engineering of Metals*. Warsaw: WNT (in Polish).

4 Kingery, W.D., Bowen, H.K., and Uhlmann, D.R. (1976). *Introduction to Ceramics*, 2e. New York, NY: Wiley.

5 Wand, J. and Guo, X. (2020). Adsorption isotherm models: classification, physical meaning, application and solving method. *Chemosphere* 258: 127279.

6 Elder, S.H., DiSalvo, F.J., Topor, L., and Navrotsky, A. (1993). Thermodynamics of ternary nitride formation by ammonolysis: application to $LiMoN_2$, Na_3WN_3 and Na_3WO_3N. *Chem. Mater.* 5 (10): 1545–1553.

7 Ashby, M.F. and Jones, D.R.H. (1991). *Matériaux, Tome 1*. Paris: Dunod (in French).

8 Callister, W.D. Jr. (1994). *Materials Science and Engineering: An Introduction*, 3e. New York, NY: Wiley.

9 Lide, D.R. (ed.) (1997). *CRC Handbook of Chemistry and Physics*, 78e. Boca Raton, FL: CRC Press.

10 Frommeyer, G., Rablbauer, R., and Schäfer, H.J. (2010). Elastic properties of B2-ordered NiAl and NiAl-X (Cr,Mo,W) alloys. *Intermetallics* 18 (3): 299–305.

11 Technical Data of catalogue Delloro Stellite, Cheney Manor Industrial Estate, Swindon SN2 2PW, UK, 2008.

12 The Engineering ToolBox available on https://www.engineeringtoolbox.com/young-modulus-d_417.html checked on 18 September 2024.

13 Ashby, M.F. and Jones, D.R.H. (1991). *Matériaux, Tome 2*. Paris: Dunod (in French).

14 Dörre, E. and Hübner, H. (1984). *Alumina*. Berlin: Springer.

15 Pawłowski, L. and Blanchart, P. (2018). *Industrial Chemistry of Oxides for Emerging Applications*. Chichester: Wiley.

16 Samsonov, G.V. (ed.) (1978). *Physical and Chemical Properties of Oxides*. Moskva, USSR: Metallurgija (in Russian).

17 Snustad, I., Roe, I.T., Brunsvold, A. et al. (2018). A review on wetting and water condensation - perspectives for CO_2 condensation. *Adv. Colloid Interf. Sci.* 256: 291–304.

18 Eustathopoulos, N., Nicholas, M.G., and Drevet, B. (1999). *Wettability at High Temperatures*. Kidlington, Oxford: Pergamon Press and Elsevier Science.

19 Carr, J., Milhet, X., Gadaud, P. et al. (2015). Quantitative characterization of porosity and determination of elastic modulus for sintered micro-silver joints. *J. Mater. Process. Technol.* 225: 19–23.

20 Klein, C.A. and Cardinale, G.F. (1993). Young's modulus and poisson's ratio of CVD diamond. *Diam. Relat. Mater.* 2: 918–923.

21 Callister, W.D. and Rethwisch, D.G. (2008). *Fundamentals of Materials Science and Engineering*, 3e. New York, NY: Wiley.

22 Stachurski, Z. (1987). *Engineering Science of Polymeric Materials*. Parkville, VIC: Polymer Division, Royal Australian Chemical Institute.

23 Wyatt, O.H. and Dew-Hughes, D. (1978). *Metals, Ceramics and Polymers*. Warsaw: PWN (In Polish).

24 Seidel, W.W. and Hahn, F. (2014). *Werkstofftechnik*, 10e. Munich: Hanser.

25 Małachowska, A. (2015). *Analysis of the cold gas spraying process and determination of selected properties of metallic and cermet coatings*. PhD Thesis. University of Limoges and Wrocław University of Technology.

26 Mercier, J.P. and Maréchal, E. (1996). *Chimie de Polymères: Synthèses, Réactions, Dégradation*. Lausanne: PPUR (In French).

27 Mattox, D.M. (1994). Surface preparation for film and coating deposition processes. In: *Handbook of Deposition Technologies for Films and Coatings*, 2e (ed. R.F. Bunshah), 82–126. Park Ridge, NJ: Noyes Publications.

28 Sonoda, T., Watazu, A., Zhu, J. et al. (2001). A surface cleaning method for sputter deposition of pure titanium film onto TiNi shape memory alloy substrate. *Vacuum* 60 (1–2): 197–199.

29 Kratz, F., Grass, S., Umanskaya, N. et al. (2015). Cleaning of biomaterial surfaces: protein removal by different solvents. *Colloids Surf. B: Biointerfaces* 128: 28–35.

30 Buck, V. and Deuerler, F. (1998). Enhanced nucleation of diamond films on pretreated substrates. *Diam. Relat. Mater.* 7: 1544–1552.

31 Zhang, Z.M., He, X.C., Shen, H.S. et al. (2000). Pretreatment for diamond coatings on free-shape WC-Co tools. *Diam. Relat. Mater.* 9: 1749–1752.

32 Shen, M.R., Wang, H., Ning, Z.Y. et al. (1997). Enhanced diamond nucleation on pretreated silicon substrates. *Thin Solid Films* 301: 77–81.

33 Couillard, J.G., Ast, D.G., Umbach, C. et al. (1997). Chemical treatment of glass substrates. *J. Non-Cryst. Solids* 222: 429–434.

34 Lubna, H., Auner, G., Patwa, R. et al. (2011). Role of cleaning methods on bond quality of Ti coated glass/imidex system. *Appl. Surf. Sci.* 257: 4749–4753.

35 Cooper, K., Gupta, A., and Beaudoin, S. (2001). Simulation of the adhesion of particles to surfaces. *J. Colloid Interface Sci.* 234: 284–292.

36 Thornton, J.A. and Greene, J.E. (1994). Plasma in deposition processes. In: *Handbook of Deposition Technologies for Films and Coatings*, 2e (ed. R.F. Bunshah), 29–77. Park Ridge, NJ: Noyes Publications.

37 Rossnagel, S.M. (1991). Glow discharge plasma and sources for etching and deposition. In: *Thin Film Processes II* (ed. J.L. Vossen and W. Kern), 11–76. San Diego, CA: Academic Press.

38 Mézerette, D., Belmonte, T., Hugon, R. et al. (2001). Surface cleaning and passivation of an iron foil by a nitrogen post-discharge surface treatment. *Surf. Coat. Technol.* 142–144: 761–766.

39 Mattox, D.M. (1998). *Handbook of Physical Vapor Deposition (PVD) Processing*. Westwood, NJ: Noyes Publications.

40 Zelm, P.C. (1985). Elementary processes in plasma-surface interaction with emphasis on ions. *Pure Appl. Chem.* 57 (9): 1253–1264.

41 Ochs, D., Schroeder, J., Cord, B., and Scherer, J. (2001). Glass substrate cleaning at low energy ion source. *Surf. Coat. Technol.* 142–144: 767–770.

42 Goto, T., Takekawa, S., Moriya, M. et al. (2000). Growth of a-axis oriented $YBa_2Cu_3O7_{7-\delta}$ films on ion-beam cleaned $LaSrGaO_4$ substrates. *Physica C* 341–348: 2373–2374.

43 Curran, J.E. (1983). The use of plasmas to deposit and etch materials. *Phys. Technol.* 14: 283–296.

44 Jundale V.A., Chorage G.Y. and Yadav A.A. Structural, morphological and optical properties of $CoFe_2O_4$ thin film by spray pyrolysis technique. *Materials Today: Proceedings*, in press, doi.org/10.1016/j.matpr.2020.05.077.

45 Loosen, P. and Treusch, H.-G. (1993). Laserstrahlquellen für industrielle Fertigungstechnik. In: *Werkstoffbearbeitung Mit Laserstrahlung* (ed. G. Herzigen and P. Loosen). Munich: Carl Hanser Verlag (in German).

46 Reitz, W. (1998). Surface cleaning and coating removal with lasers. In: *Lasers in Surface Engineering* (ed. N.B. Dahotre). Materials Park, OH: ASM International.

47 Pawłowski, L. (1999). Thick laser coatings: a review. *J. Thermal Spray Technol.* 8 (2): 279–295.

48 Lu, Y.F., Song, W.D., and Hong, M.H. (1996). Laser removal of particles from magnetic head sliders. *J. Appl. Phys.* 80 (1): 499–504.

49 Turner, M.W., Crouse, P.L., Li, L., and Smith, A.J.E. (2006). Investigation into CO_2 laser cleaning of titanium alloys for gas-turbine component manufacture. *Appl. Surf. Sci.* 252 (13): 4798–4802.

50 Turner, M.W., Crouse, P.L., and Li, L. (2006). Comparison of mechanisms and effects of Nd:YAG and CO_2 laser cleaning of titanium alloys. *Appl. Surf. Sci.* 252 (13): 4792–4797.

51 Raza, M.S., Das, S.S., Tudu, P., and Saha, P. (2019). Excimer laser cleaning of black sulphur encrustation from silver surface. *Opt. Laser Technol.* 113: 95–103.

52 Schmidt, M.J.J., Li, L., and Spencer, J.T. (1999). Characteristics of high power diode laser removal of multilayer chlorinated rubber coatings from concrete surfaces. *Opt. Laser Technol.* 31: 171–180.

53 Tian, Z., Lei, Z., Chen, X. et al. (2020). Nanosecond pulsed fiber laser cleaning of natural marine micro-biofoulings from the surface of aluminum alloy. *J. Clean. Prod.* 244: 118724.

54 Razaba, M.K.A.A., Noor, A.M., Jaafar, M.S. et al. (2018). A review of incorporating Nd:YAG laser cleaning principal in automotive industry. *J. Radiat. Res. Appl. Sci.* 11: 393–402.

55 Dupuis, P., Cormier, Y., Fenech, M. et al. (2016). Flow structure identification and analysis in fin arrays produced by cold spray additive manufacturing. *Int. J. Heat Mass Transf.* 93: 301–313.

56 Randall, J.N., Owen, J.H.G., Fuchs, E. et al. (2018). Digital atomic scale fabrication an inverse Moore's Law – a path to atomically precise manufacturing. *Micro and Nano Engineering* 1: 1–14.

57 Köhler, M. and Fritzsche, W. (2004). *Nanotechnology. An introduction to Nanostructuring Techniques*. Weinheim: Wiley-VCH.

58 Mayer, T.M. and Allen, S.D. (1991). Selected area processing. In: *Thin Film Processes II* (ed. J.L. Vossen and W. Kern), 622–667. San Diego, CA: Academic Press.

59 Sebastian, E.M., Jain, S.K., Purohit, R. et al. (2020). Nanolithography and its current advancements. *Materials Today: Proceedings* 26: 2351–2356.

60 Bliznetsov, V., Kumar, R., Lin, H. et al. (2006). Integrated process of photoresist trimming and dielectric hard mask etching for sub-50 nm gate patterning. *Thin Solid Films* 504: 117–120.

61 Lipson, S.G. (2003). Why is super-resolution so inefficient? *Micron* 34 (6–7): 309–312.

62 RP Photonics Encyclopedia, https://www.rp-photonics.com/numerical_aperture.html (accessed 4 January 2021).

63 Dreeskornfeld, L., Hainl, G., Kleinberg, U. et al. (2004). Nanostructuring of Mo/Si multilayers by means of reactive ion etching using three- level mask. *Thin Solid Films* 458: 227–232.

64 Jantawong, J., Atthi, N., Leepattarapongpan, C. et al. (2019). Fabrication of MEMS-based capacitive silicon microphone structure with staircase contour cavity using multi-film thickness mask. *Microelectron. Eng.* 206: 17–24.

65 Huang, H.-W., Kao, C.-C., Hsueh, T.-H. et al. (2004). Fabrication of GaN-based nanorod LED using self-assemble nickel nano-mask and ICP reactive ion etching. *Mater. Sci. Eng. B* 113: 125–129.

66 Fonseca Filho, H.D., Maricio, M.H.P., Ponciano, C.R., and Prioli, R. (2004). Metal layer mask patterning by force microscope lithography. *Mater. Sci. Eng. B* 112: 194–199.

67 Luo, J., Zheng, H., Chen, W. et al. (2019). Conical-shaped hexagonal barium ferrite nanodot arrays on an alumina substrate based on an ultrathin alumina mask method. *J. Magn. Magn. Mater.* 489: 165449.

68 Wagner, E., Sandu, C.S., Harada, S. et al. (2015). Fabrication of complex oxide microstructures by combinatorial chemical bean vapor deposition through stencil masks. *Thin Solid Films* 586: 64–69.

69 Park, S.J., Kim, D., Lee, S. et al. (2018). Diamond-like amorphous carbon layer film by an ICP system for next generation etching hard mask. *Thin Solid Films* 663: 21–24.

70 Liu, H.W., Gao, C.X., LI, X. et al. (2001). Selective deposition of diamond films on insulators by selective seeding with a double –layer mask. *Diam. Relat. Mater.* 10: 1573–1577.

71 Ashby, C.I.H. (1991). Laser-driven etching. In: *Thin Film Processes II* (ed. J.L. Vossen and W. Kern), 783–846. San Diego, CA: Academic Press.

72 Chen, H.-Y., Han, D., Tian, Y. et al. (2014). Mask-free and programmable patterning of graphene by ultrafast direct writing. *Chem. Phys.* 430: 13–17.

73 Drabik, M., Truchly, M., Ballo, V. et al. (2018). Influence of substrate material and its plasma pretreatment on adhesion and properties of WC/a-C:H nanocomposite coatings deposited at low temperature. *Surf. Coat. Technol.* 333: 138–147.

74 Pawłowski, L. (2008). *The Science and Engineering of Thermal Spray Coatings*. Chichester: Wiley.

75 Oukach, S., Pateyron, B., and Pawłowski, L. (2019). Physical and chemical phenomena occurring between solid ceramics and liquid metals and alloys at laser and plasma composite coatings formation: a review. *Surf. Sci. Rep.* 74 (3): 213–241.

76 Abermann, R. (1990). Internal stress of vapor- deposited aluminum on aluminum substrate films: effect of O_2 and water incorporated in the substrate. *Thin Solid Films* 188: 385–394.

77 Poppeller, M. and Abermann, R. (1997). Influence of substrate properties on the growth of titanium films. Part I. *Thin Solid Films* 295: 60–66.

78 Oberhauser, R. and Abermann, R. (1999). Influence of substrate properties on the growth of titanium films. Part III. *Thin Solid Films* 350: 59–66.

79 Voevodin, A.A. and Zabinski, J.S. (1998). Superhard, functionally gradient, nanolayered and nanocomposite diamond-like carbon coatings for wear protection. *Diam. Relat. Mater.* 7: 463–467.

80 Sasaki, Y., Nose, K., Kamiko, M., and Mitsuda, Y. (2011). High adhesion of diamond-like carbon thin film to aluminum alloy achieved by substrate sputtering and redeposition method. *Surf. Coat. Technol.* 206: 143–148.

81 Kang, M.-G. and Park, H.H. (2002). Surface preparation and effective contact formation for GaAs surface. *Vacuum* 67: 91–100.

82 Chang, Y., Ducroquet, F., Gautier, E. et al. (2004). Surface preparation and post thermal treatment effects on interface properties of thin Al_2O_3 films deposited by ALD. *Microelectron. Eng.* 72: 326–331.

83 Zhang, B. and Zhou, L. (1997). Effect of sandblasting on adhesion strength of diamond coatings. *Thin Solid Films* 307: 21–28.

84 Min, J., Wan, H., Carlson, B.E. et al. (2020). Application of laser ablation in adhesive bonding of metallic materials: a review. *Opt. Laser Technol.* 128: 16188.

85 Barletta, M., Rubino, G., and Gisario, A. (2011). Adhesion and wear resistance of diamond coatings laser treated WC-Co substrates. *Wear* 271: 2016–2024.

86 Knapp, W., Djomani, D., Coulon, J.F., and Grunchec, R. (2014). Influence of structuring by laser and plasma torch on the adhesion of metallic films on thermoplastic substrates. *Phys. Procedia* 56: 791–800.

87 Xu, C., Xu, F., Shi, L. et al. (2019). Enhancement of substrate-coating adherence of boron-doped diamond electrodes by nanosecond laser surface texturing pretreatment. *Surf. Coat. Technol.* 360: 196–204.

88 Grubova, I.Y., Surmeneva, M.A., Ivanova, A.A. et al. (2015). The effect of patterned titanium substrates on the properties of silver-doped hydroxyapatite coatings. *Surf. Coat. Technol.* 276: 595–601.

89 Gheriani, R., Halimi, R., and Bensaha, R. (2004). Effect of substrate surface ion bombardment etching on reaction between chromium thin films and steel substrates. *Surf. Coat. Technol.* 180–181: 49–52.

90 Sarto, F., Alvisi, M., Melissano, E. et al. (1999). Adhesion enhancement of optical coatings on plastic substrate via ion treatment. *Thin Solid Films* 346: 196–201.

91 Jones, B.J., Anguilano, L., and Ojeda, J.J. (2011). Argon plasma treatment techniques on steel and effects on DLC structure and delamination. *Diam. Relat. Mater.* 20: 1030–1035.

92 Flege, S., Hatada, R., Ensinger, W., and Baba, K. (2014). Improved adhesion of DLC films on copper substrates by preimplantation. *Surf. Coat. Technol.* 256: 37–40.

93 Goldman, M., Goldman, A., and Sigmond, R.S. (1985). The corona discharge, its properties and specific uses. *Pure Appl. Chem.* 57 (9): 1353–1362.

94 Danlos, Y., Costil, S., Liao, H., and Coddet, C. (2008). Combining effects of ablation laser and laser preheating on metallic substrates before thermal spraying. *Surf. Coat. Technol.* 202: 4531–4537.

95 Kromer, R., Costil, S., Cormier, J. et al. (2015). Laser surface patterning to enhance adhesion of plasma sprayed coatings. *Surf. Coat. Technol.* 278: 171–182.

96 Roseberry, T.J. and Boulger, F.W. (1977). *A Plasma Flame Spray Handbook*, Rapport no. MT – 043. Springfield, VA: U. S. Department of Commerce, National Technical Information Center.

97 Mellali, M., Grimaud, A., and Fauchais, P. (1994). Parameters controlling the sand blasting of substrates for plasma spraying. In: *Thermal Spray Industrial Applications* (ed. C.C. Berndt and S. Sampath), 227–232. Materials Park, OH: ASM International.

98 Mellali, M., Fauchais, P., and Grimaud, A. (1996). Influence of substrate roughness and temperature on the adhesion/cohesion of alumina coatings. *Surf. Coat. Technol.* 81: 275–286.

99 Wigren, J. (1988). Grit blasting as surface preparation before plasma spraying. In: *Thermal Spray: Advances in Coatings Technology* (ed. D.L. Houck), 99–104. Materials Park, OH.

100 Dong, S., Zeng, J., Li, L. et al. (2017). Significance of in-situ dry-ice blasting on the microstructure, crystallinity and bonding strength of plasma-sprayed hydroxyapatite coatings. *J. Mech. Behav. Biomed. Mater.* 71: 136–147.

101 Tan, N., Xing, Z.-G., Wang, X.-L. et al. (2017). Deposition mechanism of plasma sprayed droplets on textured surfaces with different diameter-to-distance ratios. *Mater. Des.* 133: 19–29.

102 Kromer, S., Costil, S., Verdy, C. et al. (2018). Laser surface texturing to enhance bond strength of spray coatings – cold spraying, wire-arc spraying, and atmospheric plasma spraying. *Surf. Coat. Technol.* 352: 642–653.

103 Costil, S., Lamraoui, A., Langlade, C. et al. (2014). Surface modifications induced by pulsed-laser texturing. Influence of laser impact on the surface properties. *Appl. Surf. Sci.* 288: 542–549.

104 Danlos, Y., Costil, S., Liao, H., and Coddet, C. (2011). Influence of Ti-6Al-4V and Al 2017 substrate morphology on Ni-Al coating adhesion. Impact of laser treatment. *Surf. Coat. Technol.* 205: 2702–2708.

105 Gao, Y., Wu, B., Zhou, Y., and Tao, S. (2011). A two-step nanosecond laser surface texturing process with smooth surface finish. *Appl. Surf. Sci.* 257: 9960–9967.

106 Taylor, T.A. (1995). Surface roughening of metallic substrates by high pressure pure waterjet. *Surf. Coat. Technol.* 76–77: 95–100.

107 Knapp, J.K. and Taylor, T.A. (1996). Waterjet surface analysis and bond strength. *Surf. Coat. Technol.* 86–87: 22–27.

108 Takeda, K. and Takeuchi, S. (1997). Removal of oxide layer on metal surface by vacuum arc. *Mater. Trans. Jap. Inst. Metals* 38 (7): 636–642.

109 Lukauskaite, R., Cernasejus, O., Skamat, J. et al. (2017). The effect of Al-Mg substrate preparation on the adhesion strength of plasma sprayed Ni-Al coatings. *Surf. Coat. Technol.* 316: 93–103.

2

Films and Coatings Deposition Techniques

2.1 Introduction

Films and coatings can be deposited on a substrate by a multitude of techniques. Their description follows the classification presented in Table I.1 and is related to the size species participating in the deposition. These sizes are in nanometer range for atoms, ions, and molecules up to hundreds of micrometers for solid particles. The ratio of the sizes is about 10^5, and the processes using them are very different. Consequently, the categorization of techniques of deposition follows the size of species and includes the deposits: (i) atomistic; (ii) granular; and (iii) bulk. Some of the described methods do not follow exactly the chosen classification, e.g. the deposition methods using liquid feedstock were attributed to granular techniques. Similarly, the plasma spraying physical vapor deposition method (*PS PVD*), which uses particles as initial feedstock, transforms entirely into vapors before contact with substrate, and this was attributed to the granular techniques.

The chapter describes the techniques of deposition including the generation (atomistic techniques) or injection (granular methods) of species. Similarly, the phenomena occurring in the species during the stage of their contact with the medium (see Fig. I.1), which are very complicated for some techniques such as *PS PVD* or *SPPS*, are synthetically described in this chapter.

Finally, the techniques of films and coatings deposition evolve all the time, and the list described in this chapter is not complete but rather typical. Their description should enable better understanding of the chapters dealing with the microstructure and properties of films and coatings.

2.2 Atomistic Deposition Methods

The constituent species of an atomistic deposit are molecules, atoms, and ions. These species are initially in a source of species (see Fig. I.1) which may be in solid, liquid, or gaseous phase.

The morphology of the solid source, called frequently ***target***, does not play an important role. It does not matter if the source is a powder, a wire, or a rod; only their chemical

Physical Deposition Methods for Films and Coatings, First Edition. Lech Pawłowski.

composition and the distribution of the elements within them are accounted for. This distribution is designed in such a way that the final deposited film, which condenses on the substrate, has the desired chemical composition. In some cases such as deposition by sputtering with a *dc* diode system, the source must be an electrical conductor.

A solid source, which releases the species by evaporation, may occur by:

- electric heating:
  - directly by ***Joule's effect*** in electric conductors;
  - indirectly by conduction heat transfer from electrically heated conductors;
- electric arc heating;
- bombardment by electrons (*EB PVD*) or ions (*sputtering*);
- bombardment by photons (*PLD*).

The techniques of heating by electric current and arc as well as by electrons have some similarities. That is why, they are described as thermal evaporation in one section. The evaporation caused by ions and by photons includes many different phenomena and, consequently, are described separately.

The vapors diffuse toward a substrate and are in direct contact with a medium (see Fig. I.1). The medium can be chemically inert such as vacuum or inert gas, which do not react with the species, or chemically active such as reactive gas or plasma.

A solid source becomes liquid when its temperature exceeds the melting point of elements or compounds or the ***liquidus line*** for alloys. This concerns also a deposition by spraying an atomized liquid. On the other hand, the use of liquid feedstock in the form of suspensions or solutions is carefully discussed in the section dealing with granular coatings obtained by spray pyrolysis (*SP*), solution plasma spraying (*SPS*), and solution precursor plasma sprayings (*SPPS*). These techniques are at present frequently studied before being introduced to the industrial practice.

Finally, a gaseous source corresponds to the case of injection of a gas into a reactor. This gas may:

- react with the medium being reactive gas (*CVD*) or plasma (*PECVD*);
- decompose;
- transform due to the action of photons (*LCVD*).

The products of the reaction, decomposition, or transformation condense on the surface of a substrate and form a deposit.

2.2.1 Physical Vapor Deposition (*PVD*) by Thermal Evaporation

2.2.1.1 History

The first evaporated films were made by Faraday in 1857 [1] by an explosion of metal wires under vacuum. Nahrwold discovered the deposition of metallic thin films by the Joule effect in 1887, while Kundt in 1888 applied them to determine the refractive indices. In the following years, the study of evaporation deposits had rather an academic character. Industrial applications of evaporation technology had been initiated after World War II. The textbooks by Holland [2] and Maissel and Glang [3] summarized the state of the technology in the 1960s and 1970s. More recently, Richardt and Durand [4] and Mattox [5] described the current state of the technology. An important growth of industrial applications of evaporation process is related to *EBPVD* technique, which enables high rate of deposit growth [6].

2.2.1.2 Principle

A sketch of vacuum evaporation method for thin-film deposition is shown in Fig. 2.1.

A source is heated by an *e-beam* and the atoms evaporate from its surface. The evaporation rate of species having mass m depends on the vapor pressure of the source material p^* being at temperature T and surrounded by a gas having pressure p can be found from the ***Hertz–Knudsen equation*** [5]:

$$\frac{\mathrm{d}N_e}{\mathrm{d}t} = c\left(2\pi mkT\right)^{-0.5}\left(p^* - p\right) \tag{2.1}$$

The value of the constant c depends on the evaporating surface cleanliness, and its maximum value is $c = 1$.The vapors diffuse to the surrounding atmosphere in a hemisphere above the source. The evaporated mass is not distributed in a homogeneous way and depends on the angles φ and θ shown in Fig. 2.1. The maximum evaporated mass density per unit area distribution at the distance from the source r, being the thickest part of the condensed film, is in the center of the substrate where the angles $\varphi = 0$ and $\theta = 0$. This distribution and the thickness of the obtained film from a point source are given by the ***cosine law***:

$$\frac{dm_e}{dA} = \frac{m_t}{\pi r^2} cos\phi \cos\theta \tag{2.2}$$

In order to reach a homogeneous thickness of the condensed film it is useful to move the substrate in an oscillating way. The evaporation rate increases with the difference between

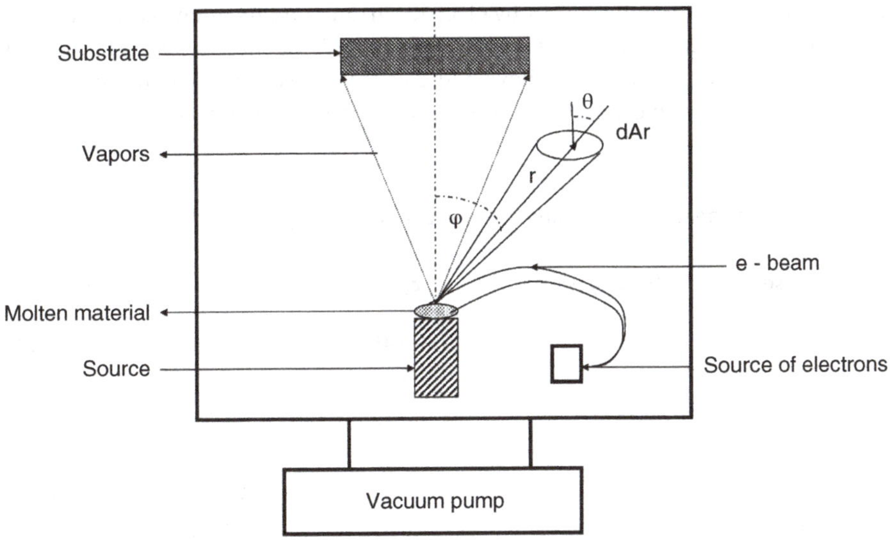

Figure 2.1 Sketch of vacuum evaporation system by electron bombardment (*EBPVD*).

the source material vapor pressure and the surrounding gas pressure. To increase this rate, the deposition takes place under vacuum being typically under pressure of $p = 10^{-6}$ hPa reached with a diffusion pump system. However, to obtain very pure deposits, pressure as low as $p = 10^{-9}$–10^{-10} hPa is sometimes applied. Such low pressure is reached using the turbomolecular pumps. The cryogenic traps between the evaporation chamber and the inlet of the diffusion pump are used to avoid the penetration of the impurities vapors.

The vapor pressure p^* depends strongly on the source temperature. The experimental values of p^* [Pa] can be described for the metals by the following equation [7]:

$$\log p^* = 5.006 + c_1 + \frac{c_2}{T} + c_3 logT \tag{2.3}$$

The constants c_1 and c_2 are collected for a few frequently evaporated metals in Table 2.1

The oxides have a much lower vapor pressure. Consequently, to reach the pressure $p^* = 100$ Pa, the source of Al_2O_3 should have a temperature of 2 395 K, and for SiO_2, a temperature of 2 641 K [7]. The vapor pressure vs. temperature for such materials can be measured using ***Knudsen effusion cells*** in which the vapors of a heated crucible containing tested material are evacuated into vacuum through an orifice, and a measured loss of mass enables to find the vapor pressure. This method was used, e.g. to find the vapor pressure of ZrO_2 [8].

2.2.1.3 Sources

The evaporated species are frequently heated using *Joule effect*. Such heating is carried out using materials, which are typically wires or powders being placed in a heating element such as spiral, or a "boat," formed of a metal sheet, sometimes covered by a ceramic layer. This layer serves as a protection against a possible chemical reaction between the source material and that of the heating element. For evaporation at temperatures below 1300 K, the metals as Pt, Fe, or Ni are used, while for higher temperatures, such metals as W, Mo, Ta are applied [4]. The heating may occur by electric current, and its intensity in a heating element is of the order of 20 to 500 A. The evaporation using *Joule effect* is easy to be applied for pure elements. The evaporation of alloys and compounds separately is more complicated because each element has its vapor pressure p^* at given temperature. Consequently,

Table 2.1 The constants enabling determination of vapor pressure by Eq. (2.3) for some metals.

Metal	Phase	Melting point, K	c_1	c_2	c_3
Al	Solid	933	9.459	−17342	−0.7927
	Liquid		5.911	−16211	0
Cu	Solid	1358	9.123	−17748	−0.7317
	Liquid		5.849	−16415	0
Ti	Solid	1941	11.925	−24991	−1.3376
	Liquid		6.358	−22747	0

Source: Adapted from Lide [7].

at the given temperature and according to ***Raoult's law*** the number of evaporated atoms belonging to the element having greater p^* is more important. Therefore, the chemical composition of the deposit may be different from that of the source. For example, in the AlMg alloy[1], aluminum has a vapor pressure p^* much lower than that of magnesium, and the film deposited with such source will therefore be richer in magnesium.

In order to avoid the change of composition, one can apply a ***flash evaporation*** in which a piece of material is completely and rapidly evaporated. Flash evaporation can be carried out using a pulse of electric current from a capacitor discharge [5]. Another problem occurs at evaporation of compounds which decompose at melting such as multi-oxides *YBCO* or *HA* [9]. The evaporation by the *Joule effect* does not enable obtaining deposits of such compounds. The intensive heating using photon pulses (*PLD* method) is more useful.

The compounds, which sublimate, e.g. Si_3N_4, SiO, BN, should be heated in such a way to avoid direct contact between heater and source. Heating by radiation or by electrons is a possible solution [1]. ***Heating by electron bombardment*** (*EB PVD* method) has two principal advantages, namely: (i) the evaporation rate is very high because of a high power density being reached by focusing an e-beam into a small source surface area; and (ii) the source is kept in a water-cooled copper container which reduces the risk of its contamination.

The electrons used in source heating in *EB PVD* method can be generated by: (i) ***thermionic electron emitters***, shown in Fig. 2.2; or (ii) ***plasma electron guns***.

The thermionic emission of electrons occurs from a tungsten filament, which is also a cathode, heated to a temperature close to that of fusion. The e-beam is then accelerated in an electric field between the cathode and the anode (voltage of about 10 to 40 kV) being focused by the electrostatic and electromagnetic lenses. To avoid oxidation of the filament, the emitter must be under a low pressure of about $p = 10^{-3}$ hPa. Typically, the emitter is located in a chamber separated from the evaporation one. Bergmann [6] used thermionic

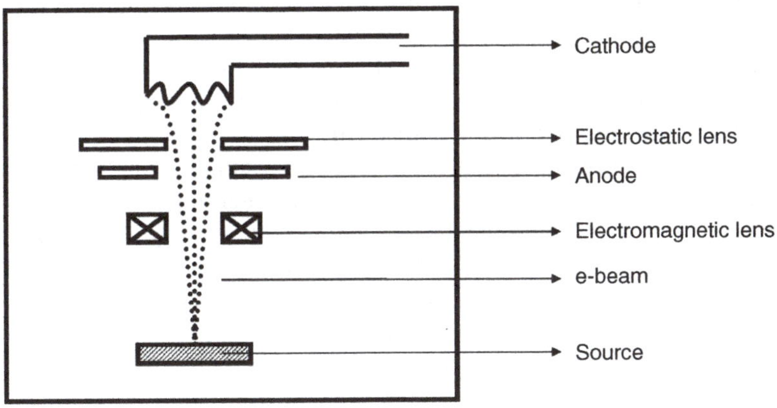

Figure 2.2 Thermionic electrons emission.

1 The vapor pressure of $p^* = 100$ Pa is reached for Al at $T = 1817$ K and for Mg at 816 K [7].

emitter to evaporate Ti, which reacted with nitrogen-containing medium enabling obtaining TiN films. The thermionic emitter consumed 12 kW electric power, and the evaporation rate was about 0.9 g/min of Ti which corresponded to film's thickness growth rate of 0.1 μm/min of TiN.

Plasma electron guns, working at a higher pressure (up to $p = 10^{-1}$ hPa), can be categorized into [1]:

- ***cold cathode electron gun***;
- ***hollow cathode electron gun***.

A cold cathode gun is a cylindrical cavity with a plasma inside emitting e-beam of intensity up to $I = 3$ A through a small aperture. The interaction between the ions and gas (such as Ar) generates plasma including electrons. The electrons are also generated by the ion bombardment of metal surface. The appropriate choice of parameters (cathode voltage, pressure, and kind of the plasma gas) makes it possible to keep the cathode at low temperature.

A hollow cathode gun is a cylinder made of a refractory metal because it works at a high temperature. A plasma gas is continuously delivered to the cylinder. Plasma is created inside it by a ***glow discharge*** between the cathode and the anode being a source. In this area, the electrons generated in plasma and emitted by ion bombardment of cylinder are collimated by magnetic field and are accelerated toward the source.

The plasma electron guns are particularly useful in *PVD* methods in which generated vapors are in contact with plasma medium, such as: (i) ***activated reactive evaporation*** (*ARE*); and (ii) ***ion plating*** (*IP*). On the contrary, the thermionic electron emitters guns are applied in the *EB PVD* facilities working under low vacuum.

The *EB PVD* technique is interesting for industrial application because it is possible to have very film high deposition rate, which can reach several tens of μm/min. The important application of *YSZ* films obtained by this technique for thermal barriers (*TBC*) is in aeronautical turbines [10]. The view of an *EB PVD* installation is shown in Fig. 2.3.

The ***electric arc heating*** of the source is also useful for the evaporation of a large quantity of material. The arc is generated as a *dc* or *ac* electric discharge between the anode and the cathode in the presence of a gas. This arc discharge is characterized by a very high current (of hundreds of amperes) and a fairly low voltage (tens of volts). The following are designs of arc source heating: (i) ***continuous arc***; and (ii) ***pulsed arc***.

In *continuous arc* heating, it is necessary to provide for a rapid consumption of the source, which must be continuously delivered to the discharge zone. Another important problem is the generation of large particles of the source material. The particles may contaminate the microstructure of the deposited film. The reduction of this contamination can be achieved by the application of particulate filters and by an appropriate choice of the processing parameters. Kourtev *et al.* [11] achieved the reduction of macro-droplets at TiN film deposition by steering the arc by rotating electromagnets.

A *pulsed arc* heating was developed in combination with pulsed laser to deposit *DLC* films by Scheibe and Schultrich [12]. The evaporation is produced by arc erosion of the source, being a cathode. The pulsed laser intensive heating enables to hook the arc to the source spot. This system is particularly useful for the deposition of multilayers and was applied to obtain Al with C multilayers [13].

Figure 2.3 *EB PVD* laboratory installation used in the research center ARCI in Hyderabad (India). *Source:* Reproduced with permission from ARCI.

2.2.1.4 Medium

The vapors from a source enter in the medium, which can be characterized by temperature and pressure. The vaporized atoms may meet other species. Their ***mean free path*** between the collisions can be expressed by a simplified relationship valid for low-pressure medium containing one gas after Groszkowski [14]:

$$\bar{l} = \frac{1}{\sqrt{2}} \frac{1}{\pi d_e^2} \frac{kT}{p} \tag{2.4}$$

The *mean free path* of N_2 molecule having diameter of $d = 364\,\text{pm}$ at $T = 300\,\text{K}$ is about $\bar{l} = 7\,\text{m}$ under low pressure of $p = 10^{-6}\,\text{hPa}$ and of $\bar{l} = 7\,\text{cm}$ at high pressure of $p = 10^{-3}\,\text{hPa}$. Typically, the distance between source and substrate is equal to a few tens of centimeters; such molecule would not have any collision at the low pressure and only a few collisions at high pressure. The collisions may lead to the chemical reactions. The number of collisions increases with the gas temperature. Consequently, the medium can be heated by an electric discharge to activate the chemical reactions with the vapors.

The evaporation under low pressure (*PVD*) is applied to obtain films of [3]:

- pure metals, such as Al, Ti, Ni, and Mg;
- metal alloys, such as Cu with Ni, Ni with Cr or Fe, Ti with Al, using different methods described in previous section (*flash evaporation*, multiples sources, *EBPVD*) to avoid change of alloy composition;
- intermetallic compounds, such as GaAs, InSb, and PbTe, typically by *flash evaporation;*
- metal ceramics composites, such as Cr + SiO and Co + SiO, also by *flash evaporation*;
- inorganic compounds, such as oxides, sulfides, or halides, using *EB PVD*.

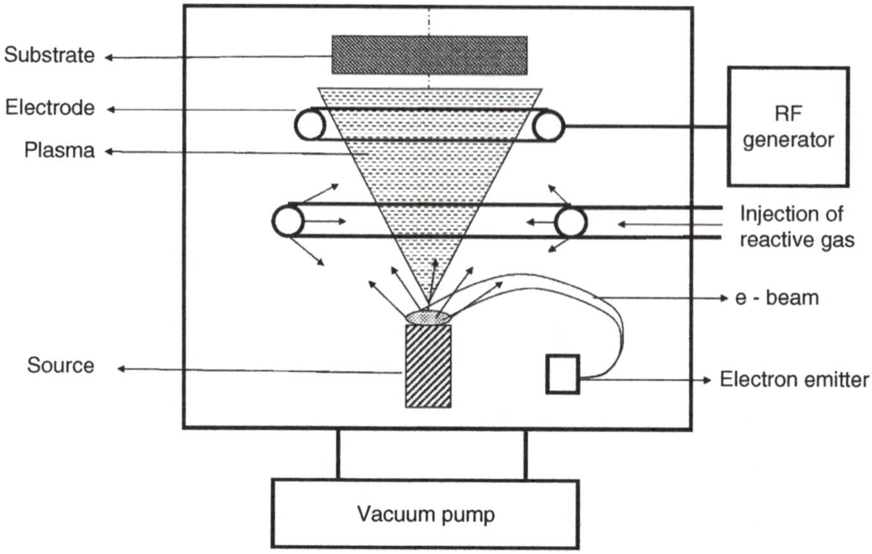

Figure 2.4 Sketch of *RE* and *ARE* installation with the source heated by *e-beam*.

The principle of installation of evaporation in a reactive atmosphere (*RE*) or in a reactive plasma (*ARE* known also as *PE PVD*) is shown in Fig. 2.4.

In this process, a reactive gas is injected into the deposition chamber. The gas can be optionally activated by an electric discharge, such as *ICP* shown in Fig. 2.4. The plasma-activated process is often called *PE PVD*. The reactive evaporation is sometimes more useful than evaporation under vacuum. An example is evaporation of an inorganic compound such as Al_2O_3. The compound can be evaporated directly, but the source needs a lot of energy to generate the vapors, which results in low deposition rate. Even worse, the evaporated film is deficient in oxygen. Consequently, it is much easier to use the metal source evaporation in gaseous atmosphere. Metallic vapors may react with oxygen or nitrogen, forming a desired compound molecule. Examples of reactions forming the films of oxide or nitride are in the equations below [1, 15]. More negative Gibbs energy indicates that the formation of alumina is easier than that of titanium nitride.

$$2Al\left(vapor\right)+\frac{3}{2}O_2\left(gas\right)\rightarrow Al_2O_3\left(film\right)\ with\ \Delta\,G^0=-1.046\frac{MJ}{mol\ O_2}\ at\ 298\ K \tag{2.5}$$

$$2Ti\left(vapor\right)+N_2\left(gas\right)\rightarrow 2TiN\left(film\right)\ with\ \Delta\,G^0=-0.307\frac{MJ}{mol\ N_2}\ at\ 298\ K \tag{2.6}$$

The reaction between evaporated atom and reactive gas can take place: (i) on the source surface; (ii) inside the volume of reactive gas or plasma; or (iii) on the surface of the substrate during film condensation (this is discussed in Chapter 3).

As the temperature of the source is high, the product of reaction with reactive gas decomposes rapidly. More probable is the reaction inside the volume of reactive gas or plasma. In order to increase the number of collisions between the species of vapors and of the gas,

Table 2.2 Experimentally determined *sputtering yield* of some oxides submitted to bombardment of Ar^+ ions having energy of $E_p^i = 5.6$ keV.

Oxide	Density, g/cm^3	Surface binding energy, E_{bs}, eV/molecule	*Sputtering yield, Y*
$\alpha - SiO_2$	2.65	18.8	1.05
TiO_2	4.25	19.5	0.9
Al_2O_3	4.0	31.6	0.9
MgO	3.58	10.3	0.82

Source: Adapted from Bach [19].

the pressure should be about $p = 10^{-4}$–10^{-3} hPa to decrease the *mean free path* and the distance between the source and substrate long enough (typically a few tens of centimeters). Finally, to increase the probability of the reaction, the reactive gas can be heated to form a plasma by electric discharges *dc*, *ac*, *RF*, or microwaves [16–18]. The activation in an evaporated atom X (or molecule X_2) in plasma can take place during the collision with an electron e^- by the way of:

- excitation: $e^- + X \rightarrow X^* + e^-$;
- ionization: $e^- + X \rightarrow X^+ + 2e^-$;
- dissociation: $e^- + X_2 \rightarrow 2X + e^-$.

Bunshah and Deshpandey [16] found that the reactive evaporation *RE* has low deposition rate (about 20 nm/min) and needs high temperature of substrate (800 to 1300 K). The use of plasma, in *ARE* processes, increases the deposition rate up to 1μm/min and does not require intensive substrate heating. Moreover, for some compounds such as TiC it is possible to condense as film only in presence of acetylene, C_2H_2, activated by electric discharge. Without such activation, the condensed film is composed of titanium and atomic carbon. Other compounds condensed using *ARE* process are shown in Table 2.2 in the section describing sputtering.

2.2.2 Physical Vapor Deposition (*PVD*) by Sputtering

2.2.2.1 History

This process is mentioned in the literature for the first time in 1852 [20] or in 1877 following Mattox [5]. The latter mentions also that Edison patented the sputter deposition of gold on cylinders used in photography. The sputtering technology using balanced magnetrons was introduced in 1970 and that of unbalanced one in 1980 after Kelly and Arnell [21]. These authors see the growing number of sputtered films being decorative, wear resistant, and having particular optical and electrical properties, such as high-temperature superconductivity mentioned also by Wasa [22]. The review of Sigmund [23] shows the major milestones in the technique of physical sputtering. The author underlined the decisive contribution of Gottfried Wehner in the 1950s in understanding of physical phenomena occurring in the source–target at sputtering. Finally, Sigmund estimated in 2011 the number of papers on the topic sputtering to be 75 000.

2.2.2.2 Principle

The vapors in the sputtering process are generated by the elastic collisions between the ions, which bombard the surface of the source, which are called the *target* in the deposition method, and vaporize the atoms from its surface (see Fig. 2.5).

In fact, the atoms exit from the target after a cascade of collisions from a depth of 5–10 nm from the surface as discussed in the following section. The sputtering chamber is initially evacuated to the pressure of $p = 10^{-6}$–10^{-10} hPa before a plasma-forming (typically Ar) gas under pressure of 10^{-3}–10^{-1} hPa is introduced. An electric glow discharge, frequently the *RF* having frequency of 13.56 MHz, creates the plasma. Consequently, the negative ions are accelerated toward the target cathode being put under negative potential of 0.5 to 5 kV [20]. The sputtered target must be an electrical conductor.

The *dc-diode-sputtering system* is the simplest one, but its price is high compared with other *PVD* systems [5]. Another possibility to sputter dielectrics using *dc* discharge is the ***triode-sputtering system***, which uses a thermionic gun, a magnetic confinement along axis emitter, and an anode put on a small positive voltage (of 50 to 100 V). The glow discharge is supplied by the electrons delivered from the emitter and it generates the ions. These ions are attracted to the high negative potential of the target, being a third electrode outside the emitter–anode axis [17]. Other sputtering systems being suitable to vaporize the dielectric targets use *ac*. These systems are discussed in Chapter 1.3.2.

Presently, the most frequently used systems are the ***magnetron-sputtering*** ones. The principal role of magnetic field, which is configured to be near to target surface, is to interact with the secondary electrons. The electrons are subjected to the ***Lorentz force*** (see Eq. 2.7):

$$\vec{F} = q\vec{E} + q\vec{v} \times \vec{B} \tag{2.7}$$

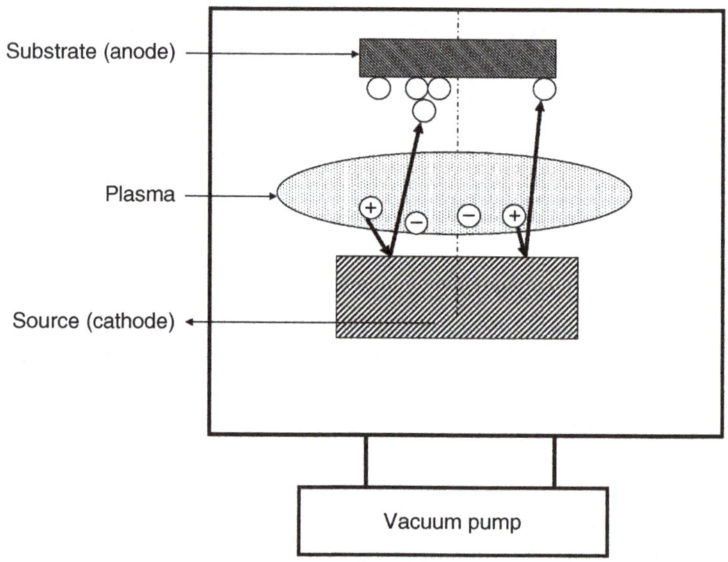

Figure 2.5 Sketch of a in a ***dc-diode-sputtering system*** in which the *target* is the cathode and the substrate – anode.

The application of a magnetic field $\vec{B}$ to the plasma modifies the trajectories of the electrons. They are subjected to the *Lorentz force,* move in a circular way around lines of a magnetic field vector, and interact with plasma increasing the number of ions generated in atom–electron collisions. The number of cations bombarding the target increases which results in an increase of sputtering rate and deposition rate. The use of magnetic field enables to decrease the working pressure of plasma gases and to lower the voltage applied to cathode from $U = -2$ to -3 kV at conventional sputtering to $U = -0.5$ kV at magnetron sputtering [21]. The *RF* can be used to avoid the charging of dielectric targets. The different techniques of magnetron sputtering are [21]: (i) ***balanced magnetron***; (ii) ***unbalanced magnetron*** with central magnetic pole being stronger compared to the outer ones; and (iii) *unbalanced magnetron* with outer magnetic poles stronger than the central one.

The magnetic field is generated between a pole being placed in the central part of the target and the outer poles. The *balanced magnetron* has the poles generating magnetic field of similar intensity. The *unbalanced magnetron* may have the outer poles generating weaker or stronger magnetic field. In this latter configuration the plasma volume is greater and may reach substrate. In this case, the ions bombard the film growing on the substrate. This bombardment densifies the film, and it is the principle of the ***ion-plating*** technique. In particular, it is desirable for making a *DLC* deposit, because it promotes the formation of diamond having sp^3-hybridized sites instead of graphite being sp^2 hybridized [24]. The ***planar magnetrons*** may be used to coat large surfaces. For example, Beister *et al.* [25] used such a system having length of 2.35 m with Si target to deposit SiO_2 and Si_3N_4 films onto windows. The *planar magnetron* systems are applied to coat planar substrates and use also planar targets. The systems with the targets having different shape, discussed by Mattox [5], include: (i) ***rotating target***[2] around the magnets; (ii) ***hollow cylinder target*** useful to coat *3D* parts; or (iii) ***hemispherical target*** used to coat hemispherical substrates. The systems can also be adapted to the substrate to be coated. An example of rotating substrates being coated using $TiAlN_x$ film is shown by Hofmann *et al* [26].

The cathodes (targets) are placed in opposition and are fitted in a set of permanent magnets, which create the plasma around them. In addition, the electromagnets create a magnetic field between two cathodes. The substrates, being cutting tools, are placed in the center on a rotating plate. The sputtering was carried out in a reactive nitrogen atmosphere using later reactive sputtering (*RS*). Sproul [27] shows a similar configuration of four unbalanced magnetron-cathodes, set around an attachment of rotating substrates holder.

2.2.2.3 Source

The analysis of the factors associated with the sputtering starts with description of ***elastic collision***[3] of a projectile with a stationary target. The initial kinetic energy of projectile is $E_p{}^i > 0$ and that of target particle is equal to $E_t{}^i = 0$. The energy transferred by an elastic

2 A system with rotating target is also called a *rotating magnetron,* and its important advantage is a more homogeneous use of the target compared to the planar configuration.

3 A collision is defined as elastic if the number and nature of the interacting particles are unchanged [28].

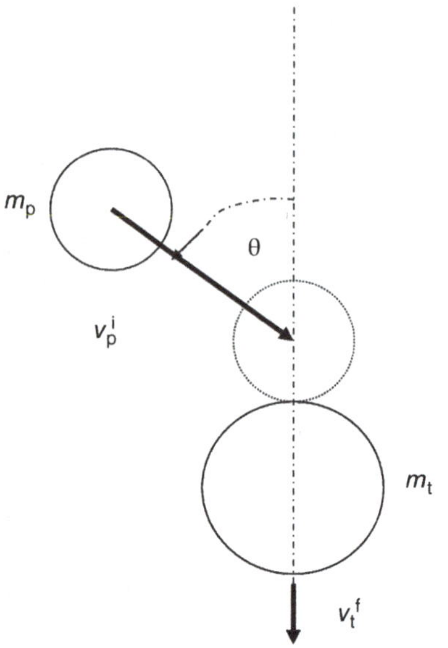

Figure 2.6 An elastic collision between a moving projectile and immobile target particles.

collision, E_t^f, can be estimated by supposing the conservation of energy and of momentum for the geometry shown in Fig. 2.6 as [5]:

$$E_t^f = E_p^i \frac{4m_p m_t \cos^2\theta}{\left(m_p + m_t\right)^2} \tag{2.8}$$

The energy transmitted to a target is maximum if the angle $\theta = 0$ and if the masses of the projectile and the target are equal ($m_p = m_t$). Typically, inert gas ions such as light argon with $m_p \approx 40$ amu or xenon with $m_p \approx 131$ amu are used to bombard a target. The incident ions enter in collision ions with atoms of the target as shown in Fig. 2.7. The collision may result in many effects such as: (i) sputtering of target atoms, frequently after a cascade of collisions, being the desired effect; (ii) implantation of incident ion inside target lattice; (iii) generation of defects in the lattice; and (iv) emission of secondary electrons. An important parameter is ***sputtering yield***, *Y*, which is defined as a number of atoms ejected from a target by an incident ion. Smentkowski [29] shows the way of its experimental measurements from the dimensions crater formed at sputtering in a target. This author presents also the following expression for *sputtering yield* useful for low-energy ion bombardment ($E_p^i < 5$ keV):

$$Y = \frac{0.3 m_p m_t \alpha E_p^i}{E_{bs}\left(m_p + m_t\right)^2} \tag{2.9}$$

Moreover, the author did theoretical calculation and experimentally determined *sputtering yield* obtained for elements ranging from as light as Be up to as heavy element as Au or U sputtered with Xe^+ ions having energy $E_p^i = 400\,\text{eV}$ under normal incidence ($\Theta = 0°$).

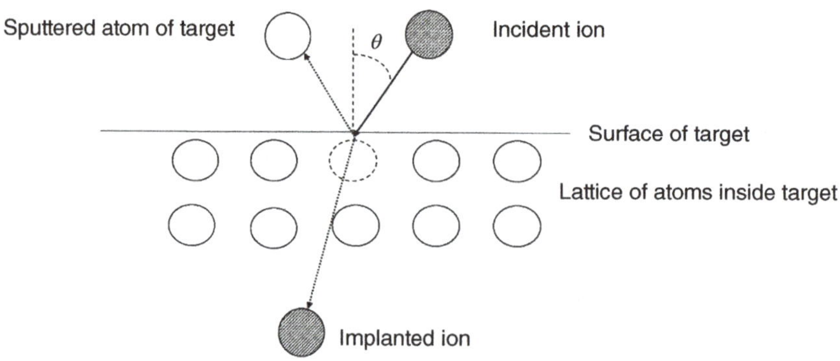

Figure 2.7 Some effects of an impact of incident ion on target surface.

Figure 2.8 The target of hafnium having diameter of 50 mm used in the laboratory of Faculty of Microsystem Electronics and Photonics at Wrocław University of Science and Technology (Poland). *Source:* Reproduced with permission of Professor W. Posadowski.

The measurements of Cu target *sputtering yield* with ions of Xe^+, Kr^+, Ar^+, and Ne^+ evidenced that the yield increased with ion energy $E_p{}^i$ as predicted by *Eq. 2.9* up to maximal value (about 20 keV for Ar^+).

Above this maximum the bombarding ions penetrate too deep into the target and get implanted inside (see Fig. 2.7). Consequently, the *sputtering yield* decreases [29].

The energies of the atoms sputtered from the target E_t^f may reach a few hundreds of eV [30]. These energies are much higher than those of evaporated species, which are generally smaller than 1 eV [5]. The species are evaporated more intensively in the sites on the target where the quantity of bombarding ions is greater. This is particularly visible at *magnetron sputtering* as seen in Fig. 2.8.

One of the greatest advantages of sputtering is the possibility of obtaining deposits of any material since the process is governed by momentum transfer and not by thermal evaporation. Nevertheless, the deposition process needs to be adapted to the target material by choosing the bombarding ions and their energy at impact. On the other hand, each element has its own *sputtering yield* for incident ions because their atomic masses are different (see *Eq. 2.9*). However, the mechanism of sputtering enables to obtain the films having the same composition even if the target is an alloy or a compound.

The simplified explanation is sputtering by ejecting the layers of atoms one after the other. If the atoms of these layers have equal or similar mass, m_t, the layers of sputtered material will condense on the substrate. If the masses are different, the *sputtering yields*, Y, are also different. Consequently, the number of atoms of a greater yield will be greater, and the remaining atomic layer in the target will become richer in atoms having lower yield. Therefore, the collisions of incident ions with the latter will be more frequent. Finally, the composition of sputtered vapors and of film condensed thereof would get closer to the target composition. This mechanism is simplified and two-phase alloys are necessary to prepare a target in the form of a cone to compensate via the incident *i-beam* angle, θ, the differences in *sputtering yield* [20].

The compounds have stronger chemical bonds and greater surface binding energy, E_{bs}, than the elements, and the *sputtering yield* of oxides shown in Table 2.2 is lower than that of the elements. Typically, when sputtering a ceramic target, a more volatile component is ejected more easily. Therefore, the sputtered films of oxides and nitrides often become deficient in oxygen or nitrogen. A possible solution is an application of deposition atmosphere rich in the deficient element such as reactive sputtering (*RS*) discussed in next section. Finally, the sputtering is particularly well suited to obtaining compounds whose structure contains atomic layers. An example of such structure is ***perovskite*** in high T_c superconductor-type Bi_2O_2.2SrO. (n-1)CaO.nCuO_2 sputtered using magnetron by Wasa [22].

2.2.2.4 Medium

The sputtering is done by neutral gas ion bombardment (frequently Ar^+). By mixing neutral gas with a reactive one such as nitrogen or oxygen, reactive atmosphere is created. The mixing is useful because the reactive gases are lighter compared to argon (m_{N2} = 14 amu, m_{O2} = 16 amu) and less useful to bombard the target. A typical mixture for *reactive spraying* (*RS*) in the presence of nitrogen is 20 vol. % N_2 and 80 % vol. of argon [5]. The *RS* deposition process is applied to synthesize many compounds shown in Table 2.3. Some of the advantages of reactive spraying are the possibility of: (i) using easy-to-prepare metallic targets (Ti, Zr, Al, etc.) to obtain oxides or nitrides; (ii) using the *dc* sputter process to obtain dielectric films; or (iii) preparing multilayered metallic and dielectric films [24].

Magnetron sputtering with the use of reactive gas may lead to its reaction with the target known as ***poisoning effect***. To avoid it, the neutral gas can be injected near the target and the reactive gas near the substrate.

The compounds obtained by reactive spraying are shown in Table 2.3. The targets are frequently elements, which enter in reaction with the gas after ejection. Sometimes, the targets are the compounds to be deposited, and the process should be carefully optimized. An example is the mixture of tin and indium oxides, known as *ITO*. The oxides were sputtered with *ITO* target at short distance to the substrate of 55 mm. The obtained films were submitted to the post-deposition annealing.

Table 2.3 Examples of compounds synthesized by thermal evaporation and sputtering in reactive atmosphere (*PEPVD*): *ARE* and *RS* [16, 26, 27].

Metallic source	Reactive gas	Compound synthesized by *ARE*	Compound synthesized by *RS*
Ti	N_2	Ti_2N, TiN	TiN
	C_2H_2	TiC	
	O_2	TiO_x ($x \leq 2$)	TiO_2
Zr	N_2	ZrN	ZrN
	C_2H_2	ZrC	
Al	O_2	Al_2O_3 (α or γ phases)	Al_2O_3
Si		SiO_2	SiO_x ($x < 2$)
Hf	C_2H_2	HfC	
Ti + V		VC + TiC	
Ti + Ni		Ni + TiC	
Ti + Al	N_2		$TiAlN_x$
In + Sn	O_2	Indium tin oxides	Indium tin oxides

2.2.3 Ion plating and *i-beam*-assisted Deposition

The techniques of ion plating and i-beam assisted deposition are associated with thermal evaporation, sputtering discussed above and with chemical vapor deposition discussed later. That is why, the discussion about source and medium is short and presented in only one section.

2.2.3.1 History

The history of the art ion plating (*IP*) began in the 1960s, and this technique was developed in the early 1970s [5]. The fundamental idea was to apply a negative voltage to a substrate in order to attract the cations coming from plasma present in the processing chamber. The ion bombardment modifies the morphology of a film deposited by sputtering or by evaporation. This bombardment is also present in the process called *IBAD*. However, the cations in this process come from an independent source, called sometimes ***ion gun***. Consequently, *IBAD* process can be realized under lower pressure than *IP*. *IBAD* is industrially used to produce the textured *YSZ* film being a part of electric cables using *YBCO* high-temperature superconductor [9, 31].

2.2.3.2 Principle

The sketch of techniques of ion plating (*IP*) and ion-beam assisted deposition (*IBAD*) associated with e-beam evaporation is shown in Fig. 2.9. The *IP* can be also associated with the sputtering techniques and is a part of the *PE PVD* processes group. The principle of *IP* is similar to that of sputtering if a target becomes a substrate. However, the negative voltage applied to the substrate and resulting thereof energy of ions should be lower than that at sputtering to enable film deposition. The plasma can be created by the discharge of a *dc* diode in an inert gas being frequently Ar under a pressure of a few Pa, and the cations Ar^+ are attracted by the negative potential of substrate. Many other types of plasma generated by *dc*, *ac*, *RF*, and microwave discharges are also possible (see Fig. 1.9). On arriving at the substrate, the cations acquired their energy from the electric field and the neutral atoms, which bombard also the surface, from the collision with the cations [32].

The *IP* techniques working with plasma enable injection of a reactive gas entering in a chemical reaction with the vapors as in the *RS, ARE,* or *PECVD* techniques. The *IBAD* technique (Fig. 2.9 b) is also characterized by the ion bombardment of substrate and growing film. The ions, however, are emitted by an *ion gun*. The direction of ions is perpendicular to the substrate in *IP* technique[4] and can be changed by the position of the gun axis in *IBAD* one. The most important parameters of both *IP* and *IBAD* processes are [33]: (i) kind of ionic species; (ii) angle of ions incidence; (iii) arrival ratio being the number of incident ions to the number of deposited atoms; and (iv) ions energy. The optimization of these parameters is necessary in order to achieve the effects desired by the process and to avoid the sputtering of the atoms of substrate [31]. In particular, too high ion energy bombarding the substrate may lead to high sputtering rate preventing the film grow-up. Mattox [5] mentioned this effect for the ion beam having energy of 30 keV.

---

4 The cations are attracted by the negative potential of the substrate. The acting force is described by Eq. (2.7).

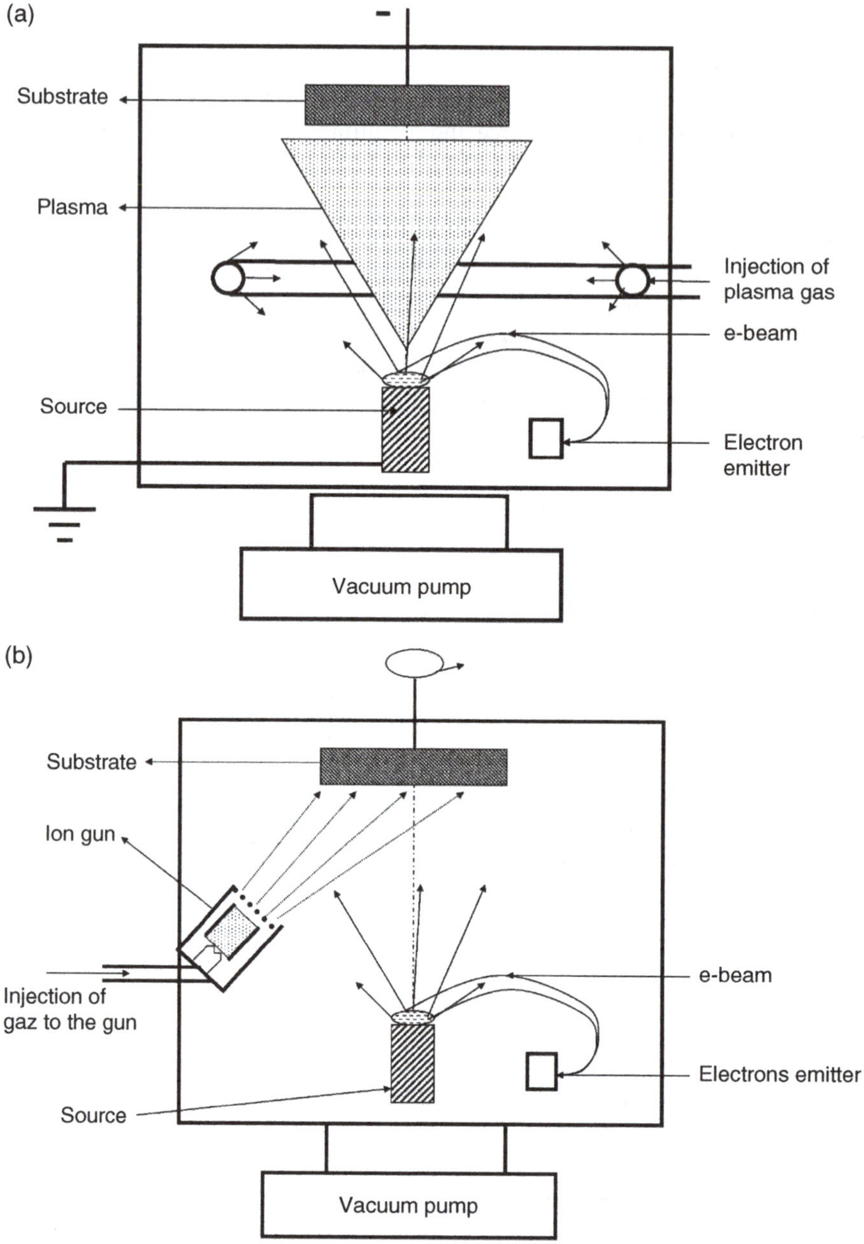

Figure 2.9 The deposition of films by evaporation using *e-beam* assisted by ions bombardment: (a) ion plating (*IP*); and (b) ion-beam assisted deposition (*IBAD*).

The *IP* and *IBAD* techniques are applied, generally, to enhance the adhesion and to modify the microstructure and properties of the films being deposited using *PVD* techniques as sputtering, thermal evaporation, or using plasma-enhanced chemical vapor deposition (*PECVD*). The modifications depend on the process parameters and on the stage of deposition. Table 2.4 describes the effects occurring in the following stages: (i) pre-deposition

Table 2.4 Most important effects of ions and neutral atoms bombardment at *IP* and *IBAD* processes [5, 32, 33].

Stage	Effects
Pre-deposition	• Volatilizing of contaminating species on the substrates' surface; • Formation of crystal lattice defects; • Densification of nucleation sites for incoming film; • Sputtering of the atoms of the substrate; • Implantation of gas atoms or surface contamination (***knock-on effect***).
During film deposition	• Modification of the interface (Interfaces can be classified as abrupt, compound, diffusion, pseudo-diffusion [32].) by implantation of incoming atoms, sputtering of substrate atoms or forming a compound from substrate and film atoms; • Decrease of porosity at the interface and densification of growing film; • Entrapping of the bombarding gas atoms leading to formation of new phase (reactive *IP*); • Sputtering of atoms of substrate and of growing film; • Generation compressive stresses in the growing film.
Post-deposition	• Densification of a film; • Generation of compressive stresses.

being, in fact, substrate cleaning described also in Chapter 1.3.2; (ii) simultaneously with film deposition; and (iii) post-deposition.

Another important effect of ion bombardment in *IBAD* processes may be texturing (grain alignment) of growing film. This effect depends strongly on the processing parameters including angle of bombarding ions and on the mobility of atoms at deposition [31].

The *IP* and *IBAD* processes may considerably improve the properties of the films, but one of the difficulties in their application is the great number of parameters to be optimized.

2.2.4 Pulsed-Laser Deposition (*PLD*)

2.2.4.1 History

The *PLD* technique was initiated in the beginning of 1960s by studies of the interaction of laser light with solids such as that of Ready [34]. These studies followed the development by Maimann of solid-state laser using ruby crystal as lasing medium in 1960 [35]. Cheung [36] reports that these initial studies were followed by experiments of condensing the films from vapors created by a target irradiated by laser. The technique was rediscovered in late 1980s by the wave of the numerous studies on the synthesis of high-temperature superconductors such as $YBa_2Cu_3O_x$ (*YBCO*). An example is the study of Bartholomew and Burton who used an excimer pulsed laser to synthesize the *YBCO* films [37]. A great advantage of *PLD*, highlighted by this occasion, is the ***congruent evaporation*** enabling obtaining the deposits of multicomponent compounds having the same composition as laser-ablated target. The industrial application of the *PLD* technique is realized, similar to *IBAD*, to production of superconducting cables, which include *YBCO* films [38].

2.2.4.2 Principle

A diagram of a PLD technique is shown in Fig. 2.10. Fig. 2.11 presents the processing chamber of a laboratory installation. The beam of a pulsed laser is focused on a target which is located in the chamber under vacuum or under reactive gas at pressure being generally lower than $p \leq 1$ hPa. The typical lasers used for *PLD* with different wavelength, λ, have the pulses duration, τ, of [39–43]:

- milliseconds – Nd: glass ($\lambda = 1054$ or 1062 nm), Nd: YAG ($\lambda = 1064$, 532, 355, or 266 nm), ruby, i.e. Cr: Al_2O_3 ($\lambda = 694$ nm);
- microseconds – CO_2 *TEA* ($\lambda = 10.6\,\mu$m);
- nanoseconds – excimer lasers[5] such as XeCl ($\lambda = 308$ nm), KrF ($\lambda = 248$ nm), or ArF ($\lambda = 193$ nm);
- picoseconds – Nd:YVO_4 ($\lambda = 1064$ nm);
- femtoseconds – sapphire, Ti:Al_2O_3 ($\lambda = 650$–1100 nm).

The angle Θ of target irradiation, shown in Fig. 2.10, is typically in the range of $\Theta = 45$–60° [41]. The following processing parameters play also an important role in interaction of laser with target:

- pulses frequency, ν, and time of deposition, t_d, being also the time interaction between laser and target, which are frequently combined with a number of pulses during deposition;
- pulse energy or power and the area of laser beam interaction, which are frequently combined as pulse energy density, called also fluency, f, or pulse power density, q.

The typical process parameters used for *PLD* deposition of films of different compounds, being very difficult to obtain using other deposition methods, are collected in Table 2.5.

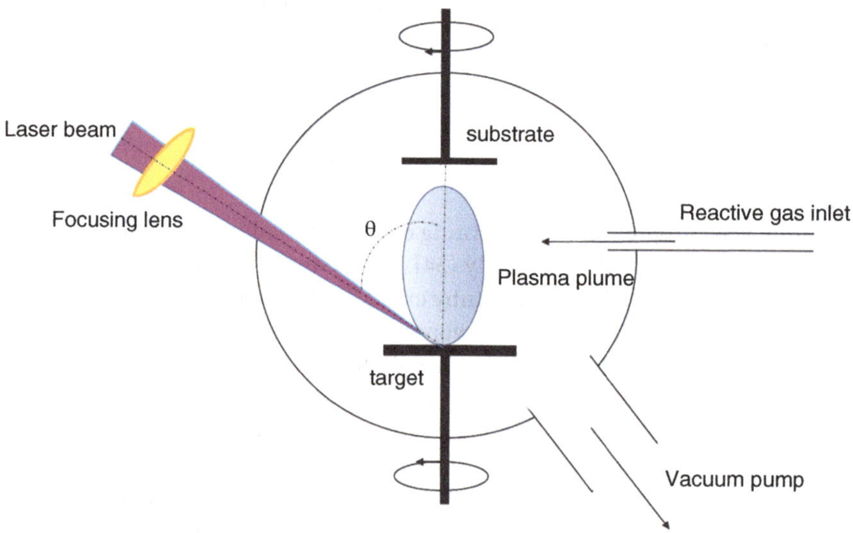

Figure 2.10 Sketch of typical pulsed laser deposition (*PLD*) processes.

5 Excimers lasers are probably the most frequently used in PLD processes.

Figure 2.11 Deposition process using *PLD* installation of the IRCER laboratory in the University of Limoges.

Table 2.5 Typical processing parameters used in different pulsed laser deposited films.

Film composition	Laser and its wavelength	Fluency, J/cm^2	Duration of laser pulse	Frequency of pulses or average power density	Gas in chamber and it pressure, Pa	References
YBCO	Excimer, KrF, $\lambda = 248\,nm$	$f = 1.5$	A few ns		O_2, $p = 13$	[41]
Perovskite, $La_xCa_{1-x}CoO_3$		$f = 7.6$	$\tau = 17\,ns$	$\nu = 10\,Hz$	Pulses of O_2 or N_2O	[44]
DLC	Excimer, KrF, $\lambda = 248\,nm$ Nd: YAG, $\lambda = 1064\,nm$	Excimer, $f = 8–20$	Excimer, $\tau = 20–25\,ns$, Nd: YAG, ms	Excimer $q = 10^{12}$ W/m^2, Nd: YAG, $q = 10^{15}$ W/m^2	High vacuum, $p = 10^{-5}$	[40, 45]
CN_x	CO_2, $\lambda = 10.6\,\mu m$	$f = 4$	$\tau = 1–3\,\mu s$	$\nu = 1\,Hz$	N_2 in post-discharge (The post-discharge is presented in Fig. 1.11e.), $p = 7–700$	[46]

2.2.4.3 Source

The photons arriving on the target are absorbed. The depth of absorption depends on the optical properties of target, namely reflectivity, R, and optical absorption depth, L. These properties are functions of the radiation wavelength. Supposing that initial laser power density is $q(0)$, the ***Lambert–Beer law*** describes its decrease, r, inside the solid target following the x axis perpendicular to the target's surface:

$$r(x) = \frac{q(x)}{q(0)} = (1 - R)\exp\left(-\frac{x}{L}\right) \tag{2.10}$$

Table 1.10 shows the values of R and L for a few metals and an oxide. The reflectivity for metal is close to $R \approx 1$ for $\lambda = 10\,\mu\text{m}$. Consequently, CO_2 laser is not useful to treat metallic targets and is used mainly on the oxide ones. The absorption of laser radiation excites the electrons, which transfer their energy to crystal lattice via electron–phonon coupling within a few picoseconds [47]. If the laser pulses duration are shorter than picoseconds (femtoseconds pulses) the energy is directly transformed into material ablation without any thermal effects. For longer pulses, the heat diffusion takes place and the material melts forming the liquid and evaporate thereafter. The heating rate of target is then very high being about 10^8–10^{10} K/s [39]. The vapors absorbs the laser pulses forming a plasma plume (Fig. 2.10) which contains electrons, ions, atoms, clusters, radicals and molecules, and larger molten droplets. The formation of the droplets results from evaporation of material below the target surface layer being still in a liquid state.

The formation of liquid particles, which are projected toward the substrate, is called ***splashing effect*** and is not desired because it lowers the quality of deposited films. Finally, the formation of plasma above target surface is associated with ablation of material, and its expansion results in a generation of waves of pressure called sometimes laser shocks [48].

2.2.4.4 Medium

The ablated target plume contains many components, and their spatial density follows the profile of cos ϕ of thermal vapors described by Eq. 2.2 or narrower profile of $\cos^n \phi$ ($4 < n < 15$) [41]. The plasma in the plume has very high degree of ionization, which may reach 80 % [39]. The energy of ions in the plasma depends strongly on experimental process parameters. The highest values may reach 2000 eV, and the average energy is about a few hundred eV [49]. The neutral atoms have lower energy of a few eV [47]. The high energy of ablated species helps in achieving high adhesion of films and in preserving the composition of target at formation of film on the substrates. The latter needs an explanation.

The ablated material is composed of molecules, which are irradiated in the plume transform in atoms and, later, on the ions. Subsequently, all these species will be found on the surface of the substrate. The energetic species, such as ions, having energy of hundreds eV, have a great mobility on the substrate. Thanks to this mobility, they can easily move toward the nucleation sites created around the molecules of the ablated compound and participate in forming the compound having initial composition. The velocity of ablated species depended on the pressure of medium and was measured to be in the range of a few km/s [41].

The *splashing effect* seems to be intrinsic to laser ablation. The size of the projected particles may reach a few micrometers, and their presence is not desirable in the condensed film. The techniques used to minimize the droplets presence include [41, 50]:

- optimization of processing parameters (lowering fluency, using target with small porosity, using laser with short wavelength);
- using rotating target;
- using vanes rotating with the speed of 3 000–10 000 rpm which sweep particulates out of the plume;
- positioning the substrate parallel to the axis of the target.

The high concentration of ablated species in plasma plume requires optimization of deposition processes to grow up the films with uniform thickness. The optimization methods were carefully analyzed for the *PLD* deposition of *DLC* films [40]. The methods include the rotation of substrate associated with:

- shifting of the substrate axis with regards to that of target;
- swinging the laser beam above the planar target;
- swinging the laser beam along the radius of cylindrical target.

The *PLD* technique can be associated with substrate polarization (*IP PLD*) and associated with sputtering [40, 45]. These two solutions were applied, e.g. to obtain the *DLC* deposits. The ion bombardment at *IP PLD* increased the content of sp^3 bonds in the diamond-like carbon film, and magnetron sputtering of Ti associated with laser ablation of graphite enabled obtaining multilayer composite Ti – TiC – *DLC*.

The medium at laser ablation may include some reactive gases as shown in Table 2.5. Consequently, the *YBCO* films synthesis is frequently realized in O_2-rich atmosphere to avoid the loss of oxygen at ablation. Similarly, the ablation of perovskites was associated with pulses of O_2 or N_2O gases synchronized with excimer laser pulses. Reactive ablation with a pulsed CO_2 laser of a graphite target in the reactive nitrogen plasma created by post-discharge was applied to obtain polycrystalline deposits of CN_x. Finally, Chitica *et al.* [51] synthesized WC deposits that were obtained by laser ablation of a tungsten, W, target in the atmosphere of methane, CH_4, under pressure of $p = 1$ Pa.

2.2.5 Chemical Vapor Deposition (*CVD*)

The *CVD* films and coatings deposition techniques include an important part of process parameters being physical. Namely, the chemical reactions take place in high temperatures in the neighborhood or on the substrate. The control of gazes fluxes and boundary layer concerning gas velocity and its temperature needs an understanding of fluid mechanics and has physical nature. On the other hand, the *PVD* processes as evaporation or sputtering include frequently chemical reactions between the species evaporated or ejected from the target. Consequently, the *CVD* and *PVD* processes have many similarities, and it makes sense to discuss them together. An important difference should be underlined in the beginning. The films deposited in the *CVD* processes grow up much slower than those in the *PVD* ones. Consequently, the films can more easily coat the substrate irregularities such as angles or holes, and the ***films conformality*** is an important advantage of *CVD* processes.

2.2.5.1 History

The *CVD* process was started to be applied at the end of XIX century to reinforce the filaments of the bulbs by coating tungsten on carbon lamp filaments by reduction of WCl_6 by H_2 [52]. The industrial ***processes of Mond*** and of ***van Arkel***, shown below, were developed in the same time [52, 53]:

$$Ni(CO)_4 \xrightarrow{423\,K} Ni + 4CO \text{ The Mond Process} \tag{2.11}$$

$$TiI_4 \xrightarrow{1473\,K} Ti + 2I_2 \text{ The van Arkel Process} \tag{2.12}$$

Until the World War II, the *CVD* technique was applied mainly to produce pure metals such as Ta, Ti, or Zr. The technique was introduced in the electronics industry to obtain semiconducting coatings in the 1960s. The plasma-enhanced chemical vapor deposition (*PE CVD*) reactors were introduced in the industry during 1970s, and the first diamond deposits were obtained by this method in the 1980s [54]. The metalorganic chemical vapor deposition (*MO CVD*) technique was developed in the 1990s.

2.2.5.2 Principle and Techniques

The chemical vapor deposition (*CVD*) process, presented schematically in Fig. 2.12, is initiated by the gaseous precursors brought to the zone of reaction. The reaction takes place in the neighborhood of the substrate and its solid-state product agglomerates on the substrate forming a film. The gaseous product of reaction is evacuated. The reactions leading to the formation of the solid phase may be activated by temperature in the fundamental process called thermal *CVD*. Some reactions, such as decomposition of ZrI_4 used to obtain film of Zr, require however as high temperature as $T \geq 1773$ K [55]. The use of such high temperatures are not useful in industrial practice. Consequently, some modifications of processes were developed to lower the reactions temperature by activating the reacting species in the different ways. An electrical discharge in the gas (mainly ***glow discharge***) may activate the species in process of plasma-enhanced chemical vapor deposition (*PE CVD*). This process can be carried out with hot cathode and is called filament chemical vapor deposition (*HF CVD*) [56, 57]. Similarly, the photons emitted by a laser may interact directly with these species in a ***photochemical*** way or by heating locally the substrate in a ***photo-thermal*** way. These processes are called laser chemical vapor deposition (*LCVD*).

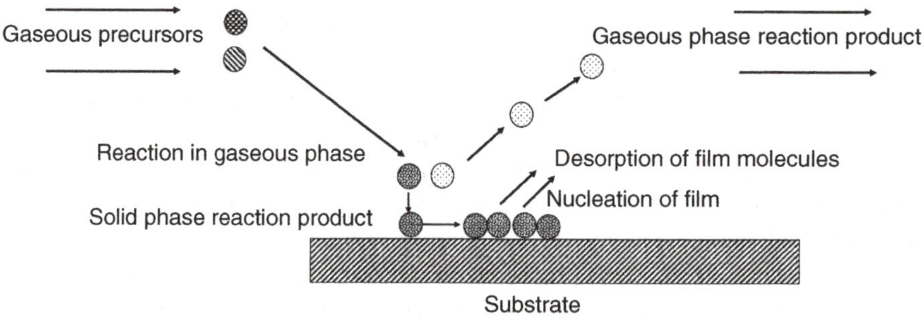

Figure 2.12 Schematic presentation of CVD processes.

The way of interaction is chosen in function to be the laser wavelength, i.e. depends on the photons energy. The local heating can be also realized by electron beam in the technique called electron beam chemical vapor deposition (*EB CVD*). Such a heating has an additional effect of secondary electrons emission. The electrons contribute in the reactions leading to the film growth [58].

Finally, the studies about gaseous precursors, which decompose at low temperature, enabled developing metalorganic chemical vapor deposition technique (*MO CVD*).

The analysis of *CVD* processes will be carried out by discussing the chemical reactions leading to generation of solid films and coatings and of their thermodynamics. Thermodynamics of the reaction enables better understanding of the process and facilitates the optimization of its parameters. Similarly, the discussion of the flow of gaseous precursors is essential to understand the particularity of generation of films on substrates having irregular surface. In particular, the *CVD* films may be created on the surfaces of powder particles suspended in ***fluidized-bed reactor*** or on the surface of micro-scale fibers [52, 58].

The technique similar to the *CVD* is atomic layer deposition (*ALD*), which uses also gaseous precursors introduced in a pulsed way to react on the substrate forming layer-by-layer deposits. This technique is discussed separately.

2.2.5.3 Sources in *CVD* Techniques

The precursors are delivered to the *CVD* reactors in gaseous phase. The quantity of gases is carefully controlled using flowmeters. The carrier gas, such as Ar or N_2, is used to transport the reacting ones.

Some precursors are initially in a liquid phase. An example is a liquid solution of $Mg(thd)_2$ in tetrahydrofuran (C_4H_8O) used by Manin *et al.* [59] to obtain MgO films by *MOCVD*. The authors used an evaporator at 473 K to gasify the liquid precursors. The mixture of N_2 with O_2 was used as carrier gas to transport the gas to the reactor. The simplified sketch of a *CVD* reactor using liquid as precursor is shown in Fig. 2.13a.

The precursors can be also a product of reaction of initially solid phase (see Fig. 2.13b). The example is the chlorination of small Ta particles described by Richardt and Durand [4] and deposition of W films described by Pierson [54]. The reaction in the chlorinator leads to formation of chloride:

$$W + 3Cl_2 \xrightarrow{1073K} WCl_6 \tag{2.13}$$

The chloride decomposes in the reaction with the carrier gas H_2 in the reactor enabling obtaining W film:

$$WCl_{6(g)} + 3H_{2(g)} \xrightarrow{773\text{–}1373K} W_{(s)} + 6HCl_{(g)} \tag{2.14}$$

2.2.5.4 Thermal *CVD*

Principle

The thermal *CVD* process includes the reaction leading to a formation of solid product (see Fig. 2.12) carried out at controlled temperature being generally higher than 1173 K [54]. The pressure in the reactor may vary from a few hundred Pascals, in a process called low-pressure chemical vapor deposition (*LP CVD*), to the atmospheric one. The *CVD* ***reactors***

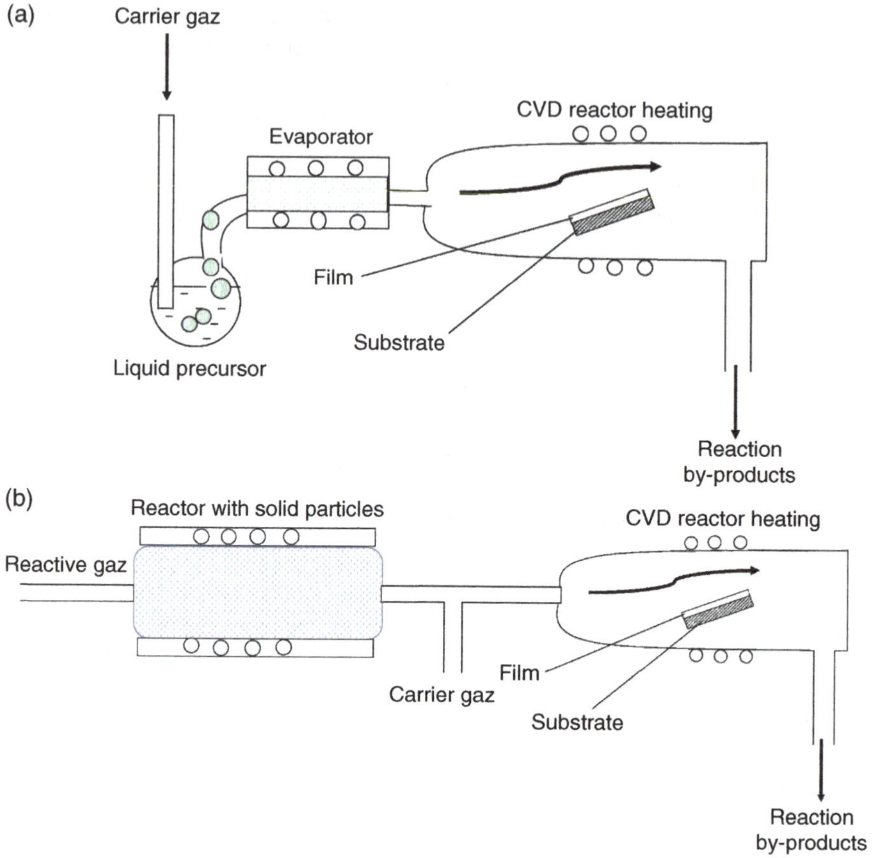

Figure 2.13 Sketch of CVD reactors delivered with: (a) liquid precursor; (b) precursor generated by reaction of solid particles.

can be either ***cold wall*** or ***hot wall*** [52]. The *hot-wall reactors* are in fact the furnaces inside which the chemical reaction occurs. The walls have the temperature of gaseous medium and their surface is consequently coated with solid films, which should be systematically removed at the reactor maintenance. An example of an industrial thermal *CVD* installation with such reactor used to obtain wear-resistant TiC films is shown in Fig. 2.14.

The *cold-wall reactors* are designed using inductive or resistive heating of the substrates only. Their drawback is the gradient of temperature resulting in gradient of concentration, which may lead to the non-uniformity of generated film. The uniformity may be improved by lowering the pressure inside the reactor [52]. The design of any *CVD* reactor has to consider recycling of the by-products. The basic steps of thermal *CVD* are as follows [55, 60]:

- mass transport and diffusion of reactants to the zone of controlled temperature;
- reaction generating the film precursor and by-products;
- diffusion of film precursor to the substrate surface;

Figure 2.14 Industrial thermal CVD installation used for deposition of titanium carbide on machine parts developed by the company *Bernex* (Olten, Switzerland). *Source:* Reproduced with permission from Bernex Bimetall AG.

- adsorption of the product by the surface;
- reactions on the surface and nucleation of the film;
- desorption of by-products of the reaction on the surface;
- mass transport and diffusion of by-products to the reactor exit.

The diffusion is shortly described in Chapter 1.2.1, but the mass transport and some fundaments of fluid dynamics will be discussed in this chapter. The possible reactions used in the thermal *CVD* techniques together with their thermodynamics being essential in this method are discussed below.

Medium

The described phenomena take place inside the reactor in the zone of reaction resulting in the formation of solid film.

Chemistry The reactions to be carried out in thermal *CVD* process depend on the solid product to be obtained. The reactions take place in very high temperature, and solid-state product should nucleate on the substrate's surface and not in the gas. The latter leads to ***homogeneous nucleation*** of powder in the reactor and should be avoided. Some typical reactions are presented in Table 2.6

Thermodynamics After having found the reaction resulting in the formation of desired solid phase, it is necessary to optimize the process. The optimization concerns the choice of: (i) temperature and pressure inside the *CVD* reactor; and (ii) molar concentration of the gaseous reactants. These variables can be determined by minimizing ***free Gibbs energy*** of

Table 2.6 Main types of chemical reactions in thermal CVD [52, 55, 61–63].

Type of reaction	Example	Reference
Thermal decomposition or *pyrolysis*	$Si(CH_3)Cl_{3\,(g)} \rightarrow SiC_{(s)} + 3HCl_{(g)}$	[52, 55, 61]
Reduction	$SiCl_{4\,(g)} + 2H_{2\,(g)} \rightarrow Si_{(s)} + 4HCl_{(g)}$	
Oxidation	$SiH_{4(g)} + O_{2(g)} \rightarrow SiO_{2(s)} + 2H_{2(g)}$	[54]
Exchange	$SiCl_{4(g)} + CH_{4(g)} \rightarrow SiC_{(s)} + 4HCl_{(g)}$	[61]
Coupled (The coupled reactions are separately as follows: $CO_{2\,(g)} + H_{2\,(g)} \rightarrow CO_{\,(g)} + H_2O_{\,(g)}$ and hydrolysis $AlCl_{3\,(g)} + 3H_2O_{\,(g)} \rightarrow Al_2O_{3\,(s)} +$ $6HCl_{\,(g)}$. These reactions were used in industrial manufacturing of films on cutting tools [63])	$2AlCl_{3(g)} + 3CO_{2(g)} + 3H_{2(g)} \rightarrow Al_2O_{3(s)} + 3CO_{(g)} + 6HCl_{(g)}$	
Carbidization	$TiCl_{4(g)} + CH_{4(g)} \rightarrow TiC_{(s)} + 4HCl_{(g)}$	[62]
Nitridation	$3SiCl_{4(g)} + 4NH_{3(g)} \rightarrow Si_{3N4(s)} + 12\ HCl_{(g)}$	

the system, which should be at its minimum to have stable reaction. The *free Gibbs energy* depends on the chemical potential, μ, of each gas participating in reaction in the following way:

$$G = \sum_i n_i \mu_i \tag{2.15}$$

The chemical potential depends on the temperature and on the partial pressure, p_i, of each gas and on p being the total pressure:

$$\mu_i = \mu_i^0 + RTln\ p_i = \mu_i^0 + RTln\ \frac{n_i}{n} p \tag{2.16}$$

Finally, the variation of *free Gibbs energy* of the entire reaction having m gaseous reactants and by-products as well as n solid products can be found from the following equation [61, 64]:

$$\Delta G = \sum_{i=1}^{m} n_{i\,(g)} \mu_{i\,(g)}^0 + RTln\ \frac{n_{i(g)}}{n} p + \sum_{i=1}^{n} n_{i\,(s)} \mu_{i\,(s)}^0 \tag{2.17}$$

For the reaction at given temperature and pressure the *Gibbs free energy* of reaction is negative $\Delta G \leq 0$ and the reaction stops if $\Delta G = 0$. It is possible to develop a computer program or to make a calculation enabling the minimization of the *free Gibbs energy* for the variable temperature; concentration of species and partial pressure of gas help in the thermal *CVD* processes optimization. The thermodynamic data required for the calculation, and in particular, free Gibbs energy of formation of a reactant or product of reaction at temperature T, i.e. $\Delta_f\, G^0{}_T$, can be found, e.g. in the handbook [7].

The major problem in such calculation is to consider the possibilities of modification of the reactants leading to many reactions having many gaseous and solid products. Pauleau [64] analyzed the reaction of carbidization similar to that shown in Table 2.6 but including hydrogen. The gaseous phases in the zone of reaction may include hydrogen, titanium chlorides, hydrocarbons and the solid-phase carbon, titanium and, the desired product, titanium carbide. Janssson *et al.* [65] made the thermodynamic calculation of free-energy minimization using program *EKVICALC* for the reactions starting from gaseous H_2, BCl_3, and CH_4, which should lead to the *CVD* nucleation of solid B_4C carbide at the temperatures from 1300 to 1500 K and total pressure of 6.7 kPa. The diagrams shown by the authors indicate the thermodynamic predictions of temperature and partial pressures of reactants being favorable to formation of boron carbide film. Choy [52] mentions other computer programs such as *SOLGAS, SOLGASMIX, FREEMIN, EKVIBASE, and MELANGE* used by many research groups, which use the minimization of Gibbs free energy to optimize the *CVD* processes.

Fluid dynamics The fluid dynamics is essential to understand the frequency of the collisions of reactants' molecules, which enables the chemical reaction to occur. Similarly, the ***laminar*** character of the gas ***flow*** is essential to coat the substrate surface irregularities. Finally, the geometry of ***boundary layer*** between gas and substrate surface influences the homogeneity of films and coatings thickness.

Consequently, the basic equations of the ***kinetic theory of gases*** will be presented supposing the flow of a gas in a cylinder having diameter d_c [9]. This theory considers, for the simplification, gas as a perfect one. The following suppositions are used: (i) n molecules of gas having mass m and diameter d_m being negligible with regard to the diameter of reactor; (ii) their number justifies statistical treatment; (iii) the molecules enter in elastic collisions with the others and with the walls of container; and (iv) interactions between molecules are negligible[6].

The distribution of the velocities of the gas molecules is given by ***Maxwell–Boltzmann distribution***, and the mean velocity of the particles is equal to:

$$\overline{v} = \frac{1}{n}\int_0^\infty v\,dn = \sqrt{\frac{8kT}{\pi m}} \tag{2.18}$$

The *mean free path*, $\overline{l}$ between the collisions is given in Eq. 2.4. The dimensionless Reynolds, *Re*, and Knudsen, *Kn*, numbers are for the flow of gases in a cylinder defined as:

$$Re = \frac{\overline{v}\rho d_c}{\eta} \tag{2.19}$$

$$Kn = \frac{\overline{l}}{d_c} \tag{2.20}$$

The gas flow regimes are generally defined using these numbers. Consequently, the principal regimes occur for: (i) $Re > 4000$ ***turbulent flow;*** (ii) $Re < 2000$ *laminar flow*; (iii) $0.01 < Kn < 0.5$ ***intermediate flow***; and (iv) $Kn > 0.5$ ***molecular flow***. As the viscosity of gas plays an important role in *turbulent* and *laminar flows*, they are called ***viscous flows***. To the latter may be added also the flows characterized by $Kn << 1$. In fact, the viscous interaction of layers of a gas with the walls of cylinder results in the formation of many vortexes at the *turbulent flow*. The laminar flow has these layers parallel to the cylinder wall. Finally, some

6 The molecules may react at the collisions.

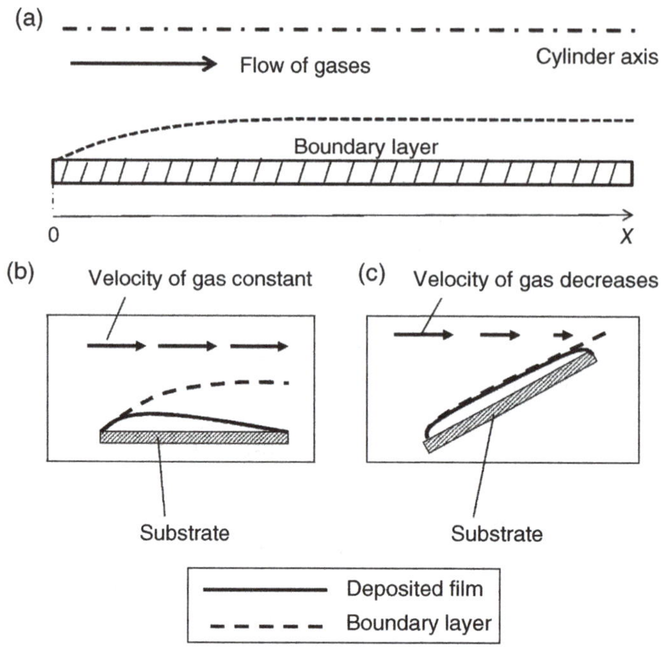

Figure 2.15 The *boundary layer* of a gas: (a) geometry in the neighborhood of cylinder walls; (b) film deposited on a substrate surface parallel to the flow of gas; and (c) film deposited on a substrate put under an angle with regard to the flow direction.

vortexes occur and some layers remain parallel to the walls for the values of *Re* numbers being between $2000 < Re < 4000$. The *molecular flow* is also called ***Knudsen flow***[7] and does not have a lot of collisions between gas molecules. The *Knudsen flow* is influenced by the collisions of molecules with the walls of cylinder.

The *boundary layer* is the region in the immediate vicinity of surface of a wall of the cylinder in which a gas flows. The thickness of the layer is defined as the distance from the wall where the thickness of the flow reaches 99% of the free stream velocity. The velocity of gas increases inside the boundary layer with the value of *x* to reach a constant value, as seen in Fig. 2.15a. The *boundary layer* for the gas velocity δ_v depends upon the flow regime. For the *laminar* and *turbulent flow regimes* and the geometry shown in the figure, the thickness is given by [64, 66]:

$$\delta_v = \frac{5x}{\sqrt{\mathrm{Re}_x}}, where\ Re_x = \frac{v\rho x}{\eta} \tag{2.21}$$

In *turbulent flow* conditions, the thickness of *boundary layer* is equal to:

$$\delta_v = \frac{0.37x}{(Re_x)^{\frac{1}{5}}} \tag{2.22}$$

7 Some authors, e.g. Pauleau [64], define *Knudsen flow* as characterized by $Kn >> 1$.

There is also a gradient of temperature between the main stream of gas and the cylinder's wall. Thermal boundary layer δ_T is smaller than the thickness of layer related to velocity, δ_v, depending on the value of dimensionless ***Prandtl number*** for the gas, *Pr*, as seen in the following equations:

$$\delta_T = \frac{0.98\delta_v}{Pr^{\frac{1}{3}}} \tag{2.23}$$

$$Pr = \frac{\eta c_p}{\lambda} \tag{2.24}$$

The boundary layer has a practical importance in the processes of *CVD* coatings deposition because of the modification of the reactions conditions (temperature and gas velocity) along the substrate surface. The thickness of the deposit depends on the mass transfer of the reaction products to the substrate, and therefore, determines its growth. Another consequence of Equation 2.21 is the mass distribution of the deposit. If the substrate surface is parallel with respect to the gas flow, the distribution is not uniform (Fig. 2.15b), and it becomes homogeneous when the substrate is put under an angle with regard to the flow (Fig. 2.15c). Consequently, the velocity of gas flow decreases gradually along the substrate, and the thickness of the boundary layer (and the thickness of the deposit) becomes more homogeneous.

Sarin [67] made numerical simulations of thermal *CVD* deposition of WC and mullite films, taking into account the thermodynamics of the reactions and the mass transfer to the substrate.

2.2.5.5 *PE CVD*

Thermal *CVD* has a major drawback, which is high temperature to be generated in the reactor to activate desired reaction. One of the methods leading to decrease the temperature is the excitation or ionization of the reactants. Let us suppose, after Reif and Kern [68], that the activation energy of chemical dissociation of species A and B, i.e. A →B requires an activation energy of ΔE. After being excited, by a collision with electrons, the species become A* and B*. The activation energy required to dissociate A*→ B* becomes ΔE^* and is, generally lower than ΔE, i.e. $\Delta E^* < \Delta E$. Consequently, the temperatures of reactions are also lower as seen in Table 2.7.

The process *PE CVD* is used frequently in electronic technology to obtain the films requiring limited deposition temperature. The major drawback is the cost of reactor being more than that of thermal *CVD*. Choy [52] mentions also a complexity of the process requiring sometimes the statistical methods of design of experiments (*DoE*) to optimize the parameters in order to obtain high-purity films. Consequently, Hössler and Ernst [69] use this method to reduce the formation of $TiSi_2$ at the interface between Ti films and Si substrate.

Principle

The reaction in the deposition process is activated mainly by a *glow discharge* at the pressure of a few kPa. This kind of discharge forms so-called **cold plasma** used also for substrates' cleaning discussed in Chapter 1.3.2. The plasma is generated in *PE CVD* reactor using *RF*

Table 2.7 Comparison of deposition temperatures of some oxide, nitrides, and carbides films by the process of thermal *CVD* and plasma-enhanced *CVD* (*PE CVD*).

	Deposition temperature, K	
Deposited film	**Thermal *CVD***	***PE CVD***
SiO_2	1073–1473	573
Si_3N_4	1173	573
TiN	1173–1327	773
TiC		
WC	1273	600–800

Source: Adapted from Pierson [54].

or microwave frequencies by inductive or capacitive coupling [70]. The remote *PE CVD* reactors generate plasma far away from the film deposition zone in order to avoid that the ionized species' bombardment of growing film [52].

Another technique, called electron cyclotron resonance (*ECR*), produces a plasma by the combination of electric and magnetic fields. The cyclotron frequency of the electron is reached when the frequency of the electric field becomes equal to that which results from a rotation of the electron around lines of the magnetic field. This effect appears for a microwave frequency of 2.45 GHz and a magnetic field with an intensity of 875 Gauss [54]. The drawback of this deposition technique is its high price and difficulty of controlling simultaneously the parameters associated with the generation of two types of field. The process of *ECR* uses a relatively low pressure of $p = 10^{-4}$ hPa [68].

The *PE CVD* technique is used commercially to obtain films of, e.g.: (i) polycrystalline and epitaxial Si or GaAs; (ii) wear-resistant TiN; and (iii) diamond-like carbon (installation is shown in Fig. 2.16).

Medium

Cold plasma is generated by different discharges under low pressure. The main difference in ***thermal plasma,*** described carefully by Boulos *et al.* [71], is the temperature of electrons, T_e, and of heavy species, T_h. These temperatures become very different at pressures lower than 10 kPa (see Fig. 2.17) and $T_e > T_h$. Arc discharge can be used at higher pressures. Consequently, *dc* arc at atmospheric pressure to obtain SiO_2 films on titanium alloy Ti6Al4V using hexamethyldisiloxane ($C_6H_{18}OSi_2$) and tetraethylorthesilicate ($(C_2H_5O)_4Si$) as precursors evaporated from liquid phase and mixed with Ar, N_2, and O_2 [73]. *Thermal plasma* formed at arc discharge is discussed in more detail in the chapter describing plasma spraying.

In addition, the mobility[8] of heavy species is much lower than that of electrons. Generally, the ions are not able to follow an electric field with a frequency higher than 100 kHz.

8 Defined as velocity of charged particle reached under action of electric field. Thornton and Greene [17] show the mobility of ions in their neutral gas.

Figure 2.16 Installation PECVD used to obtain diamond-like carbon films and developed by company *Bernex* (Olten, Switzerland). *Source:* Reproduced with permission from Bernex Bimetall AG.

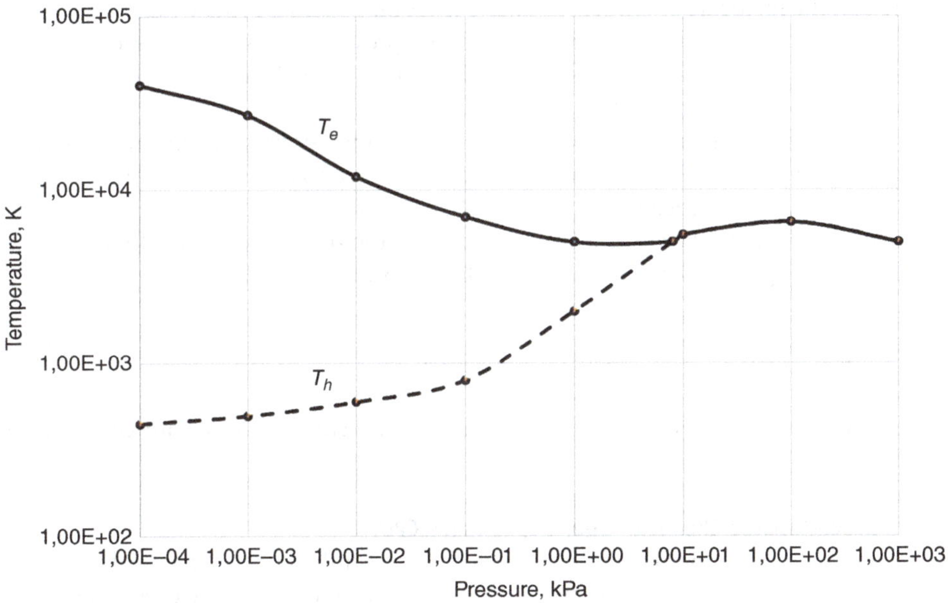

Figure 2.17 Temperatures of electrons T_e and of heavy particles T_h in a plasma created by an arc discharge vs. pressure [71, 72].

The useful discharges are generated by microwave sources having typical frequency of $\nu = 2.54$ GHz or that of *RF* with typical frequency is $\nu = 13.56$ MHz are used. The ions inertia of the ions is too great to bombard intensively the growing deposit which lowers the compressive stresses generated there.

Table 2.8 Typical parameters of plasma generated in *glow discharge* [17, 18, 57, 64].

Parameter	Value
Discharge current density	$10–10^4$ A/m^2
Radius of Debye, r_D	$r_D = 130\,\mu m$ for $n_e = 10^{16}\,m^{-3}$ and $E_e = 3\,eV$
Electron density, n_e	$10^{12}–10^{18}\,m^{-3}$
Ionization ratio	$10^{-5}–10^{-1}$
Electrons energy ($1\,eV \approx 1.602 \times 10^{-19}$ J and using expression $E = kT$ the corresponding temperature is $T \approx 11\,600\,K$), E_e	1–10 eV
Ions energy, E_i	0.02–0.1 eV
Lifetime of free radicals	about 1 ms

The electric current supplying discharge starts with the electrons being emitted from an electrode and move to another one entering in ***elastic*** and ***inelastic collisions*** with atoms and molecules leading to their excitation or ionization. The excitation and ionization occur when the energy gained in the *inelastic collision* is greater than the corresponding potential. For example, the molecule of oxygen, O_2, has excitation potential of 7.9 eV and ionization one of 12.5 eV [18]. The ionization process of species present in plasma is characterized by ***ionization rate***, which defines the ratio concentration of ionized particles (n_i or n_e) to that of all particles ($n_i + n_n = n_e + n_n$).

Plasma is electrically neutral. This neutrality occurs however only in a macro-scale. The scale of neutrality describes *Debye radius*, r_D, which characterizes the volume of the sphere where plasma is not electrically neutral. In other words, if charged particles are on the distance greater than r_D, the electric field acts to move the species in the way to restore the electrical neutral space. The radius is given by the following equation [64]:

$$r_D = \sqrt{\frac{\varepsilon_0 kT}{n_e e^2}} \tag{2.25}$$

Some typical properties of *cold plasma* generated by *glow discharge* are collected in Table 2.8.

2.2.5.6 Laser-Assisted Chemical Vapor Deposition (*LCVD*)

Radiation of light is another form of energy, which may influence the chemical reactions in *CVD* processes. Initially, the different lamps emitting *uv* radiation were used. The lamps emit the radiation resulting from discharge in vapors of such gases as mercury or xenon ones. However, since 1970s, the processes are accompanied mainly with different types of industrial lasers in the process known as laser-assisted chemical vapor deposition (*LCVD*). This process will be discussed more in detail.

Principle

The industrial lasers used in the films and coatings technologies processes are presented in Table 1.11. Their radiation can be applied in *LCVD* processes in two ways: (i) thermal *LCVD* which consists of laser heating the substrate; (ii) photochemical *LCVD* in which

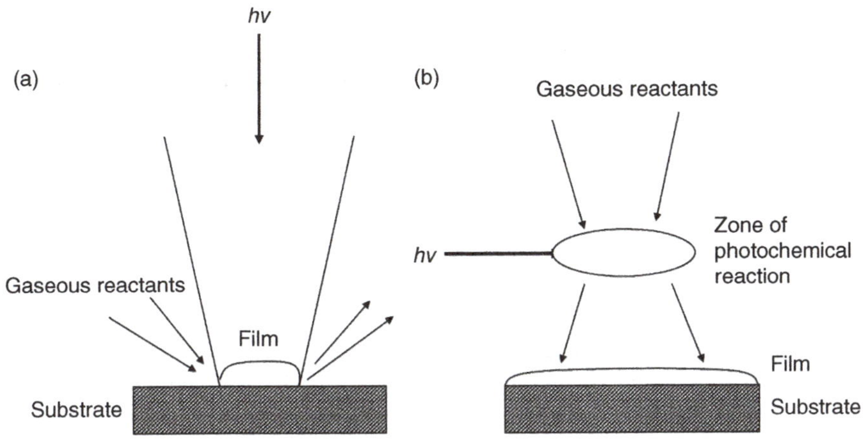

Figure 2.18 Different *LCVD* processes: (a) thermal and (b) photochemical.

laser energy influence the chemical reaction resulting in formation of solid film. The sketch of these processes is shown in Fig. 2.18.

Thermal *LCVD* requires an appropriate choice of laser to assure that the substrate absorbs well the radiation of laser. The absorption depths for common materials and different wavelengths are shown in Table 1.10. The laser light is absorbed by the substrate without interacting with the molecules of the reactive gases. The laser heats the substrate in a small area that matches the dimensions of its focal spot. The temperature of this place depends on the energy density of the laser (typically $q = 10^7$–10^{10} W/m^2). The reaction leading to the formation of a deposit takes place on the substrate. Therefore, the principle of thermal *LCVD* is similar to that of thermal *CVD* with a cold wall reactor. Some examples of films obtained by thermal LCVD are shown in Table 2.9. The lasers emitting *ir* radiation (CO_2 and Nd:YAG) are better absorbed by solid substrates and are used to heat them. The pressures are relatively high, and some processes (SiC synthesis shown in Table 2.9) are carried out at the atmospheric one.

As the reactions at photochemical *CVD* process occur in the gas, their description is made in the following section.

Table 2.9 Examples of films synthesized by thermal LCVD.

Film	Precursors	Pressure, hPa	Laser/λ, μm	*q*, power density, W/m^2 or *P*, power, W	References
$La_2Zr_2O_7$	La(dpm)$_3$, Zr(dpm)$_4$	9.5	Diode, *cw*/0.976	$q = 3.2 \times 10^5$	[74]
Al_2O_3	Al(acac)$_3$	4.7–23.3	Nd: YAG, *cw*/1.064	$P \leq 260$	[75]
SiC	$C_3H_{10}Si$	10^3	CO_2, *cw*/10.6	$6.1 \leq P \leq 7.4$	[76]
Y-doped $BaZrO_3$	Ba(dpm)$_2$, Zr(dpm)$_4$, and Y(dpm)$_3$	5	Nd: YAG/1.064	$P = 203$	[77]

Medium

In the photochemical *CVD* process, the reactive gases directly absorb laser light. The photons of laser radiation should have energy to interact with the molecules of reactants in the desired way (dissociation, excitation, or ionization). Their energy can be found knowing the laser wavelength from the *Planck–Einstein relation*:

$$E = h\nu = \frac{hc}{\lambda} \tag{2.26}$$

The relation enables to calculate the energy of photons emitted by the industrial lasers (see Fig. 2.19).

The *ir* lasers emit photons whose energy is relatively small being equal to $h\nu = 0.12\,\text{eV}$ for CO_2 laser and $h\nu = 1.2\,\text{eV}$ for Nd:YAG one. The absorption of radiation having initial intensity $q(0)$ along the reaction zone is described by *Lambert–Beer's law*, which for the gaseous medium takes the following form:

$$\frac{q(x)}{q(0)} = \exp(-n\sigma x) \tag{2.27}$$

The equation is valid only for single-photon absorption. The property of reacting gas, being essential to achieve a successful photochemical reaction, is high absorption of laser radiation. Eden [78] shows an example of platinum acetylacetonate (Pt(acac)$_2$) vapors, which absorbs well the radiation of the wavelength at about $\lambda = 300\,\text{nm}$. Consequently, the XeBr ($\lambda = 282\,\text{nm}$) or XeCl ($\lambda = 308\,\text{nm}$) excimer laser seems to be well adapted to influence photo-chemically the decomposition of this molecule [78].

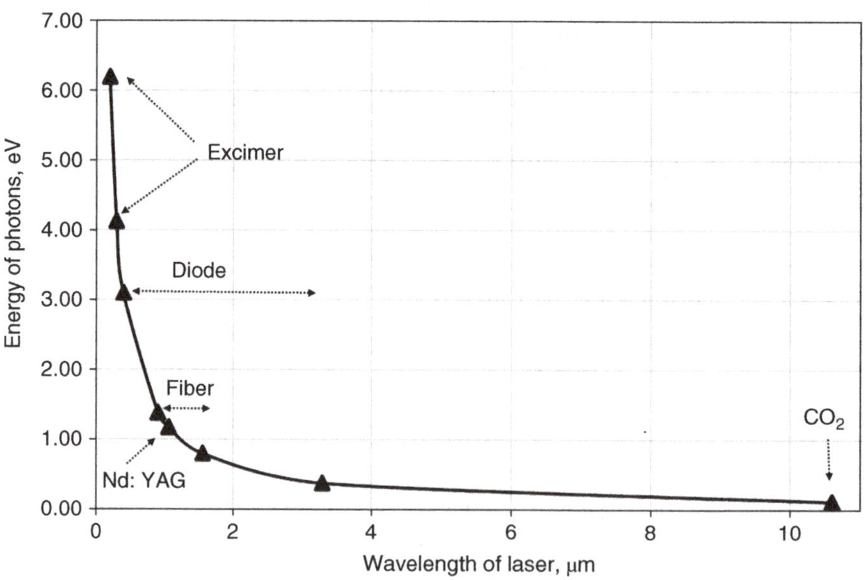

Figure 2.19 Energy of photons emitted by different industrial laser.

Table 2.10 Typical reactions in the photochemical *LCVD* process [64, 78].

Type of reaction	General equation	Example	Radiation wavelength, nm
Photodissociation	$ABC + h\nu \rightarrow A + BC$ followed by $BC \rightarrow B + C$	$Al(CH_3)_3 + h\nu \rightarrow Al(CH_3)_2 + CH_3$ followed by $Al(CH_3)_2 \rightarrow Al + C_2H_6$	$\lambda < 230$ (e.g. KrCl excimer laser)
Photoionization	$AB + h\nu \rightarrow A^+ + B^-$ or $AB + h\nu \rightarrow A^+ + B + e^-$	$InI + h\nu \rightarrow In^+ + I^-$	$\lambda < 193$ (e.g. ArF excimer laser)

The most common photochemical reactions are shown in Table 2.10. Some processes occur in two steps. Some reaction can be influenced by ***multiphoton processes***. The multiphoton absorption cross-section is different from the single-photon one.

2.2.5.7 Metal–Organic Chemical Vapor Deposition (*MO CVD*)

The metal–organic chemical vapor deposition (*MO CVD*) process is used to obtain mainly metallic films, the films of elements, oxides, carbides, and nitrides starting from liquid metal–organic precursors. Choy [52] indicates that some authors make a distinction between metalorganics (metal bonded to organic radical) and organometallic (including direct metal–carbon covalent bonds). The process including the latter is sometimes called organometallic chemical vapor deposition (*OM CVD*). The process started to be developed in the 1960s to obtain pure semiconductor coatings at lower temperature than thermal *CVD* [54].

Principle

The *MO CVD* technique uses mainly liquid precursors, which evaporate and are transported using carrier or reactive gas (H_2, O_2, or N_2) to the reactor as shown in Fig. 2.13a. Such delivery requires sufficient vapor pressure of the liquid. Kuech and Jensen [79] show the vapor pressures of common metalorganic precursors. Their pressure at $T = 298\,K$ varies from $p = 0.4$ to 488 hPa. The *MO CVD* reaction occurs at temperatures varying from $T = 573\,K$ to 1173 K under pressure varying from $p = 1\,hPa$ to the atmospheric one [54]. The high quality of precursors is a major requirement in this technology, and their low thermal instability resulting from polymerization may lead to low reproducibility of the films.

The hydrocarbons included in *MO CVD* precursors are, e.g. [52, 54, 58, 79]:

- alkyls, which include methyl (CH_3), ethyl (C_2H_5), isopropropyl (C_3H_7);
- acetylacetonates ($C_5H_7O_2$);
- alicyclic compounds, which include the carbonic groups (such as CH_2) being on the top of a triangle (cyclopropane, C_3H_6) or pentagon (cyclopentane, C_5H_{10}).

Devi [80] made a systematic study concerning the requirements and chemistry of precursors to synthesize films of different oxides. The typical precursors used in films synthesis with *MOCVD* process are shown in Table 2.11. The equipment used in the process is expensive and the quality of coating is high. Consequently, the process is frequently applied in semiconductors and in opto-electronic applications.

Table 2.11 Examples of films synthesized in *MO CVD* processes.

Film	Precursor	Carrier and reactive gas	Reaction pressure, hPa	Reaction temperature, K	Film thickness, nm	References
GaAs	$Ga(CH_3)_3$ and AsH_3	—	10^2–10^3	873–1073	—	[52]
ZrO_2	$Zr(thd)_4$	O_2 and N_2	1	973 and 1173	200 and 250	[81]
MgO	$Mg(thd)_2$	O_2 and N_2	50–10^3	673–773	About 1000 (Observed on the film section.)	[59]

2.2.5.8 Atomic Layer Deposition (*ALD*)

Finnish scientists Suntola and Antson [82] initiated the atomic layer deposition technique (*ALD*) in the 1970s as atomic layer epitaxy (*ALE*) of ZnS films. The technique was rapidly developed for polycrystalline metals and oxides, and the name *ALD* was adopted [83]. *ALD* has a lot of similarities with *CVD*. The methods were separated separated because of [84]: (i) self-limitation of precursors adsorption; and (ii) cyclic procedure of film deposition using precursors and reactants separated by purging.

The *ALD* technique included the following steps [84]: (i) exposure on precursor; (ii) purging of by-products; (iii) exposure to reactant species; and (iv) purging of final by-products. These four essential steps of *ALD* are one cycle resulting in a film thickness lower that 0.1 nm. The final film may be deposited using many cycles.

The pressure at the deposition is about $p = 1$–10^3 Pa and the substrate temperature depends on reaction between precursor and reactant but is, generally, lower than $T < 623$ K [83, 85]. The precursors used in the first cycle are not very different from those used on *MOCVD* and depend on the final film to be obtained. For example, the metal films can be obtained using: (i) inorganic precursors as metal halides, and (ii) metal organic such as alkyls, alkoxides, and amides. However, higher thermal stability of precursors for the *ALD* process is required to avoid the uncontrolled decomposition reaction [80].

The purging is made using inert carrier gases such as Ar or N_2. Finally, the typical reactants for metallic films include [85] O_2 and O_3, air, H_2, NH_3, and many others. The reactants can be activated by plasma in *PE ALD* process to become H, O, or N atoms or radicals. However, the activation by plasma renders the process more complicated by presence of ionized species and *uv* radiation.

A few examples of the *ALD* process applied to obtain metal films are shown in Table 2.12. The major advantage of the processes is *films conformality*, i.e. constant thickness on entire

Table 2.12 Description of *ALD* processes applied to obtain metal films.

Film	Substrate	Precursor	Reactant	Reaction temperature, K
Ir	SiO_2 or Al_2O_3	$Ir(acac)_3$	O_3	483–498
Ni	Si or SiO_2	$Ni(dmamb)_2$	NH_3	573
Pd	Al_2O_3	$Pd(hfac)_2$	*formalin*	473 and 373 after Pd nucleation

Source: Adapted from Emslie et al. [85].

substrate independent of its morphology and shape. This advantage renders the process useful in design of devices using nanotechnology [84, 86]. In particular, it is possible to obtain deposits onto nanostructured particles [87].

2.3 Granular Methods of Coatings Deposition

The species used in granular deposits are solid particles whose dimensions vary from nanometers to hundreds of micrometers. These particles get in contact with the substrate in the liquid, partially liquid, or solid state. The microstructure of granular deposits results from the splashing, solidification, and sintering of these particles. The granular deposits belong, generally, to the family of thermal spraying.

In thermal spraying, a solid (wires, rods, and powder) or liquid feedstock (suspensions and solutions) is introduced into the jet generated by a torch. Subsequently, many different phenomena may occur depending on deposition techniques, but a formation of molten particle and its acceleration toward a substrate is common for all of them. In most of deposition techniques the particles are sprayed on the substrate where they deform into lenticular grains called lamellae. The lamellae are attached to a substrate, solidify and sinter together, and form a deposit. Unlike atomistic deposits, the morphology of the feedstock used for thermal spraying plays an important role in reaching desired properties of coatings. Consequently, the methods of powders manufacturing will be shortly discussed.

2.3.1 Powder Fedstock

The solids used in granular methods of coating deposition are principally wires, rods, and powders [88]. The wires are generally metallic being used only in flame-spraying (*FS*) and arc-spraying (*AS*) processes. The methods of wire manufacturing are well known and are described in textbooks such as Van Vlack [89]. The wires used in practice are (i) metals, such as Al or Zn; (ii) alloys, such as Ni + 5 wt. % Al; and (iii) composites that contain metallic sheath and hard filler being frequently particles carbides such as WC or Cr_2C_3. The wires have typically an outer diameter of a few millimeters.

Another solid feedstock used in the discussed deposition methods are ceramic rods delivered to the combustion spray torches. The short rods can be inserted into a long polyethylene sheath burned out in the flame at spraying, and the deposition process is about to be continuous.

The wires and rods are introduced to plasma or flame and get molten. The resulting molten droplets are nearly monosized. This renders the resulting coating more homogeneous than that sprayed with powder feedstock. However, presently, powder feedstock is applied most frequently in granular methods of deposition.

The physical and chemical properties of powders influence considerably coatings quality. The physical characteristics include: (i) granulometry; (ii) morphology; (iii) density and flowability; and (iv) thermodynamic properties characterized by differential thermal analysis (*DTA*). The most important fundamental chemical property is chemical bond determining material properties. The bonds are: (i) ionic bonds in chemically stable and inert materials such as oxides Al_2O_3 or ZrO_2; (ii) covalent bonds of mechanically strong materials such as SiC or Si_3N_4; and (iii) metallic bonds in metals and some carbides and nitrides. The chemical properties tested typically are: (i) chemical composition tested, e.g. using *X-ray* photoelectron spectroscopy, *XPS*; (ii) phase content, characterized by *X-ray* diffraction, *XRD*; and (iii) elements distribution visualized using energy-dispersive *X-ray* spectroscopy *EDX*.

2.3.1.1 Methods of Powder Manufacturing

Following types of powders are used most frequently in thermal spraying [9, 88]: (i) metals and alloys; (ii) ceramics such as oxides, carbides, or nitrides and their alloys such as ZrO_2 + 8 wt. % Y_2O_3; and (iii) ***cermets***[9] having morphology of ceramics particles cladded with metal film or an agglomeration of metal and ceramic particles such as WC + 12 wt. % Co.

The industrial methods of the powder manufacturing depend mostly on type of the powder morphology required and on its material. Some methods of powder manufacturing are presented in Table 2.13.

Atomization, described by, e.g. Meinhardt *et al.* [90], consists of melting metal or alloy by induction heater. The liquid is then poured through a funnel having calibrated funnel and is atomized by a gas or water injected under high pressure. The liquid droplets solidify while falling down. The atomization can be carried out under vacuum, inert gas atmosphere, or in air. The atomized powder particles are generally spherical having high flowability, and their composition may include the metastable phases resulting from high cooling rates. The particles are dense and have small internal porosity.

Fusing or sintering followed by crushing is applied mainly to manufacture of ceramic or cermets powders. The process consists of bringing the raw material to high temperature being greater than melting point at fusing and by crushing the obtained solid to obtain powder. The typical powder particles are blocks as seen in Fig. 2.20 and have low flowability. The sintered powder has high internal porosity being greater than the fused one.

Table 2.13 Methods of powders manufacturing from different kinds of materials [9, 88].

Metal and alloys	Ceramics	Cermet clads	Cermet agglomerates
Atomization in vacuum, inert gas, air, or water	Fusing or sintering and crushing		Spray-drying
Fusing or sintering and crushing	Spray-drying	Cladding in: (i) solid phase (*MF*); (ii) liquid phase, and (iii) gaseous phase	*SHS*
Mechanofusion, *MF*	Self-propagating high-temperature synthesis, *SHS*		
	Powders fabrication in: (i) solid phase; (ii) liquid phase; and (iii) gaseous phase		*MF*

9 Defined as an association of ceramic and metal.

Figure 2.20 SEM micrograph (secondary electrons) of commercial fused and crushed Cr_2O_3 powder *Source:* From Pawłowski et al. [91]/DVS Verlag GmbH.

Mechanofusion, called also mechanical alloying, is a method of powders production described by, e.g. Herman *et al.* [92]. It is a laboratory-scale technique of powder manufacturing. *MF* is an ***attritor process*** in which the raw material is submitted to a friction energy in a milling machine. High energy of friction allows plastic deformation of prepared material. The internal morphology can be similar to that obtained by spray-drying or lamellar depending on the manufacturing parameters, such as rotation speed of milling. It is possible to obtain cladded powder particles. An example is stainless steel powder coated by alumina [93]. The parameters of manufacturing influence also the internal porosity and flowability of obtained powder.

Spray drying is a very popular powders production technique described in detail by, e.g. Masters [94]. The initial step is preparation of a water-based suspension, discussed in the section devoted to liquid feedstock. The suspension is subsequently atomized. The atomization leads to removal of the moisture from the droplets by evaporation, and the solid particles are formed. The obtained powders particles are spherical and have a lot of internal porosity (see Fig. 2.21a). The porosity can be reduced by, e.g. plasma densification as seen in Fig. 2.21b. The spray-dried powders have excellent flowability.

Self-propagating high-temperatures synthesis (*SHS*) is powder production technique based on exothermic combustion reaction of the powder components [88]. The reaction is initiated by an external source, and at least one component of the solid mixture should form a liquid phase. The reaction should be strongly exothermic, and minimum energy of formation of value of $\Delta_f\, G°_{298} = -70\,\text{kJ/mol}$ was mentioned [96]. The technique is used frequently to produce composite powders. The powders particles have internal porosity and require densification by plasma or laser.

Many ***other methods*** of powder manufacturing can be roughly divided following the phase in which the powder is created. Consequently, the solid-state reaction in solid state in the temperature lower than melting point is called ***calcination***. This method was, e.g. applied

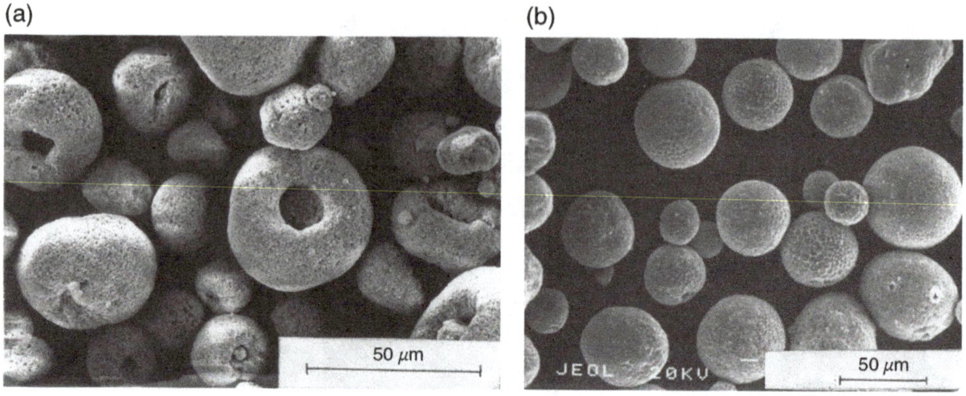

Figure 2.21 SEM micrograph (secondary electrons) of Cr_2O_3 + 5 wt. % SiO_2 powder obtained by: (a) spray drying; (b) spray drying followed by densification in plasma jet [95].

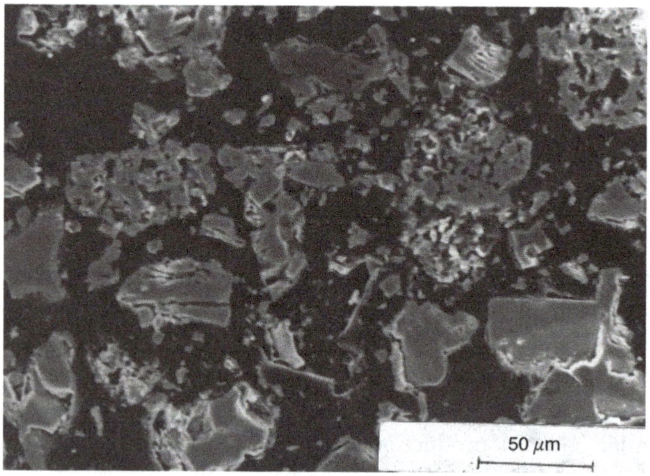

Figure 2.22 SEM micrograph (secondary electrons) of cross-section of *YBCO* powder obtained by calcination [97].

to synthesize *YBCO* powder by carrying out the reaction at $T = 1120$ K using CuO, Y_2O_3, and $BaCO_3$. The resulting product was crushed and heat treated to obtain the powder shown in Fig. 2.22. Among many liquid-phase methods, ***sol–gel*** one is frequently used to obtain metal oxide powders. The metal oxides precursors, such as alkoxides, are hydrolyzed and form polymers. The polymers are then treated at high temperature to obtain the final powder.

The details of sol – gels powders preparation for thermal spraying are presented by Scott and Woodhead [98]. The powder produced by this method have an advantage of being spherical which results in high flowability. The ***gaseous methods*** of powder manufacturing are particularly useful to obtain ***nano-powders*** having size lower than 100 nm. The synthesis consists of ***homogeneous nucleation*** of small powder particles being a product of gaseous-phase reaction similar to that used in *CVD* processes. The reaction can be activated by plasma created by glow discharge or in the zone of post-discharge. An example of CN_x powder manufactured in the zone of post-discharge of microwave-generated N_2 plasma reacting with the graphite vapors generated by CO_2 laser is shown in Fig. 2.23.

Cladding is the method of application of a film's other material onto the surface of powder particles. In thermal spraying preparation, the metallic coating is applied on ceramic, graphite, or other metal. This powder preparation technique enables, e.g. obtaining cermet coating using cold spraying. The solid-phase cladding is made using mechanical alloying, i.e. *mechanofusion*, described above. The liquid-phase cladding, described in detail by Lugscheider *et al.* [99], is most frequently applied. An example of such powder having alumina particles coated with Ni film is shown in Fig. 2.24. Finally, cladding in the

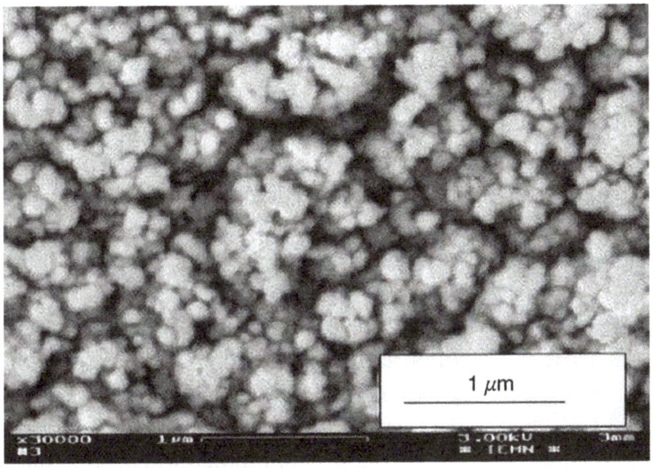

Figure 2.23 SEM micrograph (secondary electrons) of CN_x powder obtained by reaction in gaseous phase between nitrogen plasma in post-discharge and graphite vapors. The details of processes are shown in Table 2.5.

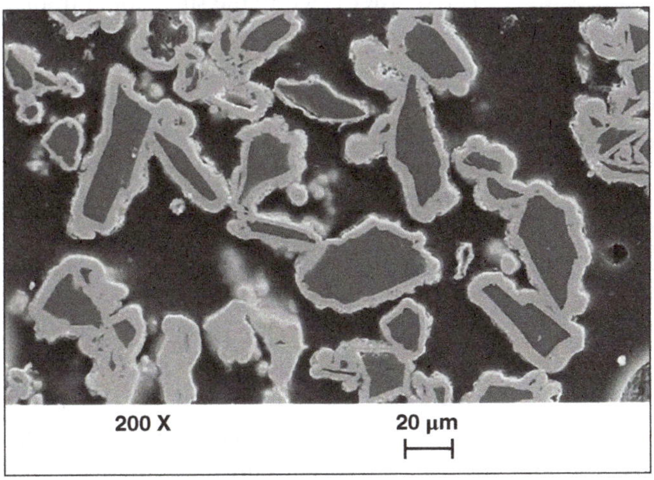

Figure 2.24 SEM micrograph (secondary electrons) of the cross-section of Al_2O_3 + 80 wt. % Ni powder cladded with Ni film. *Source:* From Winnicki et.al. [100]/with permission of ELSEVIER.

gaseous phase can be made using *CVD* (thermal *CVD* or *PE CVD*). The powder particles are coated being suspended in the *fluidized-bed reactors*. The use of sputtering, being a part of *PVD* techniques, was also reported [101].

2.3.1.2 Characterization of Powders

The powders have to be carefully characterized prior to the deposition, and the most important parameters are as follows [9, 88]:

- particles' size determines the phenomena of transfer of heat and of quantity of momentum at spraying. As the powders used in the deposition have a lot of particles of different sizes, the most frequent one used to characterize the powder size is median or average diameter corresponding to 50 % on the size distribution curve, d_{50};
- chemical and phase composition are important because of the possible transformations of particles at processing. The transformations result from: (i) the contact of liquid droplets with air; and (ii) rapid solidification at contact with substrate;
- external and internal morphology including shape and internal porosity. These parameters influence interaction with gas at spraying and coatings' microstructure;
- high-temperature behavior measured by *DSC*, *DTA*, and *TGA* enables predicting and understanding the phenomena related to powder particles melting and solidification;
- apparent density and flowability are important in predicting the transport of powder from powder feed to jet or flame. The particles density influences the transfer of quantity of momentum at spraying.

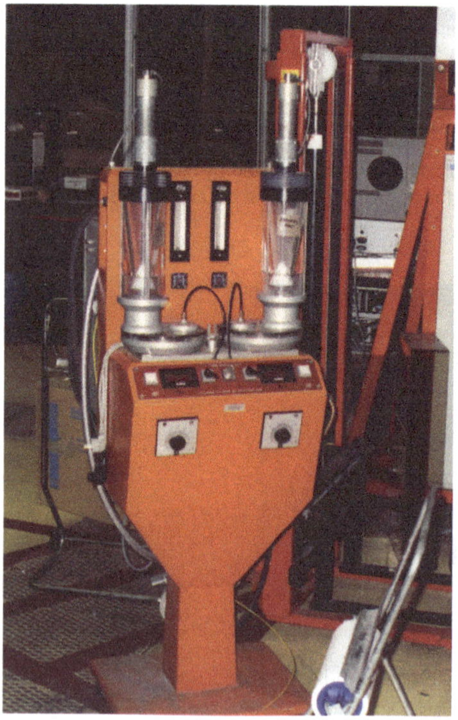

Figure 2.25 Commercial powder feeder. *Source:* Reproduced with permission of *Oerlikon Metco* (Wohlen, Switzerland).

2.3.1.3 Delivery and Injection of Powder to Flame and Plasma

The powder is fed to the spray torch from a feeder and transported by a pipeline and, finally, is injected to the flame or plasma jet. The commercial feeders deliver powder with a feed rate ranging from $q_p = 10$ to 150 g/min [88]. The commercial powder enabling simultaneous delivery of two different powders is shown in Fig. 2.25.

The sketch of pipeline and injector connecting powder feeder with the spraying torch is shown in Fig. 2.26. The geometry of the pipeline and injector determines the velocity of powder injection to the torch, v_{in}. The results of calculation of this velocity using a simple model based onto ***ballistic approach***[10] for *HA* having particles of different size are shown in Fig. 2.27. The injection velocity

10 Supposing the powder particles do not interact with each other.

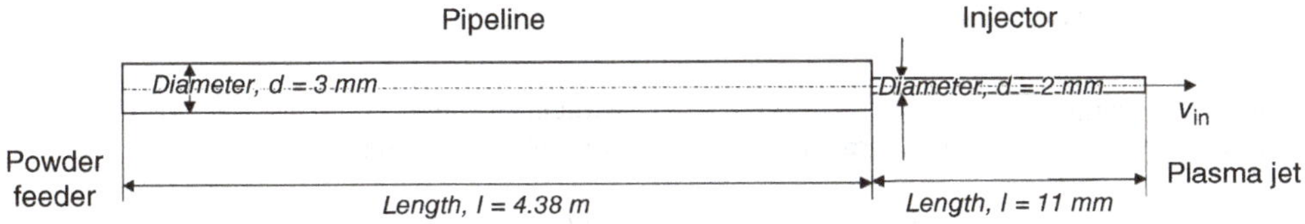

Figure 2.26 The geometry of the pipeline connecting powder feeder with the spray torch while injecting *HA* powder. *Source:* Adapted from Dyshlovenko et al. [102].

Figure 2.27 Calculation of injection velocity of HA particles of different size delivered using pipeline having geometry shown in Fig. 2.26 with Ar as carrier gas with the flow rates of q_{cg} = 3 and 3.5 slpm [102].

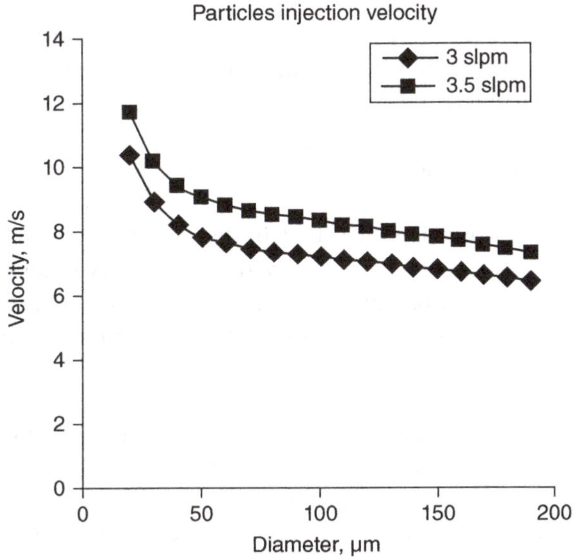

determines the trajectory of the powder particles in plasma jet or flame. The trajectory is discussed in detail in the section describing atmospheric plasma spraying (*APS*).

2.3.2 Solid Feedstock Deposition Methods

Thermal spraying techniques have been developed in a few directions starting from the initiating patents deposited by Swiss engineer, Max Ulrich Schoop, in 1909 [103]. The principal axis of this development is related to high productivity, i.e. the capacity to rapidly coat large surfaces. These techniques are described in this section as flame spraying (*FS*), arc spraying (*AS*), and atmospheric plasma spraying (*APS*). The new spray torches including water-stabilized plasma (*WSP*), one or three cathodes torch as *Axial III* can be also included in this development axis. Another axis of technology development concerns techniques enabling achievement of high kinetic energy of sprayed particles. The techniques described are high-velocity oxy-fuel (*HVOF*), detonation spraying (*DS*), and cold-gas spray method (*CGSM*). Finally, some spray techniques were developed using vacuum or intergas as spray atmosphere. This atmosphere is used in the vacuum plasma spraying (*VPS*) or in inert plasma spraying (*IPS*) to reduce the contact with air and resulting thereof oxidation of particles. On the other hand, the plasma spray physical vapor deposition (*PS PVD*) enables obtaining coating having similar to the *PVD* ones microstructure with better productivity. All the discussed techniques are shown in Table 2.14. All thermal spray techniques start

Table 2.14 Discussed thermal spraying techniques using solid feedstock.

High kinetic energy of sprayed particles	High productivity	Vacuum and inert atmosphere spraying
• *HVOF*	• *FS*	• *VPS*
• *DS*	• *AS*	• *PS PVD*
• *CGSM*	• *APS*	

with injection of feedstock to the plasma jet or flame discussed above. That is why the description of the technique will be focused on their history, principle, and the phenomena occurring in the medium.

2.3.2.1 Flame Spraying (*FS*)

History

The patents deposited by Günther and Schoop early in 20th century [104] initiated the flame spray process. Initially, the flame spray is applied for deposition of zinc against corrosion; therefore, the technique was called metallization. Later on, the process started to be used to spray the refractory metals and ceramic coatings.

Principle

The spray feedstock is introduced to the torch in the form of powder (*FS* – powder) or a wire (*FS* – wire). The sketch of the torch working with the powders is shown in Fig. 2.28. The gases inlet is axial and the powder can be introduced axially or perpendicularly to the torch. Rods and wires can be used instead of powder. The particles get molten in the flame being accelerated toward the substrate. A view of a *FS*-powder torch attached to the robot is in Fig. 2.29.

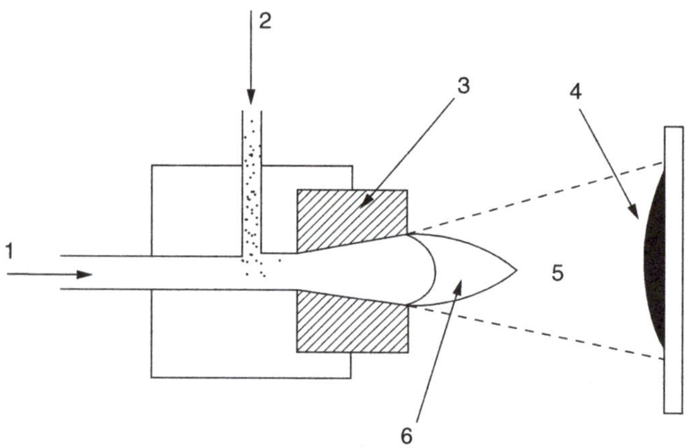

Figure 2.28 Sketch of flame spraying of powder (*FS* – powder): 1, working gases (combustion gas and oxygen); 2, injection of powder; 3, torch body; 4, sprayed coating; 5, stream of particles; 6, combustion flame.

The principal deposition parameters include: (i) combustion gas, such as acetylene, flow rate; (ii) oxygen flow rate; (iii) powder feed rate; (iv) carrier gas, such as nitrogen, flow rate; (v) heat and momentum transfer between flame and powder particles; (vi) spray distance; and (vii) kinetics of the torch with regard to the substrate.

FS is very well suited to spray metals such as Zn, Ni, Al, and Mo or alloys such as NiAl, NiCrAlY, or stainless steel. The powders have a typical size of $d_p = 5–100\,\mu m$ and their feed rate is typically equal to $q_p = 30–100\,g/min$. Another class of sprayed materials are oxide ceramics such as Al_2O_3, TiO_2, ZrO_2 [105]. Ceramics can be sprayed as a powder and as the rods being surrounded by a polymer sheath. Finally, thermoplastic polymers such as polyethylene or epoxy resins can be sprayed. A powder is injected to the flame axially as a suspension in the carrier gas such as nitrogen.

Figure 2.29 View of a torch fed with powder attached to a robot.

The wires can be used to flame spray metals and alloys. Their diameters vary between $d = 2.3$ to 6 mm and their feed is typically $q = 80–650\,g/min$. The spray distance is of the order of $l = 120–250\,mm$. Flame spraying is carried out in the ambient atmosphere, and the thickness of the deposits varies between $\delta = 100$ and $2500\,\mu m$. Often, the post-spray treatment is applied to remelt the deposits using the flame. The substrate used for flame spraying must be activated. Usually, grit blasting is used (see section 1.5.2).

Medium

The flame spraying is carried out with a torch wherein a combustion gas is burned in oxygen. The flame temperature is generally between 3000 and 3350 K, and its velocity is between 80 and 100 m/s. The volumetric ratio of combustion gas to oxygen may reach 2.5 to achieve high capacity of particles melting [105]. On the other hand, this ratio influences the atmosphere, which may be carburizing or oxidizing. The pressures and flow rates of fuel gas and oxygen vary from torch to torch. The details of interaction between the flame and the particles are discussed in the section devoted to *APS* deposition.

2.3.2.2 Arc Spraying (*AS*)

Arc spraying (*AS*) was another technique developed by Max Ulrich Schoop. The sketch of *AS* installation is shown in Fig. 2.30. The two wires, being the electrodes of the arc, are delivered from the rotating coils. The arc is initiated between the ends of these wires. These ends melt, forming liquid droplets. The droplets are blown out by the compressed gas. This gas atomizes the droplets, and the small particles are accelerated toward the substrate. The wires can be of the same material or two different materials. In the latter case, the deposits of metal/alloys mixtures are sprayed.

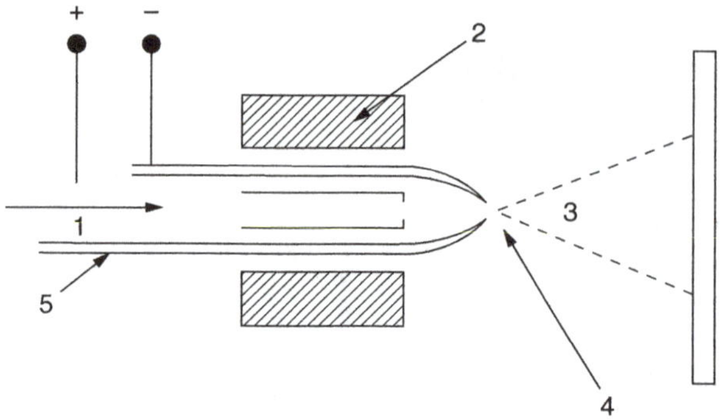

Figure 2.30 Sketch of arc spraying technique: 1, flow of atomizing gas; 2, torch; 3, sprayed particles; 4, electric arc; 5, wires – electrodes.

The quality of the sprayed coatings depends on: (i) used wire and its feed rate; (ii) electric power; (iii) momentum transfer between atomizing gas and droplets; (iv) spray distance; and (v) kinetics of the arc spray torch with regard to the substrate. The wires have to be electrically conductive materials and metals such as zinc or aluminum or alloys such as stainless steel, NiCr, or NiAl are used. It is possible to spray mixtures of metals and alloys using two different wires. Li *et al.* [106] used the wires of Al and of stainless steel being heat treated after spraying. It is also possible to obtain composites using the wire having an outer layer of a metal, which will eventually become a matrix and its internal part of a reinforcement. An example is a wire with an outer layer of Mo and internal part of $WC - W_2C$ [107]. The diameter of a wire is typically $d = 2–5\,mm$.

The electric power supplied in the arc is typically of the order of $P = 5–10\,kW$ [108]. The arc voltage is 20 to 40 V. The greater voltage leads to an increase in particle size. The spray distance is $l = 50–200\,mm$ and the wire flow rate is $q = 50–300\,g/min$. The flow rate of an atomizing gas, being often air, is $q = 1–80\,m^3/h$ and its pressure is between $p = 0.2$ and $0.7\,MPa$. The thickness of the coatings is typically $\delta = 100–1500\,\mu m$.

Medium

The medium temperature depends on the arc current, and the value of $T_g = 6100\,K$ was reported for the arc current of $I = 280\,A$. The particles created from wires can reach a speed of 150 m/s in the flame [109]. Generally, the arc spraying is carried out in open atmosphere. It is possible to arc spray under vacuum, reactive, or inert atmosphere. Consequently, the arc-spraying method of stainless steel powder production in nitrogen atmosphere was reported By Chen et al. [110].

2.3.2.3 Atmospheric Plasma Spraying (*APS*)

History

Gerdien and Lotz developed the torch used at present in plasma spraying in 1922 [111]. The plasma spraying process using electric-arc-generated plasma was patented in the USA in the early 1960s [112].

Principle

The cross section of a typical plasma torch is shown in Fig. 2.31.

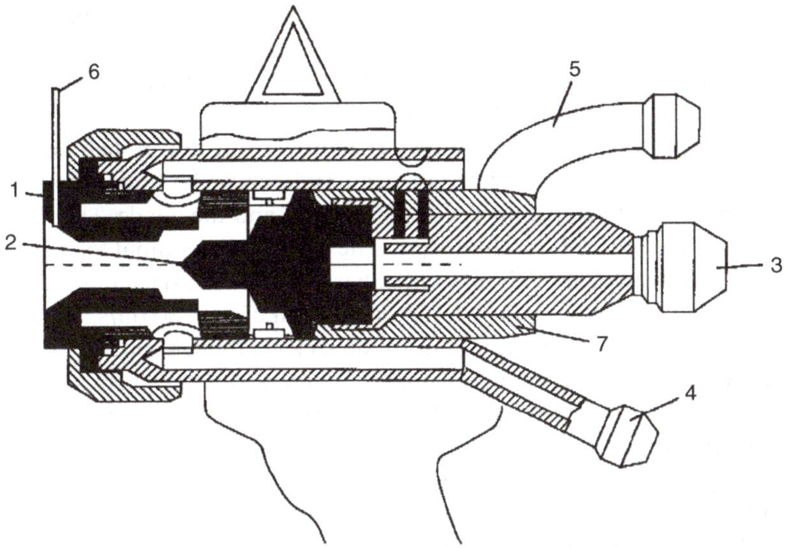

Figure 2.31 Section of a typical plasma torch: 1, anode – nozzle; 2, cathode; 3, cooling water outlet and cathode electrical connection; 4, cooling water inlet and electrical anode connection; 5, inlet of working gases; 7, electrical insulator.

Figure 2.32 A plasma torch installed on the robot [113]: 1, torch; 2, compressed air barrier; 3, compressed air cooling; 4, powder injection; 5, substrates.

The torch has a circular copper anode-nozzle and a thoriated tungsten cathode. The arc burns between the electrodes while creating plasma and moves eroding them. The plasma jet having high temperature and velocity is created from the working gases. A laboratory torch installed on the robot is shown in Fig. 2.32.

The quality of the deposits depends mainly on: (i) plasma jet; (ii) electric power input to arc; (iii) powder kind, feed rate, and its injection velocity to plasma jet; (iv) heat and momentum transfer between plasma and powder particles; (v) spray distance; and (vi) kinetics of the plasma torch with regards to the substrate.

The plasma working gases are typically mixtures of argon with hydrogen, helium, or nitrogen. Sometimes mixtures of nitrogen with hydrogen or mixtures with three gases are used. Different gases require different geometry of the cathode and the anode. The flow rate of working gases is of the order of $q = 40–80$ slpm. The electric power input to the arc may reach 100 kW in high power torches. The plasma jet has a maximum axial temperature being about $T_{\mathrm{pl}} = 14\,000$ K and the maximum axial speed of about of $v_{\mathrm{pl}} = 800$ m/s.

The powder particles size is typically $d_{\mathrm{p}} = 5–100$ μm in diameter, although generally sizes with narrow limits, such as $d_{\mathrm{p}} = 22.5–45$ μm, are used. The powders with spherical morphology having good flowability are recommended for spraying. The *APS* technology is the most versatile and all types of materials can be deposited, i.e. metals, alloys, ceramics, and composites. The carrier gas is typically Ar or N_2 with the flow rate up to $q_{\mathrm{cg}} = 10$ slpm, the powder feed rate is typically $q_{\mathrm{p}} = 50–100$ g/min, and powder injection is radial with respect to the axis of the jet. The trajectory particles in the jet can be modified using the injection angle to the plasma jet. The angle[11] is generally equal to 90° and can be modified to reach, e.g.: (i) 60° for the longer trajectories of particles in plasma jet; or (ii) 120° for the shorter trajectories. The typical thickness of the deposits varies between $\delta = 50$ and 500 μm.

Medium

The plasma jet is characterized by strong radial gradients of temperature and velocity. The axial gradients are less strong. The profiles of these two parameters for a commercial torch found using mathematical modeling are shown in Fig. 2.33. The addition to the working gases (typically Ar) of a molecular one, such as H_2, increases thermal conductivity and dynamic viscosity of plasma discussed in detail in the textbook [71]. These two parameters determine the ability of the plasma to heat particles and accelerate them.

The momentum transfer from plasma jet to a solid particle introduced results mainly from the action of drag force, which results in its acceleration. Supposing a *ballistic approach*, the equation of motion of a particle having diameter d_{p}, introduced to the plasma jet having velocity v_{pl} can be expressed as:

$$\frac{1}{6}\rho_p \pi d_p^3 \frac{dv_p}{dt} = \frac{1}{8} C_D \pi d_p^2 \rho_{pl} \left(v_{pl} - v_p\right)^2 \tag{2.28}$$

The drag coefficient, C_{D}, depends on the *Re* number, which is defined for the relative movement of particles in plasma given by *Eq. 2.29*, has the values presented in *Eqs 2.31–2.33* [88]:

$$Re = \frac{\rho_{pl} d_p \left(v_{pl} - v_p\right)}{\eta_{pl}} \tag{2.29}$$

11 The angle 180° corresponds to axial injection, which has some modern torches such as Axial III torches produced by Northwest Mettech, Surrey, Canada.

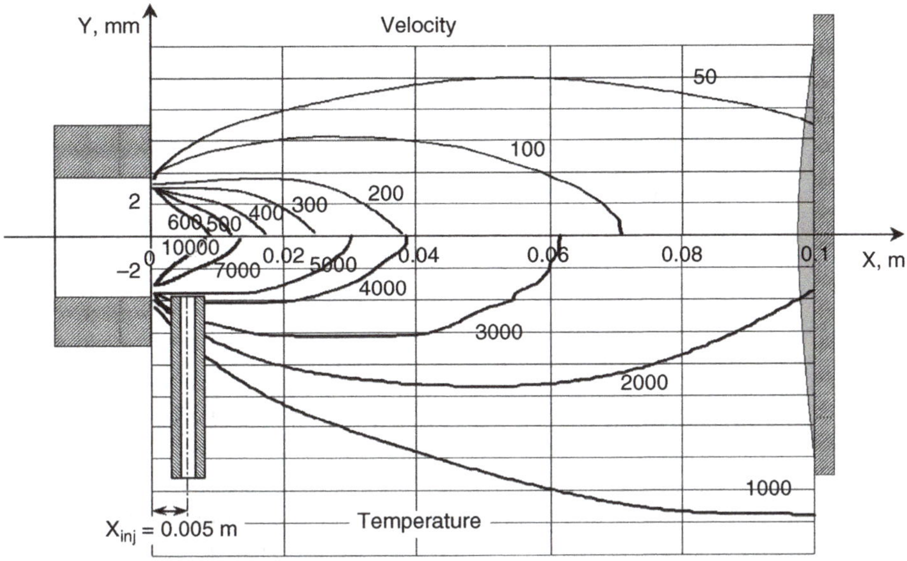

Figure 2.33 Calculated profiles of temperatures (in K) and velocities (in m/s) of a plasma jet generated by a commercial *F-4* torch (Commercialized by Oerlikon Metco, Wohlen, Switzerland) [114].

$$C_D = \frac{24}{Re}\ \textit{Stoke's regime valid for}\ Re < 0.2 \quad (2.30)$$

$$C_D = \frac{24}{Re}\left(1 + \frac{3}{16}\right)\textit{Oseen regime valid for}\ Re\ 0.2 \le Re < 2 \quad (2.31)$$

$$C_D = \frac{24}{Re}\left(1 + 0.11Re^{0.81}\right)\ \textit{valid for}\ 2 \le Re < 21 \quad (2.32)$$

$$C_D = \frac{24}{Re}\left(1 + 0.189Re^{0.632}\right)\ \textit{valid for}\ 21 \le Re < 200 \quad (2.33)$$

The values of viscosity and those of thermal conductivity in the boundary layer correspond to the mean temperature which has an average value between the particle surface temperature and that of plasma. *Eq. 2.28* does not take into account the ***effect of charge***, that is, the modification of plasma properties by a large quantity of powder particles. On the other hand, it is important while spraying small particles at atmospheric pressure (techniques *SPS* or *SPPS*) or large particles under vacuum to consider *the* ***effect of non-continuity***. The effect appears when a sprayed particle size becomes comparable to the mean free path of gas particle. This effect should be considered when spraying under low pressure.

The heat transfer from plasma jet to a sprayed particle occurs by: (i) convection, which has important contribution just after powder injection; (ii) conduction, being dominant when the particles reach the velocity close to that of the plasma jet; and (iii) radiation, which result from high temperature of plasma. The dimensionless Nusselt, *Nu*, number describes the effects of convention and conduction heat transfer. The number for a spherical particle is equal to:

$$Nu = \frac{hd_p}{\lambda_{pl}} = 2 + 0.66Re^{0.5}Pr^{0.33} \tag{2.34}$$

where *Pr* is the Prandtl number determining the thickness of boundary layer in the gas around the particle. The number is defined as:

$$\mathrm{Pr} = \frac{\eta_{pl}\left(c_p\right)_{pl}}{\lambda_{pl}} \tag{2.35}$$

The value of $Nu = 2$ can be used if the velocity of particle is equal to that of plasma which results in *Re* number, given by *Eq. 2.29*, being equal to $Re = 0$. Physically, this value of *Nu* corresponds to heat transfer by the conduction mechanism only.

The description of heat transfer from plasma jet to the particles depends strongly on their thermal conductivities and on particle's size. The well-heat-conducting metals or alloys particles of small size would have one temperature, T_p, during their trajectory in jet. For these materials, the heat balance equation[12] of a spherical particle having size, d_p, in a plasma is as follows [115]:

$$\pi d_p^2 h\left(T_{pl} - T_p\right) = \frac{1}{6}\pi\rho_p\left(c_p\right)_p d_p^3 \frac{dT_p}{dt} \tag{2.36}$$

The particles of low thermal conductivity materials, such as oxide ceramics, would have a temperature gradient. The temperature distribution inside a particle can be found from the well-known ***Fourier equation***, which, for a spherical particle, is as follows:

$$\left(c_p(T)\right)_p \rho_p(T)\frac{\partial T_p}{\partial t} = \frac{1}{r^2}\frac{\partial}{\partial r}\left(\lambda_p(T)r^2\frac{\partial T_p}{\partial r}\right) \tag{2.37}$$

This equation has initial condition $T_p = 300$ K and the following condition of symmetry:

$$\frac{\partial T_p}{\partial r} = 0 \; at \; r = 0 \tag{2.38}$$

The numerous boundary conditions result from different situations in which a particle can be found inside the plasma jet. The diagram of these situations and corresponding phases inside particle are shown in Fig. 2.34 (a, b).

2.3.2.4 High-Velocity Oxy-Fuel Spraying (*HVOF*)

History

The high-velocity oxy-fuel *HVOF* technique was developed in the late 1970s and early 1980s [116, 117]. The initial studies were made by the Thayer School of Engineering in Hanover (NH, USA) and at the Browning Engineering company in West Lebanon

12 Neglecting the radiation heat transfer.

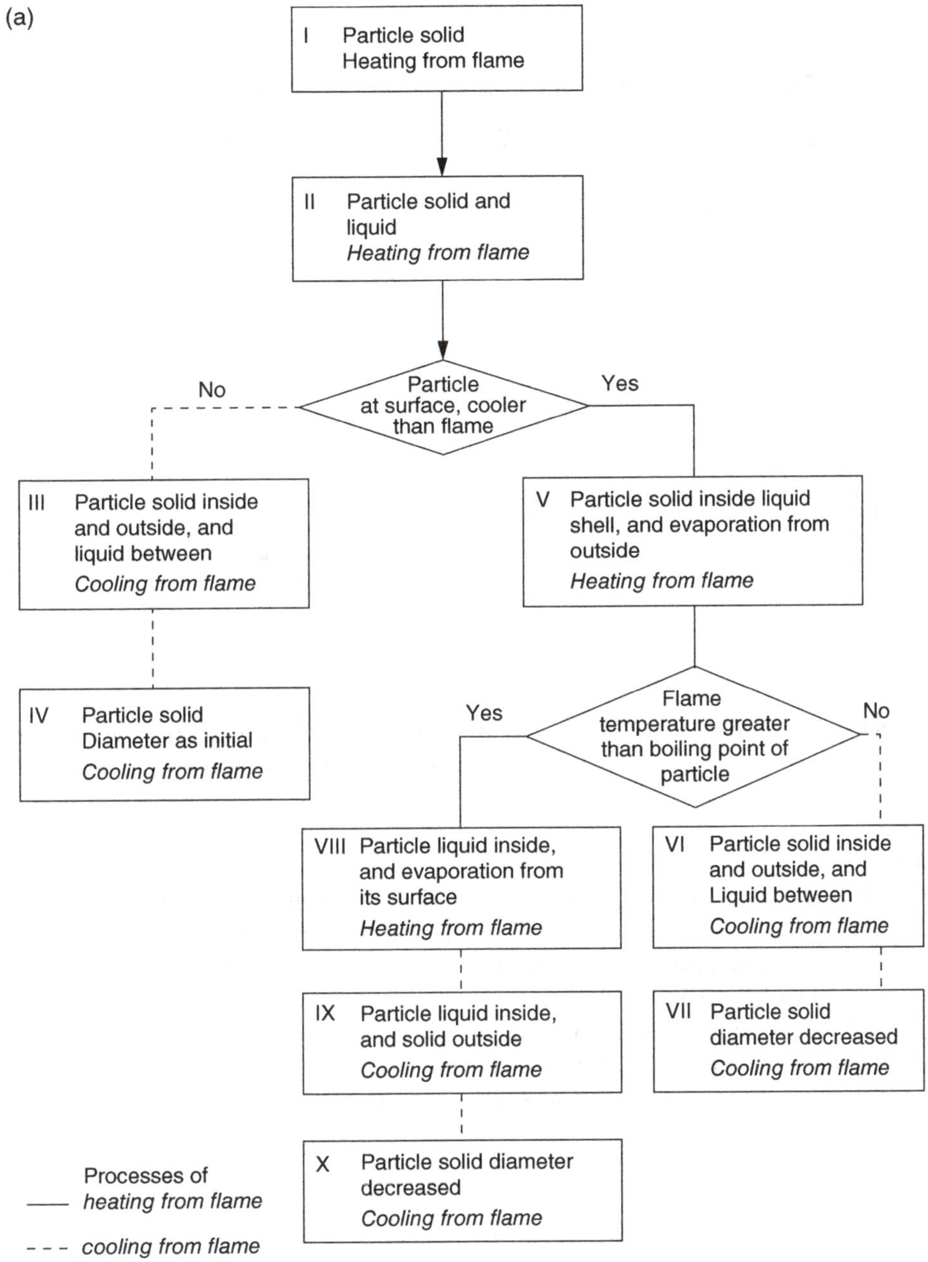

Figure 2.34 Heat transfer phenomena occurring at flight of a particle of low thermal conductivity in plasma jet [88]: (a) description of possible phenomena occurring in plasma jet; (b) sketch of phases' content in sprayed corresponding to the phenomena.

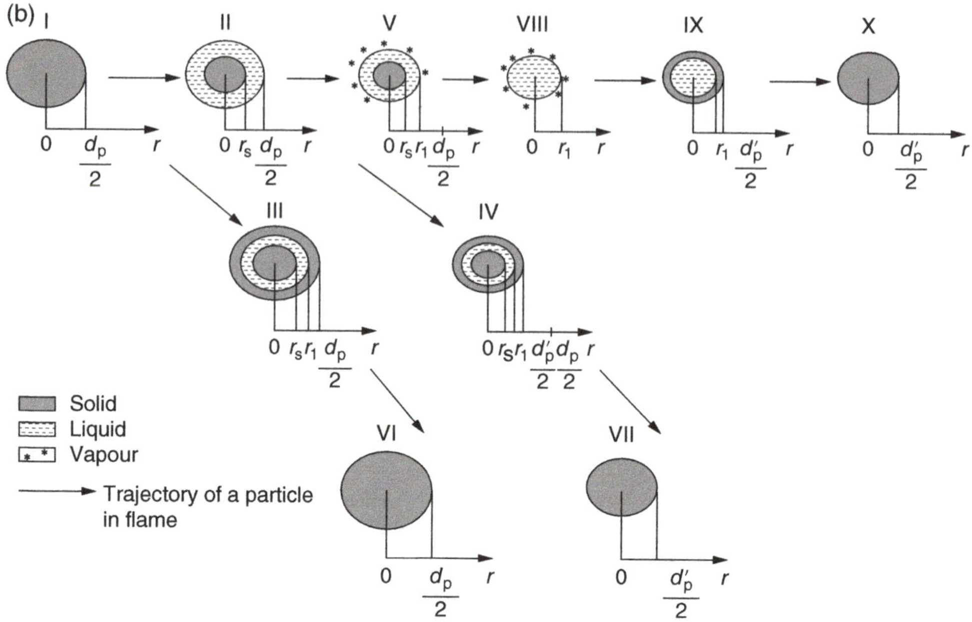

Figure 2.34 (Continued)

(NH, USA). The efforts were motivated by reaching the high quality of sprayed deposits obtained by detonation spraying, when installation was then not available commercially.

Principle

The sketch of *HVOF* torch is shown in Fig. 2.35. A combustion gas burns in oxygen in a combustion chamber under high pressure. Combustion creates a supersonic jet, which exits through a nozzle into the atmosphere. The powder is injected inside the nozzle radially or axially with regard to jet.

The *HVOF* processing parameters are similar to those of flame spraying. The important difference is the pressure in combustion chamber, which may reach $p = 1$ MPa. Vignesh *et al.* [118] realized the *DoE* statistical study, using response surface methodology, to

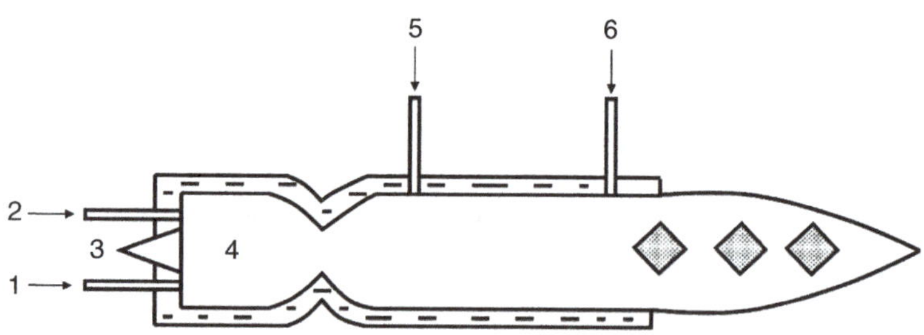

Figure 2.35 Sketch of HVOF torch: 1, oxygen inlet; 2, combustion gas inlet; 3, combustion ignition; 4, combustion chamber; 5, injection of a powder; 6, water cooling.

optimize the properties of metallic coatings and selected the following essential deposition parameters: (i) working gases, i.e. combustion and oxygen flow rates; (ii) powder feed rate; (iii) spray distance; and (iv) carrier gas flow rate.

The working gases are: (i) combustion gases such as ethylene (C_2H_2), propylene (C_3H_6), propane (C_3H_8), or natural gas having typical flow rate of $q = 50$ slpm; (ii) liquid hydrocarbons such as kerosene being atomized prior to injection in combustion chamber; and (iii) oxygen with flow rate of $q = 250$–400 slpm. The typical powder sprayed using *HVOF* are metals and alloys and composites of metals with carbides. The powder feed rate is in the range of $q_p = 20$–120 g/min, and the spray distance is in the range of $l = 150$–300 mm. The carrier gas, being typically Ar or N_2, is in the range $q_{cg} = 10$–15 slpm. Finally, the typical thickness of the coatings is in the range $\delta = 100$–$300\,\mu$m.

Medium

The reaction between the combustion gas and oxygen takes place in the combustion chamber. Taking as example propylene as combustion gas the reaction is as follows [88]:

$$C_3H_6 + \frac{9}{2}O_2 \rightarrow 3CO_2 + 3H_2 + 1936\frac{kJ}{mol} \tag{2.39}$$

The temperature of flame is greatest in stoichiometric quantity of combustion gas and oxygen flow rates [119]. The expanding gas is supersonic, and the jet has the "shock diamonds" (see Fig. 2.35). The velocities of the jets were estimated by modeling on $v_g = 1600$–2000 m/s in different HVOF torches, and their temperatures may reach $T_g = 3500$ K [88].

2.3.2.5 Detonation Spraying

History

The detonation spraying process was patented in the early 1950s, and later on the coatings sprayed using this method have been and are available from *Union Carbide* and its successor *Praxair ST,* Indianapolis, IN, USA [120]. Independently, a similar process was developed in the *Paton Institute* in Kiev, Ukraine, in the 1960s [121].

Principle

A sketch of a detonation gun is shown in Fig. 2.36 and its view in the research laboratory in Fig. 2.37. The gun has typically a long, water-cooled nozzle of 450 to 1350 mm length having internal diameter of about 25 mm [122]. A mixture of oxygen and combustion gas is injected into this nozzle as well as the powder to be sprayed. The mixture of gases explodes by ignition resulting in a detonation wave, which accelerates the powder. The detonations' frequency is $\nu = 15$ Hz, and the nozzle is purged between the detonations with N_2.

Working gases are hydrocarbons such as acetylene (C_2H_2), propane (C_3H_8), butane (C_4H_{10}) mixed with oxygen (O_2). The composition of oxygen with 45 vol. % of acetylene enabled the temperature of flame of $T = 4500$ K, and its velocity may reach 3000 m/s [123, 124]. The spray distance is reportedly about 100 mm [122]. The powders used in detonation spraying are metal, ceramics, and composites, and the sprayed coatings thickness is lower than $\delta \leq 300\,\mu$m.

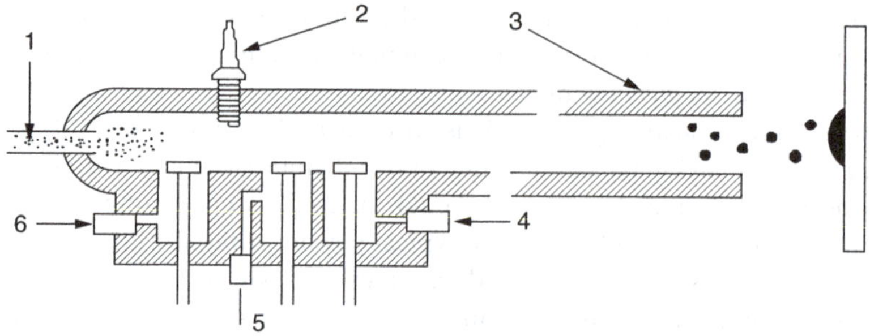

Figure 2.36 Sketch of detonation gun: 1, powder injection; 2, ignition of combustion; 3, nozzle; 4, oxygen injection; 5, acetylene injection; 6, nitrogen injection.

Figure 2.37 Detonation spray gun used in the institute ARCI in Hyderabad (India). *Source:* Reproduced with permission from ARCI.

2.3.2.6 Cold Gas Spraying Method (*CGSM*)

History

The cold gas spraying method, *CGSM,* was developed in Novosibirsk (Russia) in the end of 1980s and patented thereafter [125]. The technique was developed by a research team, which analyzed the effects of erosion of solid particles in re-entry space vehicles. The researchers found out that some particles remained on the tested surface forming a coating [126]. The process started to be applied in the industry in the beginning of XXI century.

Principle

The sketch of CGSM is shown in Fig. 2.38. The working gas in high-pressures installations is N_2 or He, and is heated up to $T = 1000\,K$. The gas is compressed up to $p = 3.5\,MPa$ and expands in the converging–diverging ***de Laval nozzle*** to reach supersonic velocities. The powder is accelerated to very high velocities and is heated up to the temperatures

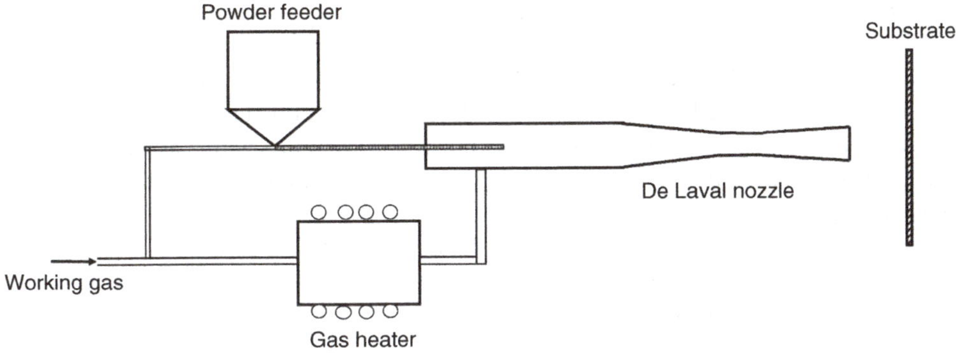

Figure 2.38 Sketch of cold gas spray method.

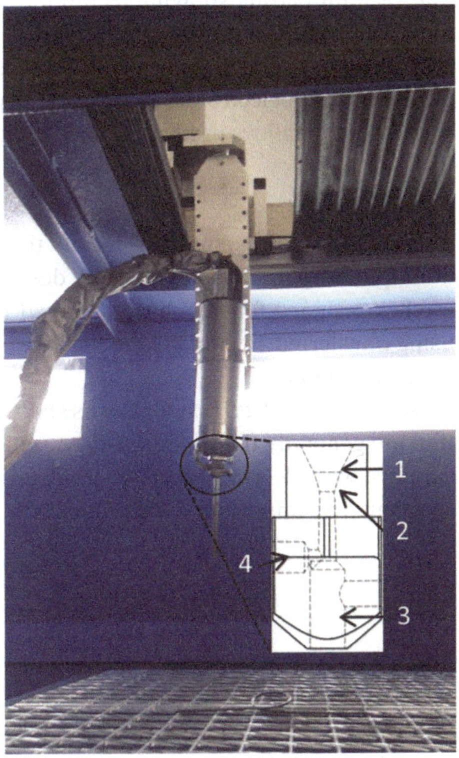

Figure 2.39 The nozzle of low-pressure CGSM installed in laboratory of Faculty of Mechanical Engineering at Wrocław University of Science and Technology (Poland): 1, de Laval nozzle; 2, convergent part, 3, divergent part; 4 – powder injection. *Source:* Reproduced with permission of Prof. Dr. M. Winnicki.

being well below the melting point. The particles get attached to the substrate surface under the effect of plastic deformation. The low-pressure cold gas spraying installation (see Fig. 2.39) uses air as working gas under the pressure up to $p = 1.2\,\text{MPa}$ and temperature up to $T = 900\,\text{K}$. The use of He as working gas is at present limited because of its limited availability, and the gas should be recycled [127]. The nozzles having rectangular section may be useful in ***cold spray additive manufacturing*** reviewed recently by Prashar and Vasudev [128].

The sprayed powders should be of the materials being able to deform plastically. Vlcek *et al.* [129] categorized the following structures of metals fulfilling this condition: (i) *FCC* structure of metals such as Al, Cu, Ag, Au, Pt, Ni, and γ-Fe; (ii) *BCC* structure of metals such as W, Ta, Mo, Nb, V, Cr, α-Fe, and β-Ti; and (iii) *HCP* structure of metals such as Cd, Zn, Co, Mg, and Ti. It is possible to spray cermets as the mixtures of ceramics with plastically deformable metal or as cladded power as shown in Fig. 2.24. The particles sizes are similar as in other spray techniques.

The cold sprayed coatings were successfully sprayed onto metallic and ceramic substrates. Małachowska *et al.* [130] also carried out the initial tests of low-pressure cold spraying onto polymer substrates.

2.3.2.7 Vacuum Plasma Spraying (*VPS*)

History

The vacuum plasma spraying technique, *VPS,* known also as low-pressure plasma spraying, *LPPS,* was initiated by Muelhlberger in the early 1970s [131]. The technique was rapidly applied to spray hot corrosion-resistant superalloy coatings [132] and metal coatings for hybrid microelectronics [133, 134]. At present, the *VPS* technique is used frequently to spray the superalloys used as bond coatings in the *TBC.*

Principle

The vacuum plasma spraying installation, shown in Fig. 2.40, is similar to that used at atmospheric plasma spraying and described in corresponding section. Sometimes the torches designed for *VPS* have *de Laval* nozzles. Another difference is the possibility of using transferred arc attached to the substrate to heat up or clean its surface, depending on the polarization.

The laboratory *VPS* installation is shown in Fig. 2.41. The working gases used in deposition are generally the mixtures of Ar with molecular gases as H_2, N_2, or He. The torch is supplied with electric power generally greater than that that used for atmospheric plasma spraying. The important differences concern also the properties of plasma jet. Namely, as

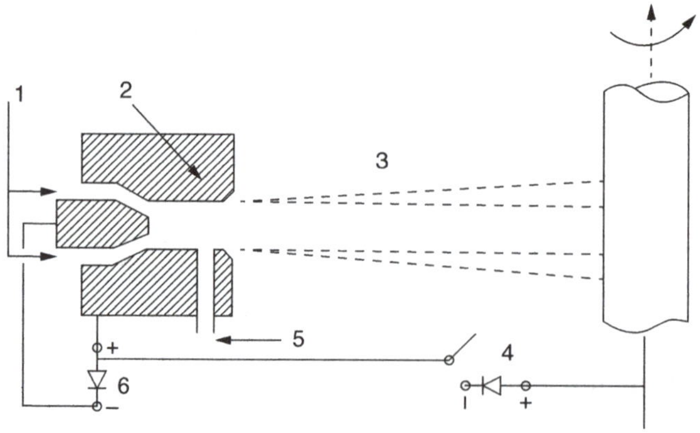

Figure 2.40 Sketch of vacuum plasma spray installation: 1, working gases inlet; 2, anode; 3, vacuum environment; 4, generator of transferred arc; 5, powder injector; 6, plasma generator.

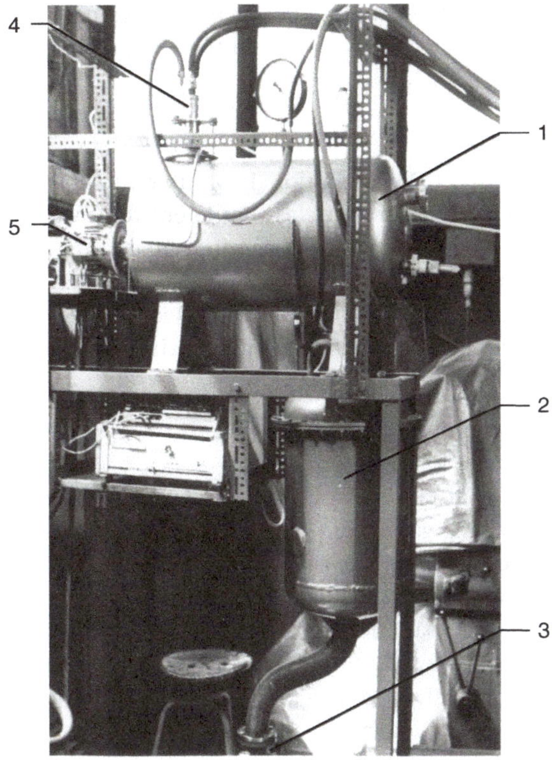

Figure 2.41 Small laboratory vacuum plasma spray installation used to spray Ta and Cu coatings [134]: 1, vacuum chamber; 2, working gases cooler; 3, vacuum pump; 4, plasma torch, 5, substrates rotation system.

shown in Fig. 2.17, the temperature of heavy species as ions or molecules diverges at lower pressure considerably from electron temperature. The values for temperature of electrons are in the range from $T_e = 10\,000$ to $15\,000$ K, and the velocity of plasma flow ranges from $v_{pl} = 1\,500$ to $3\,500$ m/s. The plasma jet is much longer than that coming from a plasma torch into open atmosphere and can reach 250 mm. The spray parameters include also:

- working pressure in the chamber being $p = 1$ Pa prior to deposition growing up to $p = 20$ kPa at spraying;
- spray distance being in the range of $l = 300$–400 mm;
- working gases flow rate is about $q = 90$–200 slpm;
- powder particles sizes are in the range of $d_p = 5$–20 μm.

2.3.2.8 Plasma Spray Physical Vapor Deposition (*PS PVD*)

History

Plasma spray physical vapor deposition, *PS PVD*, is a new deposition technique developed mainly for thermal barrier coatings technology. The initial studies seem to be made in Sulzer Metco in Wohlen (Switzerland) by K. von Niessen and M. Gindrat [135] followed by *Forschung Zentrum* (Jülich, Germany) by the group around G. Mauer and R. Vaßen [136] and the group led by M. Góral in *Technical University* of Rzeszów (Poland), [137]. The idea of the researchers was to develop the deposition technique enabling columnar microstructure of ceramic coatings similar to that obtained by *EB PVD*. The latter is used industrially to produce *TBC*s.

Principle

The *PS PVD* installation is similar to that used for vacuum plasma spraying. The injected powder particles should entirely evaporate in contact with plasma jet enabling formation of coatings from vapor phase. The torch used frequently to spray is O3CP of Oerlikon Metco (Wohlen, Switzerland) used by authors of references [135–139]. The laboratory installation of *PS PVD* is shown in Fig. 2.42. The most published research studies show the yttria-stabilized zirconia coatings (8YSZ). The main spray parameters are as follows:

- working pressure in the chamber is 50 to 200 Pa, being considerably lower than that used at *VPS*;
- spray distance ranges from $l = 300$ to 2000 mm;
- working gases are mixtures of Ar with H_2 and/or He and helium is used because of preserving its viscosity at high temperatures enabling better momentum transfer from plasma jet to the solid particles [136];
- powder particles sizes are lower than 20 μm, and with the powder manufacturing technology of *spray-drying* of fine agglomerates it is possible to obtain porous particles being easier to disintegrate after injection enabling as small as possible ceramic particles to get in contact with plasma jet [140].

Figure 2.42 Modern laboratory installation *PS PVD* used in FZ Jülich (Germany). *Source:* Reproduced with permission from Forschungszentrum Jülich GmbH.

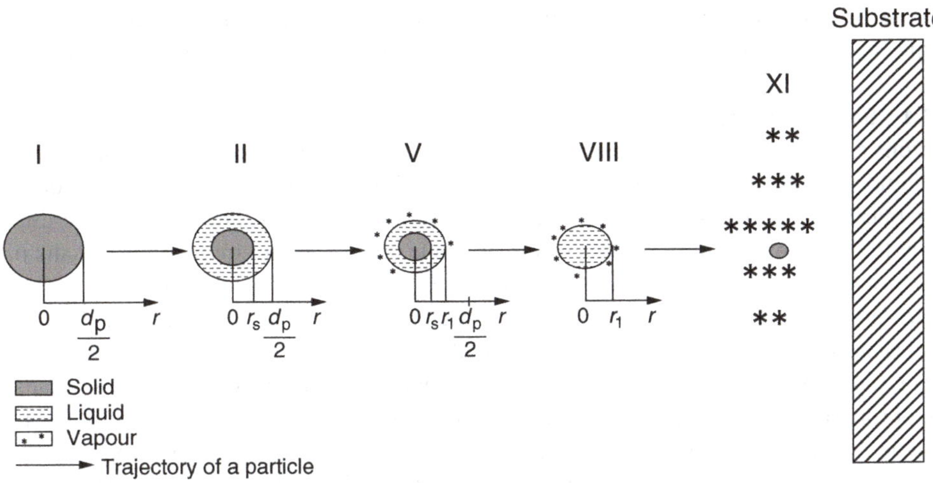

Figure 2.43 Phases content in an initially solid particle at plasma spray physical vapor deposition before contact with substrate. The numbers I, II, V, and VIII correspond to the heat transfer phenomena described in the diagram shown in Fig. 2.34a. The situation noted as XI corresponds to nearly entire evaporation of the particle with a remaining nano-dimensional liquid fine grain.

Medium

The main characteristic of *PS PVD* technique is low pressure in the operating chamber, $p \leq 200$ Pa. This pressure is still much greater than that in the conventional *PVD* techniques, but it modifies the behavior of plasma jet, which is much longer than in other thermal spray techniques and may reach 2 m [135]. The pressure in the nozzle is higher than that in the chamber being in order of $p = 40\,\text{kPa}$, its temperature $T_{pl} = 14\,000\,\text{K}$, and its velocity $v_{pl} = 6000\,\text{m/s}$ for different working gases including Ar, H_2, and He. The plasma jet expands on the nozzle exit, and its flow in the chamber has a laminar character [138]. The temperature of electrons at the spray distance of 1 to 1.5 m was estimated to be about $T_e \approx 5000\text{K}$ [141].

The agglomerated particle injected into the nozzle should disintegrate rapidly by melting the organic binder forming primary particles. The heat transfer toward this primary particles leading to its evaporation, as shown in Fig. 2.43, occurs inside the torch nozzle [142]. Knowing that entire enthalpy required to heat, melt, and evaporate zirconia is equal to $\Delta H_{ev} = 8560\,\text{kJ/kg}$, it was possible to evaporate nearly entirely only a particle having a size slightly lower than $d_p \leq 1\,\mu\text{m}$ [138, 143]. The residual clusters of zirconia, shown as situation XI in Fig. 2.43, have sizes estimated to be 5 to 20 nm.

2.3.3 Liquid Feedstock Deposition Method

The liquid feedstock used in coatings deposition are prepared as ***suspensions*** and ***solutions*** [144]. The *suspensions* include small, submicrometric or nanometric solids which participate in coatings formation at deposition after evaporation of liquid part at processing. The *solutions* include some liquid chemical products which react at processing, e.g. in flight from the source to the substrate or on the substrate (solution precursor plasma spraying, *SPPS*) or on the heated substrate (spray pyrolysis). The products of reactions are initially solids but may get molten in, e.g. plasma jet before impact with the substrate.

2.3.3.1 Solutions Preparation

Solutions are mixtures of liquids being dispersed on the molecular scale. In the mixture of two liquids, a ***solute*** is dissolved in a ***solvent***. Some solutions require mixing by mechanical or chemical energy, and some others get mixed spontaneously [9]. The *solutions* are used mainly to obtain oxides films and/or coatings. Their precursors are frequently the salts of metals such as nitrates and acetates (see Table 2.15). The appropriate mixing of these salts enables obtaining the requested stoichiometry such as 8*YSZ* shown in Table 2.15. Some chemical agents are added, e.g. acetic acid, to modify the chemistry of solution and, resulting thereof, morphology of final coatings. Water and ethanol are used as solution for these salts.

The properties of *solutions* depend strongly on their molar concentration. For example, the dynamic viscosity of aqueous salts containing yttrium and zirconium being equal to $\eta = 1.4 \times 10^{-3}$ kg/(m.s) at molar concentration of $n = 0.6$ mol/L increases to $\eta = 7.0 \times 10^{-3}$ kg/(m.s) at the concentration of $n = 2.4$ mol/L [150].

2.3.3.2 Suspension Preparation

Suspensions contain a liquid *solvent* with fine solid particles and some additives are added to improve their ***rheological properties*** such as agglomeration of fine solids, which should be avoided. In addition, the sedimentation of solids has to be reduced. Knowledge about suspension is a part of ***chemistry of colloids***, and the size of solid particles analyzed theoretically should be in the range from 1 to 500 nm [151]. However, the size of solid particles in suspensions used in practice of thermal spraying may reach a few micrometers [152]. The small particle in a liquid is submitted to the action of forces resulting from similar solids. The forces can be, e.g. electrostatic, which is repulsive or *van der Waals* one being attractive. The suspension should be designed in such a way to avoid the agglomeration of the solids, i.e. ***stabilization of suspension***, by adding the dispersing agents.

Table 2.15 Typical chemical composition of solution precursors used to spray.

Final coating	Deposition process	Precursor	Solvent	References
$La_2Zr_2O_7$	Spray pyrolysis, *SP*	Zirconyl (IV) oxynitrate hydrate and lanthanum (III) nitrate hexahydrate	Deionized water	[145]
TiO_2		Titanium (IV) isopropoxide and acetylacetone	Ethanol	[146]
Yttrium iron garnet, $Y_3Fe_5O_{12}$, *YIG*	Solution precursor plasma spray, *SPPS*	Synthesis of yttrium (III) nitrate hexahydrate and iron (III) nitrate with citric acid	Ethanol and deionized water	[147]
8 *YSZ*		Zirconium acetate and yttrium nitrate		[148]

Source: Adapted from Pawłowski [149].

The well-stabilized suspension may be used for a long time after its preparation (see Fig. 2.44).The fundamental phenomena in suspension are discussed in many textbooks on the field of industrial chemistry such as [9]. An important parameter in the preparation of suspension is the content of solid phase. This content determines the deposited coating's growth rate. However, the high solid-phase content results in high viscosity of suspension and renders it more difficult to be transported in the pipeline. Some examples of the suspensions used in thermal spraying and their properties are shown in Table 2.16.

Small solid particles used as solid phase in *suspensions* can be prepared from condensation from vapor or liquid phases. The method of fine powders manufacturing in that way is reviewed elsewhere [152, 158]. Another frequent manufacturing way is the synthesis from the solutions followed by calcining and crushing. This method was used, e.g. by Jaworski *et al.* [159], to manufacture fine *HA* powder.

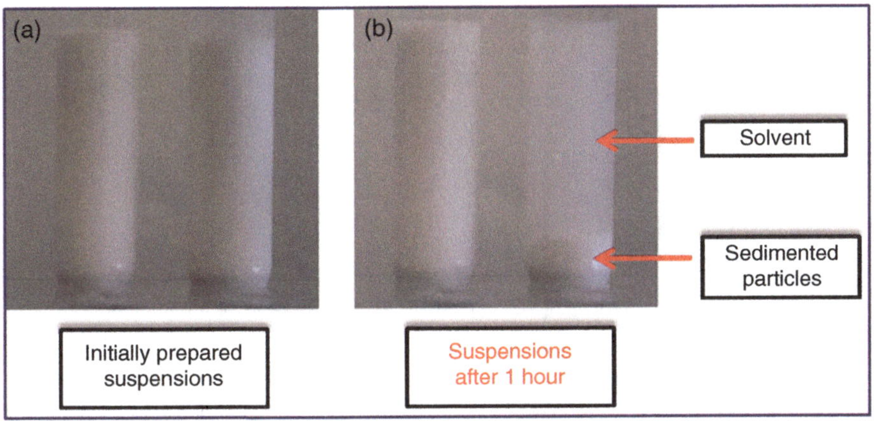

Figure 2.44 The stabilized and not-stabilized suspensions including *YSZ* solid phase: (a) immediately after preparation; (b) 1 h later [153].

Table 2.16 Typical chemical composition of aqueous suspensions used to thermal spray.

Final coating	Solid phase, size, μm	Dispersant, wt. %	Viscosity, mPa.s	Solid phase content, wt. %	Feedrate, g/min	References
TiO_2	Rutile, $d_{VS} = 0.33$	Sodium polycrylate, 0.3	—	4–10	20	[154]
YSZ	$d_{50} = 1.5$	*PAA*; *PEI*; *PBTCA*	3 up 23 depending on solid content	24	—	[155]
Al_2O_3	$\alpha - Al_2O_3$, $d_{50} = 0.4$	NaOH or HCl to control pH		20–25		[156]
HA	$Ca_{10}(PO_4)_6(OH)_2$, $d_{VS} = 1.47$	$Na_4P_2O_7$, 0.3	—	10	25	[157]

Source: Adapted from Pawłowski [152].

2.3.3.3 Delivery and Injection of Liquid to Flame or Plasma

The liquid feedstock can be delivered to the pipeline leading to the torch by peristaltic pump, which enables electronic control of the liquid feed-rate [152]. Such a feeding system does allow controlling efficiently the dynamic pressure of liquid. This pressure should be greater than that of plasma or flame, following the equation:

$$\rho_l v_l^2 > \rho_{pl} v_{pl}^2 \tag{2.40}$$

The injection velocity can be controlled in a pneumatic system in which liquid is submitted to a pressure of a compressed gas as shown in Fig. 2.45a, b. The liquid is atomized after

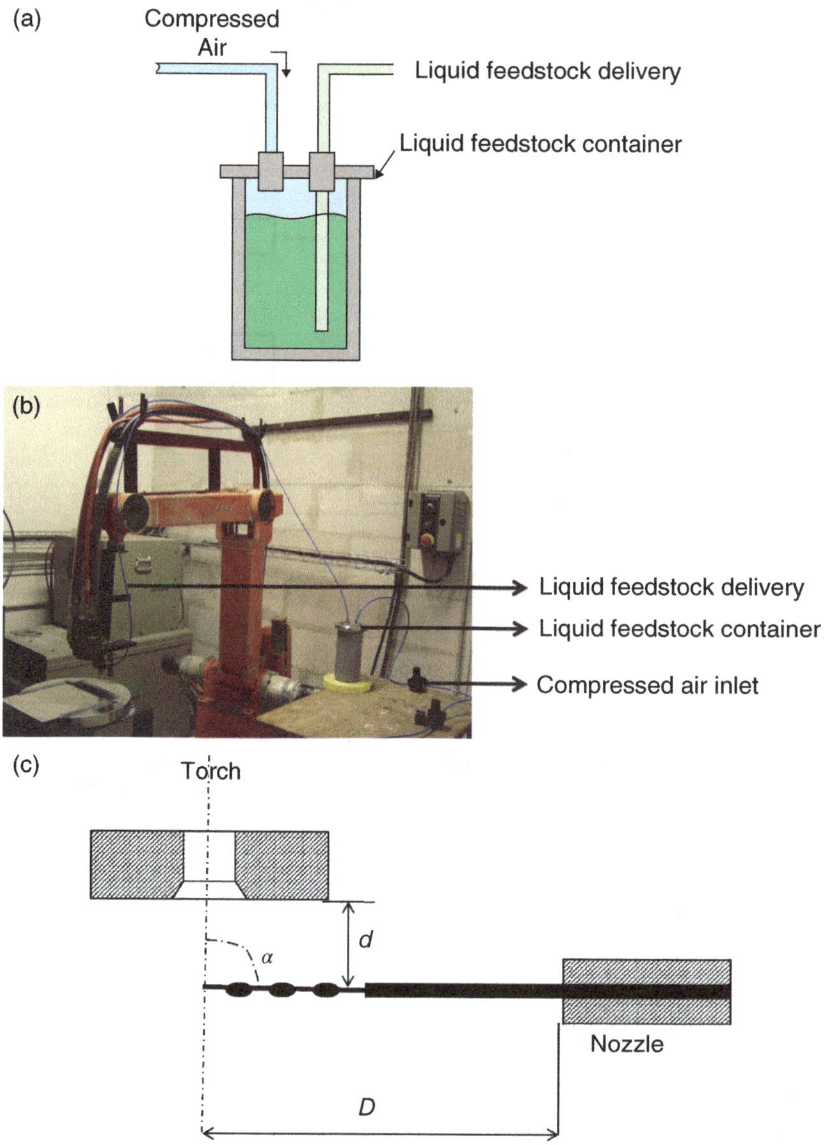

Figure 2.45 Liquid feedstock pneumatic delivery: (a) sketch of a feeder; (b) view of a feeder connected to a plasma spray torch; (c) geometry of injection to the gas/flame/plasma jet.

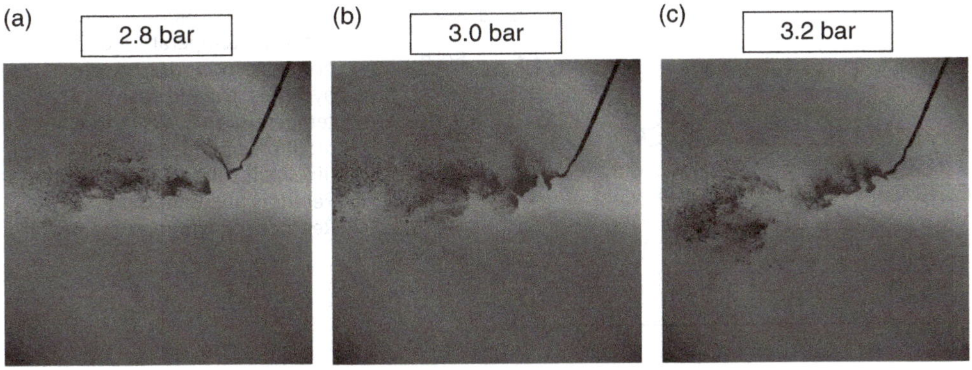

Figure 2.46 Atomization of zirconia suspension injected radially using pneumatic system into plasma jet under compressed gas pressure of: (a) 0.28 MPa; (b) 0.3 MPa; and (c) 0.32 MPa [153].

injection to a jet. The latter has geometry to be optimized as shown in Fig. 2.45c. The optimization concerns the determination of the compressed gas pressure, which corresponds to the atomization, called ***primary break up***, which results in optimal trajectory of droplets (see Fig. 2.46). The atomized droplets interact with the jet. The large droplets may disintegrate again in a ***secondary break up*** on the way to the substrate. The breaking up depends on ratio of inertial forces resulting from mechanical interaction of a droplet with the plasma jet and its surface tension forces, known as ***Weber number*** [152]. The feed-rate of liquid is significantly lower than that used in powders spraying and is typically lower than $q_l < 50$ g/min for suspension plasma spraying as in the study of Wang *et al.* [160].

2.3.3.4 Spray Pyrolysis (*SP*)

Spray pyrolysis started to be used in 1960s to obtain the films of semiconducting compounds [161]. The technique consists of spraying a solution onto heated substrate (see Fig. 2.47).

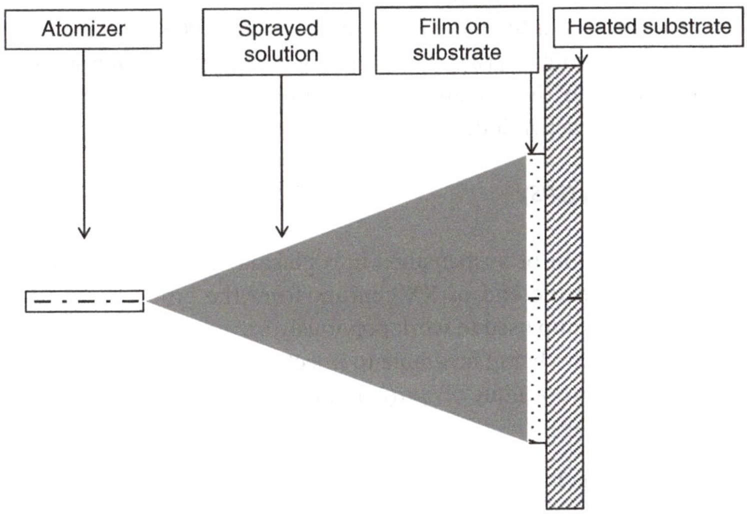

Figure 2.47 Sketch of spray pyrolysis process. *Source:* Adapted from Perednis and Gauckler [162].

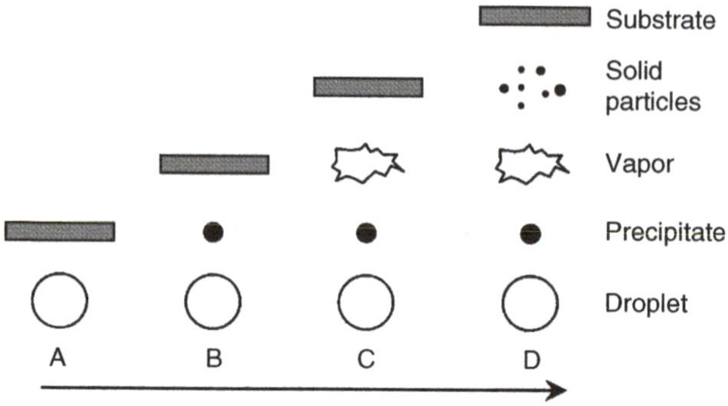

Figure 2.48 Influence of substrate temperature on the processes occurring at spray pyrolysis. The increase of temperature results in formation of: (i) precipitates; (ii) vapors; and, (iii) small solid particles. *Source:* Perednis and Gauckler [162] Reproduced with permission.

The liquid feedstock is injected through the nozzle and gets atomized. The solution flow rate ranges from $q_l = 0.1$ to 80 ml/min [163]. The droplets of solution are subsequently accelerated toward substrate. The morphology of obtained films depends mainly on precursors' chemistry and on substrate temperature.

The influence of the chemistry may be the formation of NiO films by pyrolytic decomposition of nickel chloride solution in water vapors analyzed by Yu and Kim [164]:

$$NiCl_{2(l)} + H_2O \rightarrow NiO_{(s)} + 2HCl_{(g)} \tag{2.41}$$

The authors calculated also the standard free energy of Gibbs, $\Delta G°$, in J/mol [13] for this reaction vs. temperature to be:

$$\Delta G^{\circ} = 69140.95 - 346.86\,T + 38.34 T lnT - 6.61x10^{-3} T^2 + 2.49x10^5 T^{-1} \tag{2.42}$$

The value of Gibbs free energy is negative, being about $\Delta G° = -7$kJ/mol at $T_{\text{sub}} = 700$ K, which renders possible the pyrolysis and formation of NiO film at this and higher temperatures [165]. The increase of substrate temperature modifies morphology and at 1073 K, the reaction shown in Eq. 2.41 leads to the formation of fine NiO powder described in [164]. In general, the sprayed droplets may form with increasing substrate temperature (see Fig. 2.48): (i) precipitates; (ii) vapors; and (iii) solid particles.

The substrate temperatures are frequently in the range of $T_{\text{sub}} = 600$–900 K, and the used substrates are glasses, ceramics, and metals [163]. The main advantage of spray pyrolysis is the simplicity of processes joined to the possibility of obtaining films of complicated compounds, such as magnesium or nickel-doped magnesium ferrites [166].

2.3.3.5 Suspension Plasma Spraying (*SPS*)

History

The initial works concerning injection of suspensions into plasma came probably from Sherbrooke University in Canada at the end of XX century from the group around M. Boulos and F. Gitzhofer [167]. The group used to work previously on inductively generated plasma, which has the vertical vortexes being favorable to injections of liquids. The use of the slurry could have come from the technique of *spray-drying* used frequently to produce

---

13 The authors give the free Gibbs energy in J.

the powders. The rapid development of this technology, similar to that of solution precursor plasma spraying, resulted from the possibility of obtaining the nanostructured coatings.

Principle

Killinger [168] and Fan and Bai [169] reviewed the suspension plasma spray process, *SPS*, recently. The process is carried out with a torch being similar to that used to spray powders. The three-cathode torches enabling axial injection were also reported [153]. The suspension is delivered as a continuous jet or as atomized droplets and starts to interact with a plasma jet.

Medium

The phenomena describing the physical and chemical interaction of suspension with plasma jet are shown in Fig. 2.49. The phenomena include [152]:

- *primary* and *secondary break up*;
- evaporation of liquid *solvent* with the additives;
- possible sintering of fine solids;
- melting of fine solids and of sintered agglomerates;
- evaporation of liquid particles;
- impact of droplets on the substrate.

The evaporation of liquid solvent results in the decrease of the particle mass. The time of solvent evaporation for a single droplet was analyzed for *spray-drying* powder manufacturing process by supposing that the particle is made of solvent (such as water). The analysis of Masters [94] enables to find the following differential equation, making possible to find the time of evaporation of a spherical particle having surface, A_p, mass m_p, and temperature, T_p, being immobile with regard to the gas of temperature T_{pl}:

$$-L_{ev}\frac{dm_p}{dt} = hA_p\left(T_{pl} - T_p\right) \tag{2.43}$$

The analysis of Hurevich *et al.* [170] enabled estimation of the sintering time small liquid particles of iron having size of $d_p = 1\,\mu m$ to be in the range of nanoseconds, i.e. nearly instantaneous. Finally, the melting of solid particles as well as the evaporation from their

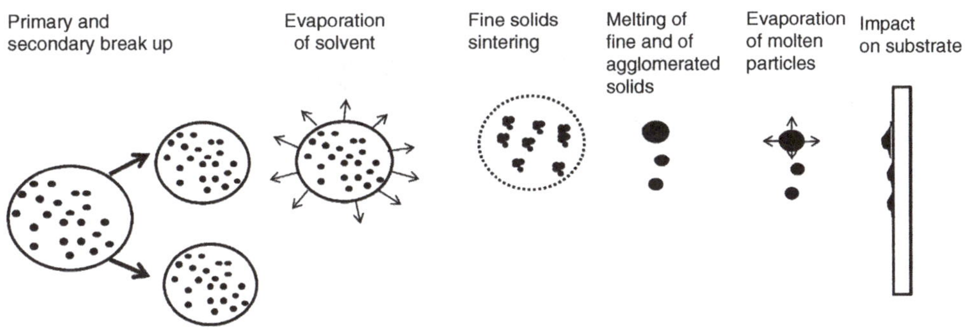

Figure 2.49 Phenomena occurring at interaction of suspension with plasma jet at *SPS*. *Source:* Pawłowski [152]/with permission of ELSEVIER.

surface correspond to the description made in the section related to *APS* and, in particular, the description of the phenomena V to X in Fig. 2.34b.

The suspension plasma sprayed coating may have columnar morphology. Such morphology is particularly interesting for top zirconia deposits of thermal barrier coatings, *TBCs*. Namely, thermal shock resistance of coatings having such morphology is similar to that obtained by the techniques *EB PVD* or *PS PVD*. Sokołowski *et al.* [171, 172] studied recently zirconia coatings having columnar microstructure deposited by *SPS*. Such coatings are about to be applied in the coatings' industry [173].

2.3.3.6 Solution Precursor Plasma Spraying (*SPPS*)

History

The application of liquid solution to thermal spraying was initiated in Stony Brook University (NY, USA) at the end of XX century in the research group of H. Herman and C.C. Berndt [174]. The research was motivated by the development of nanotechnologies. The process was rapidly developed and applied to solution precursor plasma spray of thermal barrier coatings in the University of Connecticut (CT, USA) in the research group of M. Gell and E.H. Jordan [175].

Medium

The phenomena occurring after injection of solution precursor into plasma jet are, in part, similar to those occurring at suspension plasma spraying and include [149]:

- *primary* and *secondary break up*;
- heating, vaporization, and internal precipitation;
- internal pressurization and droplet breaking-up;
- solid particle sintering, heating, and melting;
- evaporation from the melt.

The described phenomena are schematically shown in Fig. 2.50. The physical and chemical phenomena occurring at *SPPS* after injection of zirconium acetate and yttrium nitrate

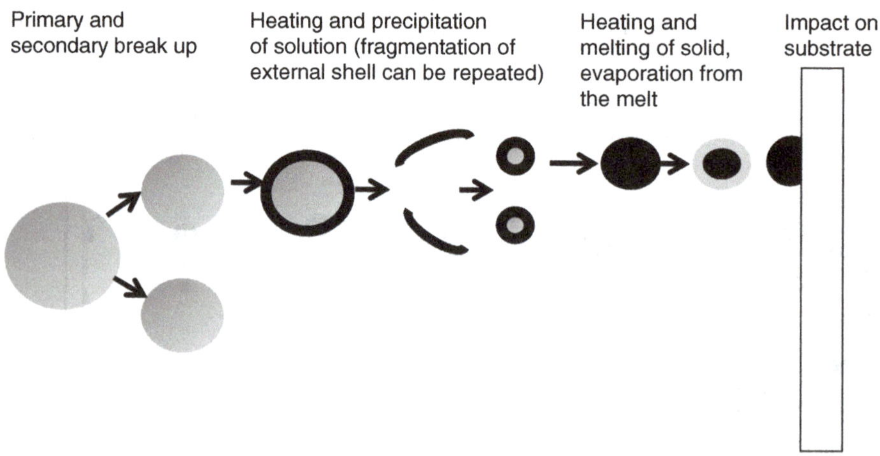

Figure 2.50 Phenomena occurring at interaction of solution precursor with plasma jet at *SPPS*. *Source:* Pawłowski [149]/with permission of ELSEVIER.

Figure 2.51 The transformation inside a droplet of solution precursor of zirconium acetate and yttrium nitrate precursor dissolved in water with ethanol starting from final break-up up to formation of liquid zirconia. *Source:* Tobon Valencia et al. [176]/with permission of ELSEVIER.

precursor solved in the mixture of water with ethanol into plasma jet were analyzed by Tobon Valencia *et al.* [176]. The transformation occurring in a solution droplet is shown in Fig. 2.51. The analysis included the evaporation, heating, and melting phenomena discussed in the section describing the processes of *APS* and *SPS*. The phenomena occurring only at the *SPPS* concern nucleation of the solid phase of solid zirconium hydroxide from liquid zirconium acetate follows the reaction:

$$C_8H_{12}O_8Zr_{(l)} + 4H_2O_{(l)} \xrightarrow{T_p=363\,K} Zr(OH)_{4(s)} + 4CH_3COOH_{(l)} \tag{2.44}$$

The above reaction is associated with ***Gibbs free energy of precipitation***, which was found to be equal to $\Delta G^* = 2.03 \times 10^{-30}$ J for the precipitation of one molecule of zirconium hydroxide.

The heating of a droplet with a solid shell outside leads to the evaporation of liquid and to the increase of the pressure inside shell. The growth of internal pressure may result in the fragmentation of shell analyzed theoretically by Basu and Cetegen [177] and shown schematically in Fig. 2.50.

2.4 Bulk Methods of Coatings Deposition

The bulk methods of coatings deposition are characterized by a large quantity of material deposited in one pass. The painting is, probably, the best known of them. This section describes the deposition characterized melting of a substrate associated with the injection of powders or introduction of a wire into the bath of a material of the deposit in the form of a wire or a powder. Another described group of methods includes the melting of already pre-deposited coating. The melting is realized using: (i) combustion flame; (ii) plasma; and (iii) laser. The melting results often in the dilution of a liquid phase in the substrate, as shown in Fig. 2.52, and the associated deposition methods are often called ***surfacing by welding*** or ***hardfacing***. The dilution, defined in Fig. 2.52, should not be greater than a few % in the well-optimized deposition process, but may reach up to 40 % [53].

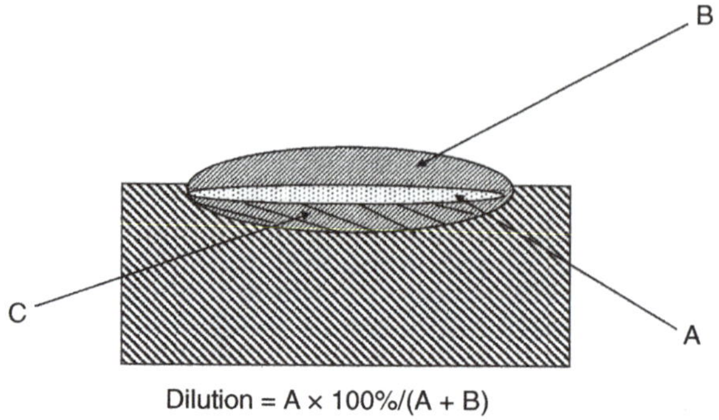

Figure 2.52 Dilution in coatings obtained using melting: (a), weight of molten zone including coatings and substrate; (b), weight of molten coatings; (c), heat-affected zone, *HAZ*.

The use of flame to melt the substrate requires longer time than the use of plasma or laser because of its relatively low temperature being lower than $T < 3350$ K for typical oxy-acetylene flame. The *hardfacing* technique using flame includes metal inert gas, *MIG*, in which a metal electrode is fed to the weld pool forming coating being protected by inert gas from the contamination. This technique is described, e.g. in the textbook of Hocking *et al.* [53] or Rudzki [109].

2.4.1 Plasma Transferred Arc (*PTA*)

The torches used in the technique of plasma transferred arc, *PTA*, are similar to those used in the technique of atmospheric plasma spraying, *APS*. The only difference is the generation of arc between cathode and the substrate. Such torches are also applied to welding in the technique known as tungsten inert gas, *TIG* [178]. The phenomena at deposition occur mainly on the substrate. The principle of *PTA* coatings is shown in Fig. 2.53. The process starts with ***pilot arc***, which is initiated between the thoriated tungsten cathode and the copper anode, cooled by water. This arc heats the plasma gas, being typically argon. Then, the torch is approached to the substrate, and the transferred arc is ignited between the

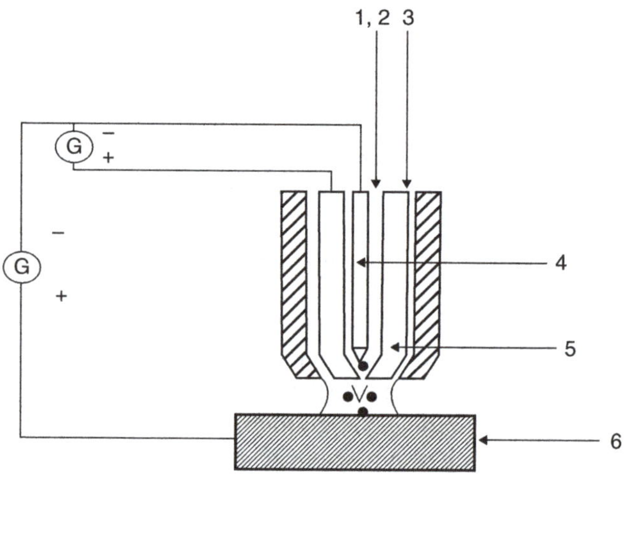

Figure 2.53 Sketch of *PTA* processes: 1, plasma gas; 2, injection of powder transported by a carrier gas; 3, protective gas; 4, cathode; 5, anode; 6, substrate.

cathode and the anode-substrate. This arc melts locally the substrate. The powder is introduced into the melt-pool to create the deposit. The pool is protected against the influence of ambient atmosphere by an inert gas such as argon [179].

The coatings obtained by the *PTA* technique are frequently steels, nickel, or cobalt-based alloys such as *Stellites* and, more recently composite coating having metallic matrix and ceramic reinforcement, *MMC* [180–182]. The quality of coatings obtained by *PTA* depends mainly on the following elements of coating's processes after Demian *et al.* [183]:

- torch (pilot arc current, transferred arc current, plasma gas flow rate, shielding gas flow rate);
- powder (chemistry, granulometry, particles velocity resulting from carrier gas flow rate);
- torch position with respect to the substrate (oscillation, distance stand-off, overlapping of the following deposition paths).

The *PTA* process can have different powder feed-rate starting from a few tenths up to a few hundreds of g/min. The latter was obtained using high-power plasma transferred arc [53]. The typical industrial torch is powered with 3 to 10 kW, and the gas flow rates are relatively low. An example of the flow rates used to spray NiCrAlY – YSZ cermets is [183]: (i) Ar plasma gas of 1.4 slpm; (ii) $Ar + H_2$ shielding gas of 14 slpm; and (iii) carrier gas of 3.5 slpm.

The dilution of a coating depends on the *pilot arc* current and on the transferred arc one. A small transferred arc current results in a small dilution, and a large pilot arc current enables better heating of the particles prior to contact with the melt-pool. The following are the typical values of these currents used, e.g. at deposition of NiAl alloy the pilot arc current is 45 A and the transferred arc 80 A [184].

The powder applied in the *PTA* process has particles having sizes in the range $d_p = 40–150\,\mu m$. The sizes are more dispersed than in the thermal spraying processes. This dispersion results from the fact that the powder is injected to be molten in melt-pool. The melting by convection in pilot arc plasma is less important, and mainly melting by conduction in melting-pool counts.

The torch moves with regard to a substrate in the following ways: (i) translation (scan speed is typically equal to a few cm/min); and ii) oscillations with an amplitude of about ten mm and its frequency about 40 to 100 min^{-1} [183, 184]. The stand-off distance between the torch and the substrate is short and about 6 to 8 cm [185] to keep the transferred arc ignited. The *PTA* coatings are deposited, formed in a couple of passes of a torch, and the final thickness of the deposits by PTA may reach 10 mm. The surface of the final coatings is not smooth (see Fig. 2.54) and requires rectification before service.

2.4.2 Laser Coatings

The techniques associated with lasers are recent. Siegman [186] reports that the beginning of laser history is in 1954 when the American (C.H. Townes, J.H. Gordon, and H. Zeiger) and, on the other side, the Iron Curtain Soviet scientists (G. Basov and A.M. Prokhorov) developed the emission of lasers representing stimulated emission at microwave frequency (called also masers). The first laser was developed using ruby as an active medium one with the wavelength of $\lambda = 0.694\,\mu m$ developed by T.H. Maiman in Hughes Research Laboratories (Malibu, CA, USA) [35].

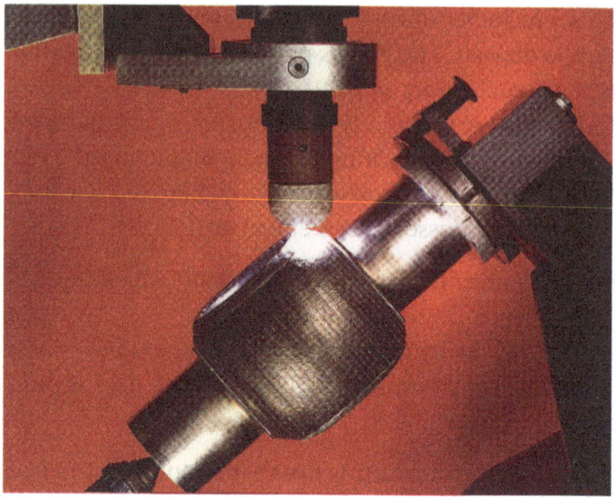

Figure 2.54 Coating deposition by PTA using a torch produced by *Saint Gobain Coating Solution*, Avignon. *Source:* Reproduced with permission from Saint-Gobain.

The laser deposits can be obtained in two processes: (i) one-step laser deposition (*1SLD*); and (ii) two-step laser deposition (*2SLD*). The *1SLD* process, similar to the *PTA*, consists of melting the substrate by a laser and injecting powder into a melt-pool. It is also possible to obtain an initial coating using with another technique, being often thermal spraying, and, then melt the pre-deposit with laser. This technique can be called the two-step laser deposition (*2SLD*). These two techniques can be categorized as (see also Fig. 2.55):

- ***cladding***, in which the deposits are chemically different from the substrate;
- ***laser surface alloying*** (*LSA*), in which the materials of the deposit and the substrate form an alloy at processing;
- ***hard-phase dispersion***, in which solid particles of a reinforcement are injected into the pool and remain solid.

Another emerging method is called ***laser prototyping*** and consists of injecting of small quantity of powder to be melted into a melt pool created by laser beam. The *laser prototyping* is a part of ***additive manufacturing.*** The latter enables creating the *3D* objects [187].

The laser deposits are mainly made using four types of lasers: CO_2, Nd: YAG, diode laser, and fiber laser. The diode and fiber lasers applications grow rapidly in *hardfacing* industry. The main properties of these lasers were discussed in the chapter describing the cleaning of substrates and are shown in Tables 1.9 and 1.11. The melting of substrate requires, however, greater laser power density than cleaning, and should be in the range of $q = 1\text{–}10^3\ \text{kW/cm}^2$ [188].

2.4.2.1 1-Step Laser Deposition (*1SLD*)

The sketch of a laser cladding is shown in Fig. 2.56. A powder is injected directly into a melt-pool in the substrate. The powder melts in contact with the pool and forms the coating. The material of the formed coating diffuses toward substrate resulting in dilution.

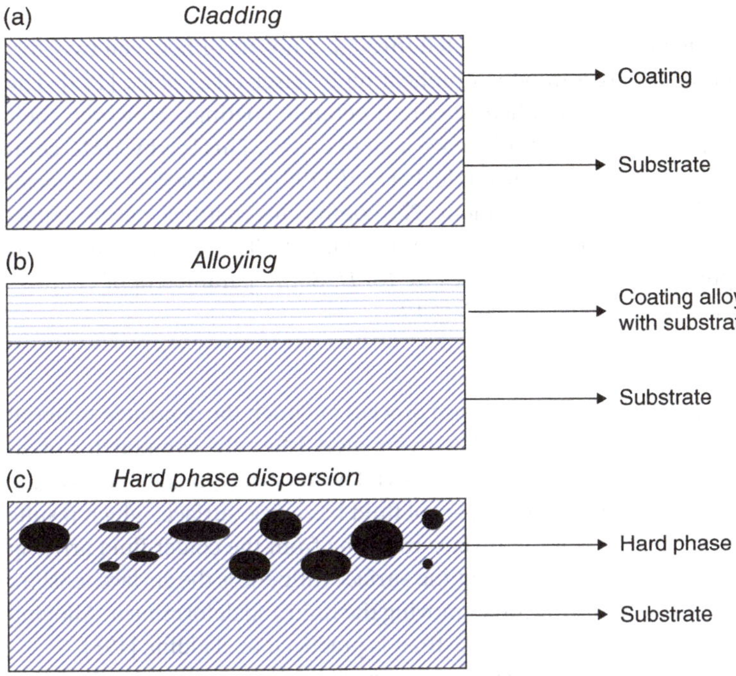

Figure 2.55 The laser-assisted coatings processes: (a) *cladding*, (b) *alloying*, and (c) *hard phase dispersion*.

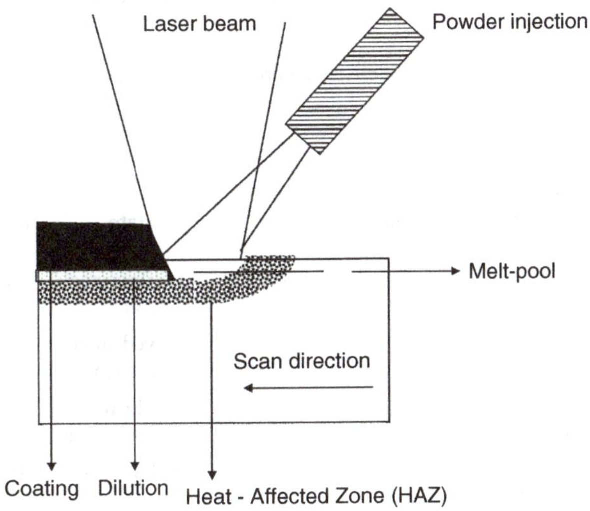

Figure 2.56 Sketch of the one-step laser deposition, *1SLD* process.

Finally, the part of the substrate, which was heated by melt-pool but remained solid, may modify its microstructure. This part is called heat-affected zone, *HAZ*. The dilution is of the order of 10% in the *cladding* process and 100% in the processes of *alloying* process and the *dispersion of hard phase*.

Cladding

In the cladding process, the powder can be injected through a nozzle or between two mirrors in a configuration a ***beam trap*** in which preheating by laser takes place [189]. The deposition by *cladding* is sometimes applied to obtain ceramic coatings. However, their adhesion to substrates is poor, and the deposits have a lot of cracks. This is why *MMC* composites are used more and more often. For example, Pei *et al.* [190] have developed a deposit with a Ni alloy matrix and the ZrO_2 reinforcement. The composites coatings may be deposited using two injectors for matrix and reinforcement separately. The *cladding* process is currently the most popular among the processes of *1SLD*. Some parameters of coatings *cladded* using different lasers are shown in Table 2.17.

Alloying

Alloying, called also laser surface alloying, *LSA*, is a process in which the deposited material is injected into the melt-pool of the substrate. The two materials are the components of an alloy which is created at the end of the deposition process. The material is made thanks to the convection movements inside the pool. The melt-pool fluid has a free surface, which results in the fluctuation of the temperature of the surface resulting from the colder fluid arriving from the bottom part and from the evaporation at the surface influences the surface tension. The gradients of surface tension create the forces moving the fluid and deforming it. Alloying requires the application of a higher-power density than is practiced in cladding [195].

LSA was frequently used to improve the wear resistance of aluminum, which has low density being attractive in many applications reviewed by Chi *et al.* [196]. The process parameters of such alloying using different lasers are shown in Table 2.18. The idea followed in several studies was the formation of intermetallic alloys from two metallic components. For example,

Table 2.17 Metal, alloys, and composite coatings obtained by laser *cladding*.

Coating composition	Laser type, power density, kW/cm^2	Wavelength, μm	Substrate	Powder particles size, μm	Powder feed rate, g/min	Scan speed, cm/s	References
Al	Diode, *cw*, 18–21	0.94	Stainless steel AISI 304	−350 + 100	1.2–1.4	0.1–0.4	[191]
Stellite 21	CO_2, 4–17	10.6	Alloyed steel, 57NiCrMoV77	−150 + 45	6–15	0.17–0.5	[192]
Alloy NiCrBSi	Fiber, *cw*, 1.3–2.2	1.07	Ductile and gray cast irons	−150 + 125	8.5	0.1–0.5	[193]
MMC composite NiCrBSi + (10–30 wt.%) WC	Nd: YAG (Power density was estimated from laser power and beam spot diameter), *cw*, 40–51	1.06	Steel S235JR	NiCrBSi: −150 + 53 WC: −210 + 40	12	2	[194]

Table 2.18 Alloying parameters of Al substrate with other metals using different lasers.

Coating composition	Laser type, power density, kW/cm^2	Wavelength, μm	Substrate	Powder particles size, μm	Powder feed rate, g/min	Scan speed, cm/s	References
Al + Zn	CO_2, *cw*	10.6	Al	−250	0.6–3.6	0.5	[197]
Al + Ni	Nd: YAG, *pulsed*, 2–10	1.06	Al	—	—	0.3	[198]

the injection of Al, Cu, Ni, or Si powders in the bath of the Mg substrate made it possible to obtain intermetallic compounds of the type: Mg_{17} Al_{12}, Mg_2Cu, Mg_2Ni, or Mg_2Si [199].

Hard-Phase Dispersion

Hard-phase dispersion is quite similar to *alloying*. The only difference is that the powder injected into the bath of the molten substrate must remain solid. The obtained coating is a composite-type *MMC*. This process of composites manufacturing was initiated in the 1970s [200, 201]. The important goal is to achieve a good wetting between liquid melt-pool and solid reinforcement. Oukach *et al.* [182] reviewed recently the problem of reinforcement wetting in the melt-pool. The thickness of the deposit depends on the process parameters, which influence the depth of penetration of a particle in the liquid bath, i.e.: (i) injection velocity determined by the flow rate of the carrier gas and (ii) viscosity of the bath liquid. Another important parameter is scan speed of the focal spot of the laser. This speed of the scan can be adjusted to enable such rapidity solidification which reduces generation of residual stresses and, resulting thereof, formation of cracks. Heating the substrate during deposition can be useful in reducing stresses. The metal matrix composites can be also obtained by cladding by injection of reinforcement and metal matrix into the melt-pool of a substrate [202, 203]. Some examples of the *MMC* coatings obtained by the dispersion of hard phase are shown in Table 2.19.

Rapid Prototyping (Additive Manufacturing)

Rapid prototyping is a modification of the cladding process that enables obtaining *3D* structures having walls formed at laser-cladding process [205]. The sketch of installation diagram for such prototyping is shown in Fig. 2.57. At the prototyping, the laser melts the powder being [206]: (i) injected through a nozzle; (ii) injected through slurry; and (iii) loosely put in container. The laser, guided by a computer, follows a path allowing creation of structures with thin walls and very good heights (z-axis direction in Fig. 2.57). Using this technique, one can form metal cylinders (Fig. 2.58) and parts of a complicated shape and unique character.

The wall thickness of structures produced by rapid prototyping varies between 100 μm (size of the focal spot of the laser) and a few millimeters. The microstructure of the parts is similar to that of sintered products.

2.4.2.2 2-step Laser Deposition (*2SLD*)

The two-step laser deposition, *2SLD*, uses laser treatment to improve the properties of a preplaced coating. The coating can be preplaced as a slurry by doctor blading and dried thereafter. Monisha *et al.* [207] presented an example of such pre-placement of the mixture of powders TiB_2 and B on Ti substrate prior to the alloying with diode laser. This section

Table 2.19 Parameters of deposits obtained by the dispersion of the hard phase using different lasers.

Coating composition	Laser type, power density, kW/cm^2	Wavelength, μm	Substrate	Powder particles size, μm	Powder feed rate, g/min)	Scan speed, cm/s)	References
WC + Ti6Al6V or TiC + Ti6Al6V	CO_2, 42–140	10.6	Ti6Al6V	−100 + 50	60–120	5–15	[200]
SiC + Ti6Al6V or TiN + Ti6Al6V or	Power density was estimated from laser power and beam spot diameter. Nd: YAG, *cw*, 14–32	1.06		—	1–3	0.8–1.7	[204]

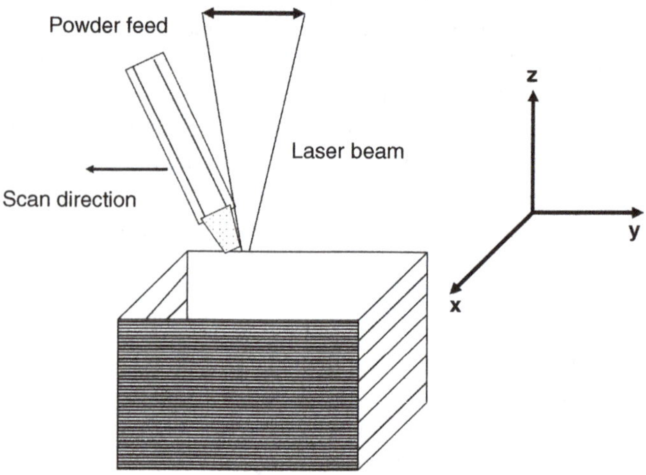

Figure 2.57 Sketch of rapid prototyping. *Source:* Pawłowski [188]/Springer Nature.

focuses mainly on the pre-deposits obtained by one of the physical deposition methods discussed previously. An example is alloying of *PVD* films of Ag, Au, Pd, Sn and Ta with on Ni substrate realized with help of Nd: YAG laser [208]. The principle of the *2SLD* method is shown in Fig. 2.59.

The laser treatment is most frequently applied to the coatings preplaced by one of the thermal spray techniques. The treatment can be realized by [188]: (i) keeping the coating in solid phase; (ii) melting the coating; and (iii) evaporating the coating material. The CO_2 laser heat treatment in solid phase was applied on plasma-sprayed *HA* coatings [209] to increase the crystallinity of the coatings. Another example is the Nd: glass laser shock treatment of Al + SiC composite obtained by *HVOF* [210]. This treatment improved the contacts

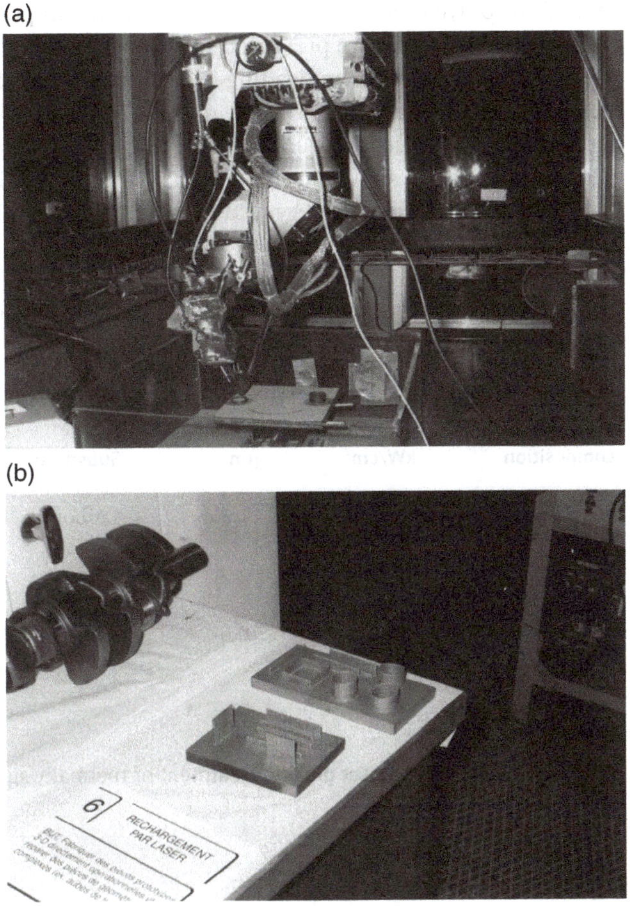

Figure 2.58 Laboratory installation of rapid prototyping: (a) preparation of a small cylinder; (b) thin walls and cyliders obtained by rapid prototyping. *Source:* From Pawłowski et.al. (188)/ Springer Nature.

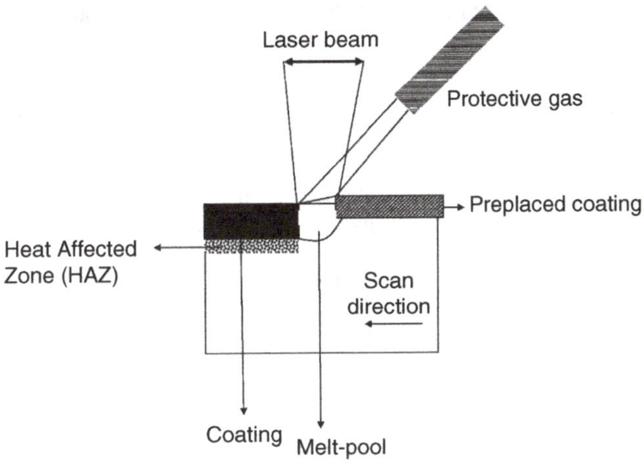

Figure 2.59 Sketch of the two-step laser deposition, *2SLD* process.

between the Al matrix and the SiC reinforcement. The parameters of the examples of plasma-sprayed coatings heat treatment in solid phase with laser are shown in Table 2.20.

The laser melting of coatings, called ***laser glazing***, is currently one of the most popular techniques for improving the properties of sprayed deposits. This treatment was applied to sprayed metals, alloys, ceramics, and composites. Examples of deposits and processing parameters are presented in Table 2.21.

Finally, the laser treatment leading to the evaporation of a part of coatings concerns the preparation of ***anilox rolls*** [217]. The rolls are first coated with plasma-sprayed Cr_2O_3 coating. The as-sprayed coatings are rectified and polished before being engraved with laser using an installation shown in Fig. 2.60. The CO_2 pulsed laser engraves the cells having a depth of tens of micrometer very next to each other. Their linear density of the cells ranges from 200 to 400 cells/cm.

Table 2.20 Parameters of heat treatment in solid phase with different lasers of plasma sprayed *HA* coatings.

Final coating phase composition	Laser type, power density, kW/cm^2	Wavelength, μm	Substrate	Coating thickness, μm	Scan speed, cm/s)	References
HA	Nd:YAG, pulsed, power up to 400 W	1.06	Carbon steel	—	0.025–2.5	[211]
HA crystalline and amorphous phase, CaP	CO_2, *cw*, 0.52–0.68	10.6	Alloy Ti6Al4V	150	2–5	[212]

Table 2.21 Example of laser glazing treatment of thermally sprayed coatings using different lasers.

Spray method of initial coating	Initial powder used to spray	Laser type, power density/ kW/cm^2	Wavelength, μm	Substrate	Coating thickness, μm	Scan speed, cm/s	References
Metal							
VPS	Ti	CO_2, *cw*	10.6	Steel 1020	300–380	30	[213]
Alloy							
HVOF	Inconel 625	Nd: YAG, *pulsed*, pulse energy 0.45 J	0.355 (third harmonic)	Mild steel	—	—	[214]
Ceramics							
APS	*8YSZ*	Diode laser, *cw*/ 2.45–7.35	0.808	Ni alloy EI 868	350	—	[215]
Composite, MMC							
APS	Fe13Cr + 55 wt. % TiC	CO_2, *cw*/ 8	10.6	Steel St 38	200–400	0.25–1	[216]

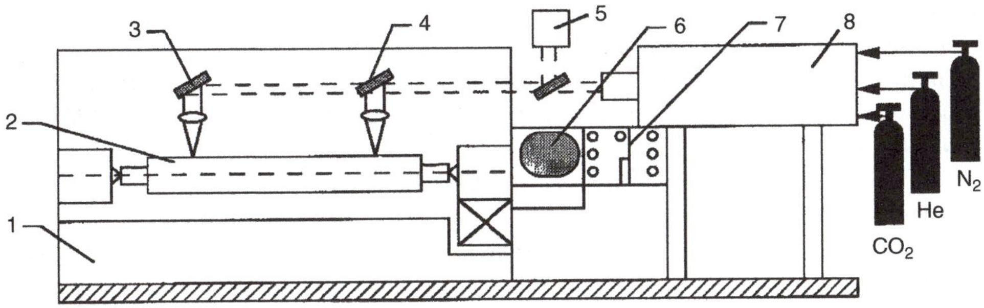

Figure 2.60 Sketch of a laser engraver: 1, engraving unit; 2, anilox roll; 3, end of engraving; 4, beginning of engraving; 5, calorimeter; 6, computer; 7, laser and engraving control unit; 8, high-power pulsed CO_2 laser.

References

1 Bunshah, R.F. (1994). Evaporation: processes, bulk microstructure and mechanical properties. In: *Handbook of Deposition Technologies for Films and Coatings*, 2e (ed. R.F. Bunshah), 131–248. Park Ridge, NJ: Noyes Publications.

2 Holland, L. (1961). *Vacuum Deposition of Thin Films*. New York, NY: John Wiley & Sons.

3 Maissel, L. and Glang, R. (1970). *Handbook of Thin Films Technology*. McGraw Hill Book Company.

4 Richardt, A. and Durand, A.-M. (1994). *Le Vide. Les couches minces – les couches dures*. Paris: In Fine (in French).

5 Mattox, D. (2010). *Handbook of Physical Vapor Deposition (PVD) Processing*, 2e. Amsterdam: Elsevier.

6 Bergmann, E. (1993). High plasma current density processes with different vapor sources. *Surf. Coat. Technol.* 57 (2–3): 133–137.

7 Lide, D.R. (ed.) (1997). *CRC Handbook of Chemistry and Physics*, 78e. Boca Raton: CRC.

8 Hoch, M., Nakata, M., and Johnston, H.L. (1954). Vapor pressured of inorganic substances. XII. Zirconium dioxide. *J. Am. Chem. Soc.* 76: 2651–2652.

9 Pawłowski, L. and Blanchart, P. (2018). *Industrial Chemistry of Oxides for Emerging Applications*. Chichester: Wiley.

10 Rigney, D.V., Viguie, R., Wortman, D.J., and Skelly, D.W. (1997). PVD thermal barrier coating applications and process development for aircraft engine. *J. Therm. Spray Technol.* 6 (2): 167–175.

11 Kourtev, J., Pascova, R., and Weissmantel, E. (1996). Arc evaporated Ti-N films with reduced macroparticles contamination. *Thin Solid Films* 287 (1–2): 202–207.

12 Scheibe, H.J. and Schultrich, B. (1994). DLC deposition by laser - arc and study of properties. *Thin Solid Films* 246: 92–102.

13 Ziegele, H., Rebholz, C., Scheibe, H.-J. et al. (1998). Mechanical and tribological properties of hard aluminium–carbon multilayer films prepared by the Laser-Arc technique. *Surf. Coat. Technol.* 107 (2–3): 159–167.

14 Groszkowski, J. (1972). *Technika Wysokiej Próżni*. Warsaw: WNT (in Polish).

15 Bunshah, R.F. (1974). Reactive evaporation. In: *Science and Technology of Surface Coating* (ed. B.A. Chapman and J.C. Anderson), 361–368. London: Academic Press.

16 Bunshah, R.F. and Deshpandey, C.V. (1986). Invited review. Plasma-assisted vapor deposition process and some applications. *Surf. Coat. Technol.* 27: 1–21.

17 Thornton, J.A. and Greene, I. (1994). Plasmas in deposition processes. In: *Handbook of Deposition Technologies for Films and Coatings*, 2e (ed. R.F. Bunshah), 29–81. Park Ridge, NJ: Noyes Publications.

18 Boenig, H.V. (1988). *Fundamentals of Plasma Chemistry and Technology*. Lancaster, PA: Technomic Publishing Co.

19 Bach, H. (1988). Ion beam sputtering of silicate glasses and oxides. *J. Non-Cryst. Sol.* 102: 36–42.

20 Thornton, J.A. and Greene, J.E. (1994). Sputter deposition processes. In: *Handbook of Deposition Technologies for Films and Coatings*, 2e (ed. R.F. Bunshah), 249–319. Park Ridge, NJ: Noyes Publications.

21 Kelly, P.J. and Arnell, R.D. (2000). Magnetron sputtering: a review of recent developments and applications. *Vacuum* 56: 159–172.

22 Wasa, K. (1993). Sputter deposition technology as a material engineering. *Bull. Mater. Sci.* 16 (6): 643–663.

23 Sigmund, P. (2012). Recollection of fifty years with sputtering. *Thin Solid Films* 520: 6031–6049.

24 Voevodin, A.A., Rebholz, C., Schneider, J.M. et al. (1995). Wear resistant composite coatings deposited by electron enhanced closed field unbalanced magnetron sputtering. *Surf. Coat. Technol.* 73: 185–197.

25 Beister, G., Dietrich, E., Schaefer, C. et al. (1995). Progress in large - area glass coatings by high rate sputtering. *Surf. Coat. Technol.* 76–77: 776–785.

26 Hofmann, D., Hensel, B., Yasuoka, M., and Kato, N. (1993). New multilayer PVD - coating technique for cutting tools. *Surf. Coat. Technol.* 61: 326–330.

27 Sproul, W.D. (1996). Physical vapor deposition tool coatings. *Surf. Coat. Technol.* 81 (1): 1–7.

28 Perez, J.P. (1995). *Mécanique*, 4e. Paris: Masson (in French).

29 Smentkowski, V.S. (2000). Trends in sputtering. *Prog. Surf. Sci.* 64: 1–58.

30 Chapman, G.E., Farmery, B.W., Thompson, M.W., and Wilson, I.H. (1972). The energy distribution of sputtered atoms from gold. *Radiat. Eff.* 13 (1–2): 121–129.

31 Matias, V. and Hammond, R.H. (2015). Ion beam induced crystalline texturing during thin film deposition. *Surf. Coat. Technol.* 246: 1–8.

32 Ahmed, N.A.G. (1994). Ion plating: optimum surface performance and material conservation. *Thin Solid Films* 241: 179–187.

33 Ensinger, W. (1998). Ion bombardment effects during deposition of nitride and metal films. *Surf. Coat. Technol.* 99: 1–13.

34 Ready, J.F. (1965). Effects due to absorption of laser radiation. *J. Appl. Phys. 36* (2): 462–468.

35 Maiman, T.H. (1960). Stimulated optical radiation in ruby. *Nature* 187: 493–494.

36 Cheung, J.T. (1994). History and fundamentals of pulsed laser deposition. In: *Pulsed Laser Deposition of Thin Films* (ed. D.B. Chrisey and G.K. Hubler), 1–19. New York, NY: John Wiley.

37 Bartholomew, C.S. and Burton, R.L. (1988). Low temperature synthesis if high T_c superconducting thin films by laser assisted reactive deposition. In: *High-Temperature Superconductors II* (ed. D.W. Capone II, W.H. Butler, B. Batlogg, and C.W. Chu), 81–84. Pittsburgh, PA: MRS.

38 https://www.fujikura.co.uk/products/energy-and-environment/2g-ybco-high-tempurature-superconductors/ (accessed 19 March 2021).

39 Metev, S.M. and Veiko, V.O. (1998). *Laser-Assisted Microtechnology*, 2e. Berlin: Springer.

40 Lu, Y., Huang, G., Wang, S. et al. (2021). A review on diamond-like carbon films grown by pulsed laser deposition. *Appl. Surf. Sci.* 541: 148573.

41 Boyd, I.W. (1996). Thin film growth by pulsed laser deposition. *Ceram. Int.* 22: 429–434.

42 Pervolaraki, M., Mihailescu, C.N., Luculescu, C.R. et al. (2015). Picosecond ultrafast pulsed laser deposition of $SrTiO_3$. *Appl. Surf. Sci.* 336: 278–282.

43 Maddi, C., Donnet, C., Loir, A.-S. et al. (2015). High N-content a-C:N films elaborated by femtosecond PLD with plasma assistance. *Appl. Surf. Sci.* 332: 346–353.

44 Lippert, T., Montenegro, M.J., Döbeli, M. et al. (2007). Perovskite thin films deposited by pulsed laser ablation as model systems for electrochemical applications. *Prog. Solid State Chem.* 35: 221–231.

45 Voevodin, A.A., Donley, M.S., and Zabinski, J.S. (1997). Pulsed laser deposition of diamond - like carbon wear protective coatings. *Surf. Coat. Technol.* 92: 42–49.

46 Alkhawwam, A., Jama, C., Ji, V. et al. (2000). Sythèse d'un dépôt CN_x polycristallin par l'ablation laser dans l'atmosphere d'azote en post-décharge. *Galvano-Organo-Traitement de Surface* 69: 199–202. (in French).

47 Shen, J., Gai, Z., and Kirschner, J. (2004). Growth and magnetism of metallic thin films and multilayers by pulsed-laser deposition. *Surf. Sci. Rep.* 52: 163–218.

48 Berthe, L., Peyre, P., Scherpereel, X. et al. (1998). Laser-shock surface processing of materials. In: *Lasers in Surface Engineering* (ed. N.B. Dahotre), 465–504. Materials Park, OH: ASM International.

49 Metev, S. (1994). Process characteristics and film properties in pulsed laser deposition. In: *Pulsed Laser Deposition of Thin Films* (ed. D.B. Chrisey and G.K. Hubler), 255–264. New York, NY: Wiley.

50 Greer, J.A. (1994). Commercial scale - up of pulsed laser deposition. In: *Pulsed Laser Deposition of Thin Films* (ed. D.B. Chrisey and G.K. Hubler), 293–311. New York, NY: Wiley.

51 Chitica, N., Gyorgy, E., Lita, A. et al. (1997). Synthesis of tungsten carbide thin films by reactive pulsed laser deposition. *Thin Solid Films* 301 (1–2): 71–76.

52 Choy, K.L. (2003). Chemical vapor deposition of coatings. *Prog. Mater. Sci.* 48: 57–170.

53 Hocking, M.G., Vantasree, V., and Sidky, P.S. (1989). *Metallic and Ceramic Coatings*. Burnt Mill: Longman Scientific & Technical.

54 Pierson, H.O. (1999). *Handbook of Chemical Vapor Deposition*, 2e. Norwich, NY: Noyes Publications.

55 T. Belmonte, Dépôts chimiques à partir d'une phase gazeuse, [in:] *Techniques de l'Ingénieur,* M1660 V2, 2010, Saint – Denis, France. (in French).

56 Muhl, S. and Méndez, J.M. (1999). A review of the preparation of carbon nitride films. *Diam. Relat. Mater.* 8: 1809–1830.

57 Rossnagel, S.M. (1991). Glow discharge plasma and sources for etching and deposition. In: *Thin Film Processes II* (ed. J.L. Vossen and W. Kern), 11–77. Boston, MA: Academic Press.

58 Thurier, C. and Doppelt, P. (2008). Platinum OMCVD process and precursor chemistry. *Coord. Chem. Rev.* 252: 155–169.

59 Manin, M., Thollon, S., Emieux, F. et al. (2005). Deposition of MgO thin film by liquid pulsed injection MOCVD. *Surf. Coat. Technol.* 200: 1424–1429.

60 Jensen, K.F. and Kern, W. (1991). Thermal chemical vapor deposition. In: *Thin Film Processes II* (ed. J.L. Vossen and W. Kern), 283–268. Boston, MA: Academic Press.

61 Carlsson, J.O. (1994). Chemical vapor deposition. In: *Handbook of Deposition technologies for Films and Coatings* (ed. R.F. Bunshah), 374–433. Park Ridge, NJ: Noyes Publications.

62 Pierson, H.O. (1996). *Handbook of Refractory Carbides and Nitrides*. Westwood, NJ: Noyes Publications.

63 Lux, B., Colombier, C., Altena, H., and Stjernberg, K. (1986). Preparation of alumina coatings by chemical vapor deposition. *Thin Solid Films* 138: 49–64.

64 Pauleau, Y. (2004). *Procédés Chimiques de Dépôt à Partir d'une Phase Gazeuse*. Paris: Lavoisier. (in French).

65 Jansson, U., Carlsson, J.-O., Stridh, B. et al. (1989). Chemical vapor deposition of boron carbides I: phase and chemical composition. *Thin Solid Films* 172: 81–93.

66 https://en.wikipedia.org/wiki/Boundary_layer_thickness consulted on 10 May 2021.

67 Sarin, V.K. (1995). Systematic development of customized CVD coatings. *Surf. Coat. Technol.* 73: 23–33.

68 Reif, R. and Kern, W. (1991). Plasma - enhanced chemical vapor deposition. In: *Thin Film Processes II* (ed. J.L. Vossen and W. Kern), 525–559. Boston, MA: Academic Press.

69 Hössler, D. and Ernst, M. (2019). Optimization of a $TiSi_2$ formation based on PECVD Ti using DoE methodology. *Solid State Electron.* 158: 51–58.

70 Cheng, L.L., Liu, M.H., Wang, M.X. et al. (2012). Preparation of SnS films using solid sources deposited by the PECVD method with controllable film characters. *J. Alloys Compd.* 545: 122–129.

71 Boulos, M.I., Fauchais, P., and Pfender, E. (1994). *Thermal Plasma: Fundamentals and Applications*. New York, NY: Plenum Press.

72 Burakowski, T. and Wierzchoń, T. (1995). *Inżynieria Powierzchni Metali*. Warsaw: WNT (in Polish).

73 Kotte, L., Althues, H., Mäder, G. et al. (2013). Atmospheric pressure PECVD based on a linearly extended DC arc for adhesion promotion applications. *Surf. Coat. Technol.* 234: 8–13.

74 Wang, D., Liu, Y., Zhu, C. et al. (2018). Preparation of lanthanum zirconate films with a widely controllable La/Zr ratio by LCVD. *Ceram. Int.* 44: 10621–10627.

75 Ito, H., Kadokura, T.K., and Goto, T. (2010). Texture and orientation characteristics of α-Al_2O_3 films prepared by laser chemical vapor deposition using Nd:YAG laser. *J. Alloys Compd.* 489: 469–474.

76 Jensen, C.J. and Chiu, W.K.S. (2006). Open-air laser-induced chemical vapor deposition of silicon carbide coatings. *Surf. Coat. Technol.* 201: 2822–2828.

77 Sakaia, T., Kato, T., Katsui, H. et al. (2020). Preparation of Y-doped $BaZrO_3$ thin film electrolyte by laser chemical vapor deposition. *Mater. Today Commun.* 24: 101184.

78 Eden, J.G. (1991). Photochemical vapor deposition. In: *Thin Film Processes II* (ed. J.L. Vossen and W. Kern), 444–496. Boston, MA: Academic Press.

79 Kuech T.F. and Jensen K.F., OMVPE of compound semiconductors, *Thin Film Processes II*, J.L. Vossen and W. Kern, (eds), Academic Press, Boston, MA, 1991. pp. 369–433.

80 Devi, A. (2013). "Old Chemistries" for new application: perspectives for development of precursors for MOCVD and ALD applications. *Coord. Chem. Rev.* 257: 3332–3384.

81 Benali, A.M., Huntz, M., Andrieux, M.I., and Poissonnet, S. (2008). Internal stresses and stability of the tetragonal phase in zirconia thin layers deposited by OMCVD. *Appl. Surf. Sci.* 254: 5807–5813.

82 Suntola T. and Antson J., Method for producing compound thin films, US Patent 4 058 430, 15 November 1977.

83 Johnson, R.W., Hultqvist, A., and Bent, S.F. (2014). A brief review of atomic layer deposition: from fundamentals to applications. *Mater. Today* 17 (5): 236–246.

84 Kim, H., Lee, H.-B.-R., and Maeng, W.-J. (2009). Applications of atomic layer deposition to nanofabrication and emerging nanodevices. *Thin Solid Films* 517: 2563–2580.

85 Emslie, D.J.H., Chadha, P., and Price, J.S. (2013). Metal ALD and pulsed CVD: fundamental reactions and links with solution chemistry. *Coord. Chem. Rev.* 257: 3282–3296.

86 Hossain, M.A., Khoo, K.T., Cui, X. et al. (2020). Atomic layer deposition enabling higher efficiency solar cells: a review. *Nano Mater. Sci.* 2 (3): 204–226.

87 van Ommen, J.R. and Goulas, A. (2019). Atomic layer deposition on particulate materials. *Mater. Today Chem.* 14: 100183.

88 Pawłowski, L. (2008). *The Science and Engineering of Thermal Spray Coatings*, 2e. Chichester: Wiley.

89 Van Vlack, L.H. (1989). *Elements of Materials Science and Engineering*, 6e. Reading, MA: Addison-Wesley.

90 Meinhardt H., Meyer B, and Müller J., Verdüste Pulver zur Herstellung von Oberflächenschichten durch thermisches Spritzen, [in:] *TS90, DVS-Berichte Band 130*, DVS, Düsseldorf, Germany, 1990, 172–177.

91 Pawłowski L., Zacchino R., Dal Maschio R. *et al.* Structure-properties relationship in plasma sprayed chromium oxide coating, [in:] *TS93, DVS-Berichte Band 152*, DVS, Düsseldorf, Germany, 1993, 132–138.

92 Herman, H., Chen, Z.J., Huang, C.C., and Cohen, R. (1992). Mechanofused powders for thermal spray. *J. Therm. Spray Technol.* 1: 129–135.

93 Ageorges, H. and Fauchais, P. (2000). Plasma spraying of stainless-steel particles coated with an alumina shell. *Thin Solid Films* 370: 213–222.

94 Masters, K. (1985). *Spray Drying Handbook*, 4e. Burnt Mill, Essex: Longman.

95 Bartnik, Z., Białucki, P., Kozerski, S. et al. (1992). Improvements in manufacturing technology of wear resistant plasma sprayed Cr_2O_3 coatings. In: *Thermal Spray: International Advances in Coatings Technology* (ed. C.C. Berndt), 983–995. Materials Park, Ohio: ASM International.

96 Blatchford, M.T., Jones, M., Horlock, A.J. et al. (2001). Improvement in HVOF sprayed cermet coatings produced from SHS powders. In: *Thermal Spray 2001: New Surfaces for a New Millenium* (ed. C.C. Berndt, K.A. Khor, and E.F. Lugscheider), 221–229. Materials Park, Ohio: ASM International.

97 Pawłowski, L. (1990). *Thermal Spraying of YBa2Cu3Ox - Materials Aspects*, unpublished Research Report. Melbourne: Monash University.

98 Scott, K.T. and Woodhead, J.L. (1982). Gel processed powders for plasma spraying. *Thin Solid Films* 95 (3): 219–225.

99 Lugscheider, E., Loch, M., and Eschnauer, H. (1991). Coated powders-the status today. In: *2nd Plasma-Technik-Symposium*, vol. 2 (ed. S. Blum-Sandmeier, H. Eschanuer, P. Huber, and A.R. Nicoll), 339–351. Wohlen Switzerland: Plasma-Technik AG.

100 Winnicki, M., Kozerski, S., Małachowska, A. et al. (2021). Optimization of ceramic content in nickel–alumina composite coatings obtained by low pressure cold spraying. *Surf. Coat. Technol.* 405: 126732.

101 Wang, B., Ji, Z., Zimone, F.T. et al. (1997). A technique for sputter coating of ceramic reinforcement particles. *Surf. Coat. Technol.* 91: 64–68.

102 Dyshlovenko, S., Pawłowski, L., Pateyron, B. et al. (2006). Modelling of plasma particle interactions and coating growth for plasma spraying of hydroxyapatite. *Surf. Coat. Technol.* 200: 3757–3769.

103 Siegmann, S. and Abert, C. (2013). 100 years of thermal spray: about the inventor Max Ulrich Schoop. *Surf. Coat. Technol.* 220: 3–13.

104 Günther, H. and Schoop, M.U. (1917). *Das Schoopsche Metallspritzverfahren: Seine Entwicklung und Anwendung*. Stuttgart: Franckh Verlag. (in German).

105 Arboleda, J.A., Serna, C.M., Cadavid, E. et al. (2018). Effect of flame spray deposition parameters on the microstructure of Al_2O_3 - 13 %TiO_2 coatings applied onto 7075 aluminum alloy. *Mater. Res.* 21 (5): 1–12.

106 Li, Q., Song, P., Ji, Q. et al. (2019). Microstructure and wear performance of arc-sprayed Al/316L stainless-steel composite coating. *Surf. Coat. Technol.* 374: 189–200.

107 Drzeniek, H. and Steffens, H.D. (1987). Cored tube wires for arc and flame spraying. In: *Thermal Spray: Advances in Coatings Technology* (ed. D.L. Houck), 33–40. Materials Park, OH: ASM International.

108 Marantz, D.R. (1974). The basic principles of electric - arc spraying. In: *Science and Technology of Surface Coatings* (ed. B.N. Chapman and J.C. Anderson), 308–321. London: Academic Press.

109 Rudzki, G.J. (1983). *Surface Finishing Systems*. Teddington, Middlesex: Finishing Publications Ltd.

110 Chen, D., Daoud, H., Scherm, F. et al. (2020). Stainless steel powder produced by a novel arc spray process. *J. Mater. Res. Technol.* 9 (4): 8314–8322.

111 H. Gerdien and A. Lotz, Über eine Lichtquelle von sehr hoher Flächenhelligkeit, *Wissenschaftliche Veröffentlichungen uas der Siemens Konzern,* vol. 2, 1922, pp. 489–496. (in German).

112 Gage R.M., Nestor D.M., and Yenni Y.M., Collimated electric arc powder deposition process, USA Patent, no. 3 016 447, 12 December 1962.

113 Pawłowski, L., Gross, A., and McPherson, R. (1991). Microstructure of plasma sprayed $YBa_2Cu_3O_x$ high - temperature superconductor. *J. Mater. Sci.* 26: 3803–3808.

114 Dyshlovenko, S., Pateyron, B., Pawłowski, L., and Murano, D. (2004). Numerical simulation of hydroxyapatite powder behaviour in plasma jet. *Surf. Coat. Technol.* 179: 110–117.

115 Borgianni, C., Capitelli, M., Cramarosa, F. et al. (1969). The behavior of metal particles injected into an argon induction plasma. *Combust. Flame* 13: 181–194.

116 Bick H. and Jürgens W., Advanced high velocity thermal spraying of metallic and ceramic powders, [in:] Proceedings 10th *ITSC*, DVS Verlag, Düsseldorf, Germany 1983, 92–94.

117 Kreye, H., Fandrich, D., Müller, H.H., and Reiners, G. (1986). Microstructure and bond strength of WC - Co by hypersonic flame spraying (Jet - Kote process). In: *Advances in Thermal Spraying* (ed. N.F. Eaton), 121–128. New York, NY: Pergamon Press.

118 Vignesh, S., Shanmugam, K., Balasubramanian, V., and Sridhar, K. (2017). Identifying the optimal HVOF spray parameters to attain minimum porosity and maximum hardness in iron based amorphous metallic coatings. *Def. Technol.* 13: 101–110.

119 Swank, W.D., Fincke, J.R., Haggard, D.C. et al. (1994). HVOF particle flow field characteristics. In: *Thermal Spray Industrial Applications* (ed. C.C. Berndt and S. Sampath), 319–324. Materials Park, OH: ASM International.

120 Poorman R.M., Sargent H.B., and Lamprey H., Method and apparatus utilizing detonation waves for spraying and other purposes, USA Patent, no. 2 714 553, 2 August 1955.

121 Borosov Y.S., Astachov E.A., and Klimenko V.S., Detonation spraying: equipment, materials and applications, [in:] *Thermische Spritzkonferenz TS90*, DVS Berichte-Band 130, DVS Verlag, Düsseldorf, Germany, 1990, 26–32.

122 Schwarz E., Detonation gun coatings for nuclear and related industries, [in:] Proceedings 9th *ITSC*, 19–23 May 1980, Hague, Pays Bas, Nederlands Instituut voor Lastechniek, Hague, Netherlands, 1980, 91–101.

123 Tucker, R.C. Jr. (1982). Plasma and detonation gun deposition techniques and coating properties. In: *Deposition Technologies for Films and Coatings* (ed. R.F. Bunshah), 454–489. Park Ridge, NJ: Noyes Publications.

124 Smith, R.G. (1974). The basic principles of detonation spraying. In: *Science and Technology of Surface Coating* (ed. B.N. Chapman and J.C. Anderson), 262–270. London: Academic Press.

125 Allkimov A.P., Papyrin A.N., Kosarev V.F. et al. Gas-dynamic spray method for applying a coating, US patent 5, 302, 414, April 12 1994.

126 van Steenkiste, T. and Smith, J.R. (2004). Evaluation of coatings produced via kinetic and cold spraying processes. *J. Therm. Spray Technol.* 13 (2): 274–282.

127 Lauricelle, F. and Jaynes, S. (2003). Helium recovery: design consideration for cold spray systems. In: *Thermal Spray 2003: Advancing the Science Applying the Technology* (ed. C. Moreau and B. Marple), 113–116. Materials Park, OH: ASM International.

128 Prashar, G. and Vasudev, H. (2021). A comprehensive review on sustainable cold spray additive manufacturing: state of the art, challenges and future challenges. *J. Clean. Prod.* 310: 127606.

129 Vlcek, J., Gimeno, L., Huber, H., and Lugscheider, E. (2003). A systematical approach to material eligibility for the cold spray process. In: *Thermal Spray 2003: Advancing the Science Applying the Technology* (ed. C. Moreau and B. Marple), 37–44. Materials Park, OH: ASM International.

130 Małachowska, A., Winnicki, M., Konat, Ł. et al. (2017). Possibility of spraying of copper coatings on polyamide 6 with low pressure cold spray method. *Surf. Coat. Technol.* 318: 82–89.

131 Muehlberger E., A high-energy plasma coating process, [in:] Proceedings 7th *ITSC*, September 10–14, 1973, London, England, The Welding Institute, Abington Hall, Cambridge CB1 6AL, Paper 58.

132 Wilms V., LPPS deposit coatings for protection against hot gas corrosion - a first evaluation, [in:] Proceedings 9th *ITSC*, 19–23 May 1980, Hague, Pays Bas, Nederlands Instituut voor Lastechniek, Hague, The Netherlands, 1980, 317–324.

133 Braguier M., Bejat J., Tueta R. et al. Improvements of plasma spraying process for hybrid microelectronics, [in:] Proceedings Conference Hybrid Microelectronics, September 25–27, 1973, Canterbury, England, 15–37.

134 Pawłowski L., The vacuum plasma sprayed copper and tantalum coatings, [in:] *Proceedings 9th ITSC*, May, 19–23, 1980, Hague, The Netherlands, Nederlands Instituut voor Lastechniek, Hague, The Netherlands, 1980, 299–305.

135 von Niessen, K. and Gindrat, M. (2011). Plasma Spray-PVD: a new thermal spray process to deposit out of the vapor phase. *J. Therm. Spray Technol.* 20 (4): 736–743.
136 Mauer, G., Hospach, A., Zotov, N., and Vaßen, R. (2013). Process conditions and microstructures of ceramic coatings by gas phase deposition based on plasma spraying. *J. Therm. Spray Technol.* 22 (2–3): 83–89.
137 Góral, M., Kotowski, S., Nowotnik, A. et al. (2013). PS-PVD deposition of thermal barrier coatings. *Surf. Coat. Technol.* 237: 51–55.
138 Mauer, G., Jarligo, M.O., Rezanka, S. et al. (2015). Novel opportunities for thermal spray by PS-PVD. *Surf. Coat. Technol.* 268: 52–57.
139 Yang, J., Cheng, Z., Zhong, X. et al. (2021). Deposition behavior of PS-PVD yttria partially stabilized zirconia coatings. *J. Therm. Spray Technol.* 30: 1136–1147.
140 Li, X., Deng, C., Niu, S. et al. (2021). Effect of calcination temperature on the microstructure, composition and properties of nanometer agglomerated 8YSZ powders for plasma spray-physical vapor deposition (PS-PVD) and coatings thereof. *Ceram. Int.* 47 (12): 16632–16640.
141 Mauer, G. and Vaßen, R. (2012). Plasma Spray – PVD: plasma characteristics and impact on coating properties. *J. Phys. Conf. Ser.* 406: 012005.
142 He, W., Mauer, G., Gindrat, M. et al. (2017). Investigations on the nature of ceramic deposits in Plasma Spray–Physical Vapor Deposition. *J. Therm. Spray Technol.* 26: 83–92.
143 Mauer, G. (2014). Plasma characteristics and plasma-feedstock interaction under PS-PVD process conditions. *Plasma Chem. Plasma Process.* 34: 1171–1186.
144 Moreno, R. and Bannier, E. (2015). Feedstock suspensions and solutions. In: *Future Development of Thermal Spray Coatings* (ed. N. Espallargas), 51–80. Cambridge, CB22 3HJ: Woodhead Publications.
145 Weber, S.B., Lein, H.L., Grande, T., and Einarsrud, M.A. (2013). Influence of the precursor solution chemistry on the deposition of thick coatings by spray pyrolysis. *Surf. Coat. Technol.* 221: 53–58.
146 Oja, A., Mere, M., Krunks, R. et al. (2006). Structural and electrical characterization of TiO_2 films grown by spray pyrolysis. *Thin Solid Films* 515: 674–677.
147 Guo, X.Z., Ravi, B.G., Devi, P.S. et al. (2005). Synthesis of yttrium iron garnet (YIG) by citrate-nitrate gel combustion and precursor plasma spray process. *J. Magn. Magn. Mater.* 295: 145–154.
148 Govindarajan, S., Dusan, R.O., and Joshi, S.V. (2011). In situ particle generation and splat formation during solution precursor plasma spraying of yttria-stabilized zirconia coatings. *J. Am. Ceram. Soc.* 94 (12): 4191–4199.
149 Pawłowski L., Application of solution precursor spray techniques to obtain ceramic films and coatings, Future Development of Thermal Spray Coatings, N. Espallargas (ed.), Woodhead Publications., Cambridge, 2015, 123–141.
150 Chen, D., Jordan, E.H., and Gell, M. (2008). Effect of solution concentration on splat formation and coating microstructure using the solution precursor plasma spray process. *Surf. Coat. Technol.* 202: 2132–2138.
151 Shaw, D.J. (1980). *Introduction to Colloid and Surface Chemistry*. London: Butterworths.
152 Pawłowski, L. (2009). Suspension and solution thermal spray coatings. *Surf. Coat. Technol.* 203: 2807–2829.

153 Sokołowski, P. and Pawłowski, L. (2016). Review of recent studies on suspension plasma sprayed ZrO_2 coatings. In: *Advances in Materials Science Research*, vol. *26* (ed. M.C. Wythers), 137–180. New york, NY: Nova.

154 Tomaszek, R., Pawłowski, L., Gengembre, L. et al. (2006). Microstructural characterization of plasma sprayed TiO_2 functional coating with gradient of crystal grain size. *Surf. Coat. Technol.* 201: 45–56.

155 Waldbillig D., Tang Z., Burgess A., and Kesler O., Effect of substrate and cathode parameters on the properties of suspension plasma sprayed solid oxide fuel cell electrolytes [in:] *ITSC 2008 Conference Proceedings*, CD-Rom, ISBN 978-3-87155-979-2, DVS-Verlag, Düsseldorf, Germany, 2008.

156 Toma F.-L., Berger L.-M., Naumann T., and Langner S., Characterization of ceramic nanostructured coatings prepared by thermal spraying of suspensions [in:] *ITSC 2008 Conference Proceedings, ibid.*

157 Tomaszek, R., Pawłowski, L., Gengembre, L. et al. (2007). Microstructure of suspension plasma sprayed multilayer coatings of hydroxyapatite and titanium oxide. *Surf. Coat. Technol.* 201: 7432–7440.

158 Pawłowski, L. (2008). Finely grained nanometric and submicrometric coatings by thermal spraying: a review. *Surf. Coat. Technol.* 202 (18): 4318–4328.

159 Jaworski, R., Pierlot, C., Pawłowski, L.L. et al. (2009). Design of the synthesis of fine HA powder for suspension plasma spraying. *Surf. Coat. Technol.* 203 (15): 2092–2097.

160 Wang, Y., Legoux, J.-G., Neagu, R. et al. (2012). Suspension plasma spray and performance characterization of half-cells with NiO/YSZ anode and YSZ electrolyte. *J. Therm. Spray Technol.* 21 (1): 7–15.

161 Tomar, M.S. and Garcia, F.J. (1981). Spray pyrolysis in solar cells and gas sensors. *Prog. Crystal Growth Charact.* 4: 221–248.

162 Perednis, D. and Gauckler, L.J. (2005). Thin film deposition using spray pyrolysis. *J. Electroceram.* 14: 103–111.

163 Patil, P.S. (1999). Versatility of chemical spray pyrolysis technique. *Mater. Chem. Phys.* 59: 185–198.

164 Yu, J. and Kim, D. (2013). The preparation of nano size nickel oxide powder by spray pyrolysis process. *Powder Technol.* 235: 1030–1037.

165 Ukoba, K.O., Eloka-Eboka, A.C., and Inambao, F.L. (2018). Review of nanostructured NiO thin film deposition using the spray pyrolysis technique. *Renew. Sust. Energ. Rev.* 82: 2900–2915.

166 Sarıtas, S., Sakar, B.C., Kundakci, M. et al. (2021). Analysis of magnesium ferrite and nickel doped magnesium ferrite thin films grown by Spray Pyrolysis. *Mater. Today: Proc.* 46 (16): 6920–6923.

167 Gitzhofer F., Bouyer E., and Boulos M.I., Suspension Plasma Spraying, US Patent 5 609 921, 3 November 1997.

168 Killinger, A. (2015). Status and future trends in suspension spray techniques. In: *Future Development of Thermal Spray Coatings* (ed. N. Espallargas), 81–122. Cambridge, CB22 3HJ: Woodhead Publications.

169 Fan, W. and Bai, Y. (2016). Review of suspension and solution precursor plasma sprayed thermal barrier coatings. *Ceram. Int.* 42: 14299–14312.

170 Hurevich, V., Smurov, I., and Pawłowski, L. (2002). Theoretical study of the powder behavior of porous particles in a flame during plasma spraying. *Surf. Coat. Technol.* 151–152: 370–376.

171 Sokołowski, P., Kozerski, S., Pawłowski, L., and Ambroziak, A. (2014). The key process parameters influencing formation of columnar microstructure in suspension plasma sprayed zirconia coatings. *Surf. Coat. Technol.* 260: 97–106.

172 Sokołowski, P., Nylen, P., Musalek, R. et al. (2017). The microstructural studies of suspension plasma sprayed zirconia coatings with the use of high-energy plasma torches. *Surf. Coat. Technol.* 318: 250–261.

173 N. Curry, *private information*, July 2021.

174 Karthikeyan, J., Berndt, C.C., Tikkanen, J. et al. (1997). Preparation of nanophase materials by thermal spray processing of liquid precursors. *Nanostruct. Mater.* 9 (1–8): 137–140.

175 Padture, N.P., Schlichting, K.W., Bhatia, T. et al. (2001). Towards durable thermal barrier coatings with novel microstructures deposited by solution precursor plasma spray. *Acta Mater.* 49: 2251–2257.

176 Tobon Valencia, V., Pawłowski, L., Lecomte-Nana, G. et al. (2020). Preliminary analysis of physical and chemical phenomena occurring in droplet at solution precursor plasma spraying of zirconia coatings. *Surf. Coat. Technol.* 397: 126059.

177 Basu, S. and Cetegen, B.M. (2007). Modeling of thermos-physical processes in liquid ceramic precursor droplets injected into plasma jet. *Int. J. Heat Mass Transf.* 50: 3278–3290.

178 Liu, Z.M., Cui, S.L., Luo, Z. et al. (2016). Plasma arc welding: process variants and its recent developments of sensing, controlling and modeling. *J. Manuf. Process.* 23: 315–327.

179 Allnat N.J., Bell G.R., and Griffith L.J., Developments in plasma arc surfacing, [in:] Proceedings 9th *ITSC*, 19–23 May 1980, Hague, The Netherlands, Nederlands Instituut voor Lastechniek, Hague, The Netherlands, 1980, 218–23.

180 Zikin, A., Hussainova, C., Katsich, E.B., and Tomastik, C. (2012). Advanced chromium carbide-based hardfacings. *Surf. Coat. Technol.* 206: 4270–4278.

181 Zikin, A., Badisch, I., Hussainova, C.T., and Danninger, H. (2013). Characterisation of TiC-NiMo reinforced Ni-based hardfacing. *Surf. Coat. Technol.* 236: 36–44.

182 Oukach, S., Pateyron, B., and Pawłowski, L. (2019). Physical and chemical phenomena occurring between solid ceramics and liquid metals and alloys at laser and plasma composite coatings formation: a review. *Surf. Sci. Rep.* 74 (3): 213–241.

183 Demian, A., Denoirjean, P.D., Pawłowski, L., and El Ouardi, R. (2016). Microstructural investigations of NiCrAlY + Y_2O_3 stabilized ZrO_2 cermet coatings deposited by plasma transferred arc (PTA). *Surf. Coat. Technol.* 300: 104–109.

184 H. Hallen and C. Herrström, PTA - Beschichtung mit Nickelalumidpulvern, [in:] *Thermische Spritzkonferenz TS 93*, DVS - Berichte Band 152, DVS Verlag, Düsseldorf, Germany, 1993, pp. 167–170. (in German).

185 Marantz D.R. and Richardson S.J., Plasma - transferred arc hard - facing an improved process for surfacing, [in:] Proceedings 9th *ITSC,* 19–23 May 1980, Hague, Netherlands, Nederlands Instituut voor Lastechniek, Hague, Netherlands, 1980, pp. 207–210.

186 A.E. Siegman, *Lasers*, University Science Books, Sausalito, CA, 1986.

187 Brandt, M. (ed.) (2017). *Laser Additive Manufacturing: Materials, Design, Technologies, and Applications*. Cambridge, Duxford, CB22 4QH: Woodhead.

188 Pawłowski, L. (1999). Thick laser coatings: a review. *J. Therm. Spray Technol.* 5: 279–295.

189 Kindler, H., Volz, R., and Huonker, M. (1996). Ein neues Verfahren zum Laserbeschichten. In: *6th Eur. Conf. Laser Treatment of Materials, ECLAT'96*, vol. 1 (ed. F. Dausinger, H.W. Bergman, and J. Sigel), 447–454. Steuttgart: AWT. (in German).

190 Pei, Y.T., Oiyang, J.H., and Lei, T.C. (1996). Laser cladding of ZrO_2 - (Ni - alloy) composite coating. *Surf. Coat. Technol.* 81: 131–135.

191 Riveiro, A., Mejías, F.L., del Val, J. et al. (2014). Laser cladding of aluminium on AISI 304 stainless steel with high-power diode lasers. *Surf. Coat. Technol.* 253: 214–220.

192 Nowottny, S., Shen, J., and Dausinger, F. (1991). Verschleiβschutz durch Auftragschweiβn von Stellite 21 mit CO_2 Laser. *Laser und Optoelektronik* 21 (6): 50–54. (in German).

193 Arias-González, F., del Val, J., Comesana, R. et al. (2016). Fiber laser cladding of nickel-based alloy on cast iron. *Appl. Surf. Sci.* 374: 197–205.

194 Deschuyteneer, F., Petit, M.G., and Cambier, F. (2015). Processing and characterization of laser clad NiCrBSi/WC composite coatings – influence of microstructure on hardness and wear. *Surf. Coat. Technol.* 283: 162–171.

195 Siddiqui, A.A. and Dubey, A.K. (2021). Recent trends in laser cladding and surface alloying. *Opt. Laser Technol.* 134: 106619.

196 Chi, Y., Gu, G., Yu, H., and Chen, C. (2018). Laser surface alloying on aluminum and its alloys: a review. *Opt. Lasers Eng.* 100: 23–37.

197 Carvalho, P.A. and Vilar, R. (1997). Laser alloying of zinc with aluminum: solidification structures. *Surf. Coat. Technol.* 91: 158–166.

198 Vaziri, S.A., Shahverdi, H.R., Torkamany, M.J., and Shabestari, S.G. (2009). Effect of laser parameters on properties of surface-alloyed Al substrate with Ni. *Opt. Lasers Eng.* 47: 971–975.

199 Galun, R., Weisheit, A., and Mordike, B.L. (1997). Surface alloying of magnesium base alloys with high power CO_2 laser. *Proceedings of the SPIE - The International Society for Optical Engineering* 3092: 744–747.

200 Schaefer, R.J., Tucker, T.R., and Ayers, J.D. (1980). Laser surface melting with carbide particle injection. In: *Laser and Electron Beam Processing of Materials* (ed. C.W. White and P.S. Percy), 754–759. Academic Press.

201 Ayers, J.D., Schaefer, R.J., and Robey, W.P. (1980). A laser processing technique for improving the wear resistance of metals. *J. Met.* 33 (8): 19–23.

202 Chang, C., Verdi, D., Garrido, M.A., and Ruiz-Hervias, J. (2016). Micro-scale mechanical characterization of Inconel cermet coatings deposited by laser cladding. *Bol. SECV* 55 (4): 136–142.

203 Abioye, T.E., Farayibi, P.K., McCartney, D.G., and Clare, A.T. (2016). Effect of carbide dissolution on the corrosion performance of tungsten carbide reinforced Inconel 625 wire laser coating. *J. Mater. Process. Technol.* 231: 89–99.

204 Kloosterman, A.B. and De Hosson, J.T.M. (1997). Ceramic particle injection during laser treatment of titanium. In: *Surface Treatment'97* (ed. M.H. Alibadie and C.A. Berbbia), 286–294. Southampton: Computational Mechanics Publications.

205 Gebhardt, A. (1996). *Rapid Prototyping*. München: Hanser Verlag.

206 Hu, Y. and Cong, W. (2018). A review on laser deposition-additive manufacturing of ceramics and ceramic reinforced metal matrix composites. *Ceram. Int.* 44: 20599–20612.

207 Monisha, K., Shariff, S.M., Manonmani, J., and Senthilselvan, J. (2021). High power diode laser alloying of TiB2 and amorphous-boron with titanium: phase, microstructure and hardness analyses. *Optik - International Journal for Light and Electron Optics* 239: 166758.

208 Draper, C.W., Preece, C.M., Jacobson, D.C. et al. (1980). Laser alloying of deposited metal films on nickel. In: *Laser and Electron Beam Processing of Materials* (ed. C.W. White and P.S. Peercy), 721–727. London: Academic Press.

209 Pawłowski, L., Rapinel, H., Tourenne, F., and Jeandin, M. (1999). Laser treatment hydroxyapatite plasma deposits. *Galvano Organo-Traitement de Surface* 676: 433–437.

210 Schnick, T., Tondu, S., Peyre, P. et al. (1999). Laser shock processing of Al-SiC composite coatings. *J. Therm. Spray Technol.* 8 (2): 296–300.

211 Khor, K.A. and Cheang, P. (1994). Laser post-treatment of thermally sprayed hydroxyapatite coatings. In: *Thermal Spray Industrial Applications* (ed. C.C. Berndt and S. Sampath), 153–157. Materials Park, OH: ASM International.

212 Ranz X., Pawłowski L., Sabatier L. et al. Phases transformation in laser treated hydroxyapatite coatings, Proceedings 15th *ITSC*, C. Coddet (ed.), ASM International, Materials Park, OH, 1998, p 1343-9.

213 Ayers, J.D. and Schaefer, R.J. (1979). Consolidation of plasma sprayed coatings by laser remelting. *Proceedings of SPIE-The International Society for Optical Engineering* 178: 57–64.

214 Yilbas, B.S., Arif, A.F.M., and Gondal, M.A. (2005). HVOF coating and laser treatment: three-point bending tests. *J. Mater. Process. Technol.* 164–165: 954–957.

215 Kobylanska Szkaradek, K. (2010). Laser melted ZrO_2–Y_2O_3 thermal barrier obtained by plasma spraying method. *J. Alloys Compd.* 505: 516–522.

216 Tondu, S., Schnick, T., Pawłowski, L. et al. (2000). Laser glazing of FeCr-TiC composite coatings. *Surf. Coat. Technol.* 123: 247–251.

217 Pawłowski, L. (1996). Technology of thermally sprayed anilox rolls: state of art, problems, and perspectives. *J. Therm. Spray Technol.* 5 (3): 317–334.

3

Nucleation, Growth, and Microstructure of Physical Deposits

3.1 Introduction

The species are generated by a source and get in contact with the medium before ending their trajectory on the surface of a substrate and beginning to form a film or coating. Starting from the first moment of contact with the substrate, the phenomena related to the adhesion of deposits take place.

The atomistic films are composed of atoms and molecules as the species, which condense on the substrate. The initial layer is formed by adhering atomes, called ***adatomes***[1], which enter in direct contact with the surface and may move before finding their final place [1]. The more energetic ones may bounce off the substrate's surface before losing their energy. Then, these species get attached to the surface forming a nucleation center. Contrarily, the atoms or molecules having low kinetic energy remain in a place at the first contact. The crystal grains that grow from the energetic species may become columnar or equiaxed depending on the deposition parameters. The microstructure of films formed by low-energy species is most frequently amorphous. Post-deposition treatment may modify the films' microstructure.

The granular process uses greater species, which have sizes ranging from tenths of nanometers in the processes as *SPS* and *SPPS* up to hundreds of micrometers in other processes. The particles are cold or warm in *CGSM* deposition method and get molten or having very high temperature in other thermal spray techniques. The cold particles may deform plastically on impact with the substrate. The molten ones splash on the substrate, solidify, and cool down rapidly. The microstructure of a sprayed coatings deposit is strongly influenced by the velocity and temperature of particles at impact. The unmelted particles arriving at the substrate lead to generation of pores in the sprayed deposits.

Finally, the microstructure of bulk deposits formed by flame, plasma, or laser heating of a substrate is strongly influenced by the solidification rate of the bath of liquid material. The microstructure of composites formed by ceramic particles injected to the bath of liquid metal or alloy is influenced by the wettability of ceramic surfaces by the liquid.

1 In the following chapter the *adatomes* can also be the molecules of the compound forming a film.

Physical Deposition Methods for Films and Coatings, First Edition. Lech Pawłowski.

The microstructures of the deposits are often designed to reach desired property. A well-known example is columnar microstructure of deposits used as a part of thermal barrier coatings (*TBC*). Such microstructure improves thermal shock resistance of the thermal barrier enabling longer service.

3.2 Atomistic Deposition Methods

3.2.1 Introduction

The factors influencing the nucleation and growth of a film in an atomistic method, shown in Fig. I.1, are related to the contact of species with the medium, which may influence their energy. The chemically reactive medium may modify also their chemical composition.

The phenomena, which occur at the contact of the species with a substrate, namely the condensation of vapors followed by the nucleation leading to deposit growth, depend on many variables, which are associated with: (i) deposition process; (ii) materials of film; and (iii) substrate material. As far as the deposition process is concerned, the variables are as follows:

- kinetic energy of the species in the moment of contact with the substrate's surface;
- growth rate of the film depending on the flux, *R*, of arriving species, which in turn depends on: (i) productivity of the source; (ii) geometry of the substrate with regards to the source; and (iii) pressure of the gas in deposition chamber;
- flow of the contamination species resulting from their partial pressure.

The variables, for the processes analyzed in this work, have the values collected in Table 3.1.

3.2.2 Phenomena at Contact and Nucleation

The elementary processes on the surface of a substrate being in contact with the arriving vapors are discussed in the following sections supposing that the species are the atoms, which become *adatoms* in contact with substrate.

The *PVD* processes described include *EB PVD*, *PLD*, sputtering, or *PS PVD* among others. In the process of sputtering, as well as *PLD*, the species have high kinetic energy at impact with substrate. Moreover, about 50% of species at *PLD* technique are ionized [7]. Their energy is reduced by the re-evaporation. The presence of ionized species may lead also to the strong interaction between the atoms of film and that of the substrate if the arriving ions form a chemical bonding. In that case, the arriving atoms adhere by ***chemisorption*** [1].

The flux, *R*, of *adatomes* arriving on the substrate's surface at thermal *CVD* deposition depends on two main phenomena [8, 9]:

- mass transport control resulting from diffusion of precursors through boundary layer above substrate's surface (see Fig. 2.15);
- surface reaction control resulting from chemical reaction of precursors occurring on this surface, being activated by the presence of ionized species in the technique *PE CVD*.

Table 3.1 Atomistic processes discussed in the chapter and variables determining the microstructure with their typical values [2–6].

Process	Kinetic energy of species, eV	Film's growth rate,nm/s	Species	Pressures at deposition, p_d, hPa	Substrate temperatures at deposition, T_d, K	Typical materials deposited
PVD – thermal evaporation	0.1–0.5	1–20 (higher for *EB PVD*)	Molecules, atoms, ions	10^{-9}–10^{-3} (higher for *ARE*)	320–600	Metals, ceramics
PVD – sputtering	1–100	1–10	Atoms, ions, molecules	10^{-3}–10^{-1}	320–600	All materials
IP and *IBAD*	1–100	1–20	Ions, atoms	10^{-6}–10^{-3}	320–600	Metals, ceramics
PS PVD	—	0.1–0.2/ torch pass	Molecules	2–10	1200–1500	Mainly *YSZ*
PLD	Up to 1000	1–20	Clusters, molecules, atoms, ions	10^{-6}–10	50–300	All materials
CVD thermal	0.1	1–5	Molecules, atoms	10^{-5}–100	1000–1500	All materials, frequently ceramics
PE CVD	1–500	1–10	Atoms, ions	10^{-4}–1	600–800	
ALD	0.1	0.1/cycle	Molecules, atoms, radicals	10^{-1}–10	500	

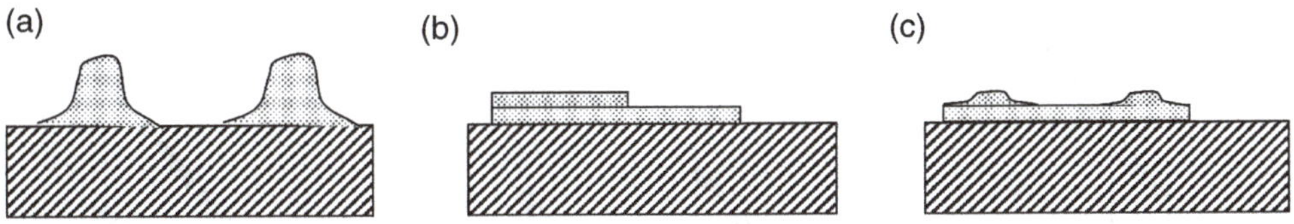

Figure 3.1 Types of films growth: (a) *3D*; (b) *2D*; and (c) mixed being partly in *2D* and partly in *3D*. *Source:* Greene [10]/John Wiley & Sons.

The above phenomena depend on the temperature in the reactor and on the total pressure of gases. The surface reaction regime is more interesting as it can be realized in lower temperatures than mass transport one. This regime can be also attained by an increase of gas low velocity and an appropriate choice of precursors [9].

The nucleated *adatoms* start to grow up forming the film. Three types of growth, shown in Fig. 3.1, are possible [7, 10]: (i) *3D* called also ***Volmer–Weber*** or island growth; (ii) *2D* called also ***Frank van der Merwe*** or layer-by-layer growth; and (iii) mixed being partly in *3D* and partly in *2D*, known as ***Stranski–Kastanov*** growth.

3.2.2.1 Growth in Three Dimensions

This type of nucleation generally concerns the deposited materials being different chemically from that of the substrate. The *adatoms,* being on the surface (see Fig. 3.2), can be adsorbed by substrates physically. Physical adsorption occurs thanks to the force of attraction between neutral species called *van der Waals* force, which decreases strongly with the distance r between interacting species, being proportional to r^{-7}. The physical adsorption mechanism results in a very low adhesion of the film to the substrate

The growth in *3D* includes the following processes shown schematically in Fig. 3.2:

- initial contact of a species (supposed atoms) with substrate's surface;
- loss of kinetic energy by re-evaporation;
- diffusion of *adatoms* toward a nucleation center, typically formed by structural defects on the surface, and creation of a cluster;
- growth of a cluster to form an island;
- coalescence of islands to cover the entire surface of the substrate.

The important step is the growth of a cluster to form an island. Greene [10] developed the thermodynamic analysis based on a model of a condensed liquid droplet, which enables to find the critical radius of a cluster, r^*. The cluster of a radius greater, i.e. $r > r^*$, can lower their volume total free energy ΔG being equal to:

$$\Delta G(r) = \frac{4}{3}\pi r^3 \Delta G_v + 4\pi r^2 \gamma \tag{3.1}$$

On the contrary, the clusters having radius lower, i.e. $r < r^*$, will disappear. The critical free energy corresponding to critical radius is equal to $\Delta G^* = \Delta G\,(r^*)$.

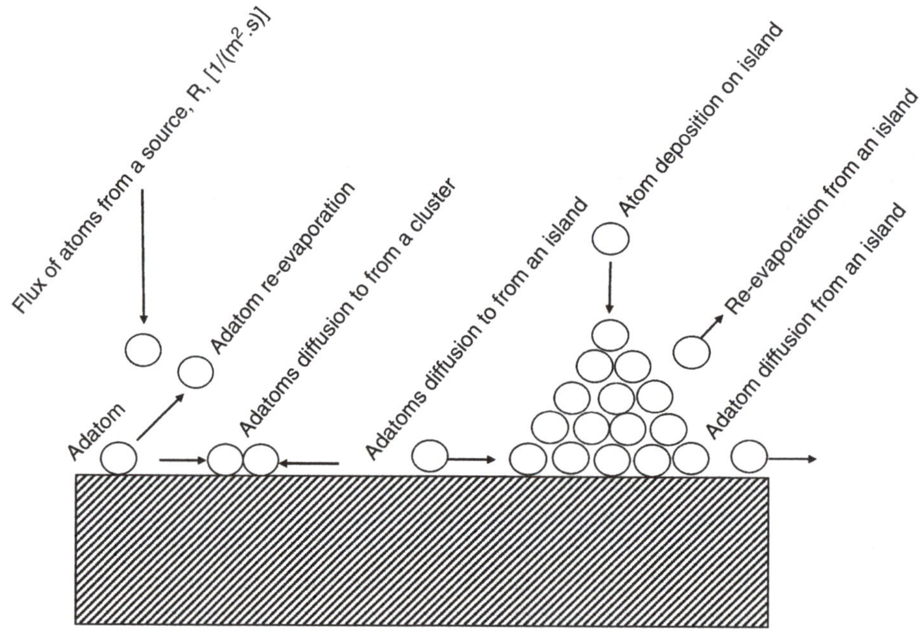

Figure 3.2 Some processes at the beginning of the condensation of a vapor on a substrate leading to the growth in *3D*. *Source:* Horwitz and Sprague [7]/John Wiley & Sons.

Some deposition techniques include ion bombardment with the ions attracted by the polarized substrate (*IP*) or by *ions gun (IBAD)*. The sketch of these techniques is shown in Fig 2.9. The following processes may occur additionally on the ion-bombarded surface:

- desorption of contaminated species;
- implantation of ions bombarding the substrate (see Fig. 2.7);
- sputtering of the substrate atoms (see Fig. 2.7);
- creation of structural defects, which can become nucleation sites.

The condensation of *adatoms* occurs when the pressure of the vapors near the surface is equal to or greater than the equilibrium pressure, i.e. when the vapors are saturated. Then the *adatoms* form clusters, which may grow up and, having radius greater than the critical one, equal to r^*, and become islands as shown in Eq. (3.1). The value of r^* depends also on the deposition temperature, T_d, and decreases with it. At low temperatures, the number of nucleation centers increases and, therefore, the film is composed of fine grains. If T_d is low enough, the microstructure of a film becomes amorphous. The high values of T_d promote the mobility of *adatoms* on the surface. At given T_d, the critical radius decreases when the flux of species, R, increases. The density of nucleation centers, which determines the growth rate of a deposit, is proportional to the [10]:

- number of atoms that have arrived at the nucleation centers having a critical radius (or greater) and, therefore, to the flux of species, R, and on their mobility;
- saturation of the vapors.

The increase of T_d decreases the vapor saturation pressure and, in the same time, increases the mobility of *adatoms* toward the nucleation centers. The nucleation results from these two actions, acting in the opposite directions. For example, if T_d becomes too great the film is no longer formed. The flux of species, R, depends on the deposition technique. This flux should be greater than the flux of impurities. The latter depends on the gas pressure at the deposition [3]. The technique with great flux of species is *PLD*. The estimation of flux of graphite at CN_x film deposition using this technique was estimated to be 1.5×10^{20} $1/(m^2.s)$ [11].

Finally, the clusters grow up to become the islands. Their growth is realized by capture of *adatoms*, other clusters, and by coalescence with other islands. The force that drives coalescence is the reduction in surface energy (see Eq. (3.1)). This is why the islands are often rounded. The islands are initially perfect crystals, their coalescence introduces structural defects such as linear and volume dislocations. The space between the islands remains in the holes and channels, which may be eventually filled by the secondary nucleation. Eventually, a film reaches its final thickness by growing islands in the direction perpendicular to the substrate creating columnar microstructure shown in Fig. 3.3. The figure shows also that the direction of grains growth depends also on the substrate's surface roughness. The average grain height is therefore comparable to the final thickness value of the deposit.

3.2.2.2 Nucleation and Growth in Two Dimensions

This type of nucleation corresponds to a film and a substrate being of quite similar pair of materials. An obvious case would be that a film is deposited onto substrate of the same material. The examples of films being different from the substrate are Cd film on a polycrystalline W or Pb film on monocrystalline Cu substrate [10]. The *chemisorption* may

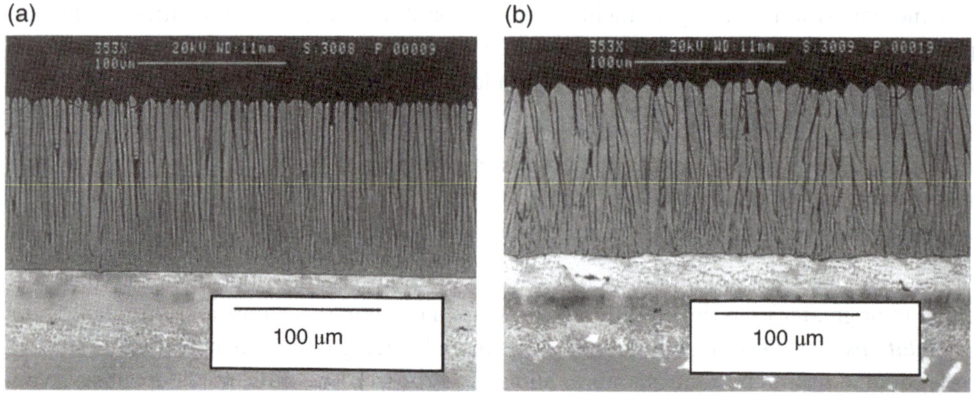

Figure 3.3 SEM micrograph of a ZrO_2 + 7YSZ deposit obtained by *EB PVD* on the substrate: (a) with the polished surface, (b) activated by sandblasting. *Source:* From Rigney *et al.* [12]/with permission of Springer Nature.

occur also in *2D* growth when the atoms of the deposit and the substrate have a high chemical affinity. If so, the atoms are not mobile on the surface and the number of nucleation sites is equal to the number of surface atoms. This type of adsorption gives good film adhesion to the substrate.

The case of two-dimensional nucleation occurs when the clusters grow up to become a monoatomic layer rather than growing in three dimensions. That is, the growth of clusters and their coalescence to become an island are the same as in the case of three-dimensional nucleation, but the thickness of the clusters and the islands does not exceed the atom size. Likewise, the influence of factors related to the deposition technique, such as T_d and R, on the nucleation rate is the same as in the case of three-dimensional growth discussed previously. After deposition of a monoatomic layer the process of growth, nucleation, and coalescence of the second one begins.

3.2.2.3 Nucleation and Growth in Two Dimensions and in Three Dimensions

The nucleation and growth of *Stranski–Kastanov* begins with nucleation in a two-dimension pattern and then the growth is in three dimensions. This mechanism may result from the modification of the first monoatomic layer, nucleated in *2D*, either by a chemical reaction of film with the substrate or by formation of mechanical stresses. The flow of species subsequently arriving on the substrate encounters a different surface and growth is in *3D*. The example of mixed growth is Ge film on Si substrate [7] or Cu film on Ag substrate [10].

3.2.3 Interface Between Film and Substrate

The interface between deposit and substrate determines the adhesion of an atomistic deposit. In general, one can distinguish following Mattox [1]:

- ***abrupt interface*** occurs if there is an abrupt change of chemical composition from that of film to that of substrate over the distance of 0.2–0.5 nm. This type of interface is characterized by very low adhesion because *adatoms* interact with those of the substrate by small *van der Waals* forces. A variation of this type of interface with improved adhesion occurs when coating an activated surface (by one of the surface activation techniques described in Section 1.5).

- ***diffusion interface*** occurs if there is a gradual change in chemical composition in the interface region without formation of a chemical compound. This mechanism takes place in the case of the solubility of the atoms of film in the substrate and inversely. Moreover, the temperature of deposition, T_d, should be high enough and its time, t_d, has to be long enough to render diffusion possible. This type of interface is characterized by gradual changes of mechanical and thermophysical properties which leads to reduction of thermal and mechanical stresses. Consequently, such interface leads to high adhesion of films after deposition and in service.
- ***compound interface*** occurs when there is a formation of a chemical compound between the deposit and the substrate. This type of interface is initiated by diffusion and can be accompanied by formation of a grain having density which may introduce mechanical stresses at the interface and lead to the generation of cracks. Generally, the formation of a compound at the interface promotes adhesion of the deposit to the substrate.
- ***pseudo-diffusion interface*** occurs if a gradual change in chemical composition from substrate to the film is generated artificially at deposition by changing the chemical composition of vapors arriving on the substrate's surface.

3.2.4 Microstructure of Films Obtained Using Different Deposition Techniques

The nucleation phase of the deposit ends when the influences of the substrate and of the interface on the growth stop to play an important role. Generally, it occurs beginning from the film thickness of a few tenths of nanometers. Then, the growth and microstructure of the deposit start to depend mainly on the deposition process and of its parameters. This phase of a film growth determines its microstructure of the deposits and, consequently, its physicochemical properties.

The frequent microstructure of atomistic films is a columnar one with the grains having shape shown in Fig. 3.3. An important factor at deposition is the flux of species, R. The deposition techniques producing a large amount of the vapors resulting in great value of R enables obtaining the microstructure including homogeneously condensed grains. Such grains are loosely deposited onto substrate's surface and do not adhere well. In particular, the *PLD* technique [11] or *PVD* with *arc evaporation* [13] may have a high degree of saturation of vapors. In particular, the *PLD* using pulsed CO_2 laser. The light from this laser is absorbed, in general, more deeply than that of other lasers and the volume of vapors is greater.

Grains having shape of columns, shown in Fig. 3.3, have been discussed in numerous microstructural studies of deposits. Their generation depends on the following parameters of the principal *PVD* deposits:

- deposition temperature, T_d, in thermal evaporation of thick layers of metals and ceramics [14, 15] and on the substrate preparation during *EBPVD* deposition of thick films of ZrO_2 stabilized with Y_2O_3 and CeO_2 [16];

- T_d and partial pressure of Ar in the film obtained by *PVD* sputtering [17];
- T_d and bombarding ion energy in the *ion-plating* process associated with the sputtering (sputter *IP*) of semiconductors and ceramics [18].

The microstructural analyses made it possible to create a structure zone model (*SZM*) shown in Fig. 3.4 for the main *PVD* processes.

The four zones of the *SZM* depend on: (i) dimensionless temperature ratio, T_d/T_m (T_m is the melting point of the substrate), for *PVD* thermal evaporation; (ii) this ratio and on partial pressure of argon during at sputtering; and (iii) this ratio as well as the energy of ions bombarding surface of the deposit during its growth. More information about the *SZM* in *PVD* films can be found in recent reviews [19, 20]. The zones in the *PVD* techniques are called generally: 1; T; 2; and 3.

Zone 1 is limited to: (i) $T_d/T_m < 0.25–0.3$ for thermal evaporation technique; (ii) $T_d/T_m < 0.3$ for sputtering depending also on the partial pressure of argon; and (iii) $T_d/T_m < 0.4$ for *ion plating* depending also on the energy of bombarding ions. The grains in the zone 1 have a columnar shape being separated by empty spaces (pores). Microscopic studies have established that the columns may be nano-dimensional having diameter ranging from 1 to 400 nm [18]. The growth mechanism in this area in thermal evaporation techniques (Fig. 3.4a) is determined by the low mobility of *adatoms*. The grains grow in the direction of the source of vapors and do not diffuse in lateral directions. The grains close to the film surface are hemispherical and larger than those at the interface with the substrate (see also Fig. 3.3). The limits of the *zone 1* have larger values of the T_d/T_m ratio at high partial pressure argon at sputtering (Fig. 3.4b). This result from a decrease in mobility of *adatoms* by the atoms of Ar adsorbed to the surface. On the contrary, the increase of bombarding ions energy at *ion plating* (Fig. 3.4c) densifies the film and decreases the *zone 1* down to the value of $T_d/T_m < 0.1$. This effect can be explained by an increase in mobility of *adatoms* due to ion bombardment [18].

Zone T (***transition zone***) exists in sputtered films in the limits of temperature ratio $0.1 < T_d/T_m \leq 0.5$, which depend on the partial pressure of argon in films obtained by *ion plating*. The zone in this deposition technique depends also on the bombarding ions energy. The *transition zone* is composed of the fibrous grains placed densely next to each other. The voids around these grains disappear as the *adatoms* have the energy transferred in lateral directions by bombarding ions. On the other hand, this energy may pulverize the clusters and re-evaporate the *adatoms* resulting in the re-deposition on the surface. Therefore, the extent of this zone increases with increasing ion energy (Fig. 3.4c).

Zone 2 was identified in the films obtained by: (i) thermal evaporation for $0.3 < T_d/T_m \leq 0.5$; and (ii) sputtering and *ion plating*, for $0.5 < T_d/T_m \leq 0.75$. This zone includes columnar grains arranged densely next to each other. The surface of the grains has an angular shape corresponding to the crystalline planes. Films that crystallize in this zone do not have pores. The *adatoms* have enough thermal energy for lateral diffusion, and the microstructure of the films does not depend either on the partial pressure of the sputtering gas or on the energy of bombarding ions. The properties of deposits obtained in this zone approximate those of massive materials [17].

Zone 3 was observed in the films: (i) thermally evaporated for $T_d/T_m > 0.5$; and (ii) sputtered and ion plated for $T_d/T_m > 0.75$. Again, the thermal energy of the *adatoms* in this zone is predominant enabling diffusion in all directions. The films crystallized in this zone have,

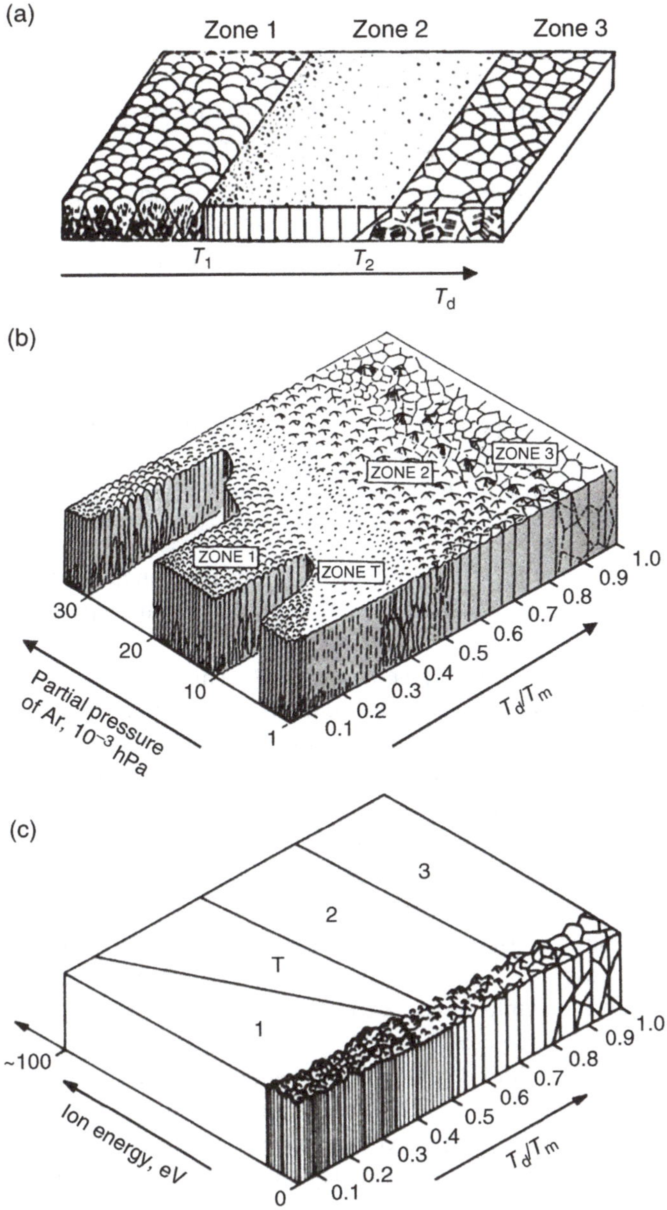

Figure 3.4 Structure zone models (*SZM*) for main atomistic *PVD* techniques: (a) thermal evaporation. *Source:* Movchan and Demchischin [14].; (b) sputtering *Source:* Thornton [17]/AIP Publishing.; and (c) *ion plating Source:* Messier *et al.* [18]/AIP Publishing.

in addition to the columnar grains, the equiaxial one, similar to that existing in bulk materials, treated at high temperature. The grains are often single crystals and their surface is flat.

In addition to the deposition parameters discussed above, the most frequent columnar microstructure of *PVD* films can be influenced by the following factors:

- ***Surface roughness*** modified at substrate activation discussed in Section 1.5.1. The columnar grains growing on the rough surface have the axes of the grains and are no longer parallel to each other as shown in Fig. 3.3b. Films synthesized on such surfaces are more porous than those obtained on the flat ones.
- ***Angle of incidence*** of vapors. The axes of the columns are generally directed toward the source of the vapors. The change of angle of incidence may modify, e.g. crystal growth direction in the textured films [21].
- ***Substrate rotation*** results in the short periods of deposition time during which the source delivers the species to the growing film under a changing angle resulting from the substrate rotation. The ***effects of shadowing*** discussed by, e.g. Schulz *et al.* [16], modify the columns' contours.

The growth of *PVD* deposits has been frequently simulated numerically using, e.g. statistical ***Monte Carlo method*** in *2D* [22] and in *3D* [23] or by a thin-film simulation program ***SIMBAD*** [24].

The films obtained by the *CVD* methods grow up generally in *2D*. This results from the flow of gas parallel to the substrate's surface pushing the *adatoms* laterally to form a step, which moves to cover the substrate (see Fig. 2.15). Cammilleri *et al.* [25] describe an example of such a lateral growth of a Ge film on the SiO_2 substrate. The possible microstructure of *CVD* films is similar to that obtained by *PVD* – thermal evaporation and shown in Fig. 3.4a. Consequently, the crystal grains can have the following shapes [26]: (i) columnar, (ii) faceted columnar; and (iii) fine and equiaxed. The microstructure varies on the materials type. Pierson [26] mentions that, in general, the metal deposits have typically one of two possible columnar structures. An example of such a columnar microstructure of W film obtained by *CVD* process at 823 K under atmospheric pressure using WF_6 precursor [27]. On the other hand, ceramic films, such as SiO_2, Al_2O_3, or Si_3N_4, have frequently a finely grained or amorphous microstructure.

The nucleation and growth in the *CVD* films depend on the phenomena occurring inside *boundary layer* above the substrate's surface (see Chapter 2.2.5). The film growth may occur in two possible regimes of: (i) mass transport controlled by diffusion of reactants through that layer; and (ii) surface reaction control. The dependence of the film growth rate on the deposition temperature and on the pressure in the *CVD* reactor, shown in Fig. 3.5, is different in each regime. The regime of surface reaction control enables obtaining film growth being less dependent on the possible variation of precursor's concentration resulting in the more homogeneous films thickness. That is why this regime is more useful in films production. The *CVD* film growth rate increases also with the gas flow velocity, and the surface reaction control can be reached at high value of this velocity. The choice of precursors enables to carry out the process in this regime even at higher temperatures [9].

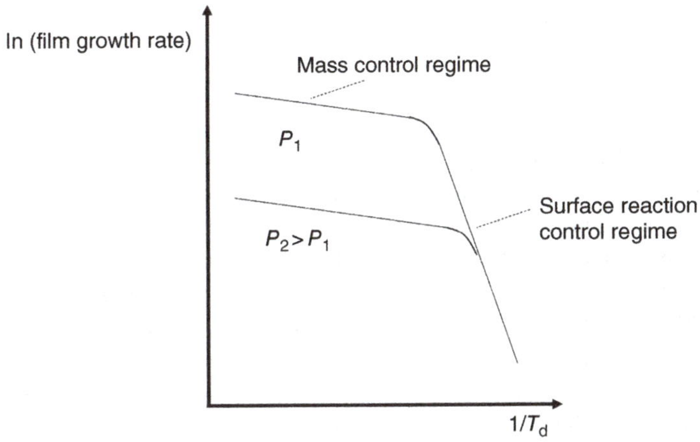

Figure 3.5 Arrhenius plot of the logarithm of growth rate vs. the reciprocal of temperature of deposition, T_d, showing possible regimes of growth of *CVD* films obtained under different reactor pressures and at different temperatures [8, 9, 28].

3.2.5 Residual Stresses and Defects in Films

The process of film deposition may lead to the generation of stresses, σ, in the deposits and in the substrate. These stresses result from the differences of the thermal and mechanical properties of film and substrate materials. The deposits on the components being in service as a part of different machines may be also submitted to the mechanical stresses resulting from contact with other surfaces. All these stresses have a three-dimensional character, which may lead to the detachment of the deposit from its substrate. The residual stresses in atomistic deposits are a sum of the following stresses [29, 30]:

$$\sigma = \sigma_{\mathrm{I}} + \sigma_{\mathrm{st}} + \sigma_{\mathrm{T}} \tag{3.2}$$

- ***Structural stresses***, σ_{st}, which results from the difference of the crystal lattices of the substrate and of the deposit. Structural stresses are generally small and can be neglected.
- ***Intrinsic stresses***, σ_{I}, generated during the growth of a film on a substrate. These stresses result from the modification (generally increase) of the volume of the deposit compared to the volume of a monocrystalline material. This modification is an effect of displacement of atoms to interstitial sites because of the collisions with energetic species arriving from the source and the generation of lattice defects. The stresses are generated by the bombardment of species such as atoms during sputtering or evaporation or ions during ion plating processes. The increase of the volume $\Delta V/V$ may result from:
  - displacement of an atom (point dislocation);
  - displacement of a line of atoms (linear dislocations);
  - creation of a gap in the crystal lattice of the deposit;
  - creation of micro-pores by the agglomeration of gaps;
  - implantation of atoms or ions of the species which bombard the deposit.

The intrinsic stress is equal to:

$$\sigma_{\mathrm{I}} = -\frac{1}{3}\frac{\Delta V}{V}\frac{E_{\mathrm{d}}}{1-\nu_{\mathrm{d}}} \tag{3.3}$$

In general, deposits obtained by sputtering or thermal evaporation at high pressure being greater than $P_{\mathrm{d}} > 7\times10^{-3}$ hPa are submitted to the anisotropic tensile stresses. The deposition using *ion plating* or sputtering at lower pressures induces compressive stresses generated by the ion bombardment [1]. These stresses depend on the potential of the substrate. An example is, discussed by Knoll and Beadley [31], the stresses measured in 8*YSZ* films deposited by *RF* sputtering on SiO_2 substrate polarized with negative potential. The deposition temperature, T_{d}, may influence also the intrinsic stresses, σ_{I}. Namely, the increase of this temperature leads to decrease of this type of stress which results from plastic deformation, cracking, and thermal relaxation of the deposit material [30].

- ***Thermal stresses***, σ_{T}, which result from the difference in the coefficients of thermal expansion between the deposited film, TEC_{d} and the substrate, TEC_{sub}. This difference may lead to the generation of stresses if the deposition temperature, T_{d}, is greater than the ambient one, T_{amb}:

$$\sigma_{\mathrm{T}} = \frac{E_{\mathrm{d}}}{1-\nu_{\mathrm{d}}}\left(\mathrm{TEC}_{\mathrm{d}} - \mathrm{TEC}_{\mathrm{sub}}\right)\left(T_{\mathrm{d}} - T_{\mathrm{sub}}\right) \tag{3.4}$$

Thermal stresses are generated during cooling of the film–substrate system after deposition. The stresses depend on the thermal expansion coefficients in the way shown in Fig. 3.6. The distribution of thermal stresses inside the deposit can be calculated theoretically [32].

The possible relaxation of residual stresses depends on the microstructure of the deposit. Stresses would not be transferred from grain to another when the grains are separated by

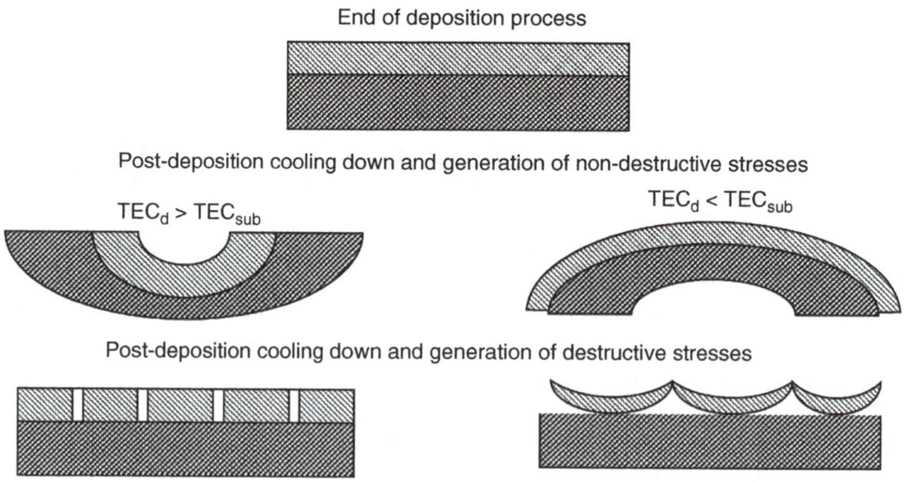

Figure 3.6 Generation of non-destructive and destructive thermal stresses in the film–substrate system.

pores or voids as the zone 1 or T of *SZM* (Fig. 3.4). Similarly, the cracks created by these stresses can stop in such pores. On the other hand, the higher mechanical resistance characterizes the dense microstructure of the zone 2 or 3 of the *SZM*. This resistance is a function of grain size and the microstructure with smaller grains is stronger [29]. There are several possibilities to reduce or cancel the residual stresses [1]:

- reduction of the thickness of the deposit;
- application of a negative potential to the substrate to keep the stresses close to zero;
- periodic ion bombardment during deposition to cancel out the stresses by creating sub-layers alternately subjected to tensile and compressive stresses inside the film;
- modification of the composition of the deposit
- generation of a porous type of microstructure as in the zone 1 or T of *SZM* which does not transmit the stresses.

The residual stresses can lead to macro-defects such as cracking or even to the detachment of a film from the substrate (see Fig. 3.6). The micro-defects in the film result from the atomistic mechanism of its growth, which includes diffusion of atoms and/or ion bombardment. Micro-defects, which are the source of intrinsic stresses, modify the electrical properties of deposits. Consequently, the electrical conductivity of films with such defects decreases. However, in some cases, their presence is beneficial – this is particularly the case with Type II superconductors in which the vectors of the magnetic field can stick to such defects resulting in an increase of the critical current [33]. The concentration of the defects may be reduced by post-deposition treatment.

3.2.6 Post-deposition Treatment

The films obtained by atomistic methods require sometimes a post-deposition treatment to improve their properties. It must be added that the quality of the atomistic deposits is generally high, and such treatment is applied less often than for the granular coatings, which are more heterogeneous. The post-deposition treatment can be carried out by:

- ***surface impregnation*** of the deposit with a polymer. The impregnation is done primarily to block open porosity near the surfaces of deposits. Typically, for better penetration into the pores, the polymers are applied as liquid solutions by brushing or spraying. Subsequently, the solution solidifies by contact with the ambient atmosphere or by *uv* heating. The epoxy resins are applied thanks to their stability, while silicone resins are used for their resistance against corrosion and for resistance in high temperatures.
- ***chemical*** or ***electrochemical treatment*** is often applied onto metallic films to protect their surface against contact with the ambient atmosphere. An example of such treatment is the anodization of aluminum or tantalum which covers the metals with a thin and dense oxide layer. Chemical treatment of a metal film by immersion in acids and salts covers the surface with a thin corrosion-resistant layer.
- ***mechanical treatment*** may be done to densify the films and to render their surface smooth. An example of such treatment is glass beads peening of the surface of CoCrAlY alloy films produced by *EB PVD* prior to oxidation being a part of *TBC* manufacturing [34].

- ***heat treatment*** is used to improve their properties. For example, the heat treatment at 1350 K in an inert atmosphere applied to superalloy deposits after the glass bead peening described above to densify the CoCrAlY films [34]. The heat treatment was used for the TiN films by *CVD* on steel substrates. The treatment was necessary to harden the substrate modified at high temperature of the *CVD* deposition [35]. Heating in an oxygen atmosphere is applied to restore oxygen lost during deposition of films of *YBCO* or *HA* [33, 36]. Finally, heat treatment by laser remelting can make it possible to create an intermetallic alloy of film with the substrate with the deposit (***laser alloying*** process) [37, 38].

3.3 Granular Deposits

3.3.1 Introduction

The species constituting a granular deposit, that is particles, have size starting generally from a few hundred of nanometers up to more than one hundred micrometers. The methods of the coarse powders manufacturing are shown in Section 2.3.1. Fine particles are generally injected to plasma jet or flame as suspension or solution and coating sprayed thereof are frequently called nanometric or nanophased [39]. The methods of liquid feedstock preparation are described in Section 2.3.3. The species used in granular deposits are 10^3 up to 10^6 times greater than the sizes of species such as atoms, molecules, or clusters being used in generation of atomistic films. Consequently, the phenomena of nucleation of crystal grains and of the coating's build-up growth resulting in the coating's microstructure are very different. Finally, the coatings microstructure depends on: (i) deposition, i.e. most often thermal spraying process; (ii) size of species entering in contact with a substrate; (iii) kind of sprayed material; and (iv) kind of substrate (metal, ceramic, or polymer).

The most important variables associated with the methods of spraying are the following:

- velocity and temperature of particles upon impact with the substrate, v_p;
- size and state of particles (liquid or solid);
- atmosphere of the spray processes;
- temperature of the coating on the end of spray processes, T_d.

The typical values for the variables of frequently used thermal spraying processes as well as sprayed materials and used substrates are presented in Table 3.2.

All materials can be thermally sprayed. However, more complicated is the processing of the materials, which: (i) decompose at high temperatures such as *YBCO* or *HA*; (ii) change their composition during spraying by, e.g. decarburization, such as WC or oxygen loss as *YBCO* or *HA*; or (iii) sublimate instead of melting, e.g. graphite[2]. The problems related to the materials decomposition can be sometimes solved by post spray heat treatment. The heat treatment was used to restore the superconducting phase in plasma sprayed *YBCO* [41]. The change of composition can be avoided by an appropriate

2 The temperature of sublimation of graphite was measured to be between 3895 and 4020 K at atmospheric pressure [40].

Table 3.2 Typical variables determining the microstructure for frequently used thermal spraying techniques described in details in sections 2.3.2 and 2.3.3.

Spray processes	v_p, m/s	Size of particles, μm	T_p/T_m	State of particle at impact	T_d, K	Typical coatings' material	Typical substrates
High kinetic of sprayed particles	Up to 1000	10–100	0.6–1	Solid at *CGSM*, molten or partly molten in the others	400–600	Metals, alloys, *MMC* with carbides reinforcement	Metals, alloys
High productivity	Up to 400	10–100	0.7 up to more than 1	Molten or partly molten	400–800	Metals, ceramics polymers as part of *MMC*	All materials
Vacuum and inert atmosphere spraying	Up to 900	5–20	More than 1	Vapors, molten or partly molten	700–900	Metals and alloys	Metal, alloys
Solution and suspension spraying	Up to 900	Up to 5	Up to 1	Molten or partly molten	Up to 400[a]	Ceramics, *MMC*	

[a] Short spray distance is generally accompanied by very intensive cooling.

preparation of powder. The cladding of WC powder with Co film protects the carbide against decarburization. Moreover, the cladding of graphite with Ni film may reduce the sublimation.

3.3.2 Contact Phenomena

The sprayed particles arrive at the substrate while being solid, partly or fully melted. The solid particles that are sprayed using cold spray method should be ductile. The particles, being fully molten, deform as the raindrops splashing on the windshield of a car. Consequently, the morphology of the splat depends on the velocity of particles prior to the impact. The molten particles solidify at impact and form lamellae. As the surface of the substrate is frequently rough, the contact between the lamellae is not perfect, and the pores may be formed inside the coating. Moreover, the shape of the substrate below the solidifying particle determines the direction of the heat removal and the shape of the crystal grains. Ultimately, all of these factors influence the adhesion of granular deposits.

3.3.2.1 Deformation of a Particle

Solid Particles

The plastic deformation of solid particles arriving at the substrate concerns mainly metals or alloys sprayed using cold gas spraying method. Vlcek *et al.* [42] recommended the metals having *FCC* structure such as Al, Cu, Ag as being easy to deform. On the contrary, the *BCC* metals, such as W, Mo, or Nb among others, have the lowest deformability. Similarly, the

Table 3.3 Estimation of critical velocities of some metals and alloys.

Material	Zn	Ta	Cu	Al	Cu + 30 wt. % Zn	Stainless steel 409[a]	Stainless steel 316 L[b]	Ti	Ti + 6 wt. % Al + 4 wt. % V
v_{cr} (m/s)	339	447	451	482	505	630	656	712	1013

[a] Stainless steel 409 has composition of Fe+(10.5–12.5) wt. % Cr + 1 wt. % Si + 1 wt. % Mn and small contents of Mo, C, S, P, Ti.
[b] Stainless steel 316 L has composition of Fe+(16–18) wt. % Cr+(10–12) wt. % Ni+(2–3) wt. % Mo and small contents of Si, P, S.
Source: Adapted from Assadi *et al.* [43].

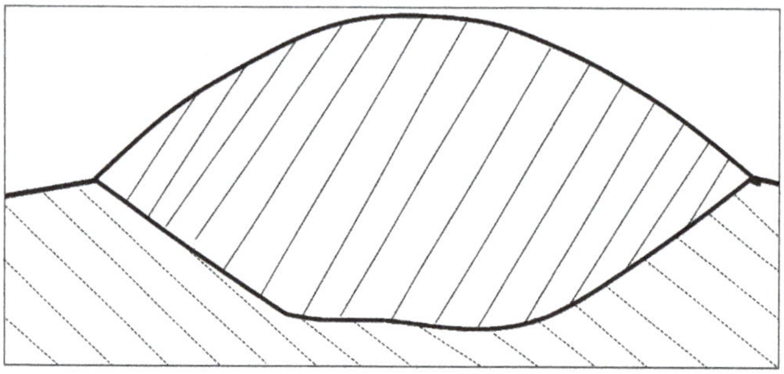

Figure 3.7 Impact of spherical particle Cu particle on Cu substrate. *Source:* Adapted from Winnicki *et al.* [46].

low melting point, T_m, is a useful indicator of the sprayability. Finally, the particles arriving on the substrate should have velocity, v_p, greater than the critical one, $v_{cr} > v_p$. The particles with lower velocity at impact would cause the abrasion of a substrate. The rough estimation critical velocities for some metals and alloys are presented in Table 3.3. The estimations of v_{cr}, measured and calculated by Assadi *et al.* [44] and Grujicic [45], show slightly different values. The impacts occurring at cold spraying deform the particle arriving and the metallic substrate. An example of Cu particle impacting Cu substrate is shown in Fig. 3.7.

Molten Particles

The ***molten sprayed particle*** deforms during impact with the substrate taking one of two morphologies categorized by Houben [47] and called ***pancake*** and ***flower***. As mentioned previously, these two shapes are similar to those left by rainwater droplets on the car windshield. The *pancake* morphology at impact occurs more frequently when spraying using less energetic jets or flames or very fine particles. The splat having this morphology is not entirely disintegrated. The splats with *flower* shape have more petals being much finer than those in *pancake* splat. The formation of a *flower* morphology is associated with the impact of the particle having high velocity and being molten. Such splats can be found in the substrate when spraying with the more energetic jets and flames. Fig. 3.8 shows the splats having both morphologies obtained while plasma spraying of *HA* powder with the use of arc with low electric power and high electric power.

Many authors, beginning from the initial study of Madejski [49], analyzed the particle impact theoretically. The analyses were carried out starting from the experimental

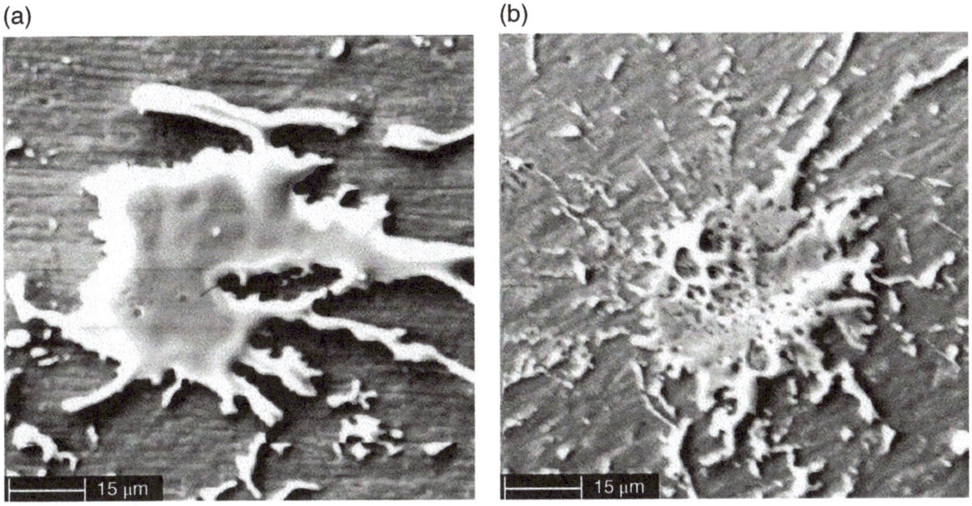

Figure 3.8 *SEM* (secondary electrons) of splat formed using *HA* particles arc plasma sprayed using: (a) low electric power resulting in a *pancake* morphology; and (b) high electric power giving *flower* splat. *Source:* From Dyshlovenko *et al.* [48]/with permission of ELSEVIER.

observation of the surface of splat, A_{sp}, and initial diameter of particle, d_p, by introducing the ***flattening parameter*** ξ defined as:

$$\xi = \frac{2}{d_p}\sqrt{\frac{A_{sp}}{\pi}} \tag{3.5}$$

The value of this parameter depends on the Re_p number describing the molten particle having velocity v_p prior to impact being defined as:

$$Re_p = \frac{\varrho_m d_p v_p}{\eta_m} \tag{3.6}$$

The value of the flattening parameter in the study was found to be equal to:

$$\xi = 1.2941\, Re_p^{0.2} \text{ for } Re_p > 100 \tag{3.7}$$

The other models of particles flattening leading to slightly different values of flattening parameter are discussed in the textbook [50]. Eq. (3.5) enables to find, by knowing the values of surface of the splat and the flattening parameter, the thickness of the splat. It is also possible to estimate the diameter of liquid droplet before impact. Such estimation was made for *YSZ* process synthesized using *SPPS* technique [51].

The size of a particle prior to the impact influences the shape of splat. The surface tension may preserve easier the integrity of splat for small particles [52]. Consequently, the splats in the coatings-sprayed techniques using particles having nanometric and submicrometric sizes, such as *SPS* or *SPPS*, have frequently *pancake* morphology. The formation of *flower* morphology starts with separation of the impacting particle in two parts. The minimal

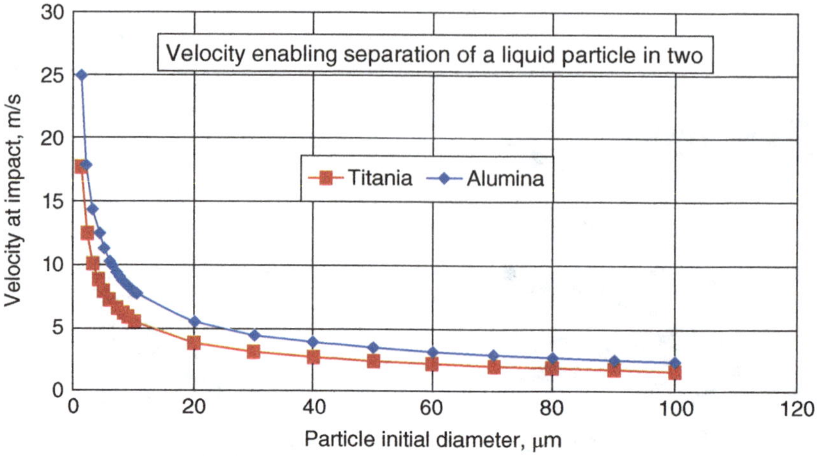

Figure 3.9 The estimation of the minimal velocity at impact of liquid particles of TiO_2 and Al_2O_3 which results in a separation of one particle in two parts of equal volume in function of initial particles' diameter. *Source:* Pawłowski [52]/with permission of Elsevier.

velocity at impact enabling such separation grows rapidly with decreasing diameter of impacting particles as seen in the calculations made for TiO_2 and Al_2O_3, shown in Fig. 3.9.

3.3.2.2 Solidification and Cooling of a Particle

The solidification and cooling of a molten particle impacting the substrate was analyzed theoretically by, e.g. Kudinov [53] and Steffens *et al.* [54]. Kurz and Fischer [55] presented fundamental theory of the solidification of a molten particle without any contact with a substrate. The important data resulting from these analyses are very useful to understand the adhesion of sprayed deposits, namely:

- contact temperature, T_{co}, which is established during the solidification time between the splat and the substrate which, depends on the thermophysical properties of the materials of the coating and substrate pair and on the initial temperature of the substrate T_0;
- solidification time, t_{sol}, where time depends on thermal contact resistance[3] and on the thickness of the splat δ_{sp}.

The calculations of contact temperature and solidification time, made for Ni particles sprayed using different methods onto the Fe substrate, are shown in Table 3.4. The thicknesses of the splats, found experimentally, depended on the deposition method. The solidification time varies between 1 and 100 μs depending on the splat thickness. The latter depends on the spray parameters. Generally, if the particle is molten on impact, the contact temperature, T_{co}, increases with the melting point temperature, T_m, of the particle material and decreases with increasing thermal conductivity, λ_p, and thermal diffusivity, a_p. It is possible that contact temperature exceeds the melting point of the substrate material. A practical example is the spraying of alumina on aluminum or copper substrate. In this case, the particle becomes embedded in the substrate and the adhesion of

3 Thermal contact resistance was not considered in the cited studies.

Table 3.4 Contact temperatures and solidification times for Ni sprayed onto the Fe substrate with different spraying techniques.

Spray technique	Process variables	T_0, K	δ_{sp}, μm	T_{co}, K	t_{sol}, μs	References
APS	—	373	18	373	23	[53]
AS	Pressure of atomization gas equal to 0.7 MPa	298	5	1232	1	[54]
	Pressure of atomization gas equal to 0.25 MPa		40		100	

the deposit improves. The solidification time, t_{sol}, decreases if thermal diffusivity of the substrate is high and the heat is removed more rapidly.

The experimental observations of the particle temperature after impact on the substrate can be made experimentally using a fast pyrometer. Moreau *et al.* [56] observed the temperatures and solidification time of Nb splats sprayed by APS onto glass and steel substrate.

3.3.2.3 Adhesion Mechanisms

Solid Particles

The solid particles deform themselves and the substrate as seen in Fig. 3.7. This can be achieved if their velocity and resulting thereof kinetic energy is greater than the critical one v_{cr}. The adhesion of particle to the substrate can be explained by ***adiabatic shear instability***, which occurs at contact of two materials [44, 45]. One of them exerts a strong stress on another. In usual non-adiabatic conditions, called *isothermal* ones, the stress is transferred resulting in the hardening of the surface of material in contact. This phenomenon occurs, e.g. at grit blasting discussed in Section 1.5.2. However, if the stress exerted in *adiabatic* conditions results in local increase of temperature in some microscopic areas, it may exceed the melting point and local softening of materials. The areas of melting can be very small, and Assadi *et al.* [44] observed that the local melting of copper impacted by copper particle having velocity of $v_p = 680\,m/s$ occurs or an mesh element used in numerical simulations having size of 0.4 μm.

The local softening of materials may lead to formation of local interfacial instabilities and adiabatic plastic flow. Fig. 3.10, based on the microscopic images of copper particle on

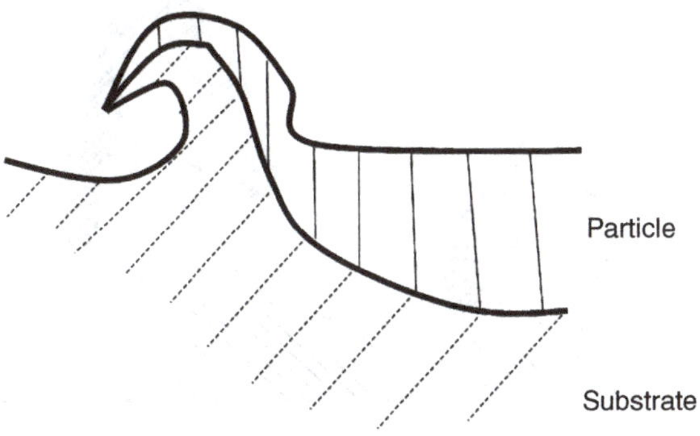

Figure 3.10 An example of interfacial instability inspired by a microscopic image of Cu particle on Cu substrate.

copper substrate shown by Assadi *et al* [44], shows a possible formation of such instability. The instability is an interfacial vortex, which explains the adhesion of cold sprayed particle to solid substrates.

Molten Particles

The adhesion of the molten particles has a mechanical nature. The liquid particles fall on the peaks of the surface of substrate activated by, e.g. sand blasting. Consequently, the splats solidify and tighten the peaks as shown in Fig. 3.11. This mechanism is sometimes called ***mechanical anchorage***.

Neither *flower* nor *pancake* splats are in contact with the substrate in their entire surface. McPherson and Cheang [57] found that the contact surface called ***contact zone*** corresponds to 20–30 % of the total surface of a splat. The contact zone is limited by:

- ***oxidation*** occurring on the surface of metallic particles at flight. To limit the oxidation of metallic powders, thermal spraying can be made in the inert atmosphere (*IPS*) or in vacuum (*VPS*). The use of alloys, having an oxygen-absorbing component, is another possibility. An example is adding of aluminum to nickel as in the Ni5Al alloy. Consequently, the coating of Ni-sprayed plasma in air contained 2% wt. of oxygen, while the one sprayed using an Ni5Al alloy contained only 10–3 wt. % of oxygen [57]. In order to avoid oxidation of the substrate's surface it is recommended to activate it shortly prior to the coating deposition.
- ***pores*** filled with air or with working gas of jet or flame. The existence of such porosity is intrinsic to the sprayed process. However, the pores are smaller if the particles are smaller. Thus, the adhesion can be improved by spraying a fine-particles powder.

The adhesion within the *contact zones* results from one of the following interaction mechanisms:

- ***physical interaction*** realized by *van der Waals forces*. The binding energy is estimated to be in the range of $E = 0.01–0.1$ eV/liaison [58]. This interaction can only exist between very clean surfaces, which allows atoms to approach to the distance smaller than 0.5 nm.
- ***metallurgical interaction***, which can be realized by one of two possible mechanisms:
  - diffusion, which is carried by the voids in the crystal lattice generated at solidification of a splat. The condition of such diffusion is mutual solubility of the coverslip materials and the substrate. Diffusion occurs at the contact temperature (T_{co}) during the

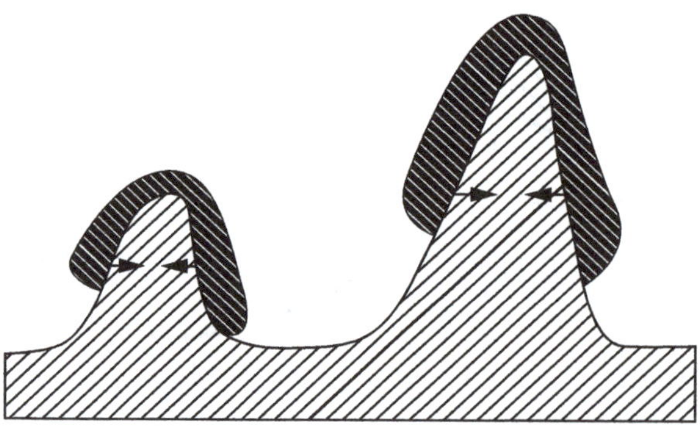

Figure 3.11 Mechanism of *mechanical anchorage* of splats on the substrate irregularities. *Source:* Pawłowski [50]/John Wiley & Sons.

solidification time (t_{sol}). T_{co} can increase due to an exothermic reaction of the elements at solidification and cooling of a splat. An example is an intermetallic reaction between molten Ni and Al, which results in formation of NiAl and/or Ni_3Al.

- chemical reaction between a splat and a substrate. In spraying practice, the chemical reaction leading to the formation of intermetallic compounds was observed when the powders of metals such as Mo and W were sprayed on the steel substrates [59].

The coatings are sprayed generally onto metal or alloy substrates. The use of polymer substrate to spray WC-Co coatings using *HVOF* technique required an initial deposition of metallic interlayers of Ni and Cu [60]. Similar substrate required an interlayer of Cu + Al_2O_3 cermet prior to the application of Cu coating by *CGSM* [61]. Finally, metallic coating of Cu and of Ta was sprayed on the Al_2O_3 substrate using VPS method. The mechanism of coatings adhesion was the *mechanical anchorage* which can be deduced from the profile of atomic concentration of elements [62].

3.3.3 Nucleation and Growth of a Crystal

The solidification of a splat begins at the interface with the substrate (or with the previously sprayed layer) having lower temperature. The latent heat of fusion is removed through this interface. The nucleation is then heterogeneous, and the grains have a columnar shape (see Fig. 3.12a). If the temperature of the substrate or that of previously deposited layer is high then the solidified grains may rather be equi-axial in shape, and the nucleation may be homogeneous. Such a microstructure was observed in the Ni coatings sprayed using VPS technique [64]. The contact between the lamellae (see the arrows in Fig. 3.12b) may also reduce the heat transfer. The thermal contact resistance, R_T, between the splat and the substrate was determined experimentally for stainless steel substrate and

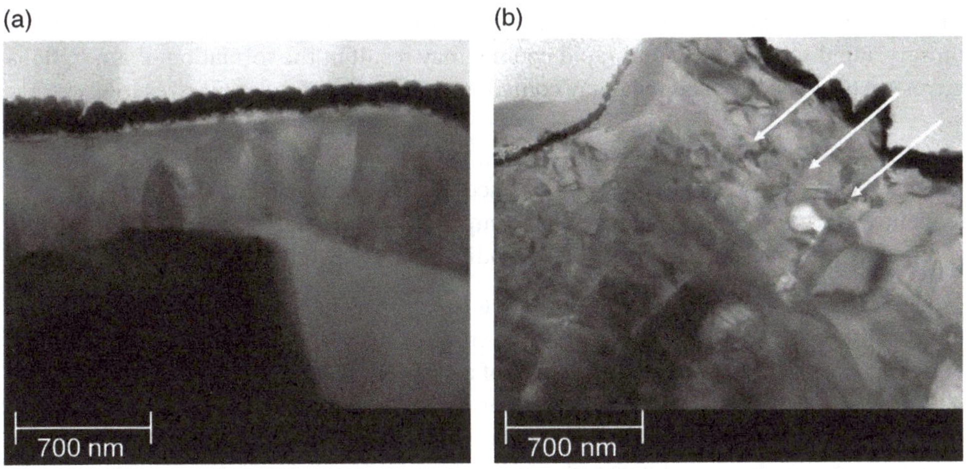

Figure 3.12 TEM micrographs of image of the of TiO_2 coatings sprayed using *SPS* method: (a) columnar grains inside a lamella close to the substrate, (b) contacts between neighboring lamellae shown by the arrows. *Source:* From Podlesak *et al.* [63]/with permission of ELSEVIER.

Table 3.5 Crystal grain sizes measured in typical sprayed coatings.

Spray technique	Substrate temperature prior to deposition, K	Method of crystal grains size determination	Coating	Substrate	Crystal grains size, nm	References
VPS	900–1020	*XRD* peaks broadening	Cu	Al_2O_3	27–37 (40 in powder feedstock)	[62]
		TEM grains observation	Ni	Cu	400–600	[64]
APS	—				100–300	
SPPS		*XRD* peaks broadening	TiO_2	Stainless steel	45 of rutile phase (58 in powder feedstock)	[65]

zirconia and alumina splats obtained by plasma spraying [64]. This resistance between two solid surfaces being in contact transferring the heat flux of q_T and having temperatures difference equal to ΔT, defined as:

$$R_T = \Delta T / q_T \tag{3.8}$$

was measured to be in the range from $R_T = 10^{-5}$ to 10^{-8} m^2.K/W. The values of the cooling rate of characterized splats were in the range of 7×10^{-6} to 3×10^{-5} K/s.

The typical size of the crystals, shown in Table 3.5, varies from a few tens to a few hundreds of nanometers. The size increases with the temperature of the coating's growth. Thus, the choice of spray parameters, which is characterized by a high temperature of the deposit, gives larger grains. Similarly, the thermal spraying of materials having low melting point and that of high latent heat of melting favor the formation of greater crystals. On the contrary, rapid solidification and rapid cooling may result in the formation of amorphous microstructure. Finally, a lamella may contain crystal grains of different sizes [66].

3.3.3.1 Phase Composition

The specification of the sprayed coatings, used in many industrial applications, includes frequently the phase composition. The composition of initial feedstock is frequently modified by thermal spray process. These modifications may result from:

- solidification and rapid cooling of a particle splat, which lead to the formation of new phases (often metastable) or even of an amorphous structure;
- modification of the chemical composition of a particle can take place during its trajectory in the jet or flame, which can be an effect of:
  - selective evaporation of one phase from a multi-phases particle;
  - oxidation or reduction of a particle.

A well-known example of the formation of a metastable phase occurs at APS deposition of powder composed of $\alpha - Al_2O_3$, having hexagonal crystal lattice. The sprayed coatings

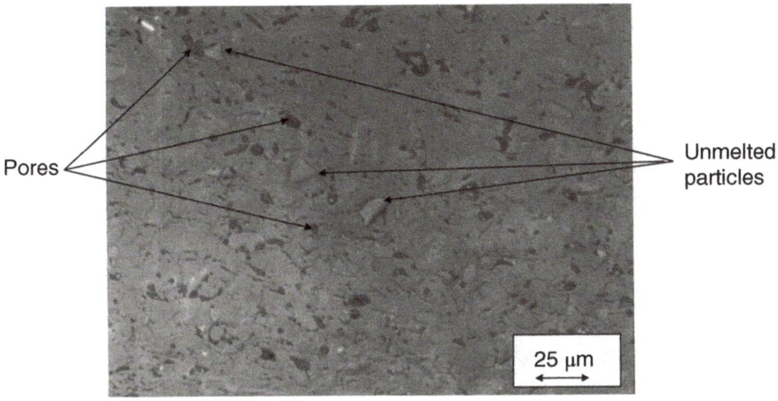

Figure 3.13 Optical microscope image of the polished cross-section of an Al_2O_3 coating obtained by APS.

crystallize in cubic $\gamma - Al_2O_3$. Its formation was explained by a critical free energy for nucleation ΔG^* energy from the liquid being more favorable to form γ – phase than α – one in the condition of *undercooling*[4] [52, 67]. Therefore, the presence of α – phase in the sprayed results from the particles, which were not molten at spraying. Fig. 3.13 shows such particles in *APS*-sprayed alumina coating.

Li and Ishigaki [68] calculated that the *undercooling* effect promotes also the formation of metastable *anatase* phase instead of *rutile* at solidification of TiO_2 melt. The size of splats resulting from the size of particles influences the undercooling effect. Consequently, small splats occurring at spraying of nanometric or submicrometric particles at *SPS* and *SPPS* deposition release less heat of solidification which increases cooling rate and promotes undercooling effect [52]. That is why the coatings *SPS* deposited using fine rutile feedstock included *anatase* phase [65].

The formation of metastable phase was also observed in atmospheric plasma sprayed *YBCO*. The coatings crystallized in the simple cubic phase and required the post-spray heat treatment to form the orthorhombic one having superconducting properties [33, 41]. Moreover, one of the oxides included in *YBCO*, namely CuO, evaporates at plasma spraying in air. Therefore, to obtain a stoichiometric composition in the deposit, the initial powder used to spray must be prepared with an excess amount of CuO.

The oxidation occurs frequently when thermal spraying of metals or alloys using techniques such as *FS*, *AS*, or *APS* in open atmosphere. Some oxides lose oxygen at high temperatures. TiO_2 may lose oxygen during spraying by *APS* and forming coatings, which include TiO_{2-x} phase [69]. Similarly, the Cr_2O_3 powder processed by this technique shows inclusion of oxide reduction to form metallic Cr as seen in Fig. 3.14.

---

4 Undercooling means that the solidification occurs at the temperature lower than melting temperature depending on the liquid cooling rate.

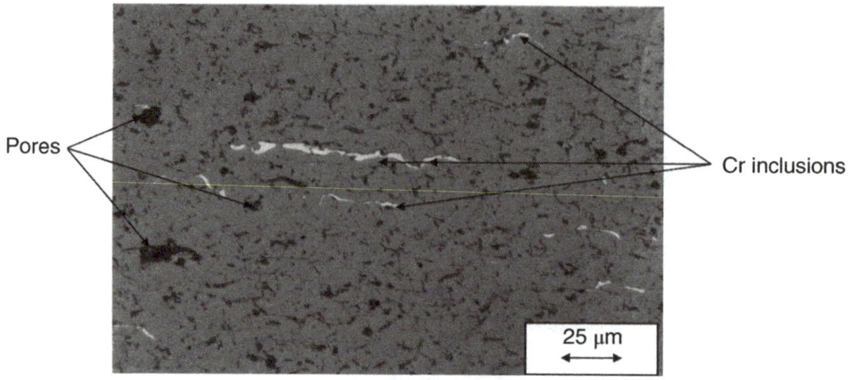

Figure 3.14 Optical microscope image of the polished cross-section of a Cr_2O_3 obtained by APS technique.

3.3.3.2 Inhomogeneity of Deposits, Composites

Thermally sprayed deposits are not homogeneous. In addition to the above-discussed chemical and physical transformations of sprayed feedstock, a deposit always contains the second phase: pores. There are small pores inside lamellas and greater, inter-lamellar pores well visible in Figs. 3.13 and 3.14. Finally, the lamellas have different sizes and shapes resulting from the inhomogeneity of initial feedstock (powder, solution, and suspension). The different shapes of splats result also from the differences in temperature and velocities of particles impacting the substrate, which is an effect of (i) radial feedstock injection used in most deposition techniques; (ii) jet or flame pulsations in techniques such as *AS*, *APS*, *VPS*, or detonation gun spraying; and (iii) pulsation of delivered feedstock.

In addition to the deposition process intrinsic inhomogeneities, the coatings can be sprayed as the composites, discussed in detail elsewhere [70]. The most frequent composite is that of metal with ceramics (*MMC*). Such composite is useful to reduce thermal stresses generated at high temperature in ceramic coating deposited onto metallic substrate. The stresses result from the difference between thermal expansion of metal and ceramics. An example of such composite coating plasma sprayed using Al and Al_2O_3 powders is shown in Fig. 3.15.

The *MMC* coatings are also used as wear-resistant coatings. In particular, the composite coatings including WC as hard ceramic phase in metallic matrix are sprayed with *HVOF* method [72]. Finally, the metallic part of *MMC* is necessary to obtain adhesion of cold sprayed deposits. The coatings can be sprayed using mixture of powders or using ceramic particles cladded with metal film (see Fig. 2.24). The resulting coatings can be used against wear resistance [73].

The reinforcements of composite deposits can be solid, namely whiskers, fibers (wires), or particles, as seen in Fig. 3.16. The liquid solution can also be used. The solutions are transformed into solid in flight, in jet, or flame at *SPPS* processing. The reinforcements are prepared as the feedstock in the following ways [50]:

- ***whiskers***, which are cylindrical in shape, typically have a diameter of 0.05–8 μm and a length of 3–80 μm. The whiskers can be used as a mixture with powder. Alternatively, the powder and the whiskers can be injected to jet/flame separately using appropriate

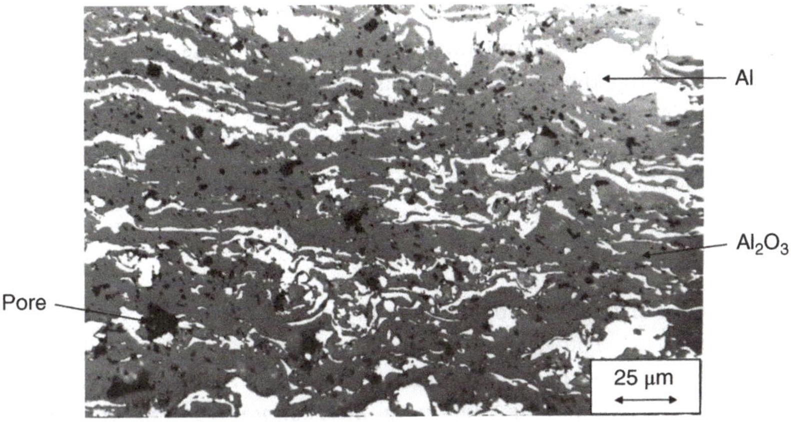

Figure 3.15 Optical microscope image of the polished cross-section of the composite Al_2O_3 + 28 vol. % Al atmospheric plasma sprayed using two powder injectors. *Source:* From Pawłowski [71]/with permission of ELSEVIER.

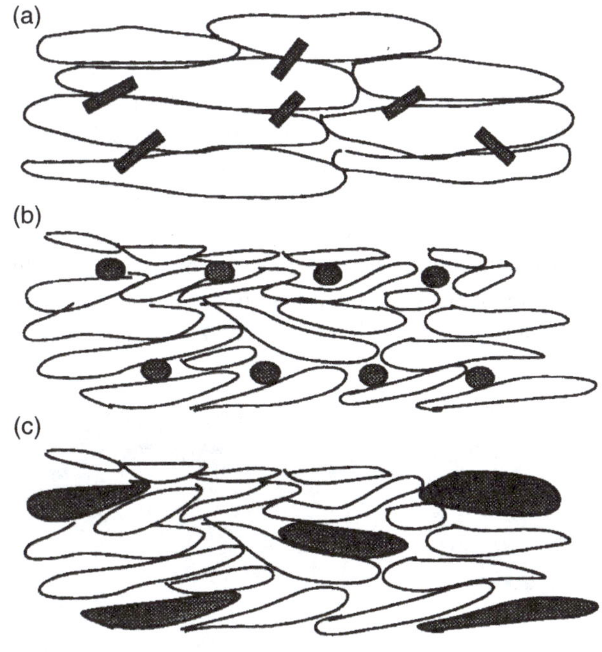

Figure 3.16 Composite deposits reinforced with: (a) whiskers; (b) fibers; and (c) particles.

distributors [74]. Finally, a composite powder can be prepared as an agglomerate with the whiskers by sintering and crushing or by spraying-drying.

- ***fibers*** or ***wires*** are wrapped around the substrate having cylindrical shape, and then the free space is filled by spray coating forming the matrix. The distance between the fibers ranges from tens of micrometers to a few millimeters [75]. The wrapping of a wire reinforcement can be repeated to create a multi-layered composite. The fiber fraction in the composite can vary from 5 to 40 vol. %.

- ***particles*** of a reinforcement can be mixed mechanically with the particles of the matrix prior to spraying. Another method of preparation is the preparation of the powders by spray-drying technique. Finally, the matrix and reinforcement powders can be injected separately in jet or flame (as shown in Fig. 3.15).
- ***solutions*** are prepared from the salts of metals such as nitrates or acetates mixed with different solvents to obtain the reinforcements of desired composition [76].

3.3.4 Build-up of a Coating

The microstructure of granular deposits depends mainly on used feedstock and on the operational spray parameters. The coatings sprayed using solid feedstock have mainly ***lamellar microstructure***. The liquid feedstock and, in particular, suspension enabled obtaining coatings having ***two-zone microstructure*** and ***columnar microstructure***.

The *lamellar microstructure*, shown in Fig. 3.17, is typical for deposits sprayed using solid feedstock resulting in micrometric size of droplets impacting the substrate. The lamellas are the splats being deposited on top of each other. Many properties of a deposit result from the quality of contacts between the lamellas. A good contact would be reached if the new lamella were sprayed onto the already-deposited one being still in liquid state. It depends, however, on the operational spray parameters such as powder feed rate and torch linear velocity discussed in detail elsewhere [50]. The final thickness of a sprayed deposit is reached after numerous passes of the spraying torch. The thickness of a layer deposited in one pass varies typically from a few μm to more than 10 μm. The microstructure of a granular deposit sprayed in open atmosphere is modified on the surface of a layer sprayed in one pass. This surface interacts with the atmosphere for some time. During this interaction, the surface can be modified chemically (oxidation) or physically (cooling and generation of thermal stresses which can lead to cracking). The resulting thereof contacts between the layers deposited in each pass are weaker than those of the lamellas inside this layer.

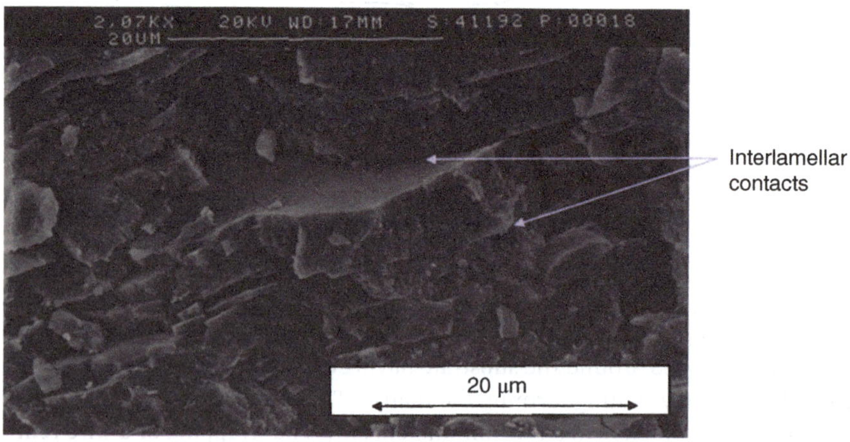

Figure 3.17 SEM (secondary electrons) image of a fractured cross-section of Cr_2O_3 coating obtained by atmospheric plasma spraying.

Finally, the microstructure depends on the distance between two passages of spraying torch over a substrate. This distance should be smaller than the diameter of the spray spot[5].

The use of suspension feedstock may result in coating, which have *two-zone microstructure* shown in Fig. 3.18. Such microstructure was observed in hydroxyapatite coatings deposited by *SPS* using suspension of sub-micrometric solids in a solvent. The solvent from droplets of this suspension evaporates at flight in plasma jet and remaining small particles get molten (see Fig. 2.49). Some of them agglomerate forming large particles, which transform later on in micrometer in size splats similar to that observed in coatings sprayed using powder feedstock. Some of the particles remain sub-micrometric. These particles may sinter inside the coating under heat flux delivered by plasma jet [78].

The *columnar microstructure*, shown in Fig. 3.19, was obtained in *APS*-deposited coatings. The deposition was realized using low- and high-energy plasma spray torches [79, 80]. The formation of *columnar microstructure* was favored by the use of suspension having low concentration of solids and by the irregularities on the substrate's surface. The columns are formed because of the lateral movement of small droplets on the substrate. The movement results from the drag force, F_D, acting by the working gas moving in lateral direction as seen in Fig. 3.20. The moving droplets may form islands around the surface irregularities and grow up as the columns as shown in Fig. 3.21.

3.3.4.1 Temperature of a Coating

The heat flux generated by working gas brought to high temperature at thermal spray deposition techniques is considerable. The deposition at atmospheric pressure allows thermal energy to be transported by convection to the coated substrate. The vacuum coating deposition favors radiative energy transfer. Finally, the heat flux generated by sprayed particles solidification and cooling plasma spraying is present in all deposition techniques with

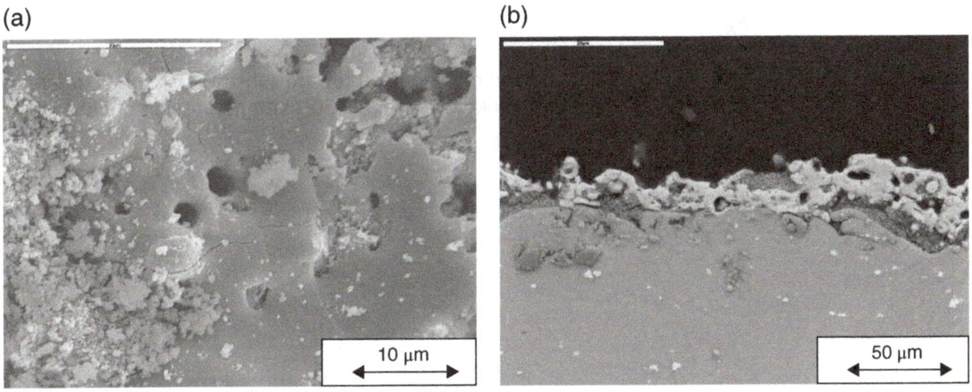

Figure 3.18 *SEM* (secondary electrons) image of suspension plasma sprayed *HA* coating: (a) surface; (b) polished cross section. *Source:* From Kozerski *et al.* [77]/with permission of ELSEVIER.

5 Spray spot diameter was estimated to be about 2.5 cm at *APS* deposition of Al_2O_3 coating [50].

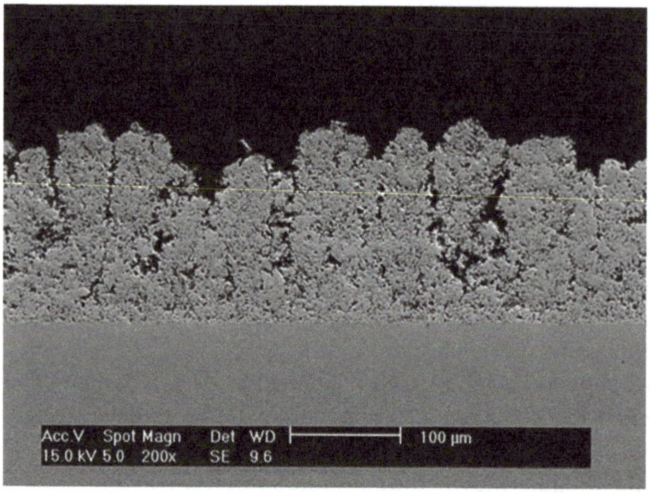

Figure 3.19 *SEM* (secondary electrons) image of *YSZ* coating sprayed using suspension feedstock with high-energy water stabilized plasma torch type WSP-H 500. *Source:* From Sokołowski et al. [79]/ with permission of ELSEVIER.

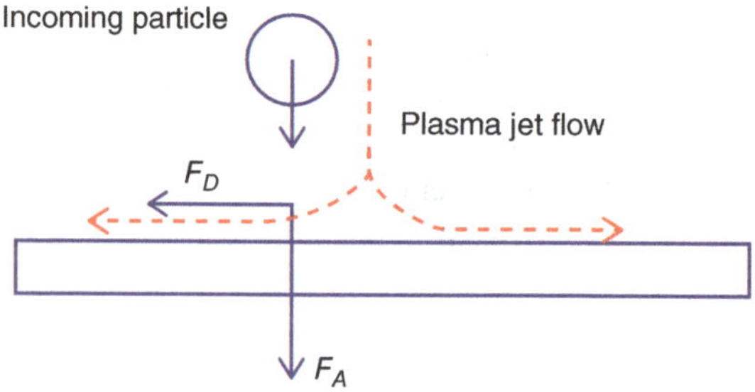

Figure 3.20 Forces acting on a small particle and moving it: (a) toward substrate, F_A; and (b) laterally on the substrate's surface, F_D, [80]. *Source:* From Sokołowski et al. [80]/with permission of ELSEVIER.

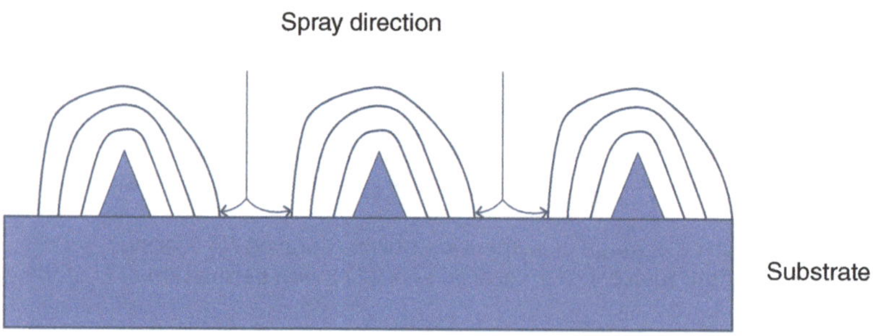

Figure 3.21 Formation of columns from small particles at the substrate's surface irregularities [80]. *Source:* From Sokołowski et al. [80]/with permission of ELSEVIER.

Table 3.6 Heat fluxes generated at thermal spraying deposition and delivered to the coating and substrate.

Heat flux	Source of the flux	Thermal spray technique
Radiation	Radiation of the plasma jet or combustion flame	Dominant in *VPS* and present in all spray techniques
Convection	Hot working gas in contact with growing coating	All techniques
Splats solidification	Liberated latent heat of fusion	All techniques with exception of *CGSM*
Lamellas cooling	Heat generated at cooling of lamellas to the local temperature of growing coating	

exception of cold spraying (Table 3.6). The total heat flux generated at *APS* deposition was estimated to be 2.6 MW/m^2 and that at *VPS* deposition, 0.62 MW/m^2 by Marynowski et al. [81] and Lee [82], respectively.

The temperature of deposition is generally controlled using thermocouple, pyrometer of *ir* camera. The cooling of deposited coating can be realized using, e.g. compressed air (see in Fig. 2.32) or liquefied cryogenic gas [83]. The cooling is necessary to reduce the thermal stresses which can lead to coating cracking.

3.3.4.2 Residual Stresses

The residual stresses can be present in the substrate activated prior to the deposition by sand or grit blasting as discussed in Section 1.5.2. These stresses are generally compressive and concentrated near the substrate's surface. Thermal spraying introduces also the residual stresses, which may influence the adhesion of coatings. Clyne and Gill [84] overviewed these stresses and showed the experimental methods of their measurements. The stresses have different origins and can be categorized as follows:

- ***Quenching stress***, σ_q, which are generated at quenching of a molten particle, having temperature T_m, on the substrate or on the previously deposited coating having temperature T_d. The quenching stress is expressed by:

$$\sigma_q = \mathrm{TEC}_d \left(T_m - T_d\right) E_d \tag{3.9}$$

  This type of stress can be measured *in situ* by determining the curvature of a thin substrate plate during deposition [85]. These stresses decrease by micro-cracking of splats occurring frequently in ceramic coatings [50].
- ***One-torch pass stress*** may be generated if spraying torch moves slowly over the substrate and/or powder feed rate is too big. Consequently, the layer deposited in one pass has a considerable thickness, and the gradient of temperatures in this layer may lead to generation of a pattern of cracks in entire coating as seen in Fig. 3.22.
- ***Thermal stresses***, σ_T, being similar to those generated in atomistic films and expressed in Eq. (3.4), are generated when the coated substrate is cooled down after deposition to room temperature. The value of the stresses in the sprayed coating and

Figure 3.22 *SEM* (secondary electron) image of the polished cross-section of *7YSZ* coating obtained by *APS* method.

their direction depend on the thermal expansion coefficients of the substrate and of the deposit in the following way [86]:
- $TEC_d < TEC_{sub}$, tensile or compressive stresses;
- $TEC_d = TEC_{sub}$, tensile stresses;
- $TEC_d > TEC_{sub}$, tensile stresses.

Thermal stresses increase also with the thickness of the deposit and with its temperature. Consequently, the intensive cooling during deposition leads to their reduction. In addition, the stresses can be reduced by applying the intermediate coating or coatings between the top-coating and the substrate to adapt the coefficients of thermal expansion. The deposits with gradient of thermal expansion coefficient belong to the family of recently reviewed functionally graded coating (*FGC*) [87].

- ***Phase transformation stresses***, σ_{pt}, generated during cooling after deposition to ambient temperature when a phase transformation in the coating material takes place. The phase transformation stresses can be calculated from the following equation supposing that the densities of the phases ρ_1 and ρ_2 before and after the transformation change, but their elasticity E_d, ν_d, remains the same [88, 89]:

$$\sigma_{pt} = \frac{E_d\left(1-\frac{\rho_1}{\rho_2}\right)}{3(2\nu_d-1)} \tag{3.10}$$

The deposition of coatings of ceramic materials such as oxides or carbides having numerous phases[6] is particularly susceptible to generate such stresses.

- ***Mechanical stresses*** are generated by high-speed particles and in particular by particles which are not melted or partly melted. Mechanical stresses were found experimentally in deposits obtained by coating sprayed using techniques *HVOF*, *DS*, and *CGSM* [90, 91]. These stresses are similar to those generated at substrate activation by sand or grit blasting being always compressive.

6 For example the phases present in plasma sprayed alumina coating, namely $\gamma-Al_2O_3$ is monoclinic and has density of $\rho = 3.56\,g/cm^3$ and α-Al_2O_3 is hexagonal and has density of $\rho = 3.98\,g/cm^3$ [89].

3.3.5 Post-deposition Treatment

Granular deposits and, in particular, thermally sprayed coatings have characteristics, such as porosity, low adhesion, presence of metastable phases, and surface roughness, which can be improved by treatment after deposition. In fact, in the industrial practice such treatment is a rule rather than an exception. There are a lot of different post spray procedures described carefully in a chapter of the textbook [50]. This section covers only most frequently used one. Generally, these procedures can be categorized as:

- ***Heat treatment***, which includes furnace heating applied mainly to ceramic coatings to eliminate metastable phases. The well-known examples are the heating in air of the alumina deposits to transform γ–Al_2O_3 into α-Al_2O_3 phase [92] or the heating in oxygen atmosphere of the *YBCO* deposits to restore the orthorhombic phase [33]. Vacuum heat treatment of the NiCoCrAlY coatings at temperature, $T = 1323$ K, is part of the standard procedure for preparing the aeronautic turbines applied by some manufacturers to coatings' adhesion. A similar treatment was made at 773–973 K by Lim *et al.* [93] to improve adhesion of WC + 17 wt. % Co coatings obtained by *APS* to the iron substrate. The furnace heating is also used to densify ceramic and metallic coatings which improves their mechanical properties or electrical properties. Laser treatment is localized heat treatment and discussed in detail in Section 3.5. This treatment densifies the metal deposits, but the ceramic may become cracked depending on laser power density [94, 95]. The pulsed CO_2 lasers are used to engrave the surface of plasma sprayed ceramic coatings used in ***anilox rolls*** industrial manufacturing [95] and tested as electron emitters [96]. Fig. 3.23 shows the surface of laser-engraved TiO_2 coating.
  Other heat treatment methods include:
  - hot isostatic pressing is applied by, e.g. Chen *et al.* [97] to ZrO_2 plasma sprayed coating;
  - spark plasma sintering used by, e.g. Khor *et al.* [98] to densify HA coatings obtained by *APS* method;
  - e-beam treatment used by, e.g. Brandl *et al.* [99] to remelt CoNiCrAlY alloy deposited by *HVOF*.

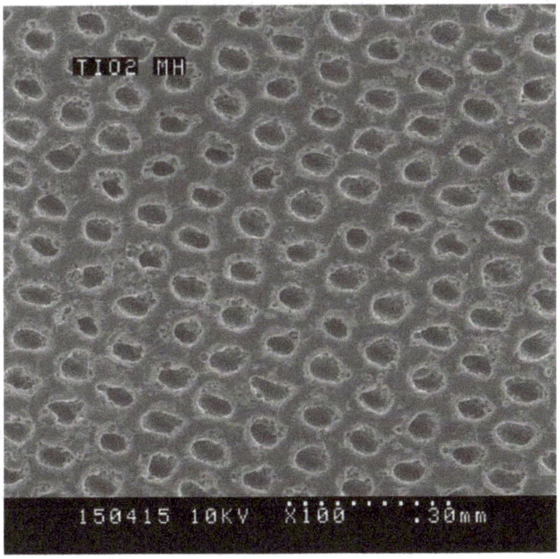

Figure 3.23 SEM (secondary electrons) image of the TiO_2 coating's surface obtained by atmospheric plasma spraying and CO_2 pulsed laser engraving. *Source:* From Znamirowski *et al.* [96]/with permission of ELSEVIER.

- ***Impregnation*** with organic and inorganic compounds is used to seal sprayed coatings' surface, which are characterized by open porosity. Knuuttila *et al.* [100] reviewed the impregnation of sprayed coatings. Impregnation with organic sealants is carried out with epoxy phenolic and silicon resins as well as with polyurethanes, polyvinyl esters, polymethacrylates, or waxes. The polymers are applied in fluid form by painting or brushing onto the surface of a deposit to better penetrate the open porosity. The liquid polymerizes in air. This type of treatment makes the deposits less permeable to gas or liquid and can be used to improve resistance to chemical corrosion. Silicone resin impregnation is part of an industrial process for manufacturing Al_2O_3 plasma sprayed coatings for corona rolls. The impregnation with inorganic sealers can be made by metal–organic precursors which are applied on the coating's surface and, later on, calcinated [100]. The molten metals can also be used to improve the mechanical properties of ceramic coatings. Ohmori *et al.* [101, 102] used molten Mn to impregnate Al_2O_3 and YSZ coatings. The treatment improved reportedly the mechanical properties of sprayed coatings. The components are applied to the surface of a deposit in liquid state. Subsequently, the impregnated deposit is heated to a temperature which exceeds the vitrification point of the compound. Mechanical treatment, that is to say, the grinding and finishing of the surface of the deposits
- ***Finishing*** is often applied in industrial practice to decrease the surface roughness of sprayed coatings. For ceramic deposits, diamond or cubic boron nitride (*CBN*) grinding wheels are applied, while for metals – silicon carbide grinding wheels [103]. The deposits are first polished and then lapped with increasingly fine grinding wheels. A good setting of the finishing parameters makes it possible to obtain a roughness of the order of $R_a = 0.2\,\mu m$ for Cr_2O_3 coatings obtained by atmospheric plasma spraying.

3.4 Bulk Deposits

3.4.1 Introduction

The species constituting a bulk deposit are, as in granular deposits, particles having size up to a few hundreds of micrometers (see Table 2.17). These particles are injected into a melt pool in a substrate in a one-step process or are already predeposited inside initial coating in two-step processes. The microstructure of the deposit depends on the way of melting surface and forming melt pool and on the coating and substrate materials. The synthesis of typical processes of bulk methods of deposition discussed in this chapter is collected in Table 3.7.

Methods such as *hardfacing*, tungsten inert gas (*TIG*), or metal inert gas (*MIG*) are conventional and used frequently in manual way in small workshops. Their microstructure is similar to that of the *PTA* coatings. In the latter, the ratio of pilot arc current to that of transferred arc current determines the dilution defined in Fig. 2.52. As far as the microstructure of the *PTA* coatings is concerned, it is similar to that obtained in one of the *1SLD* process, namely *cladding*, *laser surface alloying* (*LSA*), and *hard-phase dispersion*. In fact, the most important difference between plasma and laser processes has to do with the energy source that produces the fusion of the material. The energy in *PTA* process is produced by

Table 3.7 Variables determining the microstructure of bulk deposits.

Process	Source of energy	Power density and velocity of the source scanning	Atmosphere in contact with melt pool	Typical coating materials	Typical substrate materials
Hardfacing, i.e. treatment of predeposited coating	Combustion flame	Scan velocity up to a few cm/s	Air	Metals and alloys	Steel
Plasma transferred arc (*PTA*)	Electric arc		Air, nitrogen, argon		
Laser treatment: • One-step laser deposition (*1SLD*) with powder injection • Two steps laser deposition (*2SLD*) i.e. treatment of predeposited coating	• Lasers type: CO_2, Nd: YAG, diode, fiber. • Laser radiation: pulsed or continuous wave	• Power density up to 10^9 W/m^2 • Scan velocity up to a few hundreds cm/s	Vacuum, inert gas, air	• *Cladding*: metals, alloys, ceramics, cermets • *Laser surface alloying*: metals, alloys • *Hard phase dispersion*: oxides, carbides and nitrides	Metals, alloys

Joule effect, i.e. electric heating of conducting materials and the laser processes use the absorbed photon energy. This difference does not determine the microstructure of the deposits. Therefore, the discussion of microstructure will be done together for plasma- and laser-deposited coatings.

The laser-assisted coating processes use the energy of photons determined by the laser wavelength. Consequently, the optical properties of substrate material are important because they determine the fraction and the depth of energy absorbed by the substrate (see Table 1.11). The distribution of energy in the laser beam determines the shape and the size of the melt pool as well, described later by the ***Bénard–Marangoni convection*** effect, the movement of the liquid in the melt pool can result in coatings microstructure and shape of surface. The microstructure of *1SLD* coatings is generally less homogeneous than that of *2SLD* ones. This inhomogeneity results from a radial distribution of powder injected to the melt pool and the distribution of powder resulting from the convection movement inside the pool. Another important parameter is the viscosity of the fluid in the pool. The viscosity in turn depends on the temperature distribution and, finally, on the density of absorbed laser power absorbed. The scanning speed of the energy source (laser or arc) influences the solidification and cooling and rate of melt pool which also determines the bulk coatings' microstructure.

The *PTA* process is often carried out in the ambient atmosphere. Contact of coating being in liquid phase or cooling down with air can lead to surface oxidation. The oxide layer would have to be removed by mechanical treatment after deposition. The *PTA* or laser deposition can be made also under inert gas atmosphere or under vacuum to prevent oxidation. On the other hand, the deposition under vacuum at *2SLD* allows gas included in the treated coating open pores to be evacuated at processing. Otherwise, this gas escape through the melt pool and leave holes on the surface of a solidified deposit.

The materials for *PTA* and laser-assisted coating deposition materials are mainly stainless steel; Ni-base alloys; Co-base alloys including *Stellites* [104] and, to a lesser extent, ceramics and cermets such as Ni + WC or NiCrAlY + YSZ [105–107]. Laser-assisted cladding of pre-placed ceramic powders is seldom successful because of frequent cracking of final deposits. More useful is the preplacement of thin (200 nm) ceramic coatings obtained by the sol–gel technique, which is treated by laser treatment to enable obtaining deposits adhering to the substrate [108]. Another laser-assisted technique, namely *hard-phase dispersion,* is typically realized using carbides or nitrides. The reduction of the stresses generated at deposition is achieved by substrate pre-heating. The microstructure of bulk deposits is determined by the phenomena of rapid solidification. Consequently, the crystal grains are frequently dendrites.

The properties of the substrate material in the liquid-phase properties influence the depth of injection of the injected particles and their spatial distribution in *1SLD*. An important issue in *hard-phase dispersion* method is wettability of solid ceramic particles by liquid metal or alloy [109].

3.4.2 Phenomena in the Melt-pool

The energy source (laser, arc, or combustion flame) heats and melts the substrate very quickly. However, the temperature in the melt pool results from the energy distribution of the source and from the diffusion of heat is not homogeneous. The liquid in the pool is submitted to some forces. The forces result from [110]: (i) the substrate movement; (ii) viscosity of liquid; (iii) pressure exerted by photons and working/cooling gas; and (iv) surface tension. In particular, the radial temperature distribution of the pool leads to the convective movement of the liquid in the bath, shown in Fig. 3.24, by the effect of *Bénard–Marangoni,* called also ***thermo-capillarity***. The effect results from the fact that the surface tension, γ, depends on temperature in the following way [111, 112]:

$$\gamma(T) = \gamma_{\mathrm{m}} + \mathrm{CST}(T - T_{\mathrm{m}}) \tag{3.11}$$

The values of the surface tension at melting point, γ_{m}, and of the coefficient of surface tension (*CST*) for some metals are collected in Table 3.8. The *CST* coefficient is negative for pure metals[7]. Consequently, the liquid metal moves on the surface radially as shown in Fig. 3.24, and the velocity of its movement may reach 1 m/s [111]. After arriving at the boundary of the melt pool the liquid is redirected to the bottom of the pool forming a vortex. The convective movement leads to the formation of rippling surface of laser-treated surface analyzed, e.g. by Anthony and Cline [114] and by Joshi and Sundararajan [115]. The solidification of the melt pool freezes the moving liquid leaving the waves on its surface.

---

7 The *CST* becomes reportedly positive for stainless steel containing more than 40 ppm of sulfur [113].

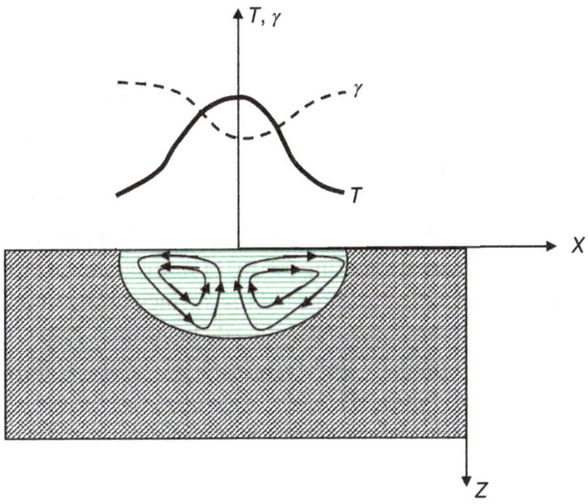

Figure 3.24 The convective movement of the liquid in the melt pool generated by laser beam having Gaussian distribution of energy (mode TEM_{00}).

Table 3.8 Experimental values of surface tension for some liquid metals.

Metal	Melting point temperature, T_{m}, K	Surface tension at melting point temperature, γ_m, N/m	Coefficient of surface tension, *CST*, N/(m.K)
Al	933	867×10^{-3}	-0.15×10^{-3}
Co	1768	1884×10^{-3}	-0.37×10^{-3}
Cr	2130	1628×10^{-3}	-0.20×10^{-3}
Cu	1357	1355×10^{-3}	-0.19×10^{-3}
Fe	1809	1844×10^{-3} and 1855×10^{-3}	-0.23×10^{-3} and -0.5×10^{-3}
Ni	1726	1838×10^{-3}	-0.42×10^{-3}

Source: Eustathopoulos *et al.* [112]/with permission of Elsevier.

The convective movement results in homogenization of the melt temperature and in an increase of cooling rate. The particles of the powder injected to the melt pool in *1SLD* process follow the lines of the vortex. Consequently, the particles are brought to the bottom part of the melt pool improving the mixing of materials which is useful in the process of *laser surface alloying* or *hard-phase dispersion*. The convective movement is not useful in the process of *cladding*. In fact, the cladding parameters are optimized so as not to melt the substrate too deeply. Moreover, the particle in *1SLD* processes may be intensively heated by laser beam prior to the injection into melt pool. For example, SiC powder particles injected to Ti substrate had reportedly temperature of about 2300 K on impact the melt pool [116]. Finally, the convective movement may lead to the phases separation, which should be avoided in *2SLD* cladding process. This separation was observed, for example, on the surface of a NiTa-alloy deposit [38].

3.4.3 Laser Shock Treatment

In general, bulk deposits are obtained by heating and melting of the substrate or coated substrate using flame, electric arc, or laser. There are, however, processes in which substrate and coating remain solid, and pulsed laser interacts only with the gas above the treated surface forming plasma which acts mechanically by the mechanical shocks generated by plasma expansion. The laser shocks are applied to the densification of pre-deposit similar to other *2SLD* processes. The laser shock treatment was applied to densify the *HVOF*-sprayed composite coatings with Al matrix and the SiC reinforcement [117]. Laser shocks deformed plastically deformed the surface of the matrix of initial coating and which generated some of structural defects (dislocations) [118]. On the other hand, the laser shocks reduced the porosity and improved the contacts between the matrix and the reinforcing particles (see Fig. 3.25).

3.4.4 Nucleation in the Melt Pool

The energy from the source is absorbed by a solid, which melts and forms a pool in *PTA* and in most of laser-assisted processes. The liquid can also contain particles of a solid phase being reinforcement of the composites. The solid particles interact with the liquid by chemical reaction, dissolution, etc. The microstructure of final deposit is determined by the solidification rate of the liquid in the pool [55, 119]. The solid grains' form of dendritic columns or a form of equiaxed dendrites is shown in Fig. 3.26. The columns are created when the heat flux is directional (contact with solid), and the equiaxed dendrites are formed in in the middle of the melt pool. The spacing of dendrite arms decreases with increased solidification rate and is about 1 μm for 10^5 K/s [110]. The solidification and germination of grains is discussed for three types of processes shown in Fig. 2.55, namely: (i) *cladding*; (ii) alloying; and (iii) *hard-phase dispersion*.

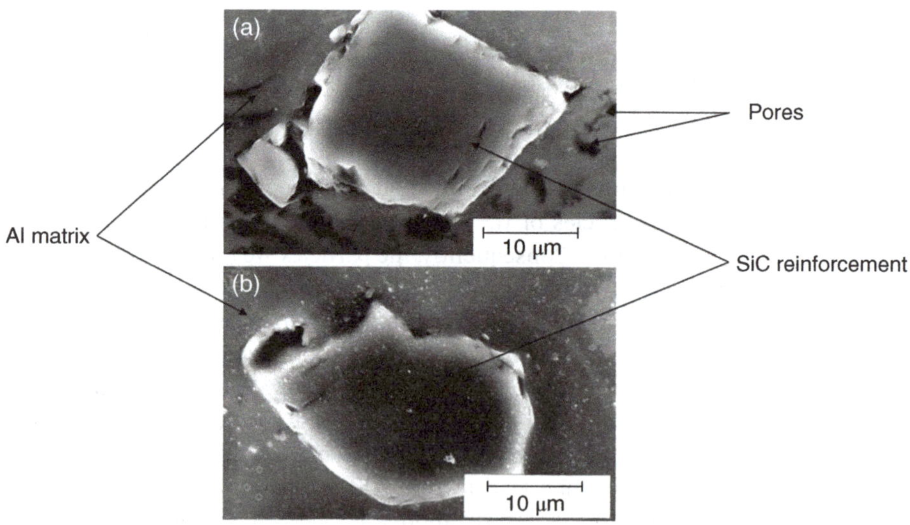

Figure 3.25 SEM (secondary electrons) image of the cross-section of an Al+15 wt. % SiC cermet coating: (a) sprayed initially using HVOF technique; (b) treated by laser shocks.

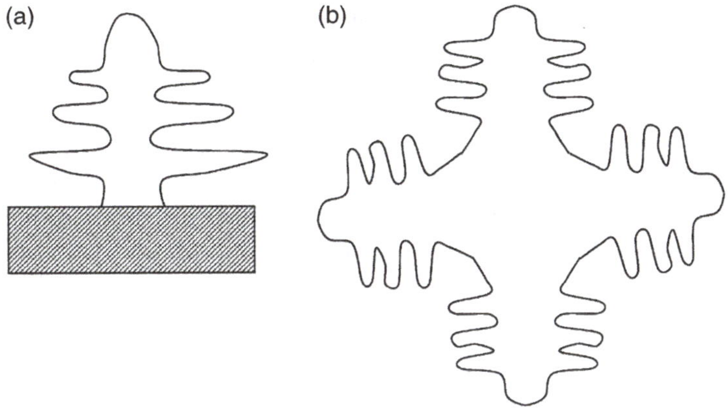

Figure 3.26 The dendrites formed at solidification of metals or alloys: (a) dendritic columns; (b) equiaxed dendrites.

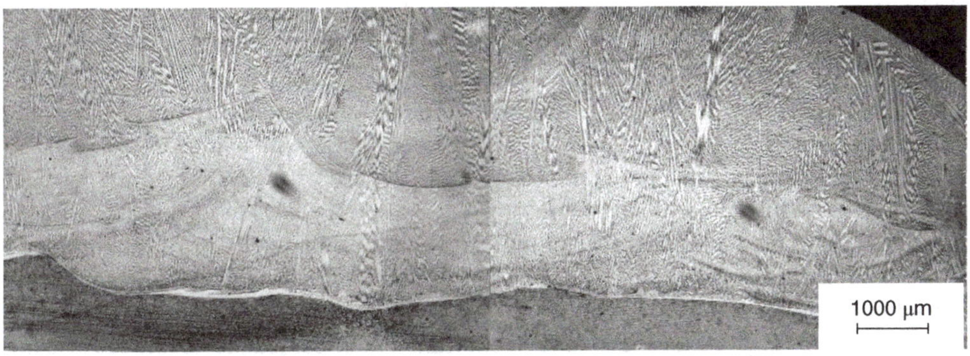

Figure 3.27 Optical microscope image of the polished cross-section of the Inconel coating obtained in one-step *cladding* using a CO_2 laser. *Source:* From Blechet [120].

Cladding consists of melting the powder injected in the melt pool (one-step process) or its pre-deposition (two-step process). The substrate remains not significantly molten, i.e. the dilution is small, as shown in Fig. 2.52. A typical microstructure of powder injection laser cladding deposits (one-step process) is shown in Fig. 3.27.

The coating includes the grains having dendritic form. The final thickness of the deposit, of about 3 mm, was reached in two laser passes. The interfaces between these two layers do not show either the important porosity or the signs of oxidation. Similarly, the contact area between the deposit and the substrate does not show the important dilution. Dendritic columns were observed in *1SLD* cladded coatings made of Ni-base, Co-base alloys reported by Hidouci *et al.* [105] as well as in Ti6Al4V alloy studied by Cottam and Brandt [121]. The *cladding* process is in two steps (*2SLD*) of TiC nanoparticles powder preplaced on Ti6Al4V substrate enabled obtaining 400 μm thick coatings having also dendritic microstructure [122]. Similarly, the *2SLD* processing of plasma sprayed Cr-50Mo alloy subsequently remelted with CO_2 laser had similar microstructure [123]. The reported studies indicated that the dendritic columns are formed during a solidification, which began at the bottom of the melt pool (see 3.26a). Consequently, the dendrites in the zone close to the interface with substrate had the arms in greater distance than that being at the coating surface. This indicates that the solidification rate was greater at the coating surface than at its interface with the substrate.

The alloying process with, e.g. laser (*LSA*), enables modifying the microstructure and properties of substrate by the elements introduced by powder, rod, or wire into the melt pool in one-step process (*1SLD*) or by their pre-deposition by thermal spraying, electrodeposition, or other method in two-step process (*2SLD*). An example of *1SLD* processes is CO_2 laser melting of the carbon steel substrate (0.2 wt. % of C) and injecting into melt pool a mixture of powders powder having composition Cr + 38 wt. % Ni + 8 wt. % Mo was presented by Anjos *et al.* [124]. The final coating, obtained after solidification, was an alloy having chemical composition of Fe + 2.7 wt. % Cr + 2.6 wt. % Ni + 0.6 wt. % Mo + 0.2 wt. % C and mainly martensitic microstructure. The crystal grains had an equiaxed dendrite (see Fig. 3.26b). The martensitic microstructure is known to occur in processes with a high solidification rate. Another example of one-step processes is *laser surface alloying* of Al with Cr by the same research group [125]. The authors used Al substrate and injected a mixture of Al + 25 wt. % Cr powder to the melt pool. The final coating required two laser remelting processes and was composed of compounds Al_7Cr and $Al_{11}Cr_2$ having microstructure including again equiaxed dendrites as crystal grains. Two-step alloying (*2SLD*) was used to obtain hard nickel aluminides ($NiAl_3$ and Ni_3Al) on the Al substrate electroplated initially with Ni layer and treated with Nd: YAG pulsed laser [126]. Similar process was used to improve the wear and corrosion resistance of stainless steel by use of Cr and CrB_2 powders predeposited as slurry and glued to the substrate [127]. The CO_2 laser melting enabled obtaining a lot of different phases including martensite, austenite, FeCr solid solution and compounds such as Cr_2B and $Cr_{23}C$ having crystal grains of dendrites, eutectics, and others depending on processing parameters.

The *hard-phase dispersion* process is different from that discussed previously because the hard phase used as *MMC* composite reinforcement being generally carbides, nitrides, or oxides is not melted at processing. The major issue is the wetting of reinforcement grains in metallic melt pool and their homogeneous distribution in the matrix. The wetting is generally better, i.e. angle of wetting, $\Theta << 90°$ for carbides reinforcement particles in metallic melt pool that for oxides ones [108]. The homogeneous distribution of the reinforcement in the melt pool requires an optimization of processing parameters. Otherwise, the hard phase particles may agglomerate as shown by Celik [128] for *MMC* coatings of Ti6Al4V matrix and WC reinforcement obtained by *PTA* with different arc currents. The same matrix and the SiC particles reinforcement were used to obtain composite coating in *1SLD* processes with the use of Nd: YAG laser by De Hosson and Teeuw [107]. The authors analyzed carefully the modification of the chemical composition of the reinforcement particles and the matrix in different positions of solidified coating and found that the SiC particles are surrounded by:

- boundary layer including TiC inside the matrix preserving its initial composition at the interface with unmolten part of the substrate;
- modified matrix which contains the TiC dendrites and the grains of the Ti_5Si_3 compound at its surface.

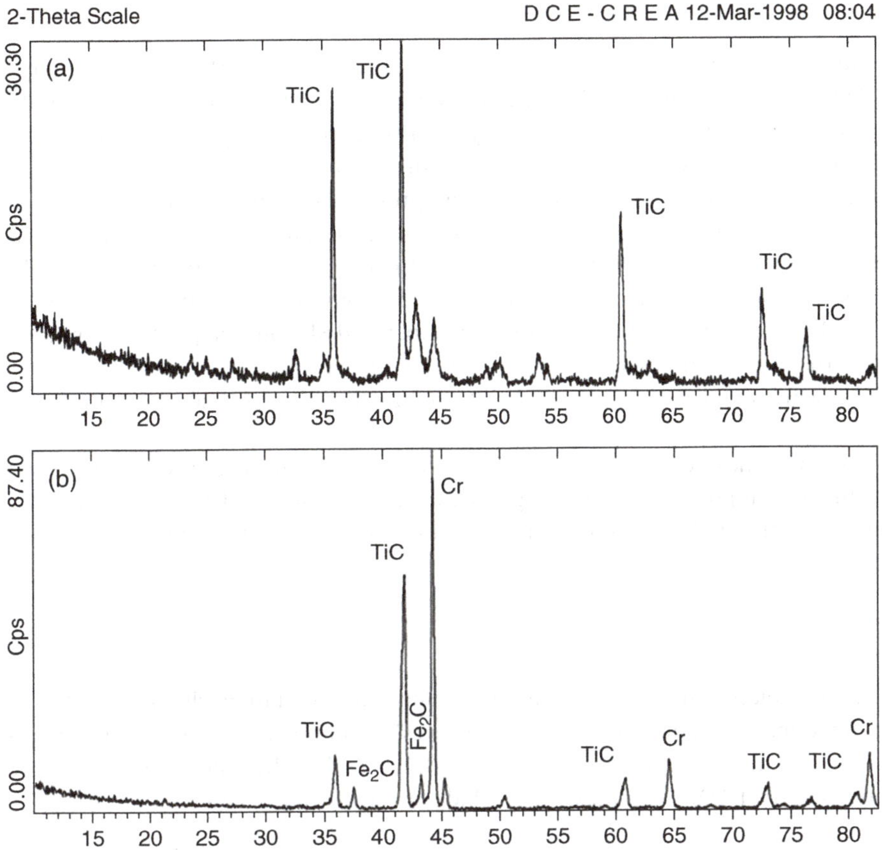

Figure 3.28 XRD of *MMC* coating having composition Fe13Cr + 55 wt. % TiC obtained in two steps by: (a) atmospheric plasma spraying, (b) CO_2 laser treatment of sprayed coating. *Source:* Tondu *et al.* [129]/with permission of Elsevier.

The *2SLD* process was used to obtain *MMC* having matrix of Fe13Cr reinforced with 55 wt. % TiC[8]. The composite coating was predeposited using atmospheric plasma and submitted to CO_2 laser treatment [129]. The XRD spectra of the coating before and after laser treatment are shown in Fig. 3.28.

The temperature of the melt pool at laser treatment was controlled to be about 2300 K being lower than melting point of TiC (T_m = 3430 K). The temperature was chosen to melt the alloy matrix while keeping the reinforcement solid. The solidification of the melt resulted in the formation of Fe_2C compound. This compound was synthesized presumably with carbon atoms dissolved from the TiC reinforcement in the melt pool at processing. Moreover, the Cr metal was found in solidified coating. Its presence could have resulted from selective evaporation of Fe from the melt pool (Fe has melting point of T_m = 1812 K).

8 The process can be also qualified as post-spray laser glazing. The discussion concerning coating's microstructure is similar for laser glazing and for *hard-phase dispersion* in two steps using laser.

3.4.5 Inhomogeneities of Deposits and Thermal Stresses

The inhomogeneities in the bulk deposits occur inside and outside the passes of energy sources (flame, arc, or laser). Inside the pass, the inhomogeneities concern the solidified melt pool, which may contain porosity resulting from the gas liberated during chemical reaction or phase transformation at cooling of the pool. In particular, the two-step treatment heats up the gas contained in the pre-deposit. The gas increases its volume and is evacuated through the liquid of the pool through the surface of deposit as shown in Fig. 3.29. The laser-treated zone shows the cracks of the surface. The cracks resulted from the stresses generated at cooling to the room temperature. Such stresses may occur at PTA deposits of metals or alloys. Cooper and Slebodnick [130] recommend to preheat processed samples to the temperatures in the range of 573–723 K and to decrease the scan velocity of energy source to reduce thermal stresses and to prevent cracking.

The surface of bulk deposits is rough because of *Bénard–Marangoni* convection and because of the overlapping of the following passes of energy sources. The theoretical study of Li and Ma [131] enabled finding the optimum overlapping to reduce the surface roughness.

3.4.6 Post-deposit Treatment

In the industrial practice, massive deposits are usually submitted to mechanical treatment (grinding, finishing) to reduce the coatings surfaces' roughness. The coatings of metals or alloys can be mechanically treated with c-BN wheels. The diamond wheels are recommended to treat *MMC* coatings with carbide or nitride reinforcements.

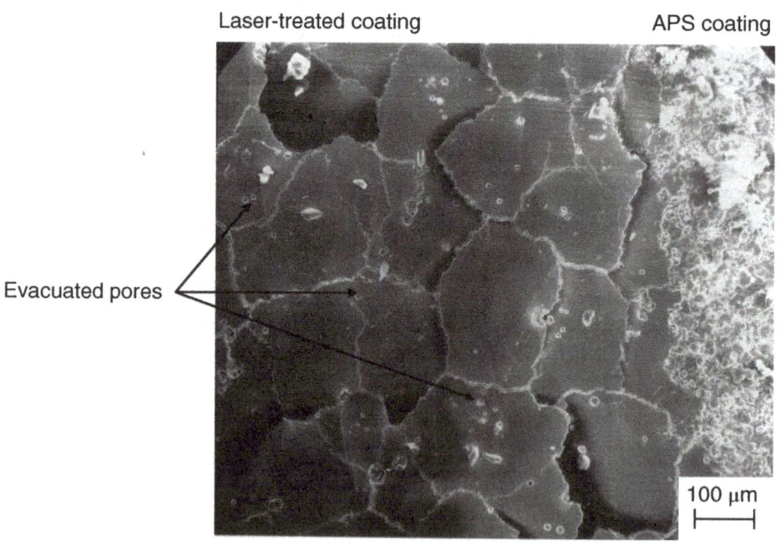

Figure 3.29 SEM (secondary electron) image of the surface of atmospheric plasma sprayed Al_2O_3 coating remelted by CO_2 laser.

References

1 Mattox, D.M. (2010). *Handbook of Physical Vapor Deposition (PVD) Processing*. Amsterdam: Elsevier.

2 Bunshah, R.F. (1982). Deposition technologies: an overview. In: *Deposition Technologies for Films and Coatings* (ed. R.F. Bunshah), 454–489. Park Ridge, NJ: Noyes Publications.

3 Hubler, G.K. (1994). Comparison of vacuum deposition techniques. In: *Pulsed Laser Deposition of Thin Films* (ed. D.B. Chrisey and G.K. Hubler), 327–354. New York, NY: Wiley.

4 von Niessen, K. and Gindrat, M. (2011). Plasma Spray-PVD: a new thermal spray process to deposit out of the vapor phase. *J. Thermal Spray Technol.* 20 (4): 736.

5 He, W., Mauer, G., Gindrat, M. et al. (2017). Investigations on the nature of ceramic deposits in plasma spray–physical vapor deposition. *J. Thermal Spray Technol.* 26: 83.

6 Johnson, R.W., Hultqvist, A., and Bent, S.F. (2014). A brief review of atomic layer deposition: from fundamentals to applications. *Mater. Today* 17 (5): 236.

7 Horwitz, J.S. and Sprague, J.A. (1994). Film nucleation and film growth in pulsed laser deposition of ceramics. In: *Pulsed Laser Deposition of Thin Films* (ed. D.B. Chrisey and G.K. Hubler), 229–254. New York, NY: Wiley.

8 Pauleau, Y. (2004). *Procédés Chimiques de Dépôt à Partir d'une Phase Gazeuse*. Paris: Lavoisier. (in French).

9 Carlsson, J.-O. (1994). Chemical vapor deposition. In: *Handbook of Deposition Technologies for Films and Coatings*, 2e (ed. R.F. Bunshah), 374–433. Park Ridge, NJ: Noyes.

10 Greene, J.E. (1994). Nucleation, film growth, and microstructural evolution. In: *Handbook of Deposition Technologies for Films and Coatings*, 2e (ed. R.F. Bunshah), 681–739. Park Ridge, NJ: Noyes.

11 Skrzypczak, M., Bertrand, P., Zdanowski, J., and Pawłowski, L. (2001). Modeling of temperature fields in the graphite target at pulsed laser deposition of CN_x films. *Surf. Coat. Technol.* 138: 39.

12 Rigney, D.V., Viguie, R., Wortman, D.J., and Skelly, D.W. (1997). PVD thermal barrier coating applications and process development for aircraft engines. *J. Thermal Spray Technol.* 6: 167.

13 Kourtev, J., Pascova, R., and Weissmantel, E. (1996). Arc evaporated Ti-N films with reduced macroparticle contamination. *Thin Solid Films* 287: 202.

14 Movchan, B.A. and Demchishin, A.V. (1969). Study of the structure and properties of thick vacuum condensates of Ni, Ti, W, Al_2O_3, ZrO_2. *Fiz. Met. Metalloved.* 28 (4): 83. (in Russian).

15 Westra, K.L. and Thomson, D.J. (1995). The microstructure of thin films observed using atomic force microscopy. *Thin Solid Films* 257: 15.

16 Schulz, U., Fritscher, K., and Peters, M. (1996). EB PVD Y_2O_3 – and CeO_2/Y_2O_3 - stabilized zirconia thermal barrier coatings-crystal habit and phase composition. *Surf. Coat. Technol.* 82 (3): 259.

17 Thornton, J.A. (1974). Influence of apparatus geometry and deposition conditions on the structure and topography of thick sputtered films. *J. Vac. Sci. Technol.* 11 (4): 666.

18 Messier, R., Giri, A.P., and Roy, R.A. (1984). Revised structure zone model for thin film physical structure. *J. Vac. Sci. Technol.* A2 (2): 500.

19 Kameneva, A.L. (2015). Models of structural zones for sputtered and evaporated thin films. *Res. J. Pharm., Biol. Chem. Sci.* 6 (5): 464.

20 Abegunde, O.O., Akinlabi, E.T., Oladijo, O.P. et al. (2019). Overview of thin film deposition techniques. *AIMS Mater. Sci.* 6 (2): 174.

21 Sohn, Y.H., Biederman, R.R., and Sisson, R.D. Jr. (1994). Microstructural development in physical vapor-deposited partially stabilized zirconia thermal barrier coatings. *Thin Solid Films* 250: 1.

22 Srolovitz, D.J. (1986). Grain growth phenomena in films: a Monte Carlo approach. *J. Vac. Sci. Technol.* A4 (6): 2925.

23 Zhang, P., Zheng, X., Wu, S., and He, D. (2004). A computer simulation of nucleation and growth of thin films. *Comput. Mater. Sci.* 30: 331.

24 Tait, R.N., Smy, T., and Brett, M.J. (1993). Modelling and characterization of columnar growth in evaporated films. *Thin Solid Films* 226: 196.

25 Cammilleri, V.D., Yam, V., Fossard, F. et al. (2008). Lateral epitaxial growth of germanium on silicon oxide. *Appl. Phys. Lett.* 93: 043110.

26 Pierson, H.O. (1999). *Handbook of Chemical Vapor Deposition*, 2e. Norwich, NY: Noyes Publications.

27 Lv, Y., Song, J., Lian, Y. et al. (2015). The thermal properties of high purity and fully dense tungsten produced by chemical vapor deposition. *J. Nucl. Mater.* 457: 317.

28 Kuiper, A.E.T., van den Brekel, C.J.H., de Groot, J., and Veltkamp, G.W. (1982). Modeling of low-pressure *CVD* processes. *J. Electrochem. Soc.* 129 (10): 2288.

29 Rickerby, D.S., Eckold, G., Scott, K.T., and Buckley-Golder, I.M. (1987). The interrelationship between internal stress, processing parameters and microstructue of physically vapour deposited and thermally sprayed coatings. *Thin Solid Films* 154: 125.

30 Oettel, H. and Wiedemenn, R. (1995). Residual stresses in PVD hard coatings. *Surf. Coat. Technol.* 76–77: 265.

31 Knoll, R.W. and Beadley, A.R. (1984). Correlation between the stress and microstructure in bias sputtered ZrO_2-Y_2O_3 films. *Thin Solid Films* 117: 201.

32 Chiu, C.-C. and Liou, Y. (1995). Residual stresses and stress-induced cracks in coated components. *Thin Solid Films* 268: 91.

33 Pawłowski, L. (2018). Synthesis properties and applications of $YBa_2Cu_3O_{7-x}$. In: *Industrial Chemistry of Oxides for Emerging Applications* (ed. L. Pawłowski and P. Blanchart), 211–253. Chichester: Wiley.

34 Shafer, S.J., Boone, D.H., Lambertson, R.T., and Pecock, D.E. (1983). The effect of deposition and processing variables on the oxide structure of M-Cr-Al coatings. *Thin Solid Films* 107: 463.

35 Kessler, O., Hoffmann, F., and Myr, P. (1994). Nachwärmebehandlung CVD- und PVD-hartstoffbeschichteter Stähle. *Härterei-Technische Mitteilungen* 49: 48. (In German).

36 Pawłowski, L. (2018). Synthesis, properties and applications of hydroxyapatite. In: *Industrial Chemistry of Oxides for Emerging Applications* (ed. L. Pawłowski and P. Blanchart), 311–352. Chichester: Wiley.

37 Draper, C.W., Preece, C.M., and Jacobson, D.C. (1980). Laser alloying of deposited metal films on nickel. In: *Laser and Electron-Beam Processing of Materials* (ed. C.W. White and P.S. Peercy), 721–727. New York, NY: Academic Press.

38 Draper, C.W., Gibson, J.M., Jacobson, D.C. et al. (1985). The microstructure of laser alloyed Ni-Ta surface layers. *J. Mater. Sci.* 20: 2303.

39 Pawłowski, L. (2008). Finely grained nanometric and submicrometric coatings by thermal spraying: a review. *Surf. Coat. Technol.* 202: 4318.

40 Abrahamson, J. (1974). Graphite sublimation temperatures, carbon arcs and crystallite erosion. *Carbon* 12: 111.

41 Pawłowski, L., Gross, A., and McPherson, R. (1991). Microstructure of plasma sprayed $YBa_2Cu_3O_x$ high-temperature superconductors. *J. Mater. Sci.* 26: 3903.

42 Vlcek, J., Gimeno, L., Huber, H., and Lugscheider, E. (2003). A systematical approach to material eligibility for cold spray process. In: *Thermal Spray 2003: Advancing the Science Applying the Technology* (ed. C. Moreau and B. Marple), 37–44. Materials Park, OH: ASM International.

43 Assadi, H., Kreye, H., Gärtner, F., and Klassen, T. (2016). Cold spraying - a materials perspective. *Acta Mater.* 116: 382.

44 Assadi, H., Gärtner, F., Stoltenhoff, T., and Kreye, H. (2003). Bonding mechanism in cold gas spraying. *Acta Mater.* 51: 4379.

45 Grujicic, M. (2007). Particle/substrate interaction in the cold-spray bonding process. In: *The Cold Spray Material Deposition Process* (ed. V.K. Champagne), 148–177. Cambridge: Woodhead Publishing Limited.

46 Winnicki, M., Małachowska, A., Dudzik, G. et al. (2015). Numerical and experimental analysis of copper particles velocity in low-pressure cold spraying process. *Surf. Coat. Technol.* 268: 230.

47 Houben, J.M. (1988). Relation of the adhesion of plasma sprayed coatings to the process parameters, size, velocity and heat contant of the spray particles. PhD Thesis. Technical University of Eindhoven, The Netherlands.

48 Dyshlovenko, S., Pawłowski, L., Roussel, P. et al. (2006). Relationship between plasma spray operational parameters and microstructure of hydroxyapatite coatings and powder particles sprayed into water. *Surf. Coat. Technol.* 200 (12–13): 3845.

49 Madejski, J. (1976). Solidification of droplets on a cold surface. *Int. J. Heat Mass Transf.* 19: 67.

50 Pawłowski, L. (2008). *The Science and Engineering of Thermal Spray Coatings.* Chichester: Wiley.

51 Tobon Valencia, V., Pawłowski, L., Lecomte-Nana, G. et al. (2020). Preliminary analysis of physical and chemical phenomena occurring in droplet at solution precursor plasma spraying of zirconia coatings. *Surf. Coat. Technol.* 397: 125059.

52 Pawłowski, L. (2009). Suspension and solution thermal spray coatings. *Surf. Coat. Technol.* 203: 2807.

53 Kudinov, V.V. (1977). *Plasma Coatings.* Moskva, USSR: Izdatelstvo Nauka (in Russian).

54 Steffens, H.-D., Wielage, B., and Drozak, J. (1991). Interface phenomena and bonding mechanism of thermally sprayed metal and ceramic composites. *Surf. Coat. Technol.* 45: 299.

55 Kurz, W. and Fischer, D.J. (1998). *Fundamentals of Solidification*, 4e. Zürich: Trans Tech Publications.

56 Moreau, C., Cielo, P., Lamontagne, M. et al. (1991). Temperature evolution of plasma sprayed niobium particles impacting on a substrate. *Surf. Coat. Technol.* 46: 173.

57 McPherson R. and Cheang P. (1989). Microstructural analysis of Ni-Al plasma sprayed coatings, [in:] *12th Int. Thermal Spraying Conf.*, London, UK, The Welding Institute, Abington Hall, paper 17.

58 Rickerby, D. (1996). Mesurement of coating adhesion. In: *Metallurgical and Ceramic Protective Coatings* (ed. K.H. Stern), 306–331. London: Chapman & Hall.

59 Kitahara, S. and Hasui, A. (1974). A study of bonding mechanism of sprayed coatings. *J. Vac. Sci. Technol.* 11: 747.

60 Grimberg, I., Soifer, K., Bouaifi, B. et al. (1997). Tungsten carbide coatings deposited by high-velocity oxy-fuel spraying on a metallized polymeric substrate. *Surf. Coat. Technol.* 90: 82.

61 Małachowska, A., Winnicki, M., Konat, Ł. et al. (2017). Possibility of spraying of copper coatings on polyamide 6 with low pressure cold spray method. *Surf. Coat. Technol.* 318: 82.

62 Pawłowski L. (1980). The vacuum plasma sprayed copper and vacuum coatings, [in:] *9th Int. Thermal Spraying Conf.*, The Hague, Netherlands, Nederlands Institut voor Lastechniek, The Hague, pp. 299–305.

63 Podlesak, H., Pawłowski, L., Laureyns, J. et al. (2008). Advanced microstructural study of suspension plasma sprayed titanium oxide coatings. *Surf. Coat. Technol.* 202: 3723.

64 Sampath S. and Herman H. (1989). Microstructural development of plasma sprayed coatings, [in:] *12th Int. Thermal Spraying Conf.*, London, UK, The Welding Institute, Abington Hall, UK, paper 53.

65 Tomaszek, R., Pawłowski, L., Gengembre, L. et al. (2006). Microstructural characterization of plasma sprayed TiO2 functional coating with gradient of crystal grain size. *Surf. Coat. Technol.* 201 (1–2): 45.

66 Bianchi, L., Denoirjean, A., Blein, F., and Fauchais, P. (1997). Microstructural investigation of plasma-sprayed ceramic splats. *Thin Solid Films* 299 (1–2): 125.

67 McPherson, R. (1973). Formation of metastable phases in flame and plasma prepared alumina. *J. Mater. Sci.* 8: 851.

68 Li, Y. and Ishigaki, T. (2002). Thermodynamic analysis of nucleation of *anatase* and *rutile* from TiO_2 melt. *J. Cryst. Growth* 242: 511.

69 Gardon, M. and Guilemany, J.M. (2014). Milestones in functional titanium dioxide thermal spray coatings: a review. *J. Thermal Spray Technol.* 23: 577.

70 Pawłowski, L. (2012). Thermal spray coatings. In: *Wiley Encyclopedia of Composites*, 2e (ed. L. Nicolais and A. Borzacchiello), 3014–3034. Hoboken, NJ: John Wiley & Sons.

71 Pawłowski, L. (1991). The properties of plasma sprayed aluminum-aluminum oxide cermets. *Surf. Coat. Technol.* 48: 219.

72 Nahvi, S.M. and Jafari, M. (2016). Microstructural and mechanical properties of advanced HVOF-sprayed WC-based cermet coatings. *Surf. Coat. Technol.* 286: 95.

73 Winnicki, M., Kozerski, S., Małachowska, A. et al. (2021). Optimization of ceramic content in nickel–alumina composite coatings obtained by low pressure cold spraying. *Surf. Coat. Technol.* 401: 126732.

74 Iwamoto, N., Umesaki, N., Endo, S., and Morimura, T. (1987). Characterization of plasma sprayed and whisker reinforced alumina coatings. *J. Mater. Sci.* 22: 1113.

75 Steffens, H.-D., Kern, H., Fischer, G. et al. (1991). Mechanical behaviour of plasma sprayed fibre reinforced composites. In: *2nd Plasma Technik Symposium* (ed. S. Blum-Sandmaier, H. Eschnauer, P. Huber, and A.R. Nicoll), 151–158. Wohlen, Lucerne: Plasma Technik, AG.

76 Unabia, R., Candidato, R. Jr., and Pawłowski, L. (2018). Current progress in solution precursor plasma spraying of cermets: a Review. *Metals* 8: 420.

77 Kozerski, S., Pawłowski, L., Jaworski, R. et al. (2010). *Two zones microstructure* of suspension plasma sprayed hydroxyapatite coatings. *Surf. Coat. Technol.* 204: 1380.

78 Łatka, L., Goryachev, S.B., Kozerski, S., and Pawłowski, L. (2010). Sintering of fine particles in suspension plasma sprayed coatings. *Materials* 3: 3845.

79 Sokołowski, P., Nylen, P., Musalek, R. et al. (2017). The microstructural studies of suspension plasma sprayed zirconia coatings with the use of high-energy plasma torches. *Surf. Coat. Technol.* 318: 250.

80 Sokołowski, P., Kozerski, S., Pawłowski, L., and Ambroziak, A. (2014). The key process parameters influencing formation of columnar microstructure in suspension plasma sprayed zirconia coatings. *Surf. Coat. Technol.* 260: 97.

81 Marynowski, C.W., Halden, F.A., and Farley, E.P. (1965). Variables in plasma spraying. *Electr. Technol.* 3: 109.

82 Lee, D. (1983). A finite elements modeling of the low pressure plasma deposition process – I. *Int. J. Mech. Sci.* 25 (8): 543.

83 Cosack T., Pawłowski L., Schneiderbanger S., and Sturlese S. (1992). Thermal barrier coatings on turbine blades by plasma spraying with improved cooling, (in:] *Proc. 37th ASME Int. Gas Turbine and Aeronautic Congress and Exposition*, Cologne, Germany, 1–4 June, Paper 92-GT-319.

84 Clyne, T.W. and Gill, S.C. (1996). Residual stresses in thermal spray coatings and their effect on interfacial adhesion: a review of recent work. *J. Thermal. Spray Technol.* 5 (4): 401.

85 Kuroda, S. and Clyne, T.W. (1991). The quenching stresses in thermally sprayed coatings. *Thin Solid Films* 200: 49.

86 Takeuchi, S., Ito, M., and Takeda, K. (1990). Modelling of residual stresses in plasma sprayed coating: effect of substrate temperature. *Surf. Coat. Technol.* 43 (44): 426.

87 Łatka, L., Pawłowski, L., Winnicki, M. et al. (2020). Review of functionally graded thermal sprayed coatings. *Appl. Sci.* 10 (15): 5153.

88 Kingery, W.D., Bowen, H.K., and Uhlman, D.R. (1976). *Introduction to Ceramics.* New York, NY: John Wiley & Sons.

89 Lamouria, S., Hamidouche, M., Bouaouadja, N. et al. (2017). Control of the γ-alumina to α-alumina phase transformation for an optimized alumina densification. *Bol. SECV* 56: 47.

90 Kuroda, S., Tashiro, Y., Yumoto, H. et al. (1998). Measurement of stress development during HVOF thermal spray. In: *Thermal Spray: A United Forum for Scientific and Technological Advances* (ed. C.C. Berndt), 805–813. Materials Park, OH: ASM International.

91 McCune, R.C. (2007). The use of cold spray coating for corrosion protection. In: *The Cold Spray Materials Deposition Process* (ed. V.K. Champagne), 302–312. Cambridge: Woodhead Publishing Limited.

92 Thompson, V.S. and Whittemore, O.J. Jr. (1968). Structural changes on reheating plasma-sprayed alumina. *Ceram. Bull.* 47: 637.

93 Lim, L.C., Lim, S.C., Lai, M.O. et al. (1996). Annealing of plasma sprayed WC-Co coating. *Surf. Coat. Technol.* 79: 151.

94 Khor, K.A. and Jana, S. (1997). Pulsed laser processing of plasma sprayed thermal barrier coatings. *J. Mater. Process. Technol.* 66: 4.

95 Pawłowski, L. (1996). Technology of thermally sprayed anilox rolls: state of art, problems and perspectives. *J. Thermal Spray Technol.* 5: 317.

96 Znamirowski, Z., Nitsch, K., and Pawłowski, L. (2013). The electric charge transport in titania–alumina composite cold cathodes made using atmospheric plasma spraying and laser engraving. *Surf. Coat. Technol.* 220: 271.

97 Chen, H.C., Haberlein, J., and Pfender, E. (1997). TEM characterization of plasma-sprayed thermal barrier coatings and ceramic-metal interfaces after hot isostatic pressing. *Thin Solid Films* 301: 105.

98 Khor K.A., Yu L.G., Chan S.H., and Cheang P. (2002). Spark plasma sintering of plasma sprayed HA coatings, [in:] *ITSC 2002 Essen*, E. Lugscheider, C.C. Berndt (eds), DVS Verlag, Düsseldorf, Germany, pp. 1024-8.

99 Brandl, W., Marginean, G., Maghet, D., and Utu, D. (2004). Effects of specimen treatment and surface preparation on the isothermal oxidation behaviour of the HVOF-sprayed MCrAlY coatings. *Surf. Coat. Technol.* 188–189: 20.

100 Knuuttila, J., Sorsa, P., and Mäntylä, T. (1999). Sealing of thermal spray coatings by impregnation. *J. Thermal Spray. Technol.* 8 (2): 249.

101 Ohmori, A., Zhou, Z., and Inoue, K. (1994). Improvement of plasma-sprayed ceramic coating properties by heat-treatment with liquid Mn. In: *Thermal Spray Industrial Applications* (ed. C.C. Berndt and S. Sampath), 543–548. Metals Park, Ohio: ASM International.

102 Ohmori, A., Zhou, Z., and Inoue, K. (1994). Liquid-Mn sintering of plasma-sprayed zirconia-yttria coatings. *Thin Solid Films* 251: 141.

103 Ibrahim, A. and Sampath, S. (1996). Effect of grinding mechanisms on surface finish and hardness of thermally sprayed WC-Co. In: *Thermal Spray: Practical Solutions for Engineering Problems* (ed. C.C. Berndt), 493–499. Materials Park, Ohio: ASM International.

104 Galvanetto, E., Adami, M., Beber, P., and Scaccia, A. (1992). Industrial applications of thermal spraying technology. *Mater. Eng.* 3 (2): 202.

105 Hidouci, A., Pelletier, J.M., Ducoin, F. et al. (2000). Microstructural and mechanical characteristics of laser coatings. *Surf. Coat. Technol.* 123: 17.

106 Demian, C., Denoirjean, A., Pawłowski, L. et al. (2016). Microstructural investigations of NiCrAlY + Y_2O_3 stabilized ZrO_2 cermet coatings deposited by plasma transferred arc (PTA). *Surf. Coat. Technol.* 300: 104.

107 De Hosson, J.T.M. and Teeuw, D.H.J. (1998). Laser deposition of ceramic coatings. In: *Lasers in Surface Engineering* (ed. N.B. Dahotre), 205–254. Materials Park, OH: ASM International.

108 Oukach, S., Pateyron, B., and Pawłowski, L. (2019). Physical and chemical phenomena occurring between solid ceramics and liquid metals and alloys at laser and plasma composite coatings formation: a review. *Surf. Sci. Rep.* 74 (3): 213.

109 Steen, W.M. (1991). *Laser Material Processing*. London: Springer.

110 Wissenbach, K., Backes, G., Drenker, A. et al. (1993). Thermisch Oberfächenbehandlung mit Laserstrahlung. In: *Werkstoffbearbeitung mit Laserstrahlung* (ed. G. Herzigen and P. Loosen), 88–134. München: Carl Hanser Verlag.

111 Vannes, A.B., Langlade-Bomba, C., and Sallamand, P. (1999). Le laser de puissance: un vecteur ideal pour réaliser des traitements superficiels. In: *Revêtement et Traitement de Surface* (ed. S. Audisio, M. Caillet, A. Galerie, and H. Mazille), 499–517. Lausanne: PPUR. (in French).

112 Eustathopoulos, N., Nicholas, M.G., and Drever, B. (1999). *Wettability at High Temperatures*. Kidlington, Oxford: Pergamon and Elsevier.

113 Bayat, M., Nadimpalli, V.K., Pedersen, D.B., and Hattel, J.H. (2021). A fundamental investigation of thermo-capillarity in laser powder bed of metal and alloys. *Int. J. Heat Mass Transf.* 166: 120766.

114 Anthony, T.R. and Cline, H.E. (1977). Surface rippling by surface-tension gradients during laser surface melting and alloying. *J. Appl. Phys.* 48 (9): 3888.

115 Joshi, S.V. and Sundararajan, G. (1998). Laser for metallic and intermetallic coatings. In: *Lasers in Surface Engineering* (ed. N.B. Dahotre), 121–179. Ohio Park, OH: ASM International.

116 Kloosterman, A.B. and De Hosson, J.T.M. (1997). Ceramic particle injection during laser treatment of titanium. In: *Surface Treatment* (ed. M.H. Aliabadi and C.A. Brebbia), 287–294. Southampton: Computational Mechanics Publications.

117 Schnick, T., Tondu, S., Peyre, P. et al. (1999). Laser shock processing of Al-SiC composite coatings. *J. Therm. Spray Technol.* 8 (2): 296.

118 Podlesak, H., Schnick, T., Pawłowski, L. et al. (2000). Microscopic study of Al-SiC particulate composites processed by laser shocks. *Surf. Coat. Technol.* 124: 32.

119 Dantzig, J.A. and Rappaz, M. (2009). *Solidification*. Boca Raton, FL: CRC Press.

120 Blechet, J.J. (1997). *Private information about studies made in DGA, F-94114.* Arcueil, France.

121 Cottam, R. and Brandt, M. (2011). Laser cladding of Ti-6Al-4V powder on Ti-6Al-4V substrate: effect of laser cladding parameters on microstructure. *Phys. Procedia* 12: 323.

122 Chen, T., Deng, Z., Liu, D. et al. (2021). Bioinert TiC ceramic coating prepared by laser cladding: microstructures, wear resistance, and cytocompatibility of the coating. *Surf. Coat. Technol.* 423: 127635.

123 Tjong, S.C. (2000). Performance of laser-consolidated plasma-spray coatings on Fe-28-7Al-1C alloy. *Thin Solid Films* 274: 95.

124 Anjos, M.A., Vilar, R., Li, R. et al. (1995). Fe-Cr-Ni-Mo-C alloys produced by laser surface alloying. *Surf. Coat. Technol.* 70: 235.

125 Almeida, A., Anjos, M.A., Vilar, R. et al. (1995). Laser alloying of aluminium alloys with chromium. *Surf. Coat. Technol.* 70: 221.

126 Vaziri, S.A., Shahverdi, H.R., Torkamany, M.J., and Shabestari, S.G. (2009). Effect of laser parameters on properties of surface-alloyed Al substrate with Ni. *Opt. Lasers Eng.* 47: 971.

127 Suna, G.F., Zhang, Y.K., Zhang, M.K. et al. (2014). Microstructure and corrosion characteristics of 304 stainless steel laser-alloyed with Cr–CrB_2. *Appl. Surf. Sci.* 295: 94.

128 Celik, O.N. (2013). Microstructure and wear properties of WC particle reinforced composite coating on Ti6Al4V alloy produced by the plasma transferred arc method. *Appl. Surf. Sci.* 274: 334.

129 Tondu, S., Schnick, T., Pawłowski, L. et al. (2000). Laser glazing of FeCr–TiC composite coatings. *Surf. Coat. Technol.* 123: 247.

130 Cooper, K.P. and Slebodnick, P. (1989). Recent developments in laser melt/particle injection processing. In: *Laser Materials Processing* (ed. G. Bruck), 3–16. Berlin: Springer Verlag.

131 Li, Y. and Ma, J. (1997). Study on overlapping in the laser cladding process. *Surf. Coat. Technol.* 90: 1.

4

Methods of Films and Coatings Characterization

4.1 Introduction

Films and film coatings are deposited to improve some properties of a substrate in view of an industrial application. These properties are of different nature and are closely related to the deposits microstructure. The characterization of properties and of microstructure takes place in a research and development (*R&D*) stage presented in Fig. I.4 being made carefully using many different techniques. At the end of this stage, the properties of deposits are compared with the initial specifications. The deposits, successfully developed and introduced to the industrial production, are generally tested less carefully at the quality control stage.

The testing microstructure requires not only the knowledge of deposited material and deposition process but also much more time to spend at the *R&D*. It is, however, useful because it enables avoiding "trial-and-error" approach being replaced by fundamental understanding as underlined by Jacobson [1]. The study of microstructure leads to the characterization of deposited material obtained by knowing its:

- chemical composition as well as the chemical bonds, such as sp^2 and sp^3 in *DLC* deposits;
- crystal structure of phases present in the deposits and their orientation;
- morphology of crystal grains including their texture as well as the defects such as voids or different phases and their distribution in the deposits.

The microstructure of films and coatings influences their properties. The latter are tested in function of the future application of the deposits. As there are many applications and their number increases steadily, it is not possible to describe all methods of testing. The following chapter describes the following properties:

- mechanical, which includes adhesion, hardness and microhardness, elasticity, fracture toughness, friction and wear resistance, and residual stresses;
- physical, such as thickness, porosity, thermophysical properties, and contact angle;
- corrosion resistance against liquids and against hot gases;
- electric and magnetic including resistivity, dielectric properties, critical current of a superconductor and magnetization;

Physical Deposition Methods for Films and Coatings, First Edition. Lech Pawłowski.

- optical such as absorption, transmission, and reflection of light;
- biological including some *in vitro* and *in vivo* characterizations.

The characterizations of the properties of the deposits are frequently discussed together with their microstructure and, in the conclusion of this discussion, the deposition process is modified or the deposits are applied to the real parts and, finally, are introduced to the production. The produced pieces are generally submitted to the non-destructive quality tests using acoustical and optical methods being shortly discussed in the following chapter.

The methods of deposits characterization cannot be categorized according to the family of deposits, i.e. atomistic, granular, or bulk. The appropriate choice of the characterization method depends rather on: (i) deposit's thickness; (ii) its chemical composition and the light elements, such as carbon, are particularly hard to analyze; and (iii) available equipment.

4.2 Methods of Microstructure Characterization

The word microstructure suggests the analysis of objects whose size is of the order of a micrometer. However, the real size of tested elements of deposit, such as atoms, molecules, grains, and defects, has a size ranging from a fraction of nanometers to millimeters.

The microstructural studies are realized by excitation of a deposit (sample) with a primary beam composed of the species or the waves. The information about the studied element is obtained by analyzing emitted or transmitted beams (see Fig. 4.1). The

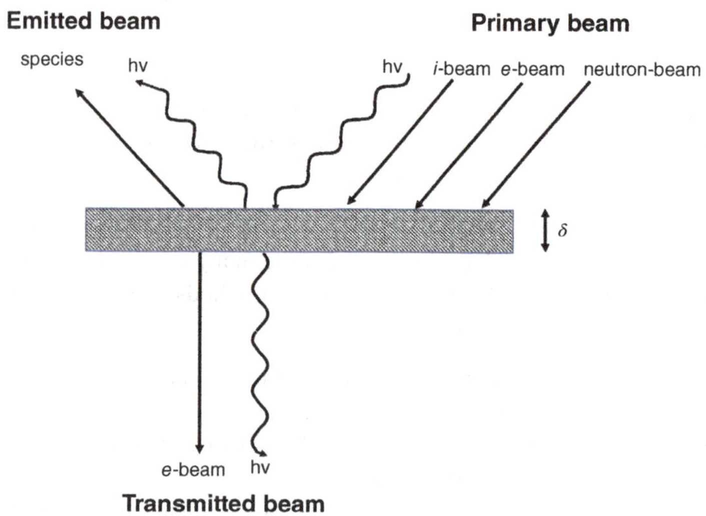

Figure 4.1 The beams of species or waves used to excite the tested sample and result in generation of emitted and transmitted beams used in characterizations of its microstructure (see Table 4.1).

wavelength of the wave in primary beam is smaller than the size of the elements. The following species are used in primary beam to excite a sample[1]:

- electrons in techniques such as *AES*, *EBSD*, *EDX*, *EELS*, *EMPA*, *SAD*, *SEM*, *WDX*, and *TEM*;
- ions in, e.g. Rutherford backscattering spectroscopy and in secondary ion mass spectroscopy;
- neutrons in, e.g. neutron diffraction technique.

The samples can be excited using electromagnetic waves of different wavelengths. The examples of the techniques using waves to excite the samples are as follows:

- visible light in techniques such as *OM* or *RS*;
- *X*-rays in techniques such as *XPS*, *XRD*, or *XRF*.

Finally, there are a couple of new techniques, in which the profile of the tested sample's surface is obtained by using the tunnel effect (*STM*) or piezoelectric effect (*AFM*) without external excitation.

It should be added that the beam of species can be associated with a wave according to the *de Broglie* hypothesis. Thus, the wavelength associated with a particle in movement having mass m, momentum p, and energy E is equal to [2]:

$$\lambda = \frac{h}{p} = \frac{h}{\sqrt{2mE}} \tag{4.1}$$

The excitation of a sample by a primary beam may result in an emission of species of the same kind as that of the primary beam or in the emission of an electromagnetic wave. The species or waves may be emitted from the surface excited by primary beam or from the other side of tested sample. The methods used to characterize the deposits categorized as analyses of chemical composition, crystal structure, and of morphology are shown in Table 4.1. Table 4.2 presents synthetically the characteristics of these methods.

4.2.1 Chemical Composition Analyses

The methods enabling to find the chemical of deposits were selected following their popularity. Some of these techniques, such as *AES*, *EELS*, or *RS*, are used more frequently to characterize thin atomistic films. The popularity of techniques such as *EDX* or *EMPA* results probably from the frequent use of *SEM* to which these techniques are often attached. Finally, the *XPS* technique is used to test all types of deposits.

4.2.1.1 Auger Electron Spectroscopy (*AES*)

The collision of energetic electrons, being a part of primary beam with an atom of the sample, can result in ejection of inner shells electrons. The atom becomes excited and its de-excitation may occur by emission of electrons by *Auger* effect [8, 9]. The kinetic energy

1 The techniques discussed in this chapter are shown using their abbreviations.

Table 4.1 Techniques used in characterization of films and coatings[a].

Chemical composition	Crystal structure	Morphology of deposits
AES	*EBSD*	*AFM*
EDX	*SAD*	*EBSD*
EELS	*XRD*	*OM*
EMPA		*SEM*
Fourier transform infrared spectroscopy [2, 3]		*STM*
Infrared spectroscopy [3]		
Rutherford backscattering spectroscopy [3]		*TEM*
RS		
WDX		Reflection high-energy electron diffraction [4]
Secondary ion mass spectroscopy [3]		
XPS		
XRF		

[a] The techniques which are not discussed in present chapter are shown with the literature source.

of these electrons is approximately equal to the difference of energy between the binding energies of three following electrons involved (see Eq. 4.2): (i) E_1 – ejected from the inner shell by the primary beam; (ii) E_2 – filling the hole from the outer shell; and (iii) E_3 – electron emitted from outer shell, called also *Auger* electron.

$$E_{\text{kin}} = E_1 - E_2 - E_3 \tag{4.2}$$

The kinetic energy is characteristic for the excited atoms. Their analysis enables their identification. It is important to mention that the ejected electrons have relatively low energy which results in the short mean free path in the sample. Consequently, the identification concerns the zone close (distance lower than a few nm) to the sample's surface.

The primary electron beam has an energy of 1–20 keV. The additional beam of ions (typically Ar^+) is used to bombard the surface in order to perform the *AES* analyzes at the depth of a sample. The analyses are is carried out under a high vacuum at a pressure less than $p < 10^{-9}$ hPa. The number of *Auger* electrons ejected from the surface of an excited solid is much smaller than that of secondary or backscattered electrons of the primary beam electrons. Their detection is made using an analyzer such as a *coaxial cylindrical* one enabling obtaining the entire spectrum of kinetic energy emitted or reemitted electrons enabling identification of atoms present in the sample [10].

All elements having more than three electrons involved in the *Auger* effect have their characteristic energies allowing elemental identification. The energies are available in the literature (see, e.g. Ref. 11) making elementary analysis an important application of *Auger* spectroscopy. Many *Auger* spectrometers allow fracturing of tested sample *in situ* in order to avoid any contact of tested surface with air [8].

Table 4.2 Synthesis of the techniques presented in Table 4.1 and discussed in detail in corresponding sections [2–7].

Technique	Primary beam of species or waves	Energy of species in primary beam, *E*, keV	Detected and analyzed beam	Maximum analyzed depth, nm	Resolution in sample's depth, nm	Lateral resolution, nm	Preparation of sample for analysis	Detection limit, ppm	Atomic numbers of analyzed atoms
AES	Electrons	1–20	Electrons reemitted	2–5	2	>10	Surface cleaning necessary	10^3	3–92
AFM	—	—	—	1	0.1	<1	Easy	—	—
EBSD	Electrons	1–30	Electrons reemitted	50	—	50	Polishing necessary	—	all
EDX		1–40	X-ray reemitted	10^2–3×10^3	typically 10^3 can reach 50		Easy	10^2	11–92
EELS		10–300	Electrons transmitted	—	1	—	Thinning to less than 100 nm	5×10^3	3–92
EMPA		1–50	X-ray reemitted	5×10^3	≤50		Easy	10^2	4–92
OM	Visible	—	Visible	—	—	4×10^2–7×10^2	Polishing necessary	—	—
RS	Visible or *ir*	—	*uv*, visible, *ir*	—	1×10^3–5×10^3	2.2×10^2–10^3	Easy	10^4	—
SAD	Electrons	100–200	X-ray diffracted	—	—	0.2	Thinning of sample	10^4	all
SEM		1–30	Electrons reemitted	10 for *SE*; ≥10^2 for *BSE*	—	typical 5–10	Easy for conducting specimens	10^3	5–92
STM	—	—	—	as *SEM*	0.01	0.1	Clean conducting surface is are necessary	—	—
TEM	Electrons	80–1000	Electrons transmitted	—	—	<0.1	Thinning to less than 100 nm required	10^3	5–92
WDX		10–50	X-ray reemitted	5×10^3	5×10^2–10^3		Cleaning of surface required	10	4–92
XPS	X-ray or *uv*	—	Electrons reemitted	0.5–10	1–20	10^4		10^2–10^3	3–92
XRD	X-ray	—	X-ray reemitted	—	2–3×10^4	10^6–10^7	Easy	10^4	all
XRF				—	—	10^5		1	5–92

The *Auger* electrons spectroscopy is used rather for electrically conducting samples such as films of metallic glasses [10]. The technique was also used to identify sp^2 and sp^3 sites[2] in amorphous films containing carbon and hydrogen (a-C:H) by Fuchs *et al.* [12]. The disadvantage of this method of characterization is the interaction of the primary beam with the film, which induces graphitization resulting in increase of sp^2 sites. In addition, diamond deposits having too low electrical conductivity could not be tested by this method [13].

4.2.1.2 Energy-dispersive X-ray Spectroscopy (*EDX*)

The sufficiently energetic primary beam of electrons absorbed by the sample may result in the emission of X-rays. The emitted spectrum includes a continuous part resulting from ***Bremsstrahlung***[3], i.e. radiation generated at deceleration of electron by atoms and other electrons and the pics resulting from deionization of atoms excited by primary beam of electrons as seen in Fig. 4.2. Peaks, such as Kα and Kβ[4], are characteristic for each atom and called also X-ray fluorescence. The wavelength of the photons moving with the light speed, *c*, corresponds to their energy, *E*, in the following way:

$$E = h\frac{c}{\lambda} \tag{4.3}$$

The continuous radiation disappears if the primary beam has too small energy and the electrons are entirely braked in the bombarded sample. That is why, there is a

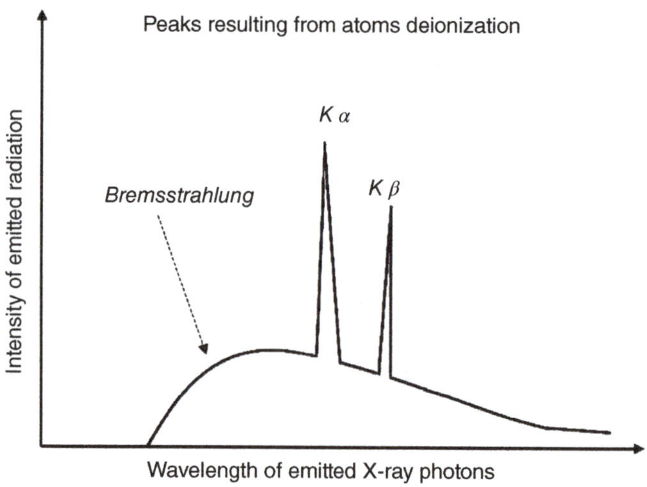

Figure 4.2 The spectrum of X-rays emitted by a sample bombarded by primary beam of electrons showing X-ray characteristic radiation Kα and Kβ.

2 sp^2 and sp^3 are the types of hybridization describing geometry of bonding between the atoms. The hybridization sp^3 corresponds to the tetrahedral pyramid.

3 *Bremsstrahlung* means, in German, braking radiation.

4 Kα or Kβ is a **Siegbahn** notation corresponding to characteristic X-ray radiation lines resulting from dropping of electron from outer shell to the inner shell without electron. The inner shell was excited initially and an electron left it. Consequently, Kα corresponds to radiation associated to electron dropping from L atomic shell to K atomic shell and Kβ – from M shell to the K one.

minimum value of the wavelength of continuous radiation, and this value decreases with the increase energy of electrons in primary beam [2].

The *EDX* spectrometers, contrarily to the *WDX* ones, analyze entire spectrum of energy emitted by tested sample. The emitted energy goes directly to the detector. The spectrometers are frequently part of a *SEM* or *TEM* microscope. The X-rays photons emitted by the sample are detected typically by a diode of Si doped with Li being cooled by liquid nitrogen. The photons produce the pairs of electron–hole in the diode. The number of pairs is proportional to the energy of arriving photons. The current generated by the diode is subsequently amplified enabling the quantification of the photon's energy [8]. The detector enables resolution of 100 eV. The entire spectrum of the photons as a function of their energy, $E = 1$–20 keV, is recorded in time of 100 s and can be observed [2]. This rapidity is an advantage of *EDX* with regard to *WDX* spectroscopy. The strong background electronic noise lowering the signal-to-noise ratio is its major drawback.

The *EDX* technique has become a routine tool for quickly determining a chemical composition in a location of the deposit visualized by a scanning microscope *SEM*. The popularity of this technique is strongly correlated with that of the microscope use and with the rapidity of *EDX* data collection. The *2D* mapping of elemental distribution in a deposit is frequently cited in the literature. Notthof *et al.* [6] presented the mapping of Al distribution in 80 nm thick GaAs/AlAs heterostructures. Cattini *et al.* [14] used *EDX* mapping to visualize the distribution of Si and P in graded coating obtained by suspension plasma spraying of bio-glass as top coating and *HA* as the coating at interface with the substrate (see Fig. 4.3).

4.2.1.3 Wavelength-dispersive X-ray Spectroscopy (*WDX*)

The X-ray emission produced by energetic electron bombardment can be detected using crystal spectrometers. The crystals having the lattice spacing, d_{hkl}, diffract the X-rays of wavelength, λ, arriving under angle, θ, following ***Braggs' law***:[5]

$$2d_{hkl} \sin\theta = n\lambda \tag{4.4}$$

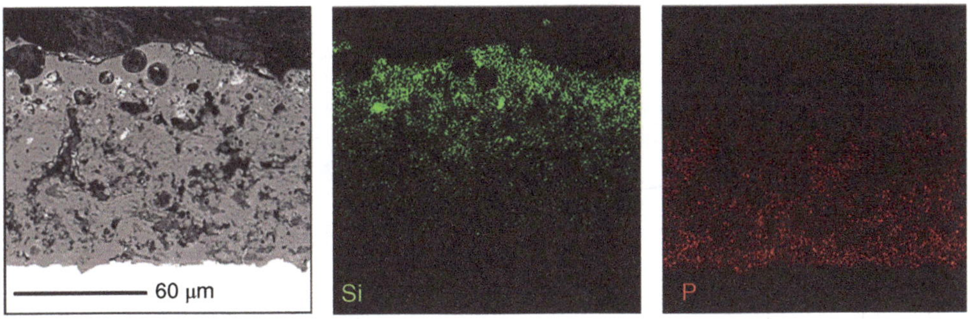

Figure 4.3 *EDX* mapping of graded *SPS* coating showing distribution of Si (representing bio-glass) and P (representing *HA*). *Source:* Cattini et al. [14]/with permission of ELSEVIER.

5 The general equation includes the orders of reflection, n. The first order $n = 1$ has much stronger intensity than the following and is generally supposed that the equation is: $2d_{hkl} \sin\theta = \lambda$.

The *WDX* includes monochromators (called also diffractors) being monocrystals prepared in the way to have the lattice spacing adapted to different X-rays wavelengths' ranges, such as [2, 8]:

- LiF of $d_{hkl} = 0.213$ nm for wavelengths range $\lambda = 0.84–0.33$ nm;
- SiO_2 of $d_{hkl} = 0.3343$ nm for wavelengths range $\lambda = 0.14–0.55$ nm;
- $KAl_3(SiO_4)_3$ of $d_{hkl} = 0.992$ nm for wavelengths range $\lambda = 0.50–1.75$ nm;
- $C_8H_5O_4K$ of $d_{hkl} = 1.306$ nm for wavelengths range $\lambda = 0.55–2.2$ nm.

The crystals are curved to focus the X-ray radiation in the detector. This curve is a part of ***Rowland circle*** where the sample emitting X-rays and their detector are placed. The latter is frequently the gas proportional counter, which counts electrically the arriving photons ionizing the gas being often argon.

The *WDX* analysis is precise enabling quantification of elements present in tested sample. For example, Kommer *et al.* [15] carried out the elemental analysis of magnetron-sputtered molybdenum nitride wear-resistant films. The authors optimized the deposition process by finding the influence of process parameters, such as working gas pressure on the content of Mo to N ratio varying from 1.05 to 1.5. The elemental *WDX* analysis was used for anti-bacterial *DLC* – Ag composite films obtained by pulsed laser deposition [16]. The content of Ag in the composite films varied from 1.07 up to 9.34 at. % depending on repetition rate of KrF laser pulses bombarding the Ag target.

4.2.1.4 Electron Energy Loss Spectroscopy (*EELS*)

The primary beam of electrons may pass through a thin sample and the loss of its energy can be analyzed. Some of the electrons traverse the sample without any interaction and preserve their initial energy. Some others have inelastic collisions with atoms in the sample losing some initial energy. Their energy loss is detected by electron spectrometer on the other side of the sample. The typical electrons energy loss spectrum is shown in Fig. 4.4.

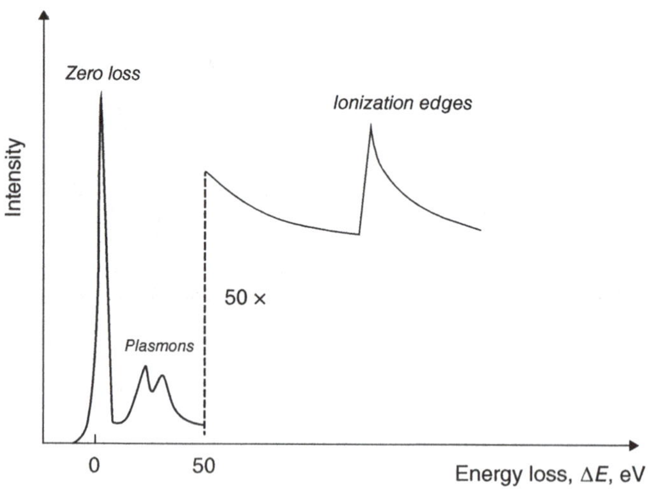

Figure 4.4 The spectrum of the energy loss of electrons traversing a sample. *Source:* Martin and George [2]/PPUR presses polytechniques.

Typical spectrum has three characteristic parts at [2, 8]:

- $\Delta E \approx 0$, there is a peak of the electrons which have not undergone any interaction with the sample;
- $\Delta E < 50\,\text{eV}$, there are the peaks for metal and alloys samples of the ***plasmons*** excitations associated with the collective oscillations of the free electrons;
- for $\Delta E > 50\,\text{eV}$, there is a zone of very small peaks, whose observation requires the increase 50× of intensity scale. The zone contains some characteristic discontinuities called ***edges***. The *edges* correspond to the energy absorbed by atomic excitation, which enables atom's identification. These *edges* are smaller for the samples including having many electrons, and the *EELS* technique i*s well adapted to light atoms, e.g. carbon,* being complementary to the *EDX* technique.

The electrons, having traversed the sample, are detected with help of magnetic field emitted by so-called ***magnetic prism*** [2]. The field interacts on the electrons with the ***Lorentz force***, which depends on their energy and which modifies their trajectory. Finally, the deviated electrons can be detected using a semiconductor. The beam of traversed electrons may excite the electrons in the valence band of the semiconductor. The excited electrons may jump to the conduction band promoting the detectable generation of ***electron–hole pair.*** The pairs can move in the electric field applied to the semiconductor by the electrodes generating electric pulse, which can be amplified. Finally, the electric pulses can be related to the energy of the electrons traversing the tested sample.

Among several applications of *EELS* to characterize deposits, the characterization of *DLC* films should be highlighted. It is important to know if there are bonds between carbon atoms of the type sp^3 (as in diamond), which render the film hard and resistant. In contrast, the bonds sp^2 render the film similar to graphite [17]. *EELS* makes it possible to discover carbon atoms with these two types of bonds using a part of the spectrum at low ΔE with a peak characteristic to interaction with *plasmons* between 20 and 34 eV [18]. The part at high energy has the edge at $\Delta E = 285\,\text{eV}$, which moves toward 290 eV with the increase of sp^3 bonds fraction.

4.2.1.5 Electron Microprobe Analysis (*EMPA*)

The electron microprobe analysis (*EMPA*) is based on the identification of the X-ray radiation caused by the deionization of an atom after an inelastic shock with an electron. The detection of the X-ray photons can be made using wavelength-dispersive (*WDX*) or electron-dispersive (*EDX*) X-ray spectroscopy. These two techniques are described separately. *EMPA* is a more general term, which is used traditionally beginning from the late 1940s when the French engineer ***Raimond Castaign*** developed this analytic technique [19].

The *EMPA* technique with a *WDX* spectrometer is often applied to obtain: (i) distribution map of the elements on an analyzed surface (see Fig. 4.3); or (ii) concentration of the elements along a line of analysis. The latter technique enabled Kitahara and Hasui [20] to visualize the diffusion of W and Mo thermally sprayed to Al substrate. Deram *et al.* [21] presented a similar example of its application by showing the profiles of Ca and P being the component of plasma sprayed *HA* coating onto Ti substrate. The authors detected also a pic of Al at the coating's interface (Fig. 4.5). This pic corresponded to the Al_2O_3 particle incrusted in the substrate at the substrate's activation by sand-blasting.

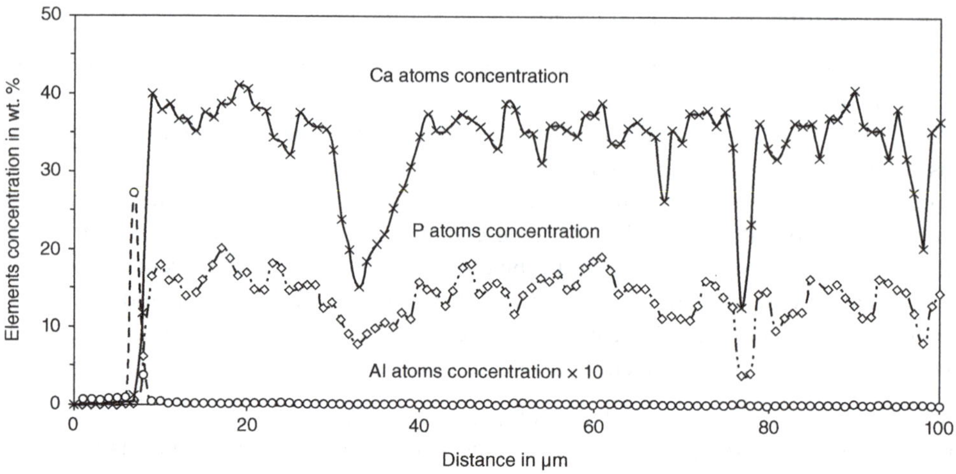

Figure 4.5 *EMPA* analysis using *WDX* detector of a line across the *HA* coating obtained by atmospheric plasma spraying onto Ti substrate being sand-blasted using Al_2O_3 prior to deposition. *Source:* Deram *et al.* [21]/with permission of Elsevier.

4.2.1.6 Raman Spectroscopy (*RS*)

The Raman spectroscopy, *RS*, is used to identify chemically the molecules or ions by identification of their vibrational of frequency ν_m by using the inelastic scattering of incident monochromatic light characterized by the photons having frequency ν_0. The radiation is generally emitted by lasers at *uv*, visible, or near-*ir* frequencies. The incident photons are diffused and get scattered by tested molecule or ion. ***Rayleigh scattering*** corresponds to elastic collisions. The frequency of incident photons, ν_0, remains the same. Inelastic collisions, corresponding to ***Raman scattering***, result in much weaker intensity of emitted radiation. The incident frequency in its spectrum is shifted by vibrating species to:

- lower values, $\nu_0 - \nu_m$, by ***Stokes scattering***[6], which consists in rendering a part of incident photon energy to the species;
- greater values, $\nu_0 + \nu_m$, by ***anti-Stokes scattering*** by taking a part the energy from the tested species.

The entire spectrum of light diffused by tested molecule, which oscillates with a frequency ν_m, includes the peak of incident light frequency ν_0 and the peaks at the frequencies $\nu_0 + \nu_m$ and $\nu_0 - \nu_m$. [22]. In the practice of *RS* wavenumber, $\tilde{\nu}$, is used rather than frequency. The wavenumber is defined as:

$$\tilde{\nu} = \frac{\nu}{c} \qquad (4.5)$$

Since the spectrum of vibrations is characteristic for chemical molecules, the *RS* allows identifying the molecular composition of tested sample. The identification is rapid. The typical spectrometer includes [23]:

6 The resulting shifts of incident frequencies are called consequently Stokes and anti-Stokes shifts, and similarly the obtained spectra of radiation are called Stokes and anti-Stokes spectra.

- incident light source such as Nd: YAG laser;
- optical system, which allows focusing the incident light on the sample and the scattered light on a spectrum analyzer;
- optical filters enabling separation of *Raman scattered* light used to the chemical analysis from the *Rayleigh scattered* light in the diffused spectrum;
- detectors such as photodiodes, or other photo-sensitive devices, such as metal oxide semiconducting (*MOS*) capacitors applied in charge-coupled devices (*CCD*) used in modern spectrometers [24].

The *RS* analysis is performed on a small area illuminated by laser, but it is also possible to obtain images by scanning successive lines on the surface of a sample [25]. Raman spectroscopy is frequently used in the analysis of *DLC* deposits. The Raman spectra enable identification of sp^2 bonds corresponding to graphite of the wavenumber about $1570\,cm^{-1}$ and called G-peak and at about 1360 called D-peak [26]. The parameters of Raman spectra corresponding to these peaks (width, intensities, ratio, etc.) enabled indirect but qualitative estimation of the content of sp^3 sites corresponding to diamond in the *DLC* + Si deposits obtained by PECVD [26] and in the *DLC* deposits obtained by ion plating [27]. Raman spectroscopy was also useful in analysis of HA coatings doped with Cu obtained by emerging SPPS technology. The analysis of the peak in Raman spectrum, corresponding to phosphate group PO_4^{3-} around $940–980\,cm^{-1}$ shown in Fig. 4.6, was interpreted by the decrease of the crystallinity of the sprayed coating with increasing Cu content [28].

4.2.1.7 X-ray Photoelectron Spectroscopy (*XPS*)

The principle of *XPS* consists of bombarding a sample with the mono-energetic X-ray photons. The X-rays used frequently are Mg $K\alpha_{12}$ having energy $h\nu = 1256.6\,eV$ or Al $K\alpha_{12}$ having energy $h\nu = 1486.6\,eV$ [29]. The photons penetrate the solid, and some of them may

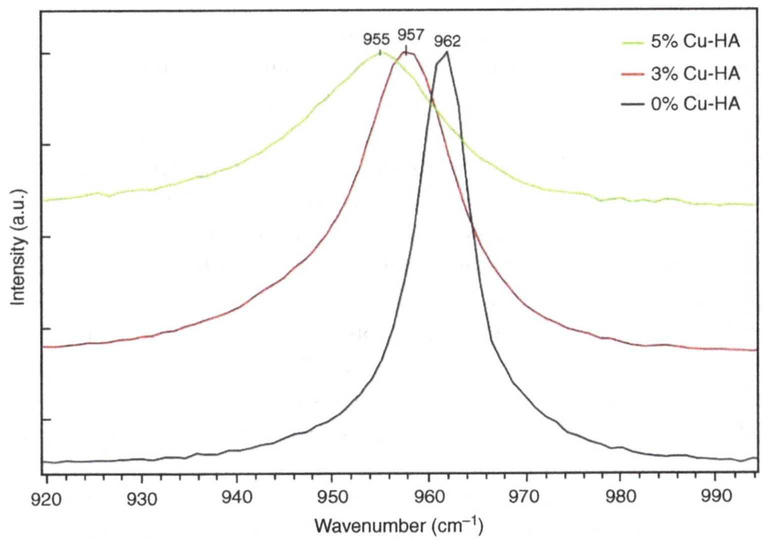

Figure 4.6 The peak corresponding to PO_4^{3-} vibrations in the Raman spectra of *HA* + Cu cermet coatings obtained by solution precursor plasma spraying. *Source:* Unabia *et al.* [28]/with permission of Elsevier.

ionize the atoms if their energy is greater than binding energy electrons at different energy levels, E_{bin}. These emitted photoelectrons have kinetic energies, E_{kin}, and may traverse the solid up to its surface and leave it. The mean free path of the photoelectrons is, generally, not greater than 10 nm which limits considerably the depth of *XPS* analyses. The balance of energy of the photoelectrons is shown in *Eq. 4.6*. The balance includes also the work function, *WF*, which can be explained as thermodynamic work necessary to bring an electron from the solid to the vacuum out of its surface [30]:

$$h\nu = E_{bin} + E_{kin} + \mathrm{WF} \tag{4.6}$$

The work function depends on the type of material (metal, semiconductor) and on its surface homogeneity, contamination, and roughness. The values of *WF* are in the range of a few eV, and the function does not influence considerably the balance of energy. The binding energy is characteristic for the elements, and one has a particular set of photoelectron transition corresponding to the atomic levels [3]. The binding energy is shifted if the atom is ionized being a part of a compound. Consequently, the analysis of kinetic energy of photoelectrons enables identification of atoms and their chemical status.

The typical *XPS* spectrometer is composed of [3, 29]:

- source of photons being frequently radiation of Mg $K\alpha_{12}$ or Al $K\alpha_{12}$ obtained by *e*-beam excitation of metallic targets and of a monochromator diffracting and focusing the emitted X-rays in order to reach the narrow spectrum of *FWHM* <1 eV;
- chamber of analysis, which requires high or very high vacuum with the pressure in the range of $p < 10^{-10}$ hPa in order to avoid contamination of tested surfaces. The chamber may be equipped in a sputtering device to clean up the surface before testing;
- electron energy analyzer being frequently hemispherical one which deviates the photoelectrons electrostatically and determines the radius of deviation. The current of photoelectrons is then multiplied before being delivered to the data acquisition system.

The *XPS* has the ability to find different levels of oxidation of the elements. The binding energy found thanks to *Eq. 4.6* and attributed to the chemical status of element is shifted when the element is a part of compound. For example, the *XPS* spectrum shows the peak of Ti $2p_{3/2}$ ranging from $E_{bin} = 454.1$ eV for atomic Ti to 458.5 eV or Ti^{+2} being a part of TiO_2 [31]. Such shifting may be useful in understanding of deposition processes. Fig. 4.7 shows the *XPS* spectra of initial powder composed of mixture of TiO_2 and TiO_x (x < 2) being initially plasma sprayed. The sprayed coatings were laser engraved. The peak Ti $2p_{3/2}$ corresponding to the binding energy of Ti ions in the initial powder was shifted toward greater binding energy in sprayed coating with regards to initial powder of $\Delta E_{bin} = 0.7$ eV. This shifting indicated the change of oxidation state, which could have resulted from the modification of oxygen content in titanium oxide at heating and cooling during plasma spraying. The post-spray laser treatment did not result in the *XPS* spectrum modification [32]. Similarly, the *XPS* analysis allowed explaining the color change of 8*YSZ* obtained by plasma spraying in open atmosphere (*APS*) and under vacuum (*VPS*) [33]. The coatings had darker surface compared to the initial powder due to the reduction of ZrO_2 to Zr_2O_3 during spraying. This reduction was more intensive at *VPS* deposition. Finally, the *XPS* analysis of chromium oxide powders enabled to identify metallic chromium inclusions in the powder prepared by the sintering and grinding method. This powder is then less suitable for spraying deposits having specifications, which exclude metallic contamination [34].

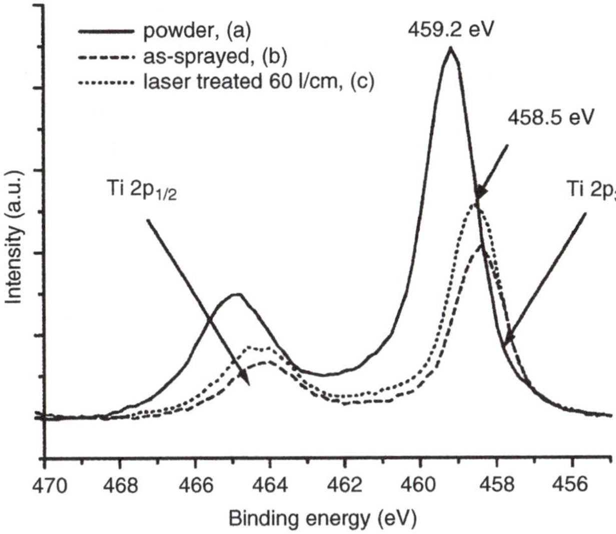

Figure 4.7 *XPS* spectra of Ti 2p energy bands corresponding to different stages of TiO_2 coating preparation using *APS* technology associated with post-spray laser engraving.: (a) initial powder; (b) as-sprayed coating; and (c) laser engraved coating. *Source:* Tomaszek et al. [32]/with permission of Elsevier.

4.2.1.8 X-ray Fluorescence Spectroscopy

The *XRF* spectroscopy uses an emission of X-rays, called secondary or fluorescent by a sample being initially excited by the power source of primary X-ray beam absorbed by the sample's material. The excitation by primary beam produces the knocking out an electron producing a hole in the shell close to atom's nucleus. Subsequently, an electron from an outer shell drops down. To restore the balance of energy, the atom emits a photon of X-ray. This emission of secondary radiation, such as Kα (associated with L→K shells transition) or Kß (M→K transition) shown in Fig. 4.2, is characteristic to each element which enables its identification. For example, excited Ti emits characteristic lines having following energies [35]: (i) Kα1, E =4.511 keV; (ii) Kα2, E = 4.505 keV; and (iii) Kß1, E = 4.932 keV. In general, the X-rays have energies varying from 100 eV to 1 MeV and the photons corresponding to this energy have wavelengths ranging from 10 nm down to 1 pm [36].

The primary X-ray beam, used to excite the samples, is emitted by generators in which the current of electrons (I = 10–100 mA) accelerated by a voltage of a few tens of kV toward an anode [2]. The primary and secondary X-ray beams can be focused by ***polycapillary optics*** down to the size of 20 μm [37]. The secondary X-ray beam is used to identify the elements using *EDX* or *WDX* spectrometers described in the corresponding sections. Such identification is advantageous, comparably to these two methods, because the X-rays irradiation of tested samples can be carried out in air, contrarily to *e*-beams used by *EDX* and *WDX* methods, which necessitate the application of vacuum chamber. Another important advantage of *XRF* spectroscopy is the fact that that the X-rays primary beam does not damage the tested specimen. These are the main reasons of using the *XRF* equipment *in situ* in medical applications [35].

Another important application of *XRF* results from the dependence of secondary X-ray beam radiation intensity on sample thickness. This dependence was used to determine the thickness of $CuGaSe_2$ film produced by chemical methods for solar cells [38]. Similarly, Zhong and Chen [39] used *XRF* equipment to determine thickness of Ti, Ni-V alloy, and Ag multi-layers sputtered on Cu substrate. The measurement of films' thicknesses can be also done *in situ* during the deposition process.

4.2.2 Crystallographic Analyses

Crystallographic analyses concern many atoms or molecules agglomerated as crystals and describe their organization. The techniques discussed in present chapter are based on the diffraction of X-rays (*XRD* and *EBSD*) or the waves associated to high-energy *e*-beam (*SAD*). The analyses to find the following characteristics of the crystals present in films and coatings:

- identification of crystalized phases;
- geometrical orientation of the crystals, i.e. their ***texture***;
- degree of their crystallization (the lowest degree is amorphous phase);
- crystals' size;
- presence of mechanical stresses generated in films and coatings during deposition.

4.2.2.1 X-ray Diffraction (*XRD*)

Bragg's law (*Eq. 4.3*) describes the diffraction of X-rays on crystal planes. The wavelength of the X-rays source generated radiation generated in an X-ray tube is essential in the analysis of the crystals. The wavelengths used in practice vary from $\lambda = 0.071$ nm (corresponding to Mo Kα) to $\lambda = 0.154$ nm (Cu Kα). The detectors measure the intensity of the diffracted radiation. The detectors are sensible on the X-ray radiation based on the phenomena of gas ionization or generation of the electrons–holes pairs in a solid-state semiconductor. The diffraction experiments are performed in different configurations for thick coatings (more than 10 μm of thickness) and for thin films. Thick coatings are tested with a so-called ***powder diffractometer*** in which the radiation source and the detector move around the circumference and the sample, placed in its center. The incidence angle is θ being equal to the angle of diffracted radiation measured by detector. The geometry of such diffractometer is known as ***Bragg–Brentano*** [2].

The X-ray diffraction enables identification of the crystal phases being present in tested sample. The experimentally obtained diffractograms, which include the intensity of peaks obtained at diffraction angle, 2θ, may be identified using *JCPDS*[7] ***Powder Diffraction Files.*** This database includes the phases identified previously in the scientific papers and in the institutes working on the standards. The new phases present in the deposits obtained using emerging method can be found directly in the literature. The latter was, e.g. necessary to identify CN_x phase obtained by laser ablation in nitrogen post-discharge atmosphere [40].

The *powder diffractometers* are useful for the crystal size being not greater than 10 μm. The bigger crystals and monocrystals can be characterized using ***Laue*** method [2].

7 The Joint Committee on Powder Standards, *JCPDS*, was initiated by American Society of Testing Materials, *ASTM*.

The coatings having privileged crystals orientation, i.e. *texture*, have the intensity of diffracted peaks radiation different from that of polycrystalline samples, which renders their identification more complicated. The size of the crystals, d_c, can be determined from the broadening of peaks in the diffractogram, $\Delta(2\theta)$[8], using the ***Scherer equation***:

$$d_c = \frac{K\lambda}{\Delta(2\theta)\cos(\theta)} \tag{4.7}$$

This equation enabled to find the sizes of TiO_2 nanoparticles having sizes of tenths of nm [41]. Another possibility is an application of ***Williamson and Hall method***; which takes into account the peaks broadening at *FWHM* and the strain in films and coatings. This method enabled to find the size of crystals in commercial *HA* coating obtained by *APS*, $d_c = 80\,\text{nm}$ [21] and $d_c = 100\,\text{nm}$ in *HA* coating obtained by *SPPS* [28].

The *powder diffractometer* geometry is not useful to very thin films or to graded films. The X-rays penetrate well inside the sample, and the diffracted signal from the top film or coating may be too weak and not sufficient for the identification. Zhang *et al.* [42] used the X-ray diffraction using small angle of incidence, *GXRD*, to analyze thin FeS films. The angles of incidence of X-ray radiation varied between 1 and 10°, which corresponded to the depth of penetration of this compound ranging from 0.14 to 1.39 µm. The X-ray penetration, limited to 10 nm, can be obtained by using technique of total reflected X-ray diffraction useful for monocrystals, *TOREXD* [43]. The incident radiation turn around for the crystal surface under an angle was smaller than the critical one, θ_{cr}, which satisfies the *Braggs diffraction* condition. This technique, using Cu Kα radiation ($\lambda = 0.154\,\text{nm}$), was applied for thin monocrystalline films of Al being 83 nm thick obtained by epitaxy.

4.2.2.2 Selected Area Diffraction (*SAD*)

The selected area diffraction (*SAD*) of electrons is frequently used together with images obtained using transmission electron microscope (*TEM*). The primary *e*-beam is the same, but the electrons *SAD* pass through a thin sample being diffracted by the crystal planes. The electrons can be treated as the waves according to *de Broglie* theory. The diffraction enables identification of crystal structure of the sample. The wave associated with electrons has very short wavelength, λ [nm], depending on the accelerating voltage accelerating, U [V], in the following way [44]:

$$\lambda = \sqrt{\frac{1.5}{U}} \tag{4.8}$$

The wavelength is as short as $\lambda = 3.9\,\text{pm}$ for $U = 100\,\text{kV}$ and only $\lambda = 1.2\,\text{pm}$ for $U = 1\,\text{MV}$. The voltage increases also the energy of electrons and renders possible longer paths inside the solid sample. Knowing that the interplanar spacings of diffracting crystal planes, d_{hkl}, are typically about a few hundreds of pm it is necessary to use a small angle of diffraction being lower than $\theta < 1°$. Consequently, the Braggs law describing diffraction is simplified with regard to *Eq. 4.4* and becomes:

$$2d_{hkl}\theta = \lambda \tag{4.9}$$

8 The instrumental peak broadening should be also considered.

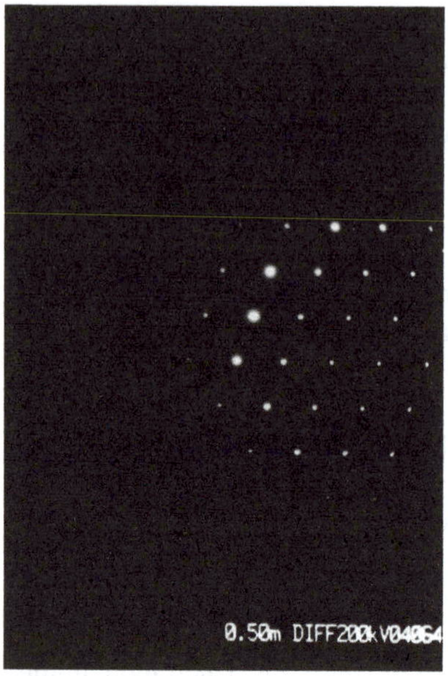

Figure 4.8 Selected area diffraction pattern of one of the Al grains shown in Fig. 4.9 and corresponding to Al (111) lattice plane. The sample was obtained by *HVOF* spraying and laser treatment of Al + 50 wt. SiC composite coating. *Source:* Podlesak *et al.* [46]/with permission of ELSEVIER.

Finally, the diffracted beams are nearly parallel to the relevant crystal planes. The diffraction pattern of a very thin sample (see, e.g. Fig. 4.8) shows typical directly transmitted primary *e*-beam and a few diffracted beams. Their geometrical analysis, knowing the crystal structure, enables determination of Miller indices of diffracting planes described by Loretto [8] and Beaunier *et al.* [44].

An advantage of short-wavelength electron diffraction, over X-ray diffraction, is the possibility of obtaining information from much smaller grains or crystals. The area of *SAD* analysis can be estimated in a few hundreds of nm compared to a few of cm at *XRD*. An important drawback is the necessity to thin down the samples. The thinning is necessary to enable electrons to go through entire sample thickness.

Diffraction patterns of small or amorphous grains do not have the peaks as in Fig. 4.8. Instead, there are the diffraction rings formed around the peak of the primary beam. In the diffracted rings of small grains, it is possible to observe small peaks corresponding to the diffracted beams. These rings become diffused for amorphous grains [3].

4.2.2.3 Electron Backscatter Diffraction (*EBSD*)

The energetic electrons bombarding a surface of a sample may enter in collision with the atoms being ***inelastically scattered*** and, later on, are diffracted by the crystal lattice of the material. The electrons backscattered to the bombarded surface are used in its imaging using scanning electron microscope, *SEM*. The backscattered and diffracted electrons are used in the technique *EBSD*. These electrons bring the information about the material structure close to the sample surface (depth of about 50 nm) [47]. The diffraction patterns known as ***Kikuchi lines*** (see Fig. 4.9a) are used in the analysis.

The first observation of *EBSD* pattern was made in 1954 [48]. The *EBSD* set up is a part of scanning electron microscope and includes the scintillating screen (frequently of phosphor) and a *CCD* camera connected to the computer helping in *Kikuchi lines* indexing (see Fig. 4.9b) described in detail by Britton *et al.* [49]. The primary *e*-beam bombards the tested sample under small angle (20° or less) with regard to the surface to increase the number of backscattered electrons.

The most frequent application of *EBSD* is the investigation of ***microtexture***, i.e. identification of the phase composition and crystallographic orientation of individual grains identified using *SEM*. The *microtexture* is not necessarily the same as the sample texture. It depends on the location of tested area and on the number of considered crystal grains. The tested samples should be metallographically polished. However, the final step of preparation should be electropolishing or etching instead of diamond polishing deforming the surface [47].

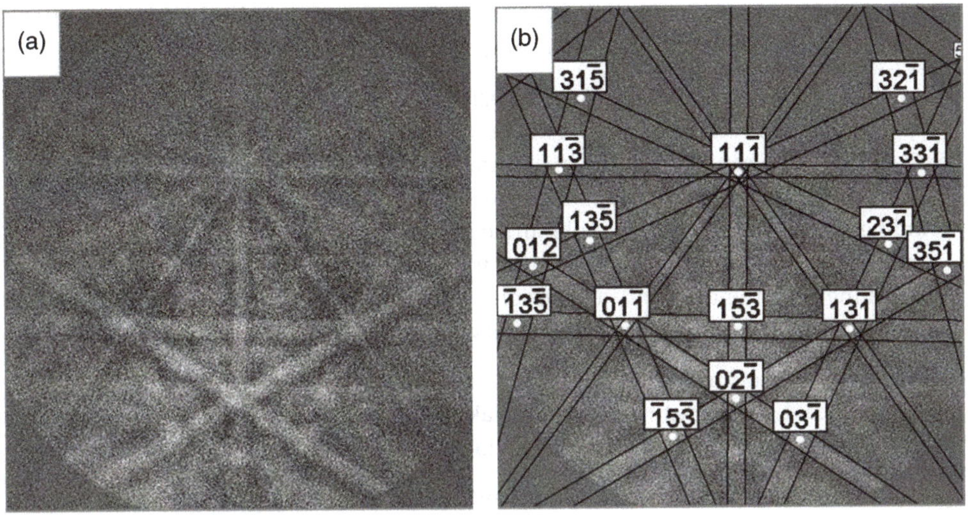

Figure 4.9 The *EBSD* pattern of tetragonal ZrO_2: (a) obtained originally; (b) indexed. *Source:* Podlesak *et al.* [45]/with permission of ELSEVIER.

The *EBSD* technique was used to analyze the *microtexture*, size, and volume of crystal grains of suspension plasma sprayed ZrO_2 stabilized using Y_2O_3 and with CeO_2 obtained using different plasma torches. The analysis helped in finding the optimal processing to obtain columnar microstructure of coatings useful for application as thermal barriers [50].

4.2.3 Analyses of Morphology

The analysis of films and coatings morphology concerns their surface and internal part. The surface is tested mainly using scanning electron microscope (*SEM*). Some studies are realized using atomic force microscope (*AFM*) or with scanning tunneling microscope (*STM*). The internal morphology of films and coatings can be studied using *SEM* observations of the deposits cross-sections or by transmission electron microscopy (*TEM*) observation of thinned samples. This morphology of thick coatings, manufactured using granular and bulk methods, is studied generally using optical metallographic microscope (*OM*). This microscope enables magnification up to 1000 × and requires very careful preparation of the tested samples. The preparation consists of polishing of the coatings' cross-section and is very important part of the morphological study. Scanning electron microscope has become the most popular tool in the development of any type of deposit.

4.2.3.1 Optical Microscope (*OM*)

The optical microscope has been used for centuries and is described in several manuals such as [51]. The microscope is most frequently used to observe the cross-sections of deposits. As the spectrum of visible light is between 400 and 700 nm, the smallest elements of the microstructure to be observed, being the microscope lateral resolution, are slightly lower than 1 μm. The longitudinal sections[9] of the samples to be characterized by

9 The section is perpendicular to the sample's surface.

optical microscope require a careful preparation. To take an example, the typical preparation steps of such section for thermally sprayed coating are as follows [52]:

- impregnation of porous or fragile sample using a low-viscosity resin;
- preparation of the section by using a diamond or SiC saw;
- embedding of the sample selected in a resin, to be done under vacuum for porous coatings to form a cylinder being useful for the next step with a typical diameter of 25–35 mm;
- grinding and polishing of the sample surface which is done gradually:
  - grinding using SiC emery paper starting from grade 120 up to 1200;
  - polishing with the diamond suspensions with the solid grains having size 6 μm and, later on, with suspension with the grains with 3 μm size.

The grinding and polishing should be carried out carefully in order to avoid the pull out of the grains being weakly bound with coating. If we take the example of granular and/or bulk deposits, such preparation of the metallographic section allows to visualize following Fowler [53]:

- porosity and size of pores and fraction and size of unmelted particles shown, e.g. in Fig. 3.13;
- thermal or mechanical deformation of the substrate at the interface with the deposit;
- the grains being reduced or oxidized during deposition (see Fig. 3.14) as well as the metallic inclusions being a part of eroded anode or cathode at plasma spraying;
- different phases and their distribution in the composite deposits as seen in Fig. 3.15;
- the dendrites present in bulk deposits as seen in Figs. 3.26 and 3.27.

The quantitative analyses of porosity, fraction of different phases or of unmelted particles in the longitudinal section of coatings are possible using free accessible software, such as ***ImageJ.***

4.2.3.2 Scanning Electron Microscope (*SEM*)

Scanning electron microscope, *SEM*, has become the most popular tool to analyze the morphology of films and coatings and, that is why, it is described here in detail. The first microscope was developed and described by von Ardenne in 1938 [54]. The microscope started to be proposed commercially in 1960s by Cambridge Instruments [55].

The primary *e*-beam used in *SEM* is generated by thermal emission by the cathodes of tungsten requiring pressure of $p = 13.3\,\text{mPa}$ or that of lanthanum hexaboride working under pressure of $p = 1.33\,\text{mPa}$. At present, the field emission cathodes have become popular because of their long lifetime. However, such cathode requires much lower working pressure of about $p = 0.133\,\mu\text{Pa}$ [56]. The electrons of the primary *e*-beam are accelerated using anode having potential of up to $U = 30\,\text{kV}$ and focused by the electro-magnetic lenses. Finally, the *e*-beam passes through the deflection coil, which scans it periodically across the observed surface. The number of lines per scan ranges from a hundred up to a few thousands [5]. The *e*-beam bombards the surface and the electrons interact with the atoms of the specimen by [44, 55, 56]:

- inelastic collisions occurring in the depth of lower than 10 nm below the surface, followed by the ionization of outer shells of atoms[10] and the emission of secondary electrons, *SE*, having generally kinetic energy $E_{\text{kin}} < 50\,\text{eV}$;

10 The emissions of electrons from inner shells of atoms, called *Auger electrons,* having kinetic energy slightly greater than that of secondary ones and emitted from the depth of a few nm is described in the section Auger Electron Emission, *AES*.

- elastic collision occurring in the depth of a few hundred nm depending on the atomic number of colliding atoms leading to the generation of backscattered electrons, *BSE*, which have the kinetic energy slightly lower than that of the bombarding electrons.

The secondary electrons come from the surface region of the deposit and enable visualizing the surface's topography. This capacity of SEM imaging is called also ***topographic contrast***. The rate of emission of backscattered electrons depends on the cross-section of collision, which depends, in turn, on the atomic number of specimen's atoms enabling to visualize the ***chemical contrast*** in the SEM images obtained using *BSE*. Consequently, the detectors of secondary and backscattered electrons are different.

The *SE* are attracted by a collector equipped with the grid being under positive potential of 200 up to 400 V before being treated by a scintillation detector. The detector provides an electrical signal that is amplified and modulates the beam of a cathode ray tube, *CRT*. The scanning of photons in *CRT* is synchronized with that of the primary *e*-beam [2, 8, 56]. The magnification of the microscope is equal to the ratio of scanning on the screen of the *CRT* to that of the sample.

The *BSE* are more energetic and do not require the electrostatic attraction by a collector. On the contrary, the grid is charged negatively to $U = -200\,\text{V}$ to push away the *SE*. The detector of *BSE* is frequently the semiconductor with the p-n junction enabling generation of electron–hole pairs. The electrical current is then treated to obtain image-forming signal.

The specimens to be observed by *SEM* may require some preparation. The tested samples should be cut to enter in the *SEM* working space. The specimens must be able to conduct electrons in order to avoid the accumulation of electrical charge. Consequently, the non-conducting specimens of ceramic or polymers or the conducting ones embedded in resin should be coated with an electrically conducting film. The films should be thin (10–25 nm) to avoid the influence on the topography of specimen. The typical film materials are Al, C, Au, Pd, or Pt. The biological specimens can be easily modified by *e*-beam at *SEM* investigation and should be prepared much more carefully [55]. The scanning electron microscope is useful for observing and characterizing the elements of deposits such as morphologies of:

- surface of as-deposited and post-deposition-treated coatings using *SE* as shown in Figs. 3.18a and 3.23 correspondingly;
- fractured sections of as-deposited coatings using *SE* shown in Figs. 3.17;
- polished cross-sections of a coating using *SE* shown in Fig. 3.18b and the *chemical contrast* using *BSE* of a deposit submitted to the test shown in Fig. 4.10. The latter was complementary to the *EDS* analysis.

4.2.3.3 Scanning Tunneling Microscope (*STM*)

The scanning tunneling microscope (*STM*) functions based on the effect of ***quantum tunneling***. This means that there is some probability that an electron has a position described by a wave function in Schrödinger equation and may exist out of the metal surface. The wave function may overlap with that of the electron from neighboring surface isolated by a small vacuum gap [58, 59]. The real microscope with the tunneling effect was developed a few years in 1980s in the *IBM* laboratories in Switzerland.

The microscope includes a metallic tip under a voltage in the range of a few mV to a few V with regards to the observed sample's electrically conducting surface [5]. The tunneling electric current may flow if the distance is smaller than 1 nm. The tunneling current decreases exponentially with the distance between tip and the surface, and its variation

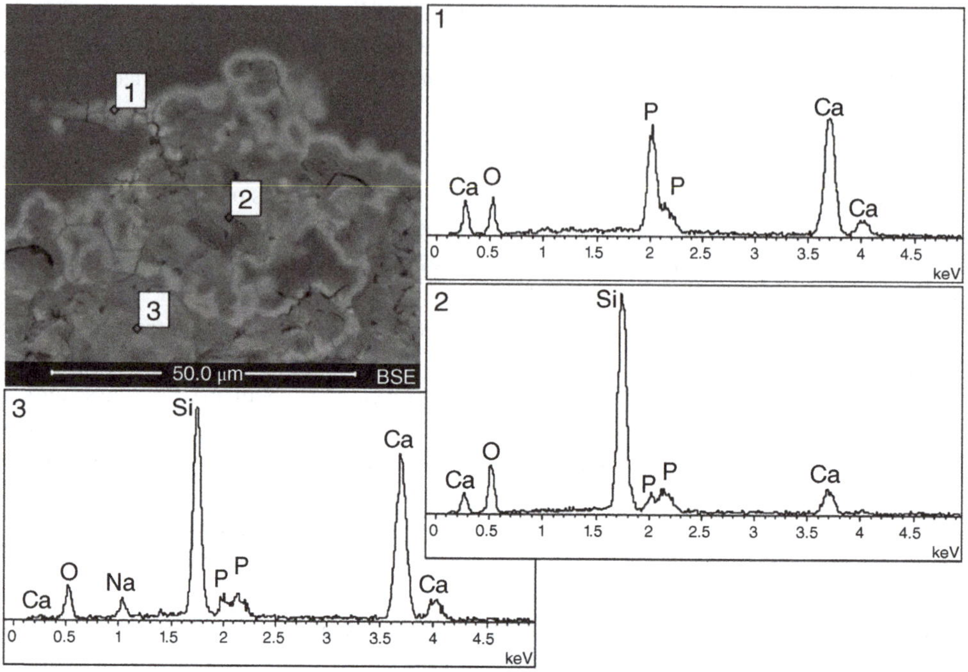

Figure 4.10 SEM micrograph (*BSE*) and *EDS* analyses of a polished cross-section of a bioactive glasses having composition in wt. % of SiO_2 + 42.3 CaO + 6.1 P_2O_5 + 4.7 Na_2O coating obtained using suspension plasma spraying and tested using simulated body fluid. *Source:* Cattini *et al.* [57]/with permission of ELSEVIER.

correlates with the variation of the surface profile [44]. The tip moves in two horizontal directions *x-y* (characterized in nm) scanning and in vertical direction *z* (characterized in pm) controlling distance of observed surface, which must be small enough to enable tunneling of electrons [59]. These movements are controlled by, e.g. piezoelectric ceramic, which is calibrated to transform a small displacement of the tip into voltage. For example, the displacement[11] of. 0.6 nm results in the generation voltage of 1 V for lead zirconate titanate. As the distance between the sample and the tip is nanometric, it is important to avoid any vibration by isolation using mechanical springs or by magnetic levitation [44].

The surface of sample and the microscope's tip should be electrically conducting and very carefully cleaned up prior to imaging. The image can be obtained if the tip is brought to the surface within the tunneling range. The distance is so small that even thermal expansion of tip or the surface irregularities may render the imaging impossible. The images may be obtained using the modes of [59]:

- *constant current* in which the voltage being related to the height on the imaged surface is recorded and the tunneling current is keeping constant;
- *constant height* in which the change of tunneling current is mapped.

11 This displacement under voltage is called also the expansion coefficient [60].

This principle makes it possible to obtain relief maps of a surface with a vertical resolution of the order of picometers. The tip can establish profiles with a maximum vertical variation of 5 μm, while the measured surface air is typically $100 \times 100\,\mu m^2$ [5]. A scanning tunneling microscope is applied to numerous studies on the atomic and molecular structures at the surfaces of semiconductors reviewed by Kubby and Boland [61] and very thin oxide films reviewed by Nilius [62]. Stout and Blunt [5] studied the bio-engineered surfaces, and Kurt *et al.* [63] observed carbonic CN_x films. Shao *et al.* [64] made recently a *STM* study of thin superconducting β-Sn film onto $SrTiO_3$ substrate.

4.2.3.4 Atomic Force Microscope (*AFM*)

The atomic force microscope (*AFM*) belongs to the family of ***scanning probe microscopes*** [60, 65]. *AFM* was developed in the same *IBM* laboratory in Switzerland as *STM* with the collaboration of Stanford University [66]. The development was a couple of years later than that of *STM*, and the motivation was the capacity of obtain topography on atomic scale of electrically insulating samples surfaces. *AFM* has undergone rapid evolution and is applied today also in nanofabrication enabling to modify the surfaces by atomic [65]: (i) modification; (ii) removal; and (iii) addition.

Habibullah [60] shows the principle of the atomic force microscope. The very fine tip, having width smaller than 10 nm, on the cantilever, being 100–200 μm long, is scanned along the tested surface. The tip may be as sharp as an atom. The tip is acting with the force in the range of μN down to nN without deforming the surface [5]. The deflection of cantilevered tip is registered by a photodetector. This deflection can be also registered on other way, e.g. using interferometer or using scanning tunnel microscope [67]. The precise scanning can be realized using ***piezoelectric tube scanners***, and a typical scanning area of commercial microscope can be $8\,\mu m \times 8\,\mu m$ [60]. The height of peaks on the tested surface, which may be observed using AFM, is the range of fraction of nm up to a few of μm [5].

AFM has been increasingly used to observe the topography of the films and coatings. Duparré *et al.* [68] studied the cross-section of MgF_2 and LaF_3 films obtained by evaporation on the silicon substrate. The authors were able to visualize the columnar grains by preparing the cross-section using ion polishing. The topography of *HA* films obtained by simultaneous evaporation of P_2O_5 (thermal evaporation) and CaO (electron beam evaporation) was studied using *AFM* by Hamdi *et al.* [69]. The authors could recognize the growth of the columnar structure with peak-to-valley distances of 0.5–1 μm size. Finally, the two-layer films generated by Nd: YAG laser ablation of Si and Al_2O_3 targets in $N_2 + O_2$ atmosphere were analyzed using AFM by Vlasova *et al.* [70]. The authors found out, e.g. the clusters in the scanned area of 400×400 nm.

4.2.3.5 Transmission Electron Microscope (*TEM*)

The first transmission electron microscope (*TEM*) was built up in *Berlin Technical Hochschule* by ***Ernst Ruska*** and ***Bodo von Borries*** in 1933 [71]. The microscopes were started to be commercialized in 1939 by Siemens in Berlin. The primary *e-beam* of such microscope is transmitted through the sample forming an image. Consequently, the tested sample should be thin enough and energy of electrons sufficiently high to make such transmission possible.

The electrons of primary beam are emitted thermally by cathodes of W or LaB_6 or by field electrons emitters similarly as in *SEM* [2]. The pressure in the chamber should be adapted to the cathode, and the lowest one is required for field emitters ($p \leq 10^{-7}$ Pa). The electrons of primary beam are accelerated by the voltage being typically in the range of 100–1000 kV. The electrons accelerated by high voltage are characterized by the wavelength of a few pm (see *Eq. 4.8*). Consequently, the resolution of the microscopes using such voltage improves. Similarly, the thickness of the sample tested using high-voltage microscope can be greater than the typical one (less than 0.1 μm). The primary beam is accelerated by an anode and then focused using one or more electromagnetic lens before bombarding the sample. Finally, the primary beam is scanned above sample's surface in the scanning transmission electron microscope (*STEM*) [8].

The electrons pass through the sample. The passage can accompanied by [3]: (i) no interaction with the atoms of sample (***bright field mode***); (ii) interaction with the atoms in elastic or inelastic way (***dark field mode***). The interaction with the atoms may also result in the diffraction of the waves associated with these electrons. The microscopes equipped to analyze the diffracted waves are described in the chapter describing selected area diffraction (*SAD*). The imaging using diffracted and primary beam electrons, e.g. combination of *bright field* and *dark field modes*, is realized using high-resolution transmission electron microscopes (*HRTEM*).

The electrons issuing from the samples are focused with the objective lenses and, subsequently, by imaging lenses, before being focused on the fluorescent screen giving final image of observed sample. The geometry of these lenses is different depending on the mode of observation, i.e. image forming or analysis of diffraction pattern [2, 8]. The final image on a fluorescent screen shows: (i) dark areas corresponding to the atoms; and (ii) structural defects and individual atoms using *HRTEM* with the highest resolution of 0.05 nm.

The diameter of samples' holder is about 3 mm and the samples should be thinner than 100 nm. Consequently, the samples preparation is complicated and time consuming. The typical procedure of such preparation is [2, 72]:

- electrolytic or chemical etching used frequently for circular metallic samples;
- ion beam milling using Ar ions accelerated to the energy of a few keV;
- cutting and polishing of the samples' cross-section in a way similar to that of optical microscope sample preparation by taking, however, into account that final sample should be sliced to form disk corresponding to the sample holder.

The transmission electron microscope makes it possible to observe the interior of crystal grains as well as the interfaces between the grains. For example, Fig. 4.11 shows the aluminum grains inside the Al-SiC composite obtained by laser shock treatment of *HVOF* coating. The axes of the grains are parallel resulting from the laser shock wave propagation. In addition, microscopic observations of this sample have revealed dislocations within the aluminum grains. These dislocations confirm the plastic deformation of the material during laser treatment [46].

Another example is the morphology of 8YSZ coating obtained by suspension plasma spraying including small and large grains of different shape (see Fig. 4.12).

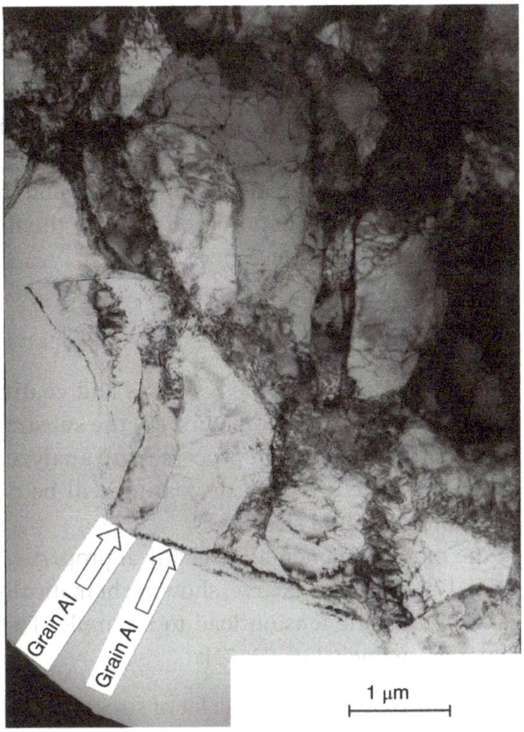

Figure 4.11 *TEM* image of the Al grains in the Al+SiC composite coating obtained by the high-velocity oxy-fuel spraying and treated by laser shocks. *Source:* Podlesak *et al.* [46]/with permission of ELSEVIER.

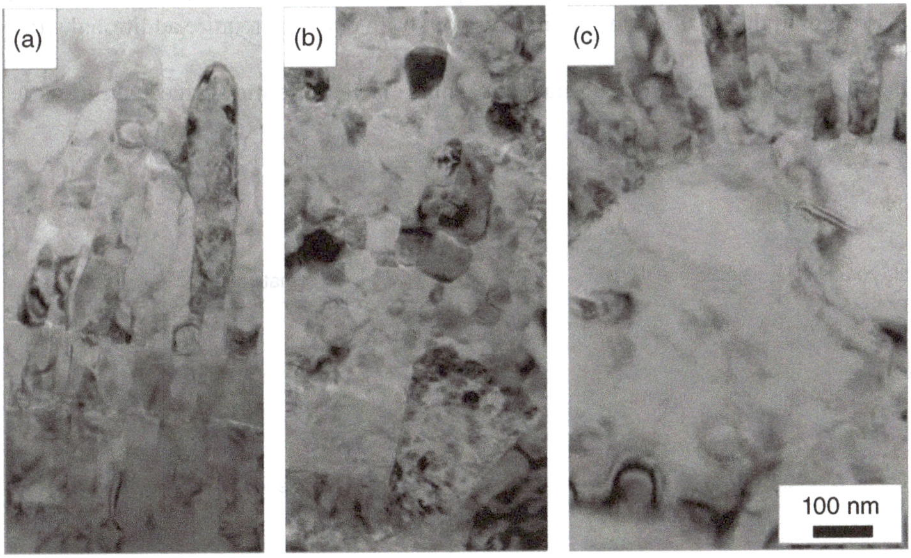

Figure 4.12 *TEM* image showing typical morphology of *8YSZ* coating obtained by suspension plasma spraying. *Source:* Podlesak *et al.* [45]/with permission of ELSEVIER. The morphology includes: (a) aligned columnar small grains within lamellae, (b) agglomerations of small equi-axed grains, and (c) large spherical particle.

4.3 Mechanical Properties of Films and Coatings

There are many different properties of films and coatings related to their practical applications. It is reasonable to start their description by mechanical properties of deposits. These properties include the adhesion of the deposit to the substrate, which is the most important property because without adhesion there is no deposit. Moreover, the wear of a surface is a very frequent reason to apply the films and coatings to extend the lifetime of many products and machines.

4.3.1 Fundamentals of Mechanical Properties

The mechanical properties of films and coatings properties depend on: (i) the deposition process and its parameters; (ii) the substrate material; and (iii) the film and coating material. The deposition processes are analyzed carefully in Chapter 2. The compatibility of substrate and film or coating will be discussed more carefully in Section 4.3.7 describing residual stresses.

The most fundamental mechanical properties of materials result from the strain–stress curve [72, 73]. Such a curve, shown schematically on Fig. 4.13, can be obtained by an application of a static tension load to a sample. The curve enables to show different class of materials, namely [36, 72, 73]:

- ceramics, which requires a lot of stress to reach small strain and get fractured for a small strain and have only elastic behavior;
- metals and alloys, which have elongated in proportion to applied stress having elastic behavior and, in the ductile zone, an increase of strain does not require additional stress prior to fracture;
- polymers, especially thermoplastic, have very small elastic zone and a large ductility zone.

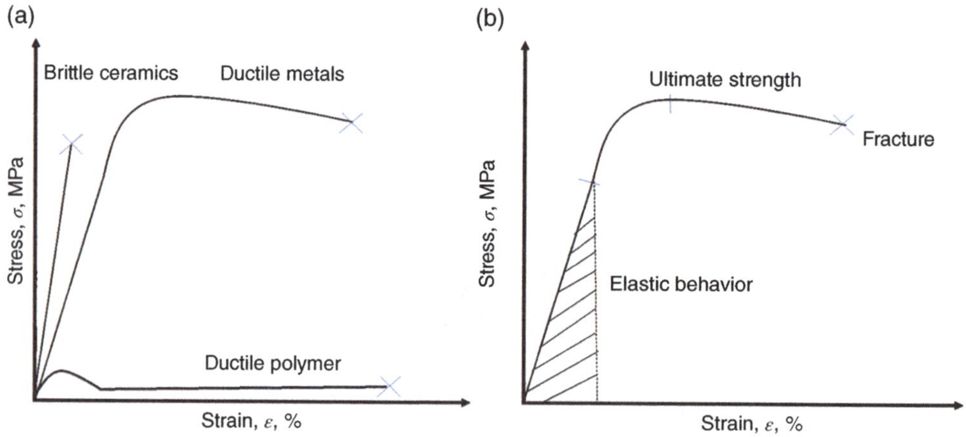

Figure 4.13 The stress–strain curves obtained by a static tension load tests for: (a) different classes of materials; (b) ductile metals with the characteristic details.

The modulus of elasticity, E, being a slope of the stress–strain curve in the elastic zone can be found from the ***law of Hooke***:

$$E = \frac{\sigma}{\varepsilon} \tag{4.10}$$

The fracture of elastic solids is based on fracture mechanics. The fracture describes ***Griffith theory*** discussed carefully by Kelly and Macmillan [74]. The theory predicts that crack created in a solid propagates if ***energy density release rate*** exceeds critical value G_{Ic}[12]. The *Griffith theory* enables to estimate the tensile stress, σ_f, corresponding to the formation of crack having length l in the solid having ***Young modulus***, E, ***Poisson modulus***, ν, and critical *energy density release rate*, G_{Ic}, as:

$$\sigma_f = \sqrt{\frac{EG_{Ic}}{(1-\nu^2)\pi l}} \tag{4.11}$$

An important material property describing its resistance to resist against crack formation is ***fracture toughness***, K_{Ic}. This property can also be found knowing these properties of the solid from the following equation:

$$K_{Ic} = \frac{G_{Ic}E}{1-\nu^2} \tag{4.12}$$

Finally, the solids can be submitted to compression or axial tension. The action of axial tension results in lateral strain, ε_{lat}, and longitudinal strain, ε_{long}. The strains are of opposite signs for uniaxial stresses as one of them corresponds to elongation and other to contraction. The *Poisson modulus* is defined as the ratio of these strains [73]:

$$\nu = -\frac{\varepsilon_{lat}}{\varepsilon_{long}} \tag{4.13}$$

The values of *Poisson modulus* vary for most metals between $\nu = 1/4$ and $1/3$ and its greatest possible value is $\nu = 1/2$.

As far as the properties of films and coatings are concerned, the property being probably most frequently characterized is adhesion. The hardness and microhardness are also frequently tested. These properties enable estimation of wear resistance and are relatively easy to test. The characterization of elasticity, i.e. *Young's modulus*, is also important because it determines the behavior of deposits submitted to different kinds of stresses discussed in Section 3.3.4.2. The stresses may lead to the deposits fracturing as shown in Fig. 3.22. That is why, the parameters related to fracture of deposits are tested. In particular, at the interface with the substrate. Finally, the great practical importance has the family of properties, which describes the contact of deposits surface in contact with different bodies. These properties are related to friction and wear. The wear may result from contacts with, e.g.: (i) other surfaces (adhesive or abrasive wear); (ii) impinging particles (erosive wear); (iii) impacting second body (impact wear) and many others.

12 I is the mode of fracture which corresponds to the forces acting in the direction of opening the crack.

Table 4.3 Tests enabling determination of mechanical properties of deposits.

Kind of test	Mechanical properties characterized
Tensile test	Adhesion of coatings, elasticity, *fracture toughness*
Indentation	Hardness and microhardness, elasticity
Scratch test	Adhesion of films
Bending test	Adhesion of coating, elasticity, *fracture toughness*
Abrasion tests	Wear resistance

Source: Adapted from Brotzen [75].

The tests leading to the characterization of mechanical properties described in following sections are synthesized in Table 4.3. Brotzen [75] described such tests for thin films.

4.3.2 Adhesion

Rickerby [76] reviewed carefully the methods of deposits adhesion measurements. The author categorizes the test of adhesion into qualitative and quantitative and, on the other hand, on mechanical and nonmechanical. Mattox [77] suggests categorizing of adhesion methods in the tests: (i) tensile; (ii) peel; (iii) shear; (iv) deformation; (v) energy deposition; (vi) fatigue; and many others. This section describes frequently used mechanical tests giving quantitative results.

In general, the quantitative mechanical tests enable to find the values of: (i) force of adhesion or eventually force per unit of area to be exerted for separation of a deposit from the substrate; or, (ii) work of adhesion to be done to detach a deposit from the substrate.

The adhesion of thick deposits, obtained by granular methods, is measured· most frequently by:

- tensile adhesion test (*TAT*), which is also called "pull off" test described in this section;
- double cantilever test (*DCB*) and other tests based on fracture mechanics, which are described in detail in section dealing with *fracture toughness*.

The tensile adhesion consists of gluing a surface of cylindrical specimen to the coated surface of similar specimen and of applying the tensile force, *F*, to the assembly (see Fig. 4.14). The fracture may occur on the interface between the deposit and specimen or inside the deposit. In the first case, by knowing the surface of the specimen, it is possible to calculate the coating ***adhesion strength***. The fracture inside coating indicates ***cohesion strength***. Thermally sprayed coatings are frequently tested by using American norm ASTM C633–2021[13], and epoxy resin is used frequently as the glue.

The norm requires the thickness of the coating to be at least 380 μm because of possible penetration of resin into its pores. The use of the epoxy resin limits the values of *adhesion strength* to 90 MPa. An example of the *adhesion strength* for ceramic coating of Cr_2O_3 plasma sprayed by plasma in air (APS) onto mild steel was reported to be in the range 21–25 MPa [78, 79].

13 Standard Test method for Adhesion and Cohesion Strength of Thermal Spray Coatings, ASTM International, West Conshohocken, PA, USA (last update November 2, 2021).

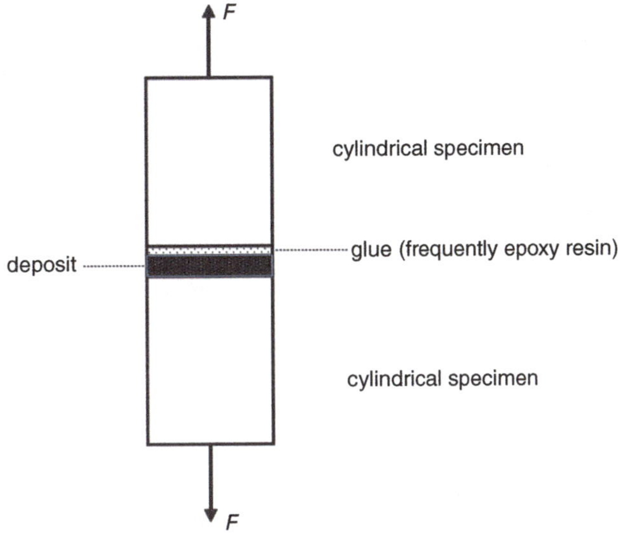

Figure 4.14 Sketch of tensile adhesion test.

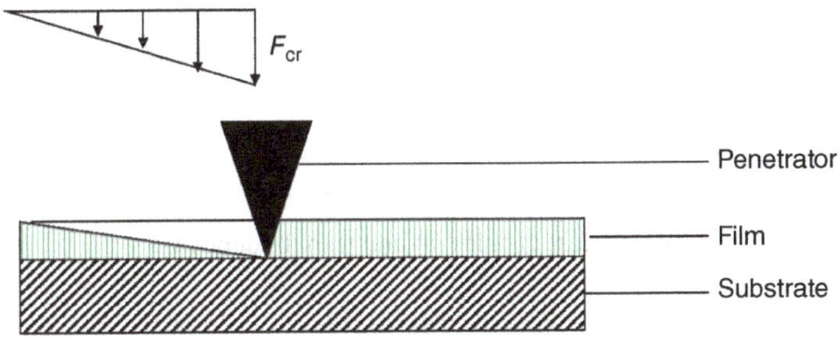

Figure 4.15 Principle of the scratch test.

The application of *HVOF* method to spray composite WC-CoCr coatings onto mild steel enables to reach much greater *adhesion strength* ranging from 55 to 65 MPa [80].

The adhesion of thin films is characterized using scratch test. This test (see Fig. 4.15) consists of scratching the surface of a film using a hard penetrator such as diamond tip. The tip is then moved along the surface applying the load, F, which gradually increases until the critical value, F_{cr}, is reached. This value corresponds to the contact of tip with the substrate (Fig. 4.16) and can be detected using acoustic emission [82]. In the case of very porous films, the help of optical microscope observation used by, e.g. Al-Rjoub *et al.* [83], or by a modification of a slope of the curve showing penetration depth vs. load [81] is more useful. The adhesion of films characterized using scratch test is expressed in Newtons.

4.3.3 Hardness and Microhardness

Hardness describes the resistance of tested material surface against a penetration of an indenter [84]. The indenter must be harder than the material. The metal surfaces can be tested using ***Brinell test***, which uses hard metal hemispheres. All materials can be tested using diamonds with a base having form of [36, 85]:

- triangle (***Berkovich test***);
- square (***Vickers test***);

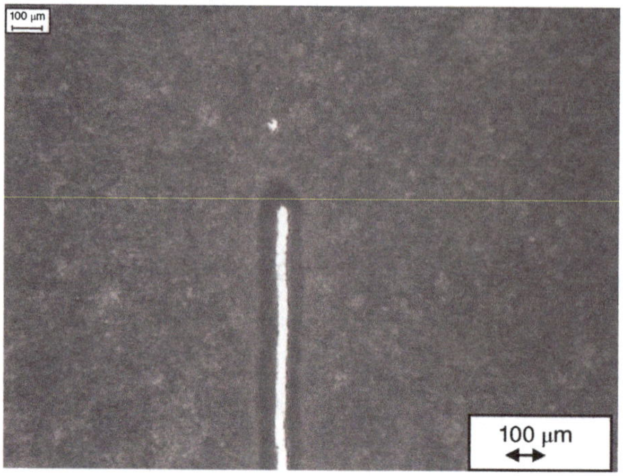

Figure 4.16 Optical microscope view of a scratch in suspension plasma sprayed (*SPS*) anatase phase of titanium oxide coating having critical load of $F_{cr} > 30$ N. *Source:* Jaworski *et al.* [81]/with permission of ELSEVIER.

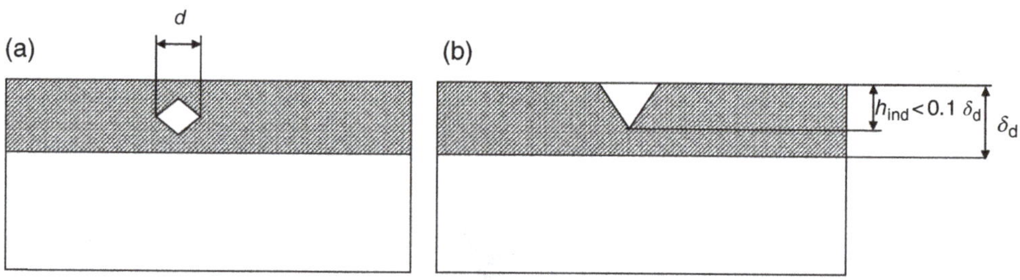

Figure 4.17 *Vickers test* of thick (a) and thin (b) deposits.

- rhombus (***Knoop test***);
- circle (***Rockwell test***)[14].

Indentations are made on the surface of the transverse cross-section or on their surface (Fig. 4.17). The values of the hardness may diverge in these cases because of difference of the deposits' microstructure in each direction. Moreover, Burnett and Rickerby [86] recommend that the thickness of a deposit, h_d, should be at least ten times greater than the depth of indentation, h_{ind}.

Hardness is expressed as a ratio of the load F to the surface are of indentation. If we consider the *Vickers test* this relation can be found by knowing the diagonal, d, in mm (see Fig. 4.17a), using following equation [36, 84]:

$$HV = \frac{0.1891F}{d^2} \tag{4.14}$$

The HV hardness values obtained indicate the use of Vickers type of indenter. Similarly, HK values indicate the use of Knoop indenter, HR – the Rockwell indenter, and finally, HB – the Brinell one. The Martens tests of hardness, HM, are increasingly popular [87].

14 Rockwell tests may use also a hard metal sphere as indenter.

The measurements are made using a pyramidal indenter similar to that used in *Vickers test*. The hardness is equal to:

$$HM = \frac{F}{26.43 h_{\mathrm{ind}}^2} \tag{4.15}$$

where F is the load and h_{ind} is the depth of indentation. The discussion of the *HM* hardness and indentation size effect is presented, e.g. by Łatka et al. [88]. The authors found *HM* value for suspension plasma sprayed hydroxyapatite coating to be about $HM = 0.5\,\mathrm{GPa}$ after deposition. The hardness values typically depend on the load, F, which is why it is important to specify it when giving the hardness value. In practice, thick deposits are tested using indenter loaded with the force which varies between $F = 0.1$ and 50 N. The values obtained with the load $F \leq 10\,\mathrm{N}$ are called microhardness, and the values obtained with greater load are called hardness [89]. Very small loads describe the microhardness of microstructural elements such as lamellae in granular deposits or reinforcements in composite ones.

The higher loads are applied to characterize the entire deposit. The granular deposits are typically tested using their longitudinal cross-sections under the loads of 1–5 N [90, 91]. The atomistic deposits are generally tested under small loads of lower than 0.5 N on their surfaces [92]. For example, the sputtered TiAlSiN films tested using Berkovich indenter under load of 15 mN had hardness of about 30 GPa [83]. It is important that the depth of indentation is not greater than 10 % of film thickness, $h_{\mathrm{ind}} < 0.1\ h_{\mathrm{d}}$ (see Fig. 4.17b). Otherwise, the properties of the substrate can influence the measured hardness value. If this condition cannot be met, it is necessary to find the hardness of film using a model developed by, e.g. Burnett and Rickerby [93].

4.3.4 Elasticity of Deposits

The modulus of elasticity can be determined using different methods, which can be categorized as [78]:

- ***destructive methods*** including:
  - indentation using deposits longitudinal cross-section (see Fig. 4.18);
  - 3-point and 4-point bending test (see Fig. 4.19);
  - compression and tensile tests;
  - cantilever beam method (used also in fracture toughness testing);
  - acoustic tests;
- ***non-destructive methods*** including:
  - indentation method using deposits surface (see Fig. 4.18);

The indentation test can be made on the deposit cross-section or on its surface. The method, developed by Oliver and Pharr [94, 95], uses generally *Berkovic*h indenter and analyzes the curve of indentation depth, d_{ind}, vs. applied load, F, shown in Fig. 4.18.

The analysis requires the sensors being able to determine precisely the depth of indenter penetration. The modulus of elasticity of the deposit, E_{d}, can be found by knowing the area

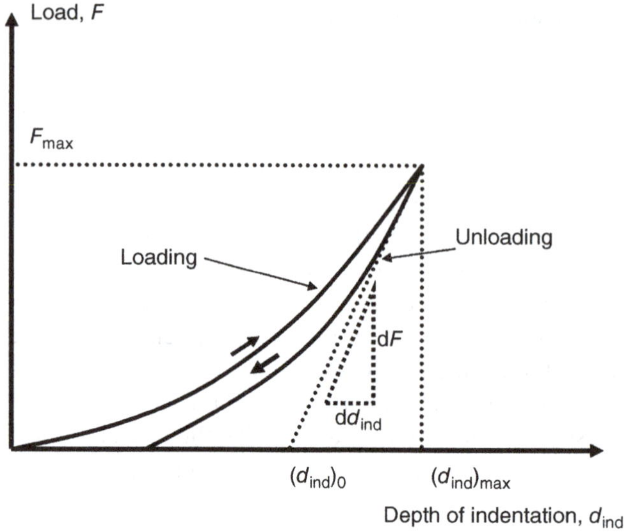

Figure 4.18 Sketch of indentation test showing loading and unloading and resulting indentation depth. *Source:* Adapted from Oliver and Pharr [94, 95].

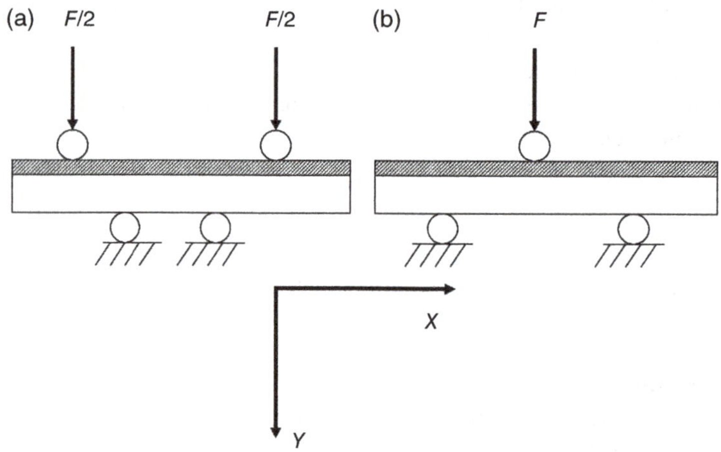

Figure 4.19 Diagram of 3-point (a) and 4-point (b) bending tests.

of indentation on the surface of deposit, A_{ind}, the mechanical properties of indenter[15], and Poisson modulus of deposit, ν_d, from the initial stiffness of unloading curve:

$$\frac{dF}{dd_{ind}} = \frac{2}{\sqrt{\pi}} E_r \sqrt{A_{ind}} \tag{4.16}$$

15 The modulus of elasticity and Poisson modulus of monocrystalline diamond are in the range of E_{ind} = 1050–1200 GPa and ν_{ind} = 0.1–0.3 [74].

The reduced elasticity modulus, E_r, appearing in this equation is defined as:

$$\frac{1}{E_r}=\frac{1-\nu_d^2}{E_d}+\frac{1-\nu_{ind}^2}{E_{ind}} \tag{4.17}$$

The indentation method can be used to find the elasticity modulus of thick and thin films obtained using different methods. Consequently, the modulus of 50–100 μm thick, suspension plasma sprayed hydroxyapatite coatings was found to be in the range of $E_d = 16–28$ GPa depending on deposition parameters [88]. On the other hand, Ivanova *et al.* [96] tested 5 μm thin Ti-Zr films obtained by ion-assisted deposition and found the elasticity modulus in the range of $E_d = 77–98$ GPa depending on the chemical composition of films.

The elasticity modulus can be found using 3-point and 4-point bending tests shown in Fig. 4.19.

The bending tests use force, *F*, acting in the *y* direction. This force results in the stress and elastic deformation in *x* direction. The deformation can be recorded precisely by the sensors and monitored acoustically to detect the micro- and macro-cracking of tested deposit. Timoshenko [97] gives the detailed description of the bending test. Young's modulus of the deposit can be found by knowing the mechanical properties of the substrates and the thicknesses of the substrate and of the deposit. For example, the 4-point bending test enabled to find the modulus of WC + 16 wt. % Co composite coating obtained using plasma spraying to be equal to $E_d = 138$ GPa.

The compression and tensile test can be carried out using free-standing deposits detached off the substrate. In order to use the standard methods used for materials (described by, e.g. Seidel and Hahn [84]), the deposits should be sufficiently thick. For example, Varis *et al.* [98] tested different deposits having thickness in the range of $\delta_d = 180–700$ μm. The deposits were thermally sprayed on Al substrate, which was chemically etched before the test. The elastic properties' free-standing samples can be also tested using acoustic methods. This method was used to characterize modulus of elasticity of alumina, which was in the range of $E = 350–380$ GPa depending on tested samples' thickness [99].

4.3.5 Fracture Toughness

The *fracture toughness*, K_{Ic}, can *be* defined as an ability to resist the growth of crack existing in pre-cracked material [100]. This parameter is related with other mechanical properties by equations 4.11–4.12.

Callister [36] presents the typical values of the toughness for the metals used as substrates and $K_{Ic} = 87.4\ \mathrm{MPa.m^{0.5}}$ for steel alloy type 4340 and $K_{Ic} = 44–66\ \mathrm{MPa.m^{0.5}}$ for titanium alloy Ti6Al4V. Alumina ceramics, which is used frequently as a coating, has $K_{Ic} = 3.0–5.3\ \mathrm{MPa.m^{0.5}}$.

The toughness of the deposits can be tested similarly to the elasticity modulus using 3-points or 4-point bending tests. For example, the 4-point bending test was used by Wiklund et al. [100] to determine cracking resistance of thin CrN ($\delta = 2.5$ μm) and TiN ($\delta = 4.3$ μm) films. The test had a qualitative character based on the microscopic and acoustic observations of cracks in the tested films.

Similarly, the uniaxial tensile test can be used to determine the toughness. The pre-cracking of the deposit can be introduced by initial tensile stress realized until the deposit

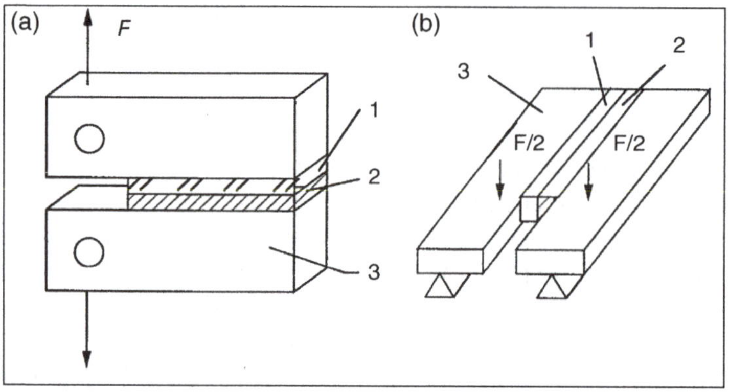

Figure 4.20 The sketches of double-cantilever beam (DCB), (a) and double torsion (DT), (b) methods used for characterization of thick coatings toughness: 1 – glue; 2 – coating; 3 – substrate; and *F* – load. *Source:* Pawłowski [78]/John Wiley & Sons.

starts to be fractured [101]. The mathematical modeling enabled the authors to find the toughness of nano-composite film of TiN/SiN_x being 3 μm on Ti6Al4V substrate of from the stress–strain curve to be about $K_{Ic} = 2.6\,\text{MPa.m}^{0.5}$.

Thick coating can be tested using double-cantilever beam (*DCB*) and double-torsion (*DT*) methods shown in Fig. 4.20. The coatings are deposited on one beam and glued to another one. The coatings should be pre-cracked before the test.

Guo and Wang [102] used *DCB* method to determine the energy density rate of plasma sprayed *8YSZ* coatings being 250–400 μm thick. The authors found the means value of $G_{Ic} = 32.5\,\text{J/m}^2$.

4.3.6 Friction and Wear

The films and coatings are very frequently applied against friction and wear. As wear leads to decrease of deposit thickness, the application is limited to thick coatings, in particular, obtained by thermal spraying or by *hardfacing*.

Friction and different kinds of wear are, following Zambelli and Vincent [103], the part of ***tribology***, which is the science of interacting surfaces of materials being in relative movement. ***Static friction*** can be defined as the resistance to movement of a solid body being initially at rest in contact with the surface of another body. Such friction results in generation of heat and results in wear of material. ***Dynamic friction*** concerns the bodies being initially in movement and is lower than the static one. The coefficient of friction is defined as:

$$\mu = \frac{T}{N} \tag{4.18}$$

where *T* and *N* are forces being tangential and normal to the surfaces in contact, respectively (Fig. 4.21). The coefficient of friction can be measured using tribometers. Its value does not depend only on the property of interacting materials but also on entire tribological system, which includes environment of testing (temperature, humidity, etc.)

Zambelli and Vincent [103] present the tribometers working on following principles:

- pin on disk;
- bloc on cylinder shown in Fig. 4.21;
- pin on bloc;
- 4-sphere tests including a rotating sphere interacting with 3 static spheres.

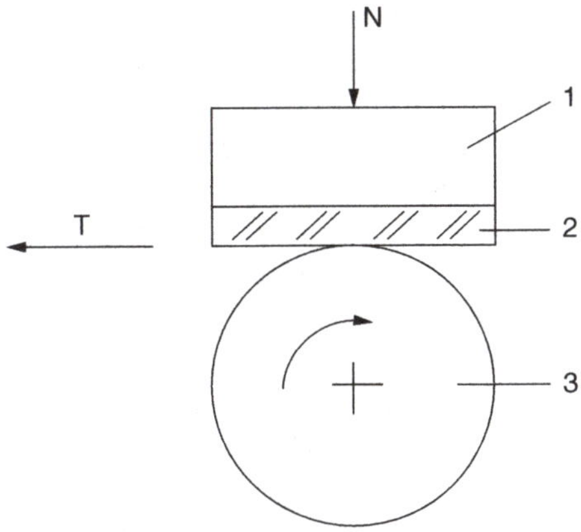

Figure 4.21 Sketch of a bloc on cylinder tribometer to determine the coefficient of *dynamic friction*: 1 – substrate; 2 – deposit; and 3 – rotating wheel. *Source:* Pawłowski [78]/John Wiley & Sons.

The friction tested by these tribometers can be generated by: (i) sliding; (ii) rotation; or (iii) rolling. The result of test is coefficient of friction or the wear rate in function of time or of the distance traveled in this time by the characterized deposit [104]. These results depend on the following main characteristics of deposit [105, 106]:

- hardness of the deposit with regard to that of tested surface;
- roughness and thickness of the deposit;
- size and hardness of debris generated at testing.

The principal mechanisms of wear of a surface submitted to the test include [107]: (i) adhesion, which results in transfer of material between contacting surfaces; (ii) abrasion, being associated with cutting of the irregularities of tested surface; and (iii) fatigue, which causes the microcracking of the deposit below tested surface.

The films and coatings are applied and tested for the following types of wear [78]:

- ***Adhesive wear***, which appears when two surfaces slide past each other. This type of wear is proportional to the distance "run" between two moving surfaces and to the load (represented by force N in Figs. 4.21 and 4.22) being inversely proportional to the hardness of the deposit. It may be determined using the tribometers discussed above or with an oscillation wear test well suited to samples of a small area (Fig. 4.22)

Figure 4.22 Wear test using oscillating sphere. *Source:* Wielage *et al.* [108]/Taylor & Francis.

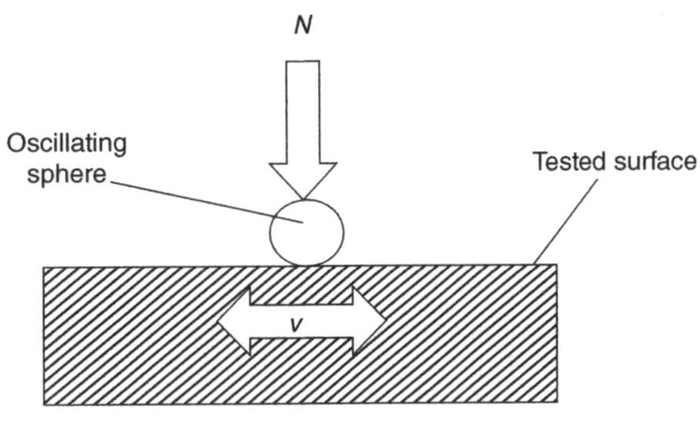

The oscillating sphere wear test used, e.g. by Wielage *et al.* [108] and shown in Fig. 4.22 applied stainless steel sphere of 8 mm diameter oscillating with frequency of 20 Hz with amplitude of 1 mm under load of N = 10 N for 1 hour over the surface of CoCrAlY coating obtained by vacuum plasma spraying and treated by laser. The tested ***linear wear*** was characterized as a depth of the cavity made by the sphere in the coating. This depth was in the range of 46–76 μm. It is possible to characterize ***volumic wear***, which is defined as a volume of worn coating divided by run distance and by load. The *adhesive wear* can be reduced by the application of a lubricant between the surfaces in contact. This kind of wear, called also ***frictional*** wear as it has similar mechanisms as friction, can be experimentally divided onto mild and severe wear. These types of adhesive wear are characterized by: (i) presence of fine debris and of the relatively reduced wear rate at ***mild wear***; and (ii) large debris which leads to the increase of the coefficient of friction and by the destruction of coatings at ***severe wear***.

The surfaces in contact may be submitted to local fusion reported, e.g. in the pistons of diesel engines by Buran and Fischer [109]. This kind of wear is called also ***scuffing wear***.

- ***Abrasive wear***, which appears in contact of deposit with the roughness of a much harder surface (***two-body abrasion***). The presence of hard particles in the contact leads to ***three-body abrasion***. The tests are shown in Fig. 4.23. The *three-body abrasion* test is normalized by ASTM G 65–04[16].

The result of this test is the volume or mass of worn coating's material divided by distance run during test.

- ***Erosive wear***, which appears when the surface of a deposit is submitted to a flow of impinging particles. The erosion rate depends on impinging particles size, velocity as well as on their impact angle [110, 111]. The mechanism of erosive wear is different for brittle and ductile materials. The brittle materials are eroded by microcracking and, resulting thereof, fracturing [112]. The erosion wear of such materials is greatest at angle of impingement[17] of 90°. Ductile materials (metals or alloys) erode by micromachining caused by impinging particles, and the erosion is the greatest at the impingement angle of 15–30°. A special kind of *erosive wear* is ***cavitation erosion*** [103]. This type of erosion is associated with turbulent flow, which results in localized pressure variations causing the formation of vapor cavities and their repetitive collapse on the coating's surface leading to material removal [113]. Cavitation first causes hollow formations or small craters, which then erode the surface [114]. This kind of wear can be tested with ultrasonic wave generators inside a water container.
- ***Fatigue wear*** occurs when the deposit is in contact with a rotating body in rotation (ball bearings).

---

16 Standard test method for measuring abrasion using dry sand/rubber wheel apparatus, G 65–04 (1 November 2004).

17 The angle with regard to eroding surface.

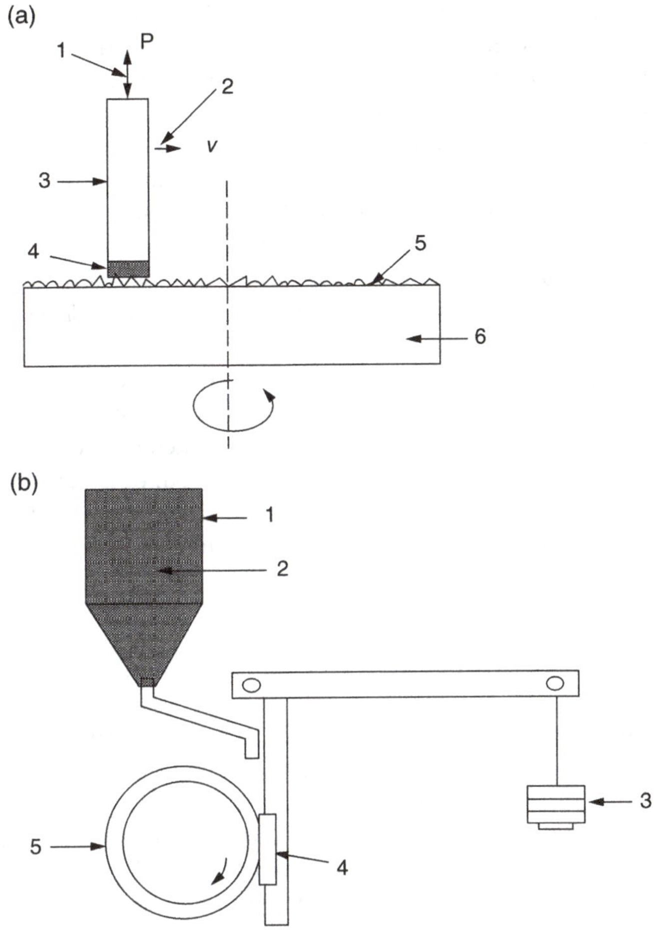

Figure 4.23 Schematic of the tests: (a) *two-body abrasion*, 1, load; 2 direction of linear motion; 3, substrate; 4, coating; 5, abrasive surface, 6, rotating disc; and (b) *three-body abrasion*, 1, sand container; 2, sand; 3, load; 4, coating with substrate; 5, rotating rubber wheel. *Source:* Pawłowski [78] / John Wiley & Sons.

- ***Fretting wear***, which appears when the deposit comes in contact with oscillating or vibrating body. Blanpain *et al.* [115] distinguish the following modes of fretting wear caused by an oscillating body:
  - mode I: linear oscillations under constant load;
  - mode II: oscillations under variable load;
  - mode III: rotational oscillations under constant normal load.
- ***Corrosive wear,*** which appears when another type of wear is accompanied by the presence of a corrosive environment such as moisture or pollution of air. Sometimes, the chemical products of corrosion are useful in protecting the surface [116].

4.3.7 Residual Stresses

The residual stresses are generated at films and coatings deposition or/and at their service. These stresses result from the difference of properties between deposit and the substrate (see Section 3.2.5). The methods for investigation of residual stresses can be divided into destructive and non-destructive ones.

Among the destructive methods, the hole drilling test is probably most frequently used. This method involves the application of gauge deposited on the deposit surface and used to monitor the strains as material of coating is removed by drilling. The drilling of a hole modifies the strains around it, and the measured values are subsequently used to calculate residual stresses in function of coating depth [117, 118].

The non-destructive methods include the X-ray one that are based on deformation of crystal lattice under action of stresses. The stresses shift the diffraction pics. This shifting enables calculation of residual stresses by ***sin²ψ method*** explained by many authors, e.g. Cullity and Stock [119] or Fitzpatrick *et al.* [120]. The method was used to test stresses in plasma sprayed ceramic and metallic coating [121, 122]. The *sin²ψ method* enables to find the residual stress in the regions of a few μm from the deposit surface because of the X-rays absorption. Moreover, an effort is necessary to select the diffracted signal only from the regions considered [123]. This renders difficult measurements of stresses distribution vs. coatings depth.

Finally, a geometrical deformation of coated relatively thin substrate can be used in the residual stresses determination. The method, described by e.g. by Gill and Clyne [124], consists of monitoring curvature of coated substrate enabling the calculation of stresses by knowing the geometry of this deformation. The method has advantage of being simple compared to the *sin²ψ* one.

4.4 Physical Properties of Films and Coatings

Among the more important physical properties is the thickness of the deposit. The mechanical and electrical properties of thin deposits are considerably influenced by their surface, interfaced with substrate, and by the substrate itself. In addition, the characterization techniques vary considerably with deposit thickness. Another important property is the porosity, which influences strongly many other characteristics of deposits. For example, high porosity worsens such mechanical properties as hardness or elasticity modulus but improves thermal shock resistance of coatings. Subsequently, the thermophysical properties such as thermal conductivity, thermal diffusivity, specific heat and emissivity, or thermal shock resistance are essential in an important application of physical deposits, which are thermal barriers. Finally, the contact angle is important for the resistance against corrosion caused by liquid.

4.4.1 Thickness

The techniques for measuring thickness can be categorized as contact techniques and non-contact techniques as seen in Table 4.4. Thickness of very thick coatings obtained by granular and bulk methods can be measured with a caliper or with a micrometer screw with a resolution of 10 μm.

4.4.2 Porosity

The pores, called also voids, are usually present in the deposits. The size of pores depends on mechanism on their generation. It is possible to distinguish:

- ***nanometric pore*s**, having size in the range of 1–10 nm, are the voids within the crystal structure. The pores can be visualized using transmission electron microscope of sample thinned by, e.g. ion bombardment (see, e.g. Fig. 3.12a);

Table 4.4 Examples of thin films thickness measurement techniques.

Technique	Measurements description	Resolution, nm	Minimum thickness, nm	Maximum thickness, μm
Contact technique				
Profilometer	Diamond stylus measures a step from substrate to film	20 (Horizontal)	10	1000
Non-contact techniques				
Optical interferometer	The light beam reflects from the surface and interface of transparent film. Their phase shifting enables thickness determination	—	30	A few
Ellipsometry	Determination of the state of polarization of light reflected from the surface and interface of transparent film (see also 4.7)	—	Lower than 0.1	5
Scanning tunneling microscope (*STM*)	Metallic tip is driven on the step of conducting film on a substrate (see also 4.2.3.3)	Less than 0.1 (lateral)	Less than 0.1	A few hundreds
Atomic force microscope (*AFM*)	Diamond tip driven by a cantilever is driven on the step of a film on a substrate (see also 4.2.3.4)		Less than 0.1	A few hundreds

Source: Stout and Blunt [5]; Chen [59]; Habibullah [60]; Mattox [77].

- ***micrometric pores***, which result from an agglomeration of defects generated during deposition or treatment after deposition. Their size varies from tens to thousands of nanometers and their visualization of possible using optical microscope observation of polished cross-section (see, e.g. Figs. 3.13–3.15). The porosity describes the fraction of the volume of voids with regard to entire volume of deposit. The porosity is often classified in a way describing the position of the pores with respect to the surface of the deposit and its interface with the substrate. Thus, we can distinguish:
- ***open porosity*** which corresponds to the pores connected with the surface of film or coating. This kind of porosity can be measured using:
  - mercury intrusion porosimetry (*MIP*) which involves placing a sample under vacuum and, subsequently, its immersion in a mercury bath. The method, described in detail by Whittemore [125], is useful for determination of porosity and of distribution of pores' size; the deposits are thicker than few tens of μm. It should be added that the use of mercury requires the strict respect of health security norms;
  - observation of changes in the acoustic wave frequencies which propagate on the surface of a deposit. The method based on the theory of Brunauer, Emmet, and Teller (*BET*) and described by Taylor *et al.* [126] requires the deposition of film or coating on piezoelectric substrate.

- ***closed porosity*** which describes the pores inside the deposit can be found using the following methods:
  - ***Archimedes' technique*** of density determination. Measurements are made with samples detached from the substrate and crushed using water immersion. The details of measurements are explained for example in the ASTM standard [127]. From the measured density and the theoretical density (typically determined with X-rays), it is possible to obtain the entire porosity of the deposit, and closed porosity could be found by knowing the open porosity.
  - microscopic image analysis technique. The images of the longitudinal section of a deposit make it possible to visualize the closed porosity having gray level very different from that of the solid material. Image analyzers are able to calculate the total area of these elements relative to the total observed area. The method is frequently applied for thick plasma sprayed or laser cladded coatings [128, 129]. The possible errors can be introduced by metallographic preparation of the cross-sections.
- ***transversal porosity*** corresponds to the pores connecting deposit's surface with its interface and its substrate. The porosity is often determined using electrochemical techniques requiring immersion in electrolyte. The method was applied, e.g. for thin films obtained by *PVD* [130, 131].

4.4.3 Thermophysical Properties

Thermophysical properties of the coatings are of importance because of the frequent industrial application of thermal barriers. Pawłowski and Fauchais [132] reviewed thermal conductivity of coatings deposited by thermal spraying. Another important property of each deposit is thermal expansion coefficient (*TEC*). This coefficient is important in design of functionally graded coatings [133]. Different carriers can transport thermal energy in solids, i.e.: (i) ***free electrons***; (ii) phonons; and (iii) photons. The emission of photons by samples' surface is described by emissivity, ε.

Each transport mechanism depends on the type of solid and on temperature range, namely [132]:

- *free electrons* occur in metals and alloys, and their number increases with temperatures at low temperatures. Consequently, this conduction mechanism is proportional to temperature in this range and is not dependent on temperature at higher range.
- ***lattice thermal conduction*** or transport by ***phonons***[18] occurs, e.g. in dielectrics with large band gap rendering very low concentration of *free electrons*. The *phonons* are scattered by structural defects in low-temperatures range and thermal conductivity depends exponentially on temperature. The scattering by other *phonons* called ***Umklapp process*** occurs at higher temperatures, and conductivity becomes inversely proportional to temperature.
- ***photon conduction*** may also occur in dielectric materials. In particular, such conduction mechanism occurs in the pores and voids at higher temperatures. Photon conduction is proportional to T^3.

18 Phonons are oscillations of crystal lattice, which can be assimilated with a quasiparticle.

Thermal energy can be transported simultaneously by many mechanisms. The transport depends mainly on [78]: (i) dimensions of crystal grains influencing mean free path of energy carriers; (ii) impurities; and (iii) porosity.

4.4.3.1 Thermal Conductivity

Thermal conductivity, λ, depends on thermal diffusivity, a; specific heat, c_p; and density, ρ in the following way:

$$\lambda = ac_p\rho \tag{4.19}$$

Conductivity can be measured directly using static or dynamic methods [134].

A frequently used static method is the hot-wire one applied for the solids of low conductivity [135–137]. In this method, the wire is put between inside two similar samples, and the determination of the conductivity is made by analysis of the temperature registered by a thermocouple. The estimation of thermal conductivity is made by supposing that the sample is infinite body and that the field of temperatures is steady.

Dynamic methods analyze the temperature fields varying with time. These methods enabled to determine thermal diffusivity, and by knowing density and specific heat, thermal conductivity can be calculated from *Eq. 4.19*. Thermal diffusivity is most frequently tested with ***laser flash method***[19] and ***modulated beam method***.

Parker et al. [138] described the principle of laser flash method. The sample is tested inside a vacuum chamber (see Fig. 4.24). The laser pulse is sent toward one of sample faces, and the temperature evolution on its parallel face is registered and analyzed [139].

Figure 4.24 Laser flash set-up used to determine thermal diffusivity. *Source:* Pawłowski *et al.* [139]/EDP Sciences.

19 The method used initially xenon lamp but at present mainly lasers to generate the energy pulses.

Temperature evolution enables to find time, which corresponds to half of maximal temperature, $t_{0.5}$[20]. Thermal diffusivity can be found by using from the following equation:

$$a = \frac{0.139\delta^2}{t_{0.5}} \tag{4.20}$$

In above equation δ is the tested sample thickness. The *laser flash method* is frequently used to test zirconia films and coatings obtained by different methods for thermal barriers, e.g. [140].

Cowan [141] developed the *modulated beam method*. The method consists of submitting a front face of a sample to sinusoidal or square heating wave. The phase shifting between the temperature oscillation of the front and the rear face of the sample enables thermal diffusivity to be determined. Brandt and Neuer [142] presented a careful analysis of the measurement made using this method.

4.4.3.2 Specific Heat and Thermal Expansion

Specific heat, c_p, is a material property defined as the quantity of heat to be delivered to a unit of mass of its sample to increase its temperature to a degree K. The samples containing n different materials with the mass fraction f_i and each of them having specific heat of $(c_p)_i$ have the specific heat determined by ***law of Kopp***:

$$c_p = \sum_{i=1}^{n} (c_p)_i f_i \tag{4.21}$$

The specific heat is frequently tested using differential scanning calorimeter (*DSC*) described in [143, 144]. Density $\rho(T)$, at a temperature T, depends on the density $\rho(300)$ at room temperature $T = 300$ K and on the coefficient of thermal expansion coefficient (*TEC*) following the equation:

$$\rho(T) = \frac{\rho(300)}{1 + 3TEC\,\Delta T} \tag{4.22}$$

The difference of temperatures ΔT is taken with regard to room temperature, and the density of the sample at room temperature can be tested using *Archimedes technique* described in Section 4.4.2 and adjusted to a given temperature with the use of thermal expansion coefficient (*TEC*) defined as:

$$TEC = \frac{1}{l}\frac{\partial T}{\partial l} \tag{4.23}$$

where l is a length of the sample. Thermal expansion measurements are made using dilatometers described by, e.g. Touloukian et al. [145]. The porosity of a sample does not influence *TEC* values.

20 Time starts with the beginning of laser flash.

4.4.3.3 Emissivity

The ratio of emission of photons by a sample's surface compared to a surface of perfect black body is described by coefficient of emissivity $\varepsilon\ (\lambda,T)$, being a function of wavelength and temperature. The coefficient varies between 0 and 1. Its greatest value, $\varepsilon = 1$, is for the perfectly black surface. The coefficient of emissivity can be measured with the use of infrared thermography applied for ceramic materials described by Martin and Fauchais [146] or using calorimetric method applied for materials considered to be used in space vehicles described by Jaworski [147].

4.4.3.4 Thermal Shock Resistance

Thermal shock resistance concerns mainly brittle materials (ceramics) and describes their ability to resist sample heating conditions, which can be convective or radiative. The resistance may be expressed as a maximum temperature variation that a sample withstands without failure [148]. The set ups enabling to test the resistance simulate closely the operating conditions of the sample in real conditions. To take an example, thermal barrier coatings including 7YSZ plasma sprayed coating for the turbine blades were tested with the use of ***burner rig*** [149]. One cycle included heating with a combustion flame up to 1760 K for 30 seconds followed by cooling down to room temperature using compressed air for 30s. The result of test was the number of cycles until spallation of coating.

4.4.4 Contact Angle

The contact angle (*CA*) corresponds, in most of the cases discussed in this book, to the contact between liquid droplet and the surface in surrounding gas. As the droplet does not move, it is called ***static contact angle,*** θ, and it depends on the interfacial surface tensions shown between (see Fig. 4.25): (i) solid and gas, γ_{sg}; (ii) solid and liquid, γ_{sl}; and (iii) liquid and gas, γ_{lg}.

The *static contact angle* depends on three interfacial surface tensions, which in equilibrium conditions can be described by *Young–Dupré law*:

$$\cos\vartheta = \frac{\gamma_{sg} - \gamma_{sl}}{\gamma_{lg}} \tag{4.24}$$

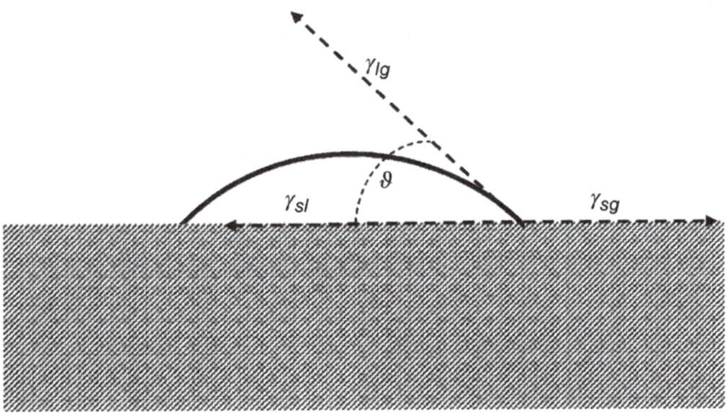

Figure 4.25 Contact angle between liquid droplet on solid surface.

The *dynamic contact angle* concerns the liquid in movement and takes into account its velocity [150].

The measurements of *static contact angle* are made generally by optical way using cameras equipped in *CCD* sensors, and the images are, subsequently, analyzed numerically [151, 152]. The angle depends on the roughness of the solid surface and on its cleaning method (discussed in Section 1.3). Moreover, recent studies evidenced that the measurements of *CA* depend on the method of droplet deposition, which include the methods of ***pendant droplet*** and ***liquid needle*** depositions techniques. The methods may result in different volume of liquid influencing the *CA* measurements [153].

4.5 Corrosion Resistance

Films and coatings obtained by physical methods are often applied to protect the surface of metals from aqueous corrosion, which can be called also low-temperature corrosion [154]. Such corrosion occurs on structural parts being in contact with water or in contact with aggressive liquids such as bases or acids. Another group of applications, discussed in this book, relates to protection against hot gases. This type of corrosion concerns the components of the turbines of aeronautical engines or gas turbines.

4.5.1 Aqueous Corrosion

Metals and alloys immersed in an aqueous solution can dissolve forming ions, for example zinc:

$$\mathrm{Zn} \xrightarrow{\text{aqueous solution}} \mathrm{Zn}^{2+} + 2e^- \tag{4.25}$$

The loss of the ions of a metal, which can be associated with oxidation of its surface, is accompanied with simultaneous release of electrons. The release ends with the creation of the negative charge in metal, which slows down the dissolution. The metals are frequently connected with the other metals and alloys as a structural part in contact with the same electrolyte. Consequently, it is easier to discuss the phenomena by treating them as two electrodes in a cell, called ***galvanic cell***. In such a cell, the electrode at which the corrosion takes place is the anode[21]. The electrons generated in the anode flow toward cathode (or cathodic region) where the reduction of another metal occurs. Consequently, the aqueous corrosion may occur in the following types of cells [155]: (i) ***dissimilar electrode cells***, which correspond to two different metals/alloys in one electrolyte; (ii) ***concentration cells***, which have identical electrodes in solution of different composition; and (iii) ***differential temperature cells***, which have similar electrodes at different temperature in the electrolyte of the same composition.

The *galvanic cell* approach enables to estimate which metal dissolves first in the solution. It can be interpreted as dissolution of one metal by another one. Such estimation can be

21 The anode or anodic region is the site where the metal loses its electrons as in Eq. 4.25. The anode has more negative potential than the cathode [155].

Table 4.5 *Standard reduction potential* of some metals at room temperature and atmospheric pressure with regard to standard hydrogen electrode.

Metal	Reaction at electrode	Standard reduction potential, V
Al	$Al \rightarrow Al^{3+} + 3e^-$	−1.662
Ti	$Ti \rightarrow Ti^{2+} + 2e^-$	−1.630
Zn	$Zn \rightarrow Zn^{2+} + 2e^-$	−0.763
Cr	$Cr \rightarrow Cr^{3+} + 3e^-$	−0.744
Fe	$Fe \rightarrow Fe^{2+} + 2e^-$	−0.447
Ni	$Ni \rightarrow Ni^{2+} + 2e^-$	−0.257
Mo	$Mo \rightarrow Mo^{3+} + 3e^-$	−0.200
Cu	$Cu \rightarrow Cu^{2+} + 2e^-$	+0.342
Ag	$Ag \rightarrow Ag^{+} + e^-$	+0.800
Au	$Au \rightarrow Au^{+} + e^-$	+1.692

Source: Lide and Frederikse [156]/Taylor & Francis.

made by comparing the ***standard reduction potential***[22]. This potential, starting from more negative up to more positive, for metals used frequently for films, coatings, and the substrates is shown in Table 4.5.

The metals having more positive potential are said to be more noble. Such metals become the cathode in a galvanic cell and dissolve the one which are less positive which become the anode. For example, Al is more reactive (sacrificial) than Cr and Al would be anode in the cell with Cr.

The corrosion-resistant films can be designed in function of its *standard reduction potential*. Consequently, the film of less noble metals than that of the substrate protects it by getting dissolved instead of it. Such a ***film*** is called ***anodic.*** An example of *anodic film* is Al deposited on Mo substrate or Cr film on Fe substrate discussed by Chiang and Wei [157]. On the contrary, ***cathodic film***, such as Ni on Ti, is of metal more noble than the substrate. Such film should be dense to protect the substrate against a contact with electrolyte. Finally, the ***inert film*** or ***inhibitive film*** does not have any electrochemical activity or slow down the ***corrosion rate,*** respectively. These films are made generally of organic materials, discussed carefully by Dhavan et al. [154]. The *corrosion rate* is defined as loss of weight by corrosion divided by the unit area and by time of submission and is measured using ***polarization cell*** [158]. A sketch of such a cell called also electrochemical cell is shown in Fig. 4.26.

The tested sample is the working electrode being the anode and the reference electrode being cathode. The electrodes are polarized by an electrometer [159]. The standard procedure is to scan the potential applied to the anode from negative to positive and to record the corrosion current (see Fig. 4.27). The *corrosion current* corresponds to metal dissolution in electrolyte. The *corrosion potential* is the point in which rate of reduction is equal to the rate

22 The reduction potential is equal to the oxidation potential with the opposite sign.

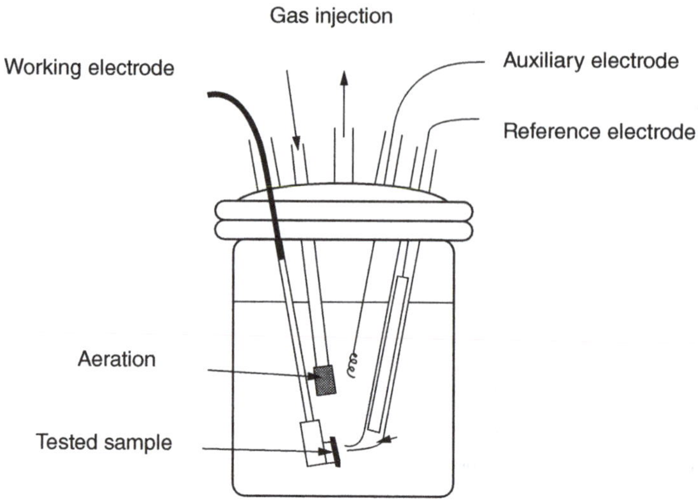

Figure 4.26 Laboratory electrochemical or *polarization cell*.

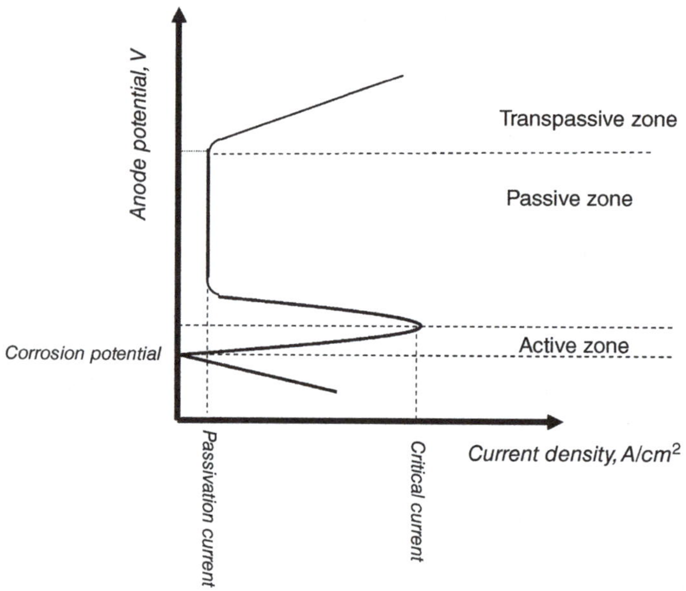

Figure 4.27 Curve of anode potential vs. current density in a polarization cell. The anode is metal or alloy with active and passives zones.

of oxidation [154, 160]. The increase of the potential above this value starts the active zone of metal dissolution until *critical current* value is reached. Then the increase of the potential is associated with reduction of current, down to *passivation current*. This zone is the passivation zone in which the oxide film is formed on the surface of anode. Further increase of potential leads to increase of the current value and of corrosion rate and the beginning of *transpassive zone* [160].

In fact, some corrosion-resistant materials require high testing potential to reach measurable corrosion current at anodic polarization test. An example is SiC ceramics tested by Cook et al. [161]. The optimization of deposition process, such as additional ion bombardment applied at magnetron sputtering of Cr films, has reduced the current density at polarization cell test down to $4\text{–}6\times10^{-6}$ A/cm^2 [157]. The testing of films and coatings requires often masking of the part of surface exposed to the electrolyte only to the deposit. For example, vacuum plasma sprayed Ta and Nb coatings were masked using resin to leave only an open surface area of 0.5 cm^2 of contact with electrolyte in electrochemical cell [162].

4.5.2 Hot Corrosion

The coatings tested against hot corrosion are mainly intermetallic compounds such as NiAl alloyed with Co, Cr, Y, or other metals which form superalloys being used mainly in aero-engines or in gas turbines. These compounds have to resist against corrosion caused by hot gases in these engines containing oxygen, carbon, or sulfur.

Oxidation is, by far, the most frequent mechanism of high-temperature corrosion. Thermodynamics of oxidation is discussed in Chapter 2 in the section describing the process of chemical vapor deposition. The kinetics of oxide film growth depends on the morphology of oxide layer. Consequently, the oxidation rate is proportional to the time of reaction, t_{ox}, for porous oxide. The dense oxide film morphology results in the square root dependence, $\sqrt{t_{ox}}$ on time. This kinetics can be measured by thermogravimetric tests. Typical results show a relative evolution of sample's mass at constant temperature in function of oxidation time.

The cyclic oxidation tests are frequently realized using *burner rigs*. The tests, being also applied to verify thermal shocks, simulate closely the real conditions in which hot-resistant films and coatings are used. Consequently, the corrosive agents are added to the fuel.

An example was the introduction of aqueous solution of NaCl to the combustor of fuel used to test duplex TBCs composed of plasma sprayed and laser glazed 8YSZ outer coatings deposited onto oxidation-resistant NiCrAlY bond coat tested at about 1320 K by Zaplatynski [163], e.g. aircraft application. Thermal cycles consisted of 55 minutes of exposition on the flame followed by 5 minutes. of cooling with air. The effects of bond coat oxidation were observed microscopically using *EMPA*.

Another example was the atomization of diluted sea-salt solution in burner ring fueled with standard aviation kerosene used by Raffaitin *et al.* [164]. The authors tested NiCoCrAlYTa coatings of different thicknesses for application in helicopter engines. The cyclic oxidation tests were carried out by heating to the temperatures of about 1170–1270 for 1 hour and cooling in air for 15 minutes. The corrosion of the coatings was observed using microscopic observation of the coating using *SEM* and their chemical analysis using *EDS*.

4.6 Electric and Magnetic Properties of Films and Coatings

Electric and magnetic properties of films and coatings are essential for their very frequent applications in microelectronics. The deposits may be used as, e.g. conducting paths, resistors, and capacitors. The films can be also a part of active semiconducting elements as diodes or transistors. Consequently, electrical resistivity as well as dielectric constant with dielectric loss factor are often tested. Thick ceramic coatings can be used to resist high voltages. Consequently, the breakdown voltage is tested and discussed in this section. The magnetic properties of the films and coatings of ferroelectric and superconducting materials deposits, such as permeability, and critical current and critical magnetic field, are also often tested. Finally, testing of deposits for an increasing application of electron emission is shortly discussed.

4.6.1 Electric Conductivity and Resistivity

Electric conductivity characterizes the transport of electric charge carriers being mainly electrons and more seldom ions[23]. In practice, the materials are characterized by their resistivity, ρ, which is equal for a wire of a section, A, and length, l, transporting current, I, under voltage U to:

$$\rho = \frac{U}{I}\frac{A}{l} \tag{4.26}$$

Electrons are weakly bound with nucleus in metals and alloys. Consequently, the resistivity of this material is low and limited mainly by the defects of the structure such as grains boundaries or pores. Moreover, the resistivity of these materials increases with temperature because the electrons are scattered by the *phonons* being oscillations of the crystal lattice or by non-conducting second phase in metal matrix composites *(MMC)*.

The electrons in superconductors, below the critical temperature, T_{cr}, are organized in *Cooper pairs*, and electric current is transported without any resistance [165]. The superconductivity effect disappears if: (i) the current is greater than the critical value, $I > I_{cr}$, critical current; or, (ii) the magnetic field generated by this current or by an external source is greater than the critical value $H > H_{cr}$.

Finally, in dielectrics and semiconductors the electrons are in two energetic bands: (i) ***valence band*** in which the electrons are closer to atom nucleus; and (ii) ***conduction band***, which has free electrons being able to move in all direction. These bands are separated by ***forbidden zone***. The electrons having enough energy may jump from valence to conduction band and participate in current transport. The energy can be delivered by, e.g. thermal energy. Consequently, the resistivity of these materials decreases with the temperature. Temperature coefficient of resistance (*TCR*), which describes the variation of resistance (and resistivity) with temperature, is rather positive for metals and alloys and rather negative for dielectrics and semiconductors.

The resistivity of the deposits can be measured by many methods described by Singh [166]. The popular test uses 4-point probe (Fig. 4.28). The current, I, is delivered by two external probes and two internal ones that measure the voltage, U.

The resistivity of the deposit can be found using Eq. 4.27. Another 4-point probe is used in ***van der Pauw test*** with the probes placed at the circumference of the sample. The method was used, e.g. to measure the resistivity of sputtered metallic glasses films on silicon substrate [167].

$$\rho = \frac{U}{I}2\pi l \tag{4.27}$$

23 Ions conduct electric current in *ionic conductors* such as solid electrolytes used, e.g. for *SOFC*.

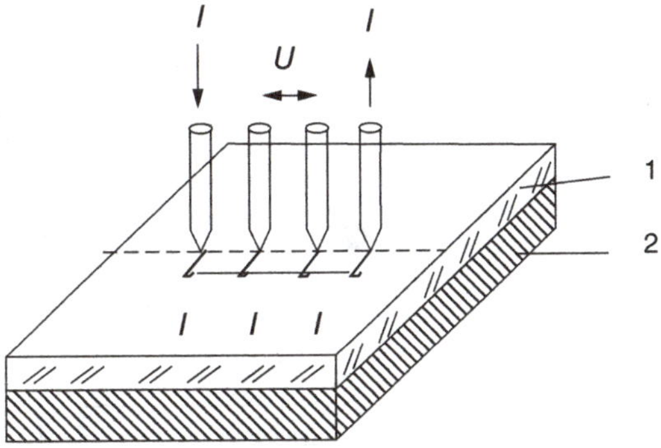

Figure 4.28 4-point resistivity probe: *1*, deposit; *2*, substrate; *I*, current; *U*, voltage; *l*, distance between the electrodes.

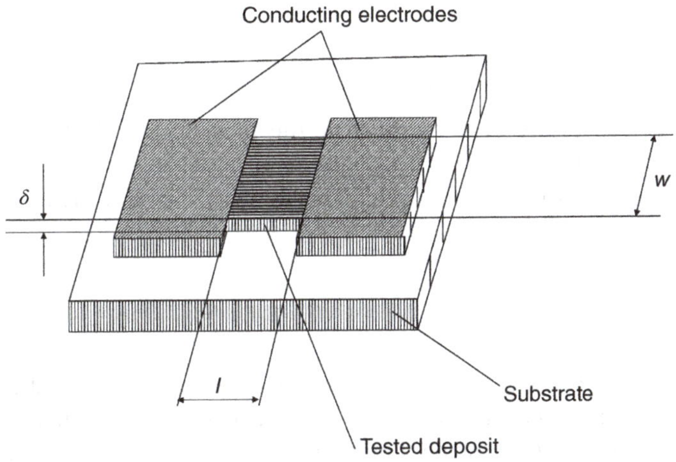

Figure 4.29 Geometry of planar resistor deposited onto insulating substrate.

The deposits used for resistors have rectangular shape and ***sheet resistance***, R_s, takes into account the number, *n*, of squares in the rectangle. This number is equal to $n = l/w$ and the resistance is equal $R_s = \rho/\delta$ by transforming *Eq. 4.26* for the geometry of tested planar resistor shown in Fig. 4.29.

Conducting electrodes, such as evaporated aluminum, are applied to pass electric current and to measure the voltage across the planar resistor. The resistance of tested deposits can be calculated as a product of *sheet resistance* and number of squares:

$$R = \frac{\rho}{\delta}\frac{l}{w} = R_s n \tag{4.28}$$

The electrodes and the planar resistor are prepared using techniques of patterning discussed in Section 1.4. The most important parameters of superconductors as critical temperature, T_{cr}, critical current, I_{cr}, and critical magnetic field, H_{cr}, can be determined using 4-point probe measurement of resistance in a cryostat of controlled temperature and external magnetic field.

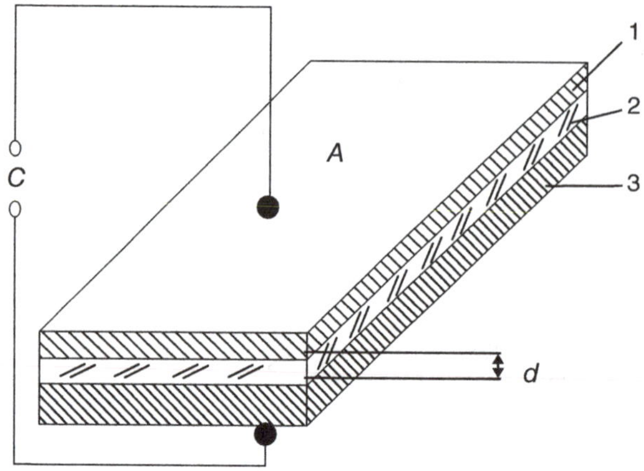

Figure 4.30 Schematic of planar capacitor: 1 – top electrode, 2 – bottom electrode, 3 – dielectric layer, C – capacitance, d – thickness of dielectric layer, A – surface area of the electrodes.

4.6.2 Dielectric Properties

Dielectrics are characterized by the presence of electric dipoles that get ordered in the external electric field. The additional electric field created by these dipoles superposes to external field. The ***dielectric constant***, ε, describes this amplification (see Eq. 4.29).

$$\varepsilon = \varepsilon_0 \varepsilon_r = \frac{Cd}{A} \tag{4.29}$$

Knowing that the *dielectric constant* of vacuum is ε_0, the dielectric can be characterized by *relative dielectric constant*, ε_r. The latter can be determined by measurement of capacitance, C, shown in Fig. 4.30. The capacitance, C, is measured using ***RC bridge***.

The dielectric constant generally decreases with the frequency of electric field. The value of ε_r remains constant in the range of frequencies, where one of the following dielectric polarization mechanisms is dominant [36]:

- polar molecules (e.g. H_2O) contribute in enhancement of electric field at the frequencies $\nu = 10^4$ to 10^8 Hz;
- ions in ionic materials contribute at the frequency range $\nu = 10^8$ to 10^{13};
- electrons contribute in the range of $\nu = 10^{13}$ to 10^{17} Hz.

To determine experimentally the dielectric constant of a deposit, with a thickness d, it is necessary to "sandwich" it between two conducting electrodes having surface area A. The *dielectric constant* can be found from the measured value of capacitance, C. At the measurements made frequently at standard frequency of $\nu = 1$ kHz, care should be taken about environmental factors such as humidity. Some dielectrics, e.g. γ-Al_2O_3 occurring frequently in thermally sprayed alumina coatings, are highly hygroscopic [168]. Consequently, the deposit must be heated up to a temperature exceeding $T \geq 100\ ^\circ C$ before starting the measurements.

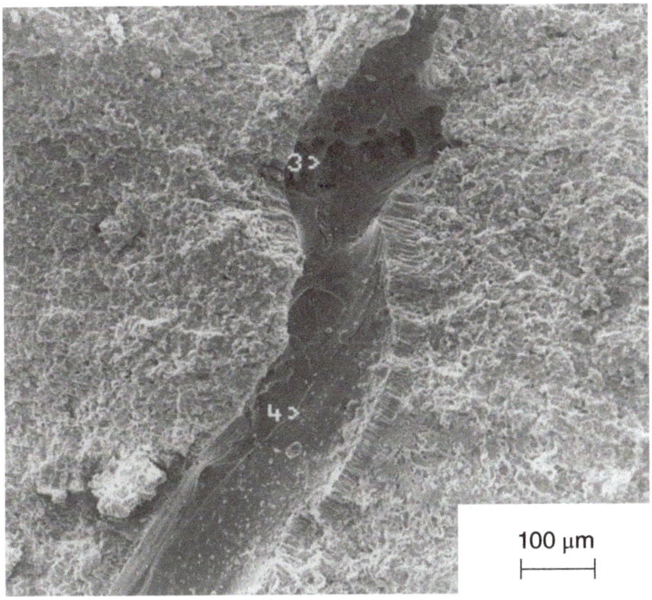

Figure 4.31 *SEM* micrograph (*SE*) of a channel left by an electrical breakdown in atmospheric plasma sprayed Al_2O_3 coating. The width of the channel indicates thermal character of the breakdown.

The ***dielectric strength*** (expressed in kV/cm) is equal to the highest electric field intensity, which can be applied to a dielectric without causing an electric breakdown. There are two principal mechanisms of such breakdown in dielectrics:

- ***electronic breakdown***, which occurs when the electrons form and avalanche is accelerated by external electric field; this avalanche creates a narrow (size of nanometers) discharge channel remaining after breakdown in the dielectric;
- ***thermal breakdown***, which occurs when a dielectric melts by electric current under the ***Joule effect***; this mechanism is characterized by a large (tenths of micrometers in size) discharge channel (see Fig. 4.31).

Thermal breakdown mechanism is most frequent in the ceramic deposits. Porosity is the major factor in reducing their *dielectric strength* [168, 169]. This results from much lower strength of the gases included in the pores compared to solid dielectric[24].

The experimental setup, similar to that shown in Fig. 4.30, can be used to determine *dielectric strength*. The voltage applied to the electrodes grows gradually until the breakdown occurs. The breakdown voltage divided by the thickness of the deposit is equal to the *dielectric strength*. At the measurements, care should be taken to avoid a corona discharge that may occur in the air around the electrodes. The immersion of the deposit with the electrodes in the oil of a transformer is a possible way in avoiding such discharge.

24 For example dense alumina has dielectric strength of 700 kV/cm compared to strength of air equal to 30 kV/cm [169].

4.6.3 Magnetic Properties

Magnetic field can be, similar to the electric one in dielectrics, reinforced in magnetic materials. This reinforcement occurs by the magnetic dipoles, which can be created by the movements of electrons around [36]: (i) the nucleus; and (ii) own axis (spin). The strength of external magnetic field, H, is reinforced in magnetic materials to obtain magnetic induction, B, by factor, μ, called permeability:

$$\mu = \mu_0 \mu_r = \frac{B}{H} \tag{4.30}$$

Magnetic materials such as ***ferromagnetic*** metals (e.g. Fe, Co, or Ni) or ***ferrimagnetic*** ionic oxides MFe_2O_4 are composed of magnetic domains and exhibit permanent magnetization, which is added to the external field, H, resulting in induction, B. The parameters characterizing the magnetic materials are related to their magnetization in external magnetic field and can be found from a hysteresis curve shown in Fig. 4.32.

The magnetization starts for $H = 0$ and reaches the maximal value at H_{max}. These materials may remain magnetized, showing ***remanence***, B_r, even if the external magnetic field is zero, $H = 0$. To eliminate the *remanence* it is necessary to apply the external field in the opposite direction called ***coercive field***, H_c. The magnetic materials with large H_c values are difficult to magnetize and demagnetize. They are called ***hard magnetic*** materials. An example *is Alnico 5* alloy (Fe + 24 wt. % Co + 14 wt. % Ni + 8 wt. % Al° + 3 wt. % Cu) having *remanence* equal to $B_r = 1.25\,\text{T}$ and the maximum energy product of $BH_{max} = 40\,\text{kJ/m}^3$. On the contrary, ***soft magnetic*** materials have small *coercive field*. Their example is, e.g. silicon iron alloy (Fe + 4 wt. % Si) having $H_c = 0.8\,\text{A/m}$ [156]. Temperature influences the magnetic properties by thermal motion of atoms and magnetization drops to zero at ***temperature Curie***, T_c. This temperature for mentioned alloys is equal to $T_c = 793\,\text{K}$ for *Alnico 5* and $T_c = 1291\,\text{K}$ for silicon iron.

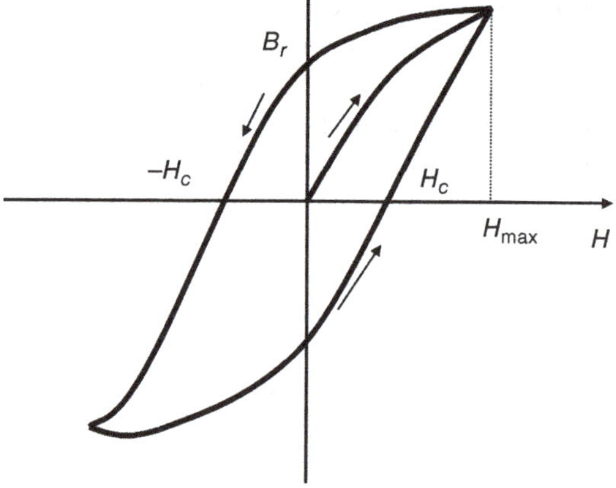

Figure 4.32 The magnetization of *ferromagnetic* or *ferrimagnetic* materials in external magnetic field.

The strength of magnetic field and resulting magnetization of solids can be measured using magnetometers. Romalis and Dang [170] described very precise magnetometers enabling discovery of magnetic induction as small as $B = 10^{-15}$ T. The vibrating sample magnetometer (*VSM*), described by Foner [171], has been frequently used to characterize the magnetic deposits. For example, thin SmCo magnetic films obtained by *PLD* on Si substrate were tested with such magnetometer [172]. The *VSM* set up includes a small internal pickup coil around the tested sample and external coil generating external magnetic field. The sample vibrates perpendicular to the external magnetic field which induces the voltage in the internal pickup coil by ***Faraday's law of induction***. The voltage is proportional to the magnetization of tested sample. The magnetization of the sample, *B*, together with external magnetic field generated by external coil, *H*, enable obtaining a hysteresis $B - H$ curve similar to that shown in Fig. 4.32.

4.7 Optical Properties of Films and Coatings

The optical properties of material result from the interaction of light with the atoms. The light is the visible spectrum of electromagnetic wave. The electromagnetic wave has magnetic and electric components. Each of them is perpendicular to the other. The phenomena resulting from these interactions are described by the ***Maxwell equations***, which enable to predict precisely the behavior of each component in the solid [173]. The light can also be presented as a flux of particles. The particles, called photons, have energy depending on the wave frequency. Their properties and the properties of electromagnetic waves in the visible spectrum and in close to these spectrum ultraviolet (*uv*) and infrared (*ir*) ones were calculated using Eq. 4.3 and are presented in Table 4.6.

The light shining on a surface of a deposit can be: (i) reflected by its surface; (ii) absorbed inside it; or (iii) transmitted through the deposit. Generally, these phenomena are described

Table 4.6 The properties of electromagnetic waves and of associated photons in the visible light and in neighboring *uv* and *ir* spectra [36, 43, 173][a].

Colors	Wavelengths, nm	Photon energies, eV	Frequencies, Hz
uv	10–380	124–3.26	3.0×10^{16}–7.9×10^{14}
Violet	400	3.10	7.9×10^{14}
Blue	470	2.64	7.5×10^{14}
Green	520	2.38	6.4×10^{14}
Yellow	580	2.14	5.8×10^{14}
Red	620	2.00	5.2×10^{14}
ir	780–1000	1.59–1.24	3.8×10^{14}–3.0×10^{14}

[a] The spectrum of visible light is continuous and the colors in the table correspond to some discrete values in the spectrum.

by the coefficients characterizing deposit material, being the ratio of corresponding light radiation intensity to the incident radiation intensity, I_0, namely:

- reflectivity, $R = I_R/I_0$;
- absorptivity, $A = I_A/I_0$;
- transmissivity, $T = I_T/I_0$.

The sum of these coefficients is equal to:

$$R + A + T = 1 \tag{4.31}$$

It is therefore sufficient to know two of them to find the value of the third one. The physical interactions of the light, being electromagnetic wave, with the solid results from their band structure and are mainly [36]: (i) ***electronic polarization***, which shifts the electron cloud with regard to nucleus; and (ii) ***electron transition***, which occurs when the photon energy corresponds to the electron transition, ΔE, from the ground state to the excited energy state. These phenomena are different for:

- ***metals***, which have free electrons. These electrons absorb the light radiation in a thin outer layer close to the surface and reemit visible light radiation. Consequently, their reflectivity is close to 1 being in the range of $R = 0.90$–0.95 [77];
- ***nonmetals***, such as ceramics and polymers, which are more transparent to light rays. The velocity of light in the nonmetals, v, is smaller than that in vacuum, c. The ratio of these velocities, n, is called ***index of refraction***, being in general a complex number[25], which depends on the light's wavelength. The index can be also found knowing *relative dielectric constant*, ε_r, and relative magnetic permeability, μ_r [36]:

$$n = \frac{c}{v} = \sqrt{\varepsilon_r \mu_r} \tag{4.32}$$

The light is absorbed gradually and the intensity of electromagnetic radiation decreases with the distance, x, from the surface of light arrival. The coefficient of absorption, a, is defined as:

$$a = -\frac{1}{I_0}\frac{dI}{dx} \tag{4.33}$$

The coefficient of absorption depends on coefficient of extinction, k, and the wavelength of light in vacuum, λ, as follows [173]:

$$a = \frac{4\pi k}{\lambda} \tag{4.34}$$

Finally, the coefficient of reflection, r, being a complex number, is defined as the ratio of reflected electric/magnetic waves amplitudes to that of incident ones. Similarly, the coefficient of transmission, t, is the ratio of transmitted electric/magnetic waves amplitudes to that of incident ones. In general, these coefficients are function of the temperature of the material and the light's wavelength.

The optical data measurements based on the analysis of reflected (reflective methods) or transmitted (transmission methods, when the deposit is detached from its substrate) light

25 Index of refraction describes the materials properties resulting for penetration of electromagnetic wave which may modify its amplitude and phase it can be also described as complex number [3, 43].

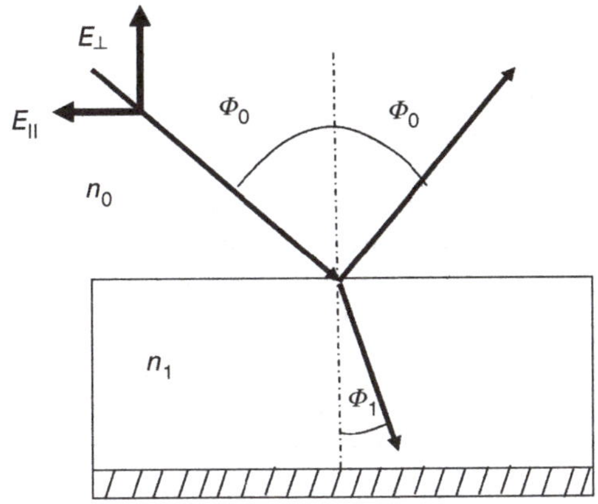

Figure 4.33 The sketch of trajectories of polarized incident light reflected on the surface and absorbed by the deposit in ellipsometric measurements.

by the sample. The techniques vary depending on the wavelength, λ, and on the polarization of the light. The transmittance can be tested using spectrophotometer [174].

The optical properties of films and coating are tested frequently with the ellipsometers [3]. The ellipsometric analysis uses a beam of polarized light shining on the solid surface being in contact with air. The technique consists of measuring the state of polarization after reflection at oblique incidence angle, as seen in Fig. 4.33. The light is reflected under the same angle, Φ_0, as the incident beam. The beam penetrates the solid deposit under refraction angle, Φ_1. This angle depends on the indexes of refraction of air n_0 and of the solid, n_1, following ***Snell's law***:

$$n_1 \sin\phi_1 = n_2 \sin\phi_2 \tag{4.35}$$

The incident light is a plane wave with linear polarization, i.e. having electric field wave being in one plane, which can be represented by the electric field components $E_\perp$ (perpendicular of the plane of incidence) or $E\|$ (perpendicular to this plane). Each component of the incident wave is reflected differently, resulting in a reflected wave that has elliptical polarization. The reflection coefficients of each component are the complex numbers, $r_\perp$ and $r_\perp$, and their ratio can be expressed as [3, 175]:

$$\frac{r_{\|}}{r_\perp} = \tan\psi e^{-i\Delta} \tag{4.36}$$

The parameter tan Ψ describes the amplitude ratio upon the reflection and the angle Δ is the phase shift of each component resulting from the reflection. The angles Ψ, Δ are obtained experimentally, and the models describing the electromagnetic light reflection enable to find optical constants of the reflecting material [176]. The measurements can be made using one wavelength associated with laser emission or with several wavelengths; this is the case of spectral ellipsometers (*SE*).

An ellipsometer with rotating polarizer and analyzer enables to find the polarization after reflection. Such ellipsometer, described in detail by McGuire [3], includes:

- incident light source emitting one or more wavelengths;
- rotating polarizer introducing linear polarization to incident ray;
- sample with the surface reflecting arriving and creating reflected ray;

- rotating analyzer of reflected ray which enables to find the polarization of wave after reflection;
- detector being generally a photomultiplier enabling to find the intensity of light issuing from analyzer and to find out its polarization.

The ellipsometers enable to find optical constants *n*, *k* as well as the thicknesses of mono- and multilayers for one or more wavelengths (*ES*). Such measurements are steadily improved and further developed especially in semiconductor industry to test the deposits growing up at manufacturing [175].

In addition to the optical constants, it is sometimes important to determine and quantify the color of a deposit. This parameter counts above all in decorative deposits. Color can be quantified using a system standardized by the Commission International de l'Éclairage described by Beck *et al.* [177].

4.8 Characterization of Biomaterials

4.8.1 Introduction

Fantozzi *et al.* [178] defined the biomaterials as *natural or synthetic materials being compatible, as a part of a medical device, with the living tissues*. The interaction with the living tissue enables also the categorizing the biomaterials as [178–180]:

- *bioinert* materials being mechanically interlocked by living tissue;
- *bioactive* materials, which form interfacial bonding with the tissue;
- *bioresorbable* materials being replaced by the surrounding tissue.

Biomaterials, similar to materials, can be grouped into the main categories of:

- metals and alloys such as bioinert Ti, stainless, steel, CoCr alloy;
- ceramics such as bioinerts Al_2O_3 or ZrO_2 and a large family of bioactive calcium phosphates as *HA* or *TCP* and bioglasses as frequently used 45S5;
- polymers being generally bioactive such as natural collagen or synthetic *PMMA*;
- composites, e.g. bioinert Al_2O_3 matrix including antibacterial Ag particles.

Films and coating of biomaterials are frequently used in the implants such as *hip joints prostheses* or *dental implants* [178, 181]. These medical devices need to withstand high level of mechanical stresses and are made of Ti stainless steel or high-strength alloys. The bioactive *HA* coating can be deposited on the part of artificial hip being in contact with human bone to improve the bio-integration of the prosthesis. The most important properties of biomaterials are (i) chemical; (ii) physical; (iii) mechanical; and (iv) biological. The important part of these parameters is presented in previous sections of this chapter or, for the interested reader, in the book of Bandyopadhyay and Bose [180]. Consequently, only the part of this characterization determining biocompatibility is described in the following sections.

The evaluation of biocompatibility starts by *in vitro*, i.e. testing "in glass" before, more complicated, *in vivo* testing, which is testing including living animals [182].

Table 4.7 The ions concentrations in *SBF* and in human blood [186–188].

	Ion concentration [mM]							
	Na^+	K^+	Mg^{2+}	Ca^{2+}	Cl^-	HCO^-_3	HPO^{2-}_4	SO^{2-}_4
Human blood plasma	142	5	1.5	2.5	103	27	1	0.5
SBF	142	5	1.5	2.5	103–147.8[a]	4.2–27[31]	1	0–0.5[a]

[a] In the different types of *SBF* [186].

4.8.2 *In vitro* Characterization

The *in vitro* method applied frequently for biomaterials is ***cell culture method***. The method consists in evaluating the biocompatibility of tested materials in contact with cells [183, 184]. The cells modifies their morphology in contact with tissue, and the methods of testing are standardized by ASTM [185]. The standards concern the details of testing such as: (i) the cells type and size; (ii) duration of contact between cell and tissue; and (iii) geometry and surface of test sample. A typical cell used in many characterizations of biomaterials is L-929 mouse fibroblast one [183, 185].

Another method consists in evaluation of the biomaterials by corrosion testing. The corrosion test, being frequently applied, uses with simulated body fluid (*SBF*) with the ions concentrations being similar to those of human blood plasma [186, 187]]. The ions and their concentration of *SBF* are shown in Table 4.7. The commercial reagents, which can be used to prepare this fluid, are shown by Hench [188]. The contact with *SBF* modifies the surface of biomaterial being in contact and forms apatite observed in solution plasma sprayed *HA* coatings and their composites with Cu by Unabia *et al.* [189]. The formation of apatite was also observed on the surface of bioglass 45S5 [186]. The modifications of the surface depend on the time of immersion and concern the morphologies of this surface and the chemical composition of apatite. The corrosion-tested biomaterials are characterized using typical methods including *SEM*, *XRD*, *ir* spectroscopy, or corrosion potential checking. Finally, the mechanical properties of biomaterials belong also to important ones to be tested [190].

4.8.3 *In vivo* Characterization

In vivo characterization of biomaterials is used mainly for testing of new materials in medical devices. The second reason is testing of entire medical device in conditions similar to those occurring at final implantation [191]. This kind of testing takes into account metabolic transformations in living body being in contact with biomaterial [192].

The medical devices including films and coatings are generally in contact with skin, blood paths, bones, and dents, and the duration of contact can be categorized as: (i) short, up to 12 weeks; and (ii) long, more than 12 weeks. The procedures including the animal require approval of organizations such as Institutional Animal Care and Use Committee in the USA [193]. The animals used in vivo experiments concerning films and coating are shown in Table 4.8.

Table 4.8 The animals used in testing different types of medical devices including films and coatings.

Medical device	Animals
Cardiovascular, such as heart valves, stents	Sheep, dog, pig, calf, monkey
Orthopedic/bone, such as implants of hips or knees	Rabbit, dog, pig, mouse, rat, sheep, monkey
Ophthalmological, e.g. contact lenses	Rabbit, monkey

Source: Anderson and Schoen [191]/with permission of Elsevier.

The biomaterials used in *in vivo* testing should fulfill the specification concerning mainly [192]: (i) physical and mechanical properties should mimic closely the natural bone; (ii) infection should be avoided; (iii) only one surgery is recommended; and (iv) corrosion resistance against physiological liquids has to be assured.

The methods of testing medical devices used in in vivo testing include X-ray computed microtomography, SEM, TEM depending on the device. The testing *in vivo* is normalized by ASTM [194].

References

1 Jacobson, B.E. (1982). Microstructures of PVD-deposited films characterized by transmission electron microscopy. In: *Deposition Technologies for Films and Coatings* (ed. R.F. Bunshah), 268–334. Park Ridge, NJ: Noyes Publications.

2 Martin, J.L. and George, A. (1998). *Caractérisation Expérimentale des Matériaux II*. Lausanne: PPUR.

3 McGuire, G.E. (1994). Characterization of thin films. In: *Handbook of Deposition Technologies for Films and Coatings* (ed. R.F. Bunshah), 763–821. Park Ridge, NJ: Noyes Publications.

4 Greene, J.E. (1994). Nucleation, film growth, and microstructural evolution. In: *Handbook of Deposition Technologies for Films and Coatings* (ed. R.F. Bunshah), 681–739. Park Ridge, NJ: Noyes.

5 Stout, K.J. and Blunt, L. (1995). Nanometres to micrometres: 3-D surface measurement in bio - engineering. *Surf. Coat. Technol.* 71: 69.

6 Notthoff, C., Winterer, M., Beckel, A. et al. (2013). Spatial high resolution energy dispersive X-ray spectroscopy on thin lamellas. *Ultramicroscopy* 129: 30.

7 Mattox, D.M. (1994). Non-elemental characterization of films and coatings. In: *Handbook of Deposition Technologies for Films and Coatings* (ed. R.F. Bunshah), 643–680. Park Ridge, NJ: Noyes Publications.

8 Loretto, M.H. (1984). *Electron Beam Analysis of Materials*. London: Chapman and Hall.

9 Agui, B., Froment, M., Abel, F. et al. (1990). *Surfaces Interfaces et Films Minces*. Paris: Dunod. (in French).

10 Thanka Rajan, S., Nandakumar, A.K., Hanawa, T., and Subramanian, B. (2017). Materials properties of ion beam sputtered Ti-Cu-Pd-Zr thin film metallic glasses. *J. Non-Crystal. Solids* 461: 104.

11 Watts, J.F. and Wolstenholme, J. (2003). *An Introduction to Surface Analysis by XPS and AES*. Chichester: J. Wiley.

12 Fuchs, A., Scherer, J., Jungn, K., and Erhardt, H. (1993). Determination of sp^2/sp^3 carbon bonding ration in a-C: H including irradiation damage by factor analysis of Auger electron spectra. *Thin Solid Films* 202: 51.

13 Dowling, D.P., Ahern, M.J., Kelly, T.C. et al. (1992). Characterization study of diamond and diamond-like carbon. *Surf. Coat. Technol.* 53: 177.

14 Cattini, A., Bellucci, D., Sola, A. et al. (2013). Suspension plasma spraying of optimized functionally graded coatings of bioactive glass/hydroxyapatite. *Surf. Coat. Technol.* 236: 118.

15 Kommer, M., Sube, T., Richter, A. et al. (2018). Enhanced wear resistance of molybdenum nitride coatings deposited by high power impulse magnetron sputtering by using micropatterned surfaces. *Surf. Coat. Technol.* 333: 1.

16 Písarík, P., Jelíneka, M., Remsa, J. et al. (2017). Antibacterial, mechanical and surface properties of Ag-DLC films prepared by dual PLD for medical applications. *Mater. Sci. Eng. C* 77: 955.

17 Valencia, F.J., Santiago, J., González, R.I. et al. (2021). Nanoindentation of amorphous carbon: a combined experimental and simulation approach. *Acta Mater.* 203: 116485.

18 Silva, S.R.P. and Stolojan, V. (2005). Electron energy loss spectroscopy of carbonaceous materials. *Thin Solid Films* 488: 283.

19 Llovet, X., Moy, A., Pinard, P.T., and Fournelle, J.H. (2021). Electron probe microanalysis: a review of recent developments and applications in materials science and engineering. *Prog. Mater. Sci.* 11: 100673.

20 Kitahara, S. and Hasui, A. (1974). A study of bonding mechanism of sprayed coatings. *J. Vac. Sci. Technol.* 11: 747.

21 Deram, V., Minichiello, C., Vannier, R.-N. et al. (2003). Microstructural characterizations of plasma sprayed hydroxyapatite coatings. *Surf. Coat. Technol.* 166: 153.

22 Turrell, G. (1996). The Raman effect. In: *Raman Microscopy, Development and Applications* (ed. G. Turrell and J. Corset), 1–23. London: Academic Press.

23 Delhaye, M., Barbillat, J., Aubard, J. et al. (1996). Instrumentation. In: *Raman Microscopy, Development and Applications* (ed. G. Turrell and J. Corset), 27–48. London: Academic Press.

24 Website of Horiba Scientific, Palaiseau, France, https://www.horiba.com/fra/technology/spectroscopy/detectors/what-is-a-ccd-detector/consulted (accessed 25 September 2024).

25 Barbillat, J. (1996). Raman imaging. In: *Raman Microscopy, Development and Applications* (ed. G. Turrell and J. Corset), 175–200. London: Academic Press.

26 Fayed, S.M., Chen, D., Li, S. et al. (2022). Corrosion behavior and passive stability of multilayer *DLC*-Si coatings. *Surf. Coat. Technol.* 431: 128001.

27 Cui, W.G., Lai, Q.B., Zhang, L., and Wang, F.M. (2010). Quantitative measurements of sp^3 content in *DLC* films with Raman spectroscopy. *Surf. Coat. Technol.* 205, 1995.

28 Unabia, R.B., Bonebeau, S., Candidato, R.T. Jr. et al. (2019). Investigation on the structural and microstructural properties of copper-doped hydroxyapatite coatings deposited using solution precursor plasma spraying. *J. Eur. Ceram. Soc.* 39: 4255.

29 Hollinger, G. and Pertosa, P. (1990). Spectroscopies d'électrons. In: *Surfaces, Interfaces et Films Minces* (ed. B. Agius and M. Froment) and co-authors, 129–235. Paris: Dunod (in French).

30 Kim, J.W. and Kim, A. (2021). Absolute work function measurement by using photoelectron spectroscopy. *Curr. Appl. Phys.* 31: 52.

31 Website of Thermo Fisher Scientific, Waltham, MA, USA, Titanium X-ray photoelectron spectra, titanium electron configuration, and other elemental information, https://www.jp.xpssimplified.com/elements/titanium.php (accessed 25 September 2024).

32 Tomaszek, R., Pawłowski, L., Zdanowski, J. et al. (2004). Microstructural transformations of TiO_2, Al_2O_3+13TiO_2 and Al_2O_3+40TiO_2 at plasma spraying and laser engraving. *Surf. Coat. Technol.* 185: 137.

33 Ingo, G. (1991). Origin of darkening in 8 wt. % yttria-zirconia plasma sprayed thermal barrier coatings. *J. Amer. Ceram. Soc.* 74: 381.

34 Pawłowski, L. and Zacchetti, N. (1995). Microstructural and microanalytical study of Cr_2O_3 plasma sprayed coatings. In: *Advances in Inorganic Films and Coatings* (ed. P. Vincenzini), 381–390. Faenza, Italy: Techna.

35 Uo, M., Wada, T., and Sugiyama, T. (2015). Applications of X-ray fluorescence analysis (XRF) to dental and medical specimens. *Jpn. Dent. Sci. Rev.* 51: 2.

36 Callister, W.D. Jr. (1994). *Materials Science and Engineering: An Introduction*, 3e. New York, NY: Wiley.

37 Formica, S.P. and Lee, S.M. (2005). X-ray fluorescence system for thin film composition analysis during deposition. *Thin Solid Films* 491: 71.

38 Klenk, M., Schenker, O., Probst, U., and Bucher, E. (1999). X-ray fluorescence measurements of thin film chalcopyrite solar cells. *Sol. Energy Mater. Sol. Cells* 58: 299.

39 S.-H. Zhong and L.-P. Chen, 2021, Thickness measurement of multi-layer thin film alloy by X-ray fluorescence spectrometer, [in:] *22nd International Conference on Electronic Packaging Technology*, August 11-14, Xiamen, China, DOI: https://doi.org/10.1109/ICEPT52650.2021.9567975.

40 Skrzypczak, M., Bertrand, P., Zdanowski, J., and Pawłowski, L. (2001). Modeling of temperature fields in the graphite target at pulsed laser deposition of CN_x films. *Surf. Coat. Technol.* 138: 39.

41 Seetharaman, A. and Dhanuskodi, S. (2014). Micro-structural, linear and nonlinear optical properties of titania nanoparticles. *Spectrochim. Acta A Mol. Biomol. Spectrosc.* 127: 543.

42 Zhang, N., Zhuang, D.M., Liu, J.J. et al. (2000). Microstructure of iron sulfide as solid lubrication coating by low temperature ion sulfurization. *Surf. Coat. Technol.* 132: 1.

43 Agius, R., Bosio, L., and Le Goff, H. (1990). Interaction photons-surface. In: *Surfaces, Interfaces et Films Minces* (ed. B. Agius and M. Froment) and co-authors, 349–433. Paris: Dunod (in French).

44 Beaunier, L., Froment, M., and Klein, J. (1990). Microscopie électronique. In: *Surfaces, Interfaces et Films Minces* (ed. B. Agius and M. Froment) and co-authors, 31–128. Paris: Dunod. (in French).

45 Podlesak, H., Łatka, L., Dietrich, D. et al. (2013). Electron microscopy and diffraction studies of suspension-plasma-sprayed ZrO_2+8 wt. % Y_2O_3 coatings. *Surf. Coat. Technol.* 220: 67.

46 Podlesak, H., Schnick, T., Pawłowski, L. et al. (2000). Microscopic study of Al–SiC particulate composites processed by laser shocks. *Surf. Coat. Technol.* 124: 32.

47 Randle, V. (2009). Electron backscatter diffraction: strategies for reliable data acquisition and processing. *Mater. Charact.* 60: 913.

48 Wilkinson, A.J. and Britton, T.B. (2012). Strains, planes, and EBSD in materials science. *Mater. Today* 15 (9): 366.

49 Britton, T.B., Jiang, J., Guo, Y. et al. (2016). Tutorial: Crystal orientations and EBSD — Or which way is up? *Mater. Charact.* 117: 113.

50 Sokołowski, P., Pawłowski, L., Dietrich, D. et al. (2016). Advanced microscopic study of suspension plasma-sprayed zirconia coatings with different microstructures. *J. Thermal Spray Technol.* 25 (1-2): 94.

51 Hadni, A. (1971). *Éléments d'Optique: Optique Géométrique*. Paris: Dunod.

52 Pawłowski, L. (1998). Mikrogefügeuntersuchungen an thermisch gespritzten Schichten. *Structure* 33: 11. (in German).

53 D.B. Fowler, 1991, Metallographic evaluation of thermally sprayed coatings, *ASTM Standardization News*, 19, May, 54.

54 von Ardenne, M. (1938). Das elektronen-rastermikroskop: theoretische grundlagen. *Z. Phys.* 109 (9–10): 553. (in German).

55 Postek, M.T., Howard, K.S., Johnson, A.H., and McMichael, K.L. (1980). *Scanning Electron Microscope: A Student Handbook*. Williston, VT: Ladd Research Industries.

56 Gabriel, B.L. (1985). *SEM: A User's Manual for Materials Science*. Metal Park, OH: ASM.

57 Cattini, A., Łatka, L., Bellucci, D. et al. (2013). Suspension plasma sprayed bioactive glass coatings: effects of processing on microstructure, mechanical properties and in-vitro behavior. *Surf. Coat. Technol.* 220: 52.

58 Binning, G., Rohrer, H., Gerber, C., and Weibel, E. (1982). Tunneling through a controllable vacuum gap. *Appl. Phys. Lett.* 40 (2): 178.

59 Chen, C.J. (2008). *Introduction to Scanning Tunneling Microscopy*, 2e. Oxford: Oxford University Press.

60 Habibullah, H. (2020). 30 Years of atomic force microscopy: creep, hysteresis, cross-coupling and vibration problems of piezoelectric tube scanners. *Measurement* 159: 107776.

61 Kubby, J.A. and Boland, J.J. (1996). Scanning electron microscopy of semiconductor surfaces. *Surf. Sci. Rep.* 26: 61.

62 Nilius, N. (2009). Properties of oxide thin films and their adsorption behavior studied by scanning tunneling microscopy and conductance spectroscopy. *Surf. Sci. Rep.* 64: 595.

63 Kurt, R., Sanjines, R., Karimi, A., and Levy, F. (2000). Structural and mechanical properties of CN_x thin films prepared by magnetron sputtering. *Diam. Relat. Mater.* 9: 566.

64 Shao, Z., Zhang, Z., Yuan, H. et al. (2018). Scanning tunneling microscopic observation of enhanced superconductivity in epitaxial Sn islands grown on $SrTiO_3$ substrate. *Sci. Bull.* 63: 1332.

65 Tseng, A.A. (2011). Advancements and challenges in development of atomic force microscopy for nanofabrication. *Nano Today* 6: 483.

66 Binning, G., Quate, C.F., and Gerber, C. (1986). Atomic force microscope. *Phys. Rev. Lett.* 56 (9): 930.

67 Linneman, R., Gotszalk, T., Hadiijski, L., and Rangelow, I.W. (1995). Characterization of a cantilever with an integrated deflection sensor. *Thin Solid Films* 264: 159.

68 Duparré, A., Ruppe, C., Pischow, K.A. et al. (1995). Atomic force microscopy of cross-sections of optical coatings: a new method. *Thin Solid Films* 261: 70.

69 Hamdi, M., Hakamata, S., and Ektessabi, A.M. (2000). Coating of hydroxyapatite thin film by simultaneous vapor deposition. *Thin Solid Films* 377–378: 484.

70 Vlasova, M., Márquez Aguilar, P.A., Reséndiz-González, M.C. et al. (2010). Laser treatment of Si–Al2O3 two-layer target. *Opt. Laser Technol.* 42: 8.

71 Hawkes, P.W. (1990). Ernst Ruska. *Phys. Today* 43: 84.

72 Roeder, R.K. (2013). Mechanical characterization of biomaterials. In: *Characterization of Biomaterials* (ed. A. Bandyopadhyay and S. Bose), 49–104. Waltham, MA: Elsevier.

73 Philpot, T.A. (2011). *Mechanics of materials, An Integrated learning System*, 2e. New York, NY: Wiley.

74 Kelly, A. and Macmillan, N.H. (1986). *Strong Solids*. Oxford: Clarendon Press.

75 Brotzen, F.R. (1994). Mechanical testing of thin films. *Int. Mater. Rev.* 39: 24.

76 Rickerby, D. (1996). Measurement of coating adhesion. In: *Metallurgical and Ceramic Protective Coatings* (ed. K.H. Stern), 306–333. London: Chapman & Hall.

77 Mattox, D.M. (2010). *Handbook of Physical Vapor Deposition (PVD) Processing*, 2e. Amsterdam: Elsevier.

78 Pawłowski, L. (2008). *The Science and Engineering of Thermal Spray Coatings*, 2e. Chichester: Wiley.

79 Pawłowski, L. (1996). Technology of thermally sprayed anilox rolls: state of art, problems, and perspectives. *J. Thermal Spray Technol.* 5: 317.

80 Górnik, M., Jonda, E., Łatka, L. et al. (2021). Influence of spray distance on mechanical and tribological properties of HVOF sprayed WC-Co-Cr coatings. *Mater. Sci. Pol.* 39: 545.

81 Jaworski, R., Pawłowski, L., Roudet, F. et al. (2008). Characterization of mechanical properties of suspension plasma sprayed TiO_2 coatings using scratch test. *Surf. Coat. Technol.* 202: 2644.

82 Steinman, P.A., Tardy, Y., and Hintermann, H.E. (1987). Adhesion testing by scratch test: the influence of intrinsic an extrinsic parameters on the critical load. *Thin Solid Films* 154: 333.

83 Al-Rjoub, A., Bin Yaqub, T., Cavaleiro, A., and Fernandes, F. (2022). The influence of V addition on the structure, mechanical properties, and oxidation behaviour of TiAlSiN coatings deposited by DC magnetron sputtering. *J. Mater. Res. Technol.* 20: 2444.

84 Seidel, W.W. and Hahn, F. (2014). *Werkstofftechnik: Werkstoffe, Eigenschaften, Prüfung, Anwendung*, 10[t]e. Munich: Carl Hanser Verlag.

85 Sakharova, N.A., Fernandes, J.V., Antunes, J.M., and Oliveira, M.C. (2009). Comparison between Berkovich, Vickers and conical indentation tests: a three-dimensional numerical simulation study. *Int. J. Solids Struct.* 46: 1095.

86 Burnett, P.J. and Rickerby, D.S. (1987). Assessment of coating hardness. *Surf. Eng.* 3: 69.

87 Shahdad, S.A., McCabe, J.F., Bull, S. et al. (2007). Hardness measurements with traditional Vickers and Martens hardness method. *Dent. Mater.* 23: 1079.

88 Łatka, L., Pawłowski, L., Chicot, D. et al. (2010). Mechanical properties of suspension plasma sprayed hydroxyapatite coatings submitted to simulated body fluid. *Surf. Coat. Technol.* 205: 954.

89 Chicot, D. and Lesage, J. (1995). Absolute hardness of films and coatings. *Thin Solid Films* 254: 123.

90 Lin, C.K. and Berndt, C.C. (1995). Statistical analysis of microhardness variations in thermal spray coatings. *J. Mater. Sci.* 30: 111.

91 Valente, T. (1997). Statistical evaluation of Vickers indentation test results for thermally sprayed materials. *Surf. Coat. Technol.* 90: 14.

92 Fridrich, C., Berg, G., Broszeit, E., and Berger, C. (1996). Measurements of the hardness of hard coatings using a force indentation function. *Thin Solid Films* 29-291: 216.

93 Burnett, P.J. and Rickerby, D.S. (1987). The mechanical properties of wear - resistant coatings I: modeling of hardness behavior. *Thin Solid Films* 148: 41.

94 Oliver, W.C. and Pharr, G.M. (1992). An improved method for determining hardness and elastic modulus using load and displacement sensing indentation experiments. *J. Mater. Res.* 7: 1564.

95 Oliver, W.C. and Pharr, G.M. (2004). Measurement of hardness and elastic modulus by instrumented indentation: advances in understanding and refinements to methodology. *J. Mater. Res.* 19: 3.

96 Ivanova, A.A., Surmeneva, M.A., Shugurov, V.V. et al. (2018). Physico-mechanical properties of Ti-Zr coatings fabricated via ion assisted arc-plasma deposition. *Vacuum* 149: 129.

97 Timoshenko, S. (1983). *Strength of Materials, Part 1 and Part 2*, 3e. Malabar, FL: Kreiger Publishing Company 1976.

98 Varis, T., Rajamäki, E., and Korpiola, K. (2001). Mechanical properties of thermal spray coatings. In: *Thermal Spray 2001: New Surfaces for a New Millennium* (ed. C.C. Berndt, K.A. Khor, and E.F. Lugscheider), 993–997. Materials Park, OH: ASM International.

99 Glandus, J.C., Platon, F., and Boch, P. (1979). Measurements of the elastic moduli of ceramics. *Mater. Eng. Appl.* 1: 243.

100 Wiklund, U., Bromark, M., Larsson, M. et al. (1997). Cracking resistance of thin hard coatings estimated by four-point bending. *Surf. Coat. Technol.* 91: 57.

101 Zhang, S., Sun, D., Fu, Y., and Du, H. (2004). Toughness measurement of ceramic thin films by two-step uniaxial tensile method. *Thin Solid Films* 469–470: 233.

102 Guo, D.Z. and Wang, L.J. (1992). Measurement of the critical strain energy release rate of plasma-sprayed coatings. *Surf. Coat. Technol.* 56: 19.

103 Zambelli, G. and Vincent, L. (1998). *Matériaux et Contact. Une Approche Tribologique.* Lausanne: PPUR.

104 Li, S., Yu, C., Deng, C. et al. (2022). Optimized mechanical and tribological properties of thermally sprayed ceramic coatings by constructing crystalline–amorphous heterojunctions. *Appl. Surf. Sci.* 604: 154552.

105 Holmberg, K. and Matthews, A. (1994). Coatings tribology: a concept, critical aspects and future directions. *Thin Solid Films* 253: 173.

106 Holmberg, K. (1992). A concept for friction mechanism of coated surfaces. *Surf. Coat. Technol.* 56: 1.

107 Föhl, J., Weißenberg, T., and Wiedemeyer, J. (1988). General aspects for tribological applications of hard particle coatings. In: *1st Plasma Technik Symposium*, vol. 3 (ed. S. Blum-Sandmeier, H. Eschnauer, P. Huber, and A.R. Nicoll), 23–34. Wohlen: Plasma Technik.

108 Wielage, B., Steinhäuser, S., Pawłowski, L. et al. (1998). Laser treatment of vacuum plasma sprayed CoCrAlY alloy. *Surf. Eng.* 14: 391.

109 Buran, U. and Fischer, M. (1988). Properties of plasma spray coatings for piston ring running surfaces. In: *1st Plasma Technik Symposium*, vol. 2 (ed. S. Blum-Sandmeier, H. Eschnauer, P. Huber, and A.R. Nicoll), 25–36. Wohlen: Plasma Technik.

110 Taylor, M.L., Murphy, J.G., and King, H.W. (1987). Brittle/ductile erosion of plasma sprayed ceramic and metallic coatings. In: *Thermal Spray: Advances in Coatings Technology* (ed. D.L. Houck), 143–149. Materials Park, Ohio: ASM International.

111 Shipway, P.H. and Hutchings, I.M. (1995). Measurements of coating durability by solid particle erosion. *Surf. Coat. Technol.* 71: 1.

112 Alonso, F., Fagoaga, I., and Oregui, P. (1991). Erosion protection of carbon-epoxy composites by plasma sprayed coatings. *Surf. Coat. Technol.* 49: 482.

113 Tiwari, R., Sampath, S., Herman, H., and Anekawa, Y. (1993). Cavitation-erosion of plasma sprayed nickel aluminides. In: *Thermal Spray Coatings: Research, Design and Application* (ed. C.C. Berndt and T.F. Bernecki), 423–428. Materials Park, Ohio: ASM International.

114 Karimi, A. (1989). Cavitation erosion of austenitic stainless steel and effect of boron and nitrogen ion implantation. *Acta Metall.* 37: 1079.

115 Blanpain, B., Mohrbacher, H., Liu, E. et al. (1995). Hard coatings under vibrational contact conditions. *Surf. Coat. Technol.* 74-75: 953.

116 Askwith, T.C. (1980). The basic mechanisms of wear. *Surf. J.* 11: 2.

117 Wenzelburger, M., Lopez, D., and Gadow, R. (2006). Methods and application of residual stress analysis in thermally sprayed coatings and composite. *Surf. Coat. Technol.* 201: 1995.

118 Riggs, W. (2004). Testing of coatings. In: *Handbook of Thermal Spray Technology* (ed. J.R. Davis), 261–272. Materials Park, Ohio: ASM International.

119 Cullity, B.D. and Stock, S.R. (2001). *Elements of X-Ray Diffraction*, 3e. Upper Saddle River, NJ: Pearson Prentice Hall.

120 Fitzpatrick, M.E., Fry, A.T., Holdway, P. et al. (2005). *Determination of Residual Stresses by X-ray Diffraction*. Teddington, Middlesex: National Physical Laboratory.

121 S. Węglowski and L. Pawłowski, 1981, A study of residual stresses in plasma sprayed alumina coating, [in:] *Proc. 5th Int. Symp. Plasma Chem.*, B. Waldie (ed.), Edinburgh, UK, 1–14 August, pp. 592–601.

122 Noutomi, A., Kodama, M., Inoue, Y. et al. (1989). Residual stresses measurement on plasma sprayed coatings. *Weld. Int.* 11: 947.

123 Kesler, O., Matejicek, J., Sampath, S. et al. (1998). Measurements of residual stresses in plasma-sprayed metallic, ceramic and composite coatings. *Mater. Sci. Engn, A* 257: 215.

124 Gill, S.C. and Clyne, T.W. (1994). Investigation of residual stresses generation during thermal spraying by continuous curvature measurements. *Thin Solid Films* 250: 172.

125 Whittemore, O.J. (1981). Mercury porosimetry of ceramics. *Powder Technol.* 29: 167.

126 Taylor, D.J., Fleig, P.F., and Hietala, S.L. (1998). Technique for characterization of thin film porosity. *Thin Solid Films* 332: 257.

127 ASTM B328 (2003). *Standard Test Method for Density and Interconnected Porosity of Sintered Powder Metal Structural Parts and Oil Impregnated Bearings*, published January. West Conshohocken; PA: ASTM International.

128 Sokołowski, P., Łatka, L., Pawłowski, L. et al. (2015). Characterization of microstructure and thermal properties of YCSZ coatings obtained by suspension plasma spraying. *Surf. Coat. Technol.* 268: 147.

129 Zeng, C., Tian, W., Liao, W.H., and Hua, L. (2016). Microstructure and porosity evaluation in laser-cladding deposited Ni-based coatings. *Surf. Coat. Technol.* 294: 122.

130 Celis, J.P., Drees, D., Maesen, E., and Roos, J.R. (1993). Quantitative determination of through-coating porosity in thin ceramic physically vapour–deposited films. *Thin Solid Films* 224: 58.

131 Vega, J., Scheerer, H., Andersohn, G., and Oechsner, M. Evaluation of the open porosity of PVD coatings through electrochemical iron detection. *Surf. Coat. Technol.* 350: 453.

132 Pawłowski, L. and Fauchais, P. (1992). Thermal transport properties of thermally sprayed coatings. *Int. Mater. Rev.* 37: 271.

133 Łatka, L., Pawłowski, L., Winnicki, M. et al. (2020). Review of functionally graded thermal sprayed coatings. *Appl. Sci.* 10 (15): 5153.

134 Touloukian, Y.S., Powell, R.W., Ho, C.Y., and Klemens, P.G. (1970). Thermophysical properties of matter. In: *Thermal Conductivity: Metallic Elements and Alloys*, vol. I. New York, NY: IFI Plenum.

135 Boer, J., Butter, J., Grosskopf, B., and Jeschke, P. (1980). Hot wire technique for determining high thermal conductivities. *Refractories J.* 55: 22.

136 Cintra, J.S. Jr. and Santos, W.N. (2000). Numerical analysis of sample dimensions in hot wire thermal conductivity measurements. *J. Eur. Ceram. Soc.* 20: 1871.

137 Penazzi, L., Jannot, Y., Meulemans, J. et al. (2022). Influence of radiation heat transfer on parallel hot-wire thermal conductivity measurements of semi-transparent materials at high temperature. *Int. J. Therm. Sci.* 179: 107690.

138 Parker, W.J., Jenkins, R.J., Butler, C.P., and Abbott, G.L. (1961). Flash method for determining thermal diffusivity, heat capacity and thermal conductivity. *J. Appl. Phys.* 32: 1679.

139 Pawłowski, L., Fauchais, P., and Martin, C. (1985). Analysis of boundary conditions and transient signal treatment in diffusivity measurements by laser pulse method. *Rev. Phys. Appl.* 20: 1.

140 Sokołowski, P., Björklund, S., Musalek, R. et al. (2017). Thermophysical properties of YSZ and YCeSZ suspension plasma sprayed coatings having different microstructures. *Surf. Coat. Technol.* 318: 28.

141 Cowan, R.D. (1963). Pulse method of measuring thermal diffusivity at high temperatures. *J. Appl. Phys.* 34: 926.

142 Brandt, R. and Neuer, G. (1979). Thermal diffusivity of solids-analysis of a modulated heating - beam technique. *High Temper. High Press.* 11: 59.

143 Touloukian, Y.S. and Buyco, E.H. (1970). Thermophysical properties of matter. In: *Specific Heat: Metallic Elements and Alloys*, vol. 4. New York, NY: IFI Plenum.

144 Richardson, M.J. (1984). Application of differential scanning calorimetry to the measurements of specific heat. In: *Compendium of Thermophysical Properties Measurements Methods* (ed. K.D. Maglic, A. Cezairliyan, and V.E. Peletsky), 519–545. New York, NY: Plenum Press.

145 Touloukian, Y.S., Kirby, R.K., Taylor, R.E., and Lee, T.Y.R. (1978). Thermophysical properties of matter. In: *Thermal Expansion: Nonmetallic Solids*, vol. 13. New York, NY: IFI Plenum.

146 Martin, C. and Fauchais, P. (1981). Mesure par thermographie infrarouge de l'émissivité de materiaux céramiques. *Rev. Int. Hautes Tempér. Refract.* 18: 95. (In French).

147 Jaworski, D.A. (1993). Thermal modeling of a calorimetric technique for measuring the emittance of surfaces and coatings. *Thin Solid Films* 236: 146.

148 Hasselman, D.P.H. (1970). Thermal stress resistance parameters for brittle refractory ceramics: a compendium. *Ceram. Bull.* 49: 1033.

149 T. Cosack, L. Pawłowski, S. Scheidenbanger, and S. Sturlese, 1992, Thermal barrier coatings on turbine blades by plasma spraying with improved cooling, [in:] Proceedings of *37th ASME Int. Gas Turbine & Aeroengine Congress and Exposition*, Cologne, Germany, June 1–4, 1992, Paper 92–GT-319.

150 Delcheva, I., Ralston, J., Beattie, D.A., and Krasowska, M. (2015). Static and dynamic wetting behavior of ionic liquids. *Adv. Colloid Interf. Sci.* 222: 162.

151 Król, D.J., Wymysłowski, A., and Allaf, K.N. (2015). Adhesion work analysis through molecular modeling and wetting angle measurement. *Microelectron. Reliab.* 55: 758.

152 Akbari, R. and Antonini, C. (2021). Contact angle measurements: From existing methods to an open-source tool. *Adv. Colloid Interf. Sci.* 294: 102470.

153 Abrar, A., Sanedrin, R., Willers, T., and Waghmare, P.R. (2022). The effect of dynamic wetting pressure on contact angle measurements. *J. Colloid Interface Sci.* 608: 1086.

154 Dhavan, S.K., Bhandari, H., Ruhi, G. et al. (2020). *Corrosion Preventive Materials and Corrosion Testing*. Boca Raton, FL: CRC Press.

155 Bunshah, R.F. (1994). Metallurgical Applications. In: *Handbook of Deposition Technologies for Films and Coatings* (ed. R.F. Bunshah), 740–762. Park Ridge, NJ: Noyes Publications.

156 Lide, D.L. and Frederikse, H.P.R. (ed.) (1997). *CRC Handbook of Chemistry and Physics*. Boca Raton, FL: CRC Press.

157 Chiang, K.-T.K. and Wei, R. (2011). Growth morphology and corrosion resistance of magnetron sputtered Cr films. *Surf. Coat. Technol.* 206: 1660.

158 Dean, S.W., France, W.D. Jr., and Ketcham, S.J. (1971). Electrochemical methods. In: *Handbook on Corrosion Testing and Evaluation* (ed. W.W. Ailor), 171–215. New York, NY: John Wiley.

159 ASTM G 3-89, reapproved 2010, Standard Practice for Convention Applicable to Electrochemical Measurements in Corrosion Testing, 3 February 1989, ASTM International. West Conshohocken; PA.

160 Viramontes-Gamboa, G., Rivera-Vasquez, B.F., and Dixon, D.G. (2007). The active-passive behavior of chalcopyrite: comparative study between electrochemical and leaching responses. *J. Electrochem. Soc.* 154: C299.

161 Cook, S.G., Little, J.A., and King, J.E. (1994). Corrosion of silicon carbide ceramics using conventional and electrochemical methods. *British Corrosion J.* 29: 183.

162 Lugscheider, E., Eschnauer, H., Häuser, B., and Jäger, D. (1985). Vacuum plasma spraying of tantalum and niobium. *J. Vac. Sci. Technol. A* 3: 2469.

163 Zaplatynski, I. (1982). Performance of laser glazed zirconia thermal barrier coatings in cyclic oxidation and corrosion burner rig test. *Thin Solid Films* 95: 275.

164 Raffaitin, A., Crabos, F., Andrieu, E., and Monceau, D. (2006). Advanced burner-rig test for oxidation–corrosion resistance evaluation of MCrAlY/superalloys systems. *Surf. Coat. Technol.* 201: 3829.

165 Rose-Innes, A.C. and Rhoderick, A.C.E.H. (1978). *Introduction to Superconductivity*, 2e. Oxford: Pergamon Press.

166 Singh, Y. (2013). Electrical resistivity measurements: a review. *Int. J. Modern Phys.: Conf. Series* 22: 745.

167 Bishara, H., Kontis, P., Dehm, G. et al. (2022). Effect of hybridization in PdAlY-(Ni/Au/Ir) metallic glasses thin films on electrical resistivity. *Scr. Mater.* 214: 114681.

168 Pawłowski, L. (1988). The relationship between structure and dielectric properties in plasma sprayed alumina coatings. *Surf. Coat. Technol.* 35: 285.

169 Morse, C.T. and Hill, G.J. (1970). The electric strength of alumina: the effect of porosity. *Proc. Brit. Ceram. Soc.* 18: 23.

170 Romalis, M.V. and Dang, H.B. (2011). Atomic magnetometers for materials characterization. *Mater. Today* 14: 258.

171 Foner, S. (1959). Versatile and sensitive vibrating-sample magnetometer. *Rev. Scientific Instr.* 30: 548.

172 Baiga, M.K., Atiqa, S., Bashir, S. et al. (2016). Pulsed laser deposition of SmCo thin films for MEMS applications. *J. Appl. Res. Technol.* 14: 287.

173 Wooten, F. (1972). *Optical Properties of Solids.* New York, NY: Academic Press.

174 Xu, J., Yang, Z., Wang, H. et al. (2014). Effect of growth temperature and coating cycles on structural, electrical, optical properties and stability of ITO films deposited by magnetron sputtering. *Mater. Sci. Semicond. Process.* 21: 104.

175 Vedam, K. (1998). Spectroscopic ellipsometry: a historical overview. *Thin Solid Films* 313–314: 1.

176 Suchaneck, G., Chernova, E., Kleiner, A. et al. (2017). Vacuum-ultraviolet ellipsometry spectra and optical properties of Ba(Zr,Ti)O_3 films. *Thin Solid Films* 621: 58.

177 Beck, U., Reiners, G., Kopacz, U., and Jehn, H.A. (1993). Decorative hard coatings: interdependence of optical, stoichiometric and structural properties. *Surf. Coat. Technol.* 60: 389.

178 Fantozzi, G., Le Gallet, S., and Nièpce, J.-C. (2011). *Science & Technologies Céramiques.* France: Les Ulis.

179 Hench, L.L. and Wilson, J. (2013). Introduction. In: *An Introduction to Bioceramics*, 2e (ed. L.L. Hench), 1–27. London: Imperial College Press.

180 Bandyopadhyay, A. and Bose, S. (2013). *Characterization of Biomaterials.* Waltham, MA: Elsevier.

181 Rather, B.D., Hoffman, A.S., Schoen, F.J., and Lemons, J.E. (2004). Biomaterials science: a multidisciplinary endeavor. In: *Biomaterials Science* (ed. B.D. Ratner, A.S. Hoffman, F.J. Schoen, and J.E. Lemons), 1–10. San Diego, CA: Elsevier.

182 J.M Anderson, R.W. Bianco, J.F. Grehan, B.C. Grubbs, S.R. Hanson, K.D. Hauch, M. Lahti, J.P. Mrachek, S.J. Northup, B.D. Rather, F.J. Schon, E.L. Schroeder, C.W. Schumacher, and C.A. Svendsen, 2004, [in:] *Biomaterials Science*, B.D. Ratner, A.S. Hoffman, F.J. Schoen, and J.E. Lemons (eds), Elsevier, San Diego, CA., 355-409.

183 Northup, S.J. (2004). *In vitro* assessment of tissue compatibility. In: *Biomaterials Science* (ed. B.D. Ratner, A.S. Hoffman, F.J. Schoen, and J.E. Lemons), 356–360. San Diego, CA: Elsevier.

184 Thasneem, Y.M. and Sharma, C.P. (2013). *vitro characterization of cell-biomaterials interactions, [in:] Characterization of Biomaterials* (ed. A. Bandyopadhyay and S. Bose), 175–205. Elsevier, Waltham: MA.

185 ASTM F813-20, Standard Practice for Direct Contact Cell Culture Evaluation of Materials for Medical Devices, Published (updated), July, 2020, ASTM Internatinal. West Conshohocken; PA.

186 Kokubo, T. and Takadama, H. (2006). How useful is SBF in predicting in vivo bone bioactivity? *Biomaterials* 27: 2907.

187 Kokubo, T. (2013). A/W glass-ceramics: processing and properties. In: *An Introduction to Bioceramics*, 2e (ed. L.L. Hench), 171–187. London: Imperial College Press.

188 Hench, L.L. (2013). Characterization of bioceramics. In: *An Introduction to Bioceramics*, 2e (ed. L.L. Hench), 521–539. London: Imperial College Press.

189 Unabia, R.B., Candidato, R.T. Jr., Pawłowski, L. et al. (2020). *In vitro* studies of solution precursor plasma-sprayed copper-doped hydroxyapatite coatings with increasing copper content. *J. Biomed. Mater. Res.* 108 (6): 2579.

190 Thanka Rajan, S., Subramanian, B., and Arockiarajan, A. (2022). A comprehensive review on biocompatible thin films for biomedical application. *Ceram. Int.* 48: 4377.

191 Anderson, J.M. and Schoen, F.J. (2004). *In vivo* assessment of tissue compatibility. In: *Biomaterials Science* (ed. B.D. Ratner, A.S. Hoffman, F.J. Schoen, and J.E. Lemons), 360–367. San Diego, CA: Elsevier.

192 Nandi, S.K. and Biswas, S. (2013). In vivo characterization of biomaterials. In: *Characterization of Biomaterials* (ed. A. Bandyopadhyay and S. Bose), 255–297. Waltham, MA: Elsevier.

193 Kabene, S. and Baadel, S. (2019). Bioethics: a look at animal testing in medicine and cosmetics in the UK. *J. Med. Ethics Hist. Med.* 12 (15): 1.

194 ASTM F1904-14, Standard practice for testing the biological responses to particles in vivo, Published (updated), January, 2014, ASTM International. West Conshohocken; PA.

5

Properties of Films and Coatings

5.1 Introduction

The specification of the deposits follows the customer's requirements, which includes the family of deposition processes, i.e. atomistic, granular, or bulk methods (Table I.1). However, each of these processes includes many particular methods enabling to obtain deposits having different thickness, microstructure, and properties resulting thereof. Consequently, the discussion of the deposit properties should be made separately for every deposition process. It is important to underline that each of these processes has a lot of particular parameters. Their optimization in view to reach desired properties of deposits can be made using *statistical design of experiments* (*DoE*) described by, e.g. Lazic [1] and reviewed for granular coatings by Pierlot *et al.* [2].

The deposition techniques are adapted to the deposited material. The general classification of materials includes: (i) metals and alloys; (ii) ceramics, especially carbides, oxides, and nitrides; and (iii) polymers being less frequently applied as deposits comparing to other materials. The composites of ceramics with metals, oxides, and ceramics are also frequently used for the deposits. Particular attention has been paid, since the end of last century, on the carbonic films and especially having some mechanical properties similar to diamond (*DLC*). Consequently, the mechanical properties of carbonic films are discussed separately.

The specification of the deposit may include the microstructure, but the most important are the properties related to the practical application, such as mechanical, electrical, optical, and others. The microstructure has an important influence of these properties. The properties shown in this chapter are presented generally together with the phase content, found generally using X-ray diffraction (*XRD*). The deposit's morphology is closely related to the deposition technique discussed in Chapter 3.

The presented properties of the deposits are typical for the deposition technique and for the used material. The essential deposition parameters include used feedstock and the substrate may help to achieve the similar properties. It must be also mentioned that some properties of bulk deposits are less frequently tested. This family of coatings is the most frequently applied for mechanical application and, in particular, as wear-resistant coatings.

Physical Deposition Methods for Films and Coatings, First Edition. Lech Pawłowski.

5.2 Mechanical Properties

Among the mechanical properties, one can distinguish those being characteristic for the films and coatings, such as adhesion and, on the other hand, the properties of materials manufactured by some conventional methods, e.g. sintering. Such characteristic properties are hardness or modulus of elasticity. In general, the deposits have worse mechanical properties than solid materials. On the other hand, atomistic deposition techniques make it possible to synthesize materials that are difficult to obtain by conventional methods such as diamond-like carbon (*DLC*).

5.2.1 Hardness and Microhardness

Hardness and microhardness are tested with a multitude of indentation techniques described in Section 4.3.3. The indentation is made with the diamond's indenters on the surface of films or on the longitudinal cross-section of the coatings. The results of typical hardnesses are presented in the using GPa as the unit. The interested reader can find more data in the recent review of Al-Asadi and Al-Tameemi [3].

5.2.1.1 Atomistic Films

Atomistic techniques are frequently applied to obtain hard carbonic films (diamond deposits), ceramics, and composites. Sometimes superhard deposits are called those whose microhardness exceeds 40 GPa.

Carbonic Films

The microhardness of carbonic deposits varies from 8 to 80 GPa (Fig. 5.1) depending on the technique and the parameters of the deposit (see Table 5.1). The greatest microhardness, close to 100 GPa, corresponds to crystalline diamond (Fig. 5.1a), being the hardest known material. High values of microhardness are also associated with the compressive stresses applied to the lattice of the carbon deposit. These stresses may be generated by ion bombardment associated with the deposition techniques such as pulsed laser deposition, ion plating, or magnetron sputtering (see Fig. 5.1b–e). The atoms in carbonic deposits may have sp^2 or sp^3 atomic bonding. The sp^2 bonding corresponds to graphite and sp^3 one correspond to diamond. Consequently, the ratio of sp^3/sp^2 sites is strongly associated with the film microhardness. Savvides and Bell [4] found the correlation of this ratio with the energy of the ions at magnetron sputtering deposition with the use of the Ar^+ ions, which bombarded the surface of the growing deposit. These authors found that the microhardness and the ratio of the sites were:

- $H = 13$ GPa and $sp^3/sp^2 = 3$ for the energy of the ions equal to 30 eV/(atom of C);
- $H = 30$ GPa and $sp^3/sp^2 = 6$ for the energy of the ions equal to 62 eV/(atom of C).

The microhardness of diamond deposits also depends on the concentration of hydrogen in the deposit being amorphous, called a–C:H. This concentration depends also on energy of the ions bombarding the surface during film deposition. The following possibilities concerning hydrogen in hard deposits can be distinguished [8, 9]:

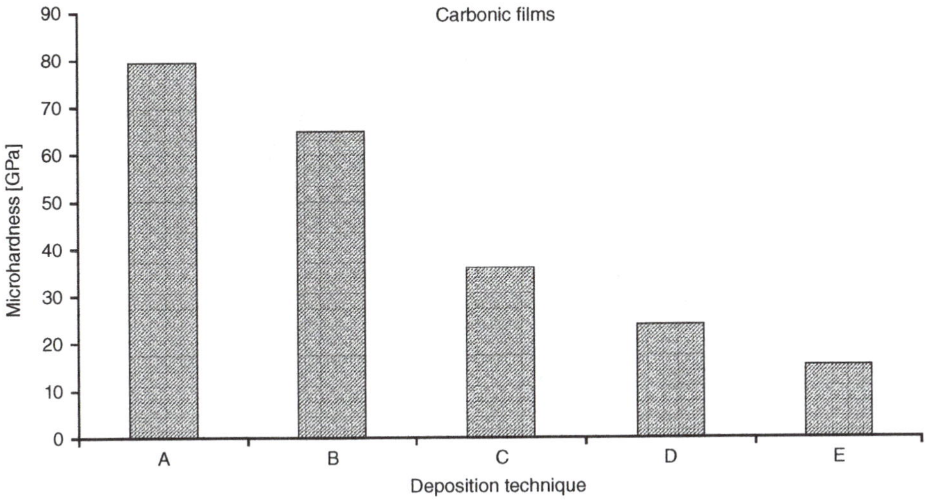

Figure 5.1 Microhardness of carbonic films obtained by atomistic techniques. The details of the technology and of the films are presented in Table 5.1.

- hydrogen is dissolved in the material without bonding with other atoms;
- most hydrogen atoms are bonded with carbon atoms;
- the film's material is a mixture of hydrocarbon phases.

Consequently, the increase of hydrogen concentration in the deposits may reduce its hardness [10].

Ceramic Films

Among the hard ceramic deposits obtained by atomistic techniques, there are above all nitrides, carbides, and composites of carbides or nitrides with metals.

The microhardnesses of typical nitrides films are shown in Fig. 5.2. The description of deposition process and of the characterization methods is explained in Table 5.2

The hardest nitride film reported in the literature is cubic boron nitride (CBN), which may reach microhardness of 60 GPa [14]. The microhardness value of this film presented with the deposition parameters is slightly lower and equal to $H = 35$ GPa (Fig. 5.2b). The frequently used and tested nitride is TiN whose microhardness varies between 44 GPa and 21 GPa (Fig. 5.2a,d, respectively). This deposit is often synthesized with sputtering techniques using nitrogen as working gas. Another tested nitride is CrN film having similar hardness similar to that of TiN (Fig. 5.2c).

The microhardness of some carbides is shown in Fig. 5.3.

Among the carbides reviewed, the hardest film was of boron carbide (B_4C), having $H = 72.3$ GPa and amorphous structure prepared by *IP* sputtering technique with ion bombardment of argon ion energy optimized at 75 eV [15]. The WC and SiC films, prepared by similar technique, had smaller hardness being slightly greater than $H = 20$ GPa [16, 17]. It must be mentioned that the hardness of amorphous SiC measured by Ulrich et al. [15]

Table 5.1 Atomistic techniques and parameters of carbon deposits having microhardnesses shown in Fig. 5.1.

		Deposition process				Deposit characteristics		
Symbol	References	Type	Target or precursor gas	Pressure, hPa	Substrate temperature, K	Crystalline phases	Thickness, δ, μm/ substrate	Test method
A	[4]	*PECVD*, microwave	precursor: $H_2 + CH_4$	—	—	diamond	3/Si	Berkovich indenter ($H = 80$–100 GPa)
B	[5]	*PLD*	target: graphite	10^{-7}	573–673	mainly sp^3 atomic bonding	0.5/steel 440 C	Nanoindentation ($H = 65$ GPa)
C	[6]	Ion plating	—	—	—	—	2.2/tool steel	($H = 36$ GPa)
D	[4]	*PVD*, magnetron sputtering	target: graphite	—	—	ratio $sp^3/sp^2 \approx 5$	5/Si	Berkovich indenter ($H = 24$ GPa)
E	[7]	*PVD*, magnetron sputtering	target: graphite	0.2×10^{-2}	473	high concentration of sp^3	multilayer composed of soft and hard layers/steel M2	Nanoindentation ($H = 15.4$ GPa]

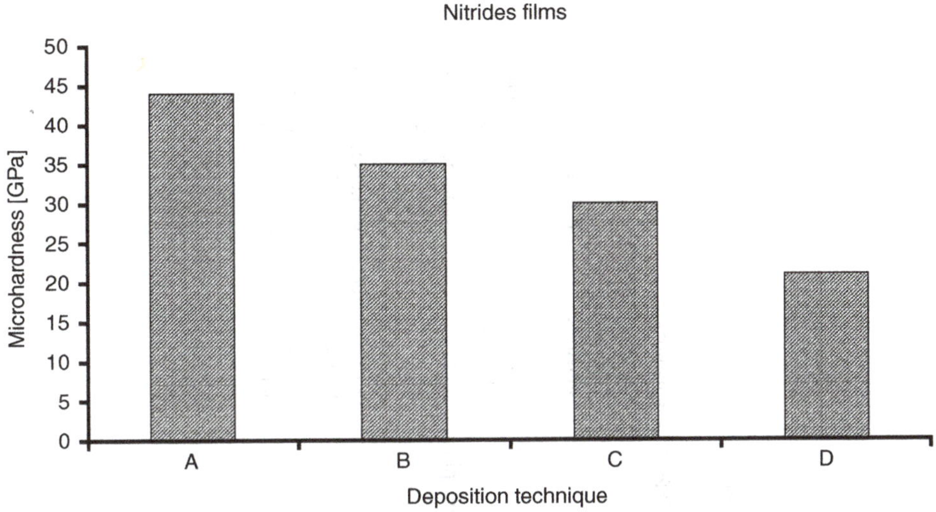

Figure 5.2 Microhardness of nitride films obtained by atomistic techniques. The description of the technology and properties of the films are presented in Table 5.2.

was considerably greater (H = 45–55 GPa). Finally, hardness of tantalum carbide film prepared by *CVD* was measured to be about 15 GPa [18].

Composite Deposits

Two types of composite hard deposits are used in practice (see Fig. 5.4):

- ***nanocomposites***, being the films in which nanocrystalline reinforcements are inside the amorphous matrix. Vepřek [19] developed the film including W_2C nanocrystals having size of about 3–7 nm included in amorphous Si_3N_4 matrix. The 4–5 μm thick films were synthesized using plasma *CVD* and ***superhard materials***[1] hardness of H = 50 GPa was reached for an optimum content of 7 at. % Si in amorphous matrix. Deng *et al.* [20] developed another design of nanocomposite. The authors optimized the power in reactive sputtering technique using target of W in $Ar + N_2$ atmosphere to obtain nanocrystalline W and W_2N grains in amorphous phase matrix. The greatest hardness of H = 23 GPa was reached at the sputtering deposition power of 7 kW.
- ***superlattices*** are defined by Sproul [21] as a thin film that is "formed by alternately depositing two different components to form a layered structure." The author used reactive sputtering of different targets, such as Ti, Nb, V. The superlattice films of TiN/NbN had hardness depending on:
  - superlattice period, λ;
  - chemical composition and stoichiometry of each layer;
  - operational sputtering parameters such as ion density and substrate bias voltage.

1 This hardness was defined by the authors as $H \geq 40$ GPa.

Table 5.2 Atomistic techniques and deposition parameters of nitrides films having microhardnesses shown in Fig. 5.2.

		Deposition process				Deposits characteristics		
Symbol	References	Type	Target or gas precursor	Working pressure, (hPa	Substrate temperature, K	Nitride	Thickness, μm/ substrate	Test method
A	[11]	Evaporation using cathodic arc (*PVD*)	Ti target and $Ar + N_2$ mixture as arc gas	10^{-3}–2×10^{-3}	623	TiN	2/stainless steel 316LVM	Berkovich indenter ($H = 44$ GPa)
B	[12]	Magnetron sputtering	Sintered hexagonal BN and $Ar + N_2$ as working gases	8×10^{-3}	up to 1173	BN (cubic phase)	up to 0.372/Si	Nanoindentation ($H \leq 35$ GPa)
C	[13]	*IP* sputtering	Cr and N_2 as working gas	2×10^{-3}	653	CrN	5.12/tool steel Vanadis	Vickers microhardness ($H = 30$ GPa)
D			Ti and N_2 as working gas	5×10^{-3}		TiN	2.95//tool steel Vanadis	Vickers microhardness ($H = 21$ GPa)

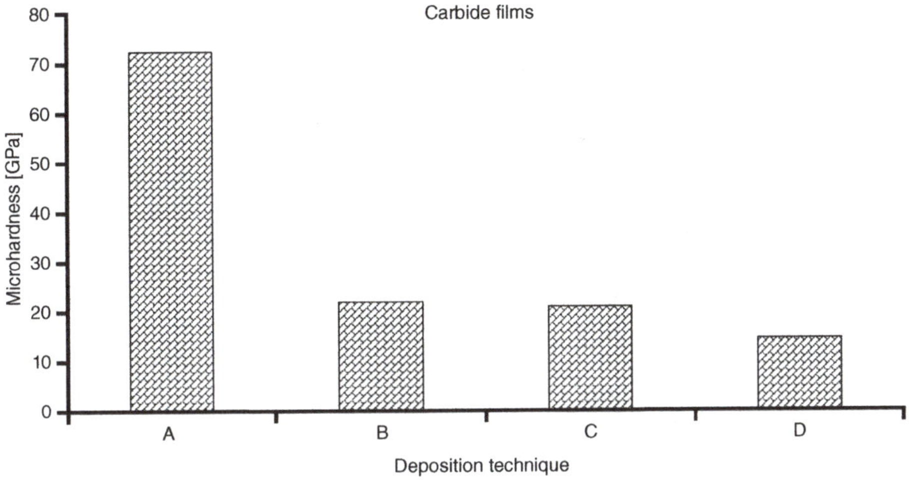

Figure 5.3 Microhardness of carbide films obtained by atomistic techniques. The description of the technology and properties of the films are presented in Table 5.3.

Consequently, for the period of $\lambda = 4–8\,\text{nm}$, hardness of TiN/NbN film reached the value of $H = 55\,\text{GPa}$. Similar technology of unbalanced sputtering using the targets of Cr and C enabled to obtain superlattice composite coating of CrN/CN_x in which carbon nitride layers were amorphous [22]. The authors obtained 1 μm thick films having the greatest hardness of $H = 32\,\text{GPa}$ for the period of about 6 nm and bias voltage of $U = -100\,\text{V}$.

5.2.1.2 Granular Coatings

The hard granular coatings are thermally sprayed using powders of carbides, oxides, alloys, and metals. The most applications are realized using carbide powders. It must be however stressed up that the carbide powders are used in the form of composites in which the carbide (WC, Cr_3C_2) is a reinforcement, and the matrix is typically a metal (Co, Ni) or an alloy (NiCr). Such powders are used as the agglomerates or as the cladded powders (see Section 2.3). The cladded powders are the carbides covered with metal or alloy. Microhardness is typically measured with the Vickers technique and sometimes Knoop one under the 3N load on the longitudinal cross-section of the coatings.

Carbide Coatings

The microhardness of the carbide coatings varies from 4 to 12 GPa (see Fig. 5.5). Tungsten carbide composite coatings including cobalt or nickel matrix have been most frequently sprayed. Also, chromium carbide composite coating with nickel chromium alloy as matrix has been often used. The microhardness depends mainly on the coatings porosity being dependent on the spray technique. The *HVOF* method enables obtaining carbide hard coatings having small porosity (Fig. 5.5a,b). Plasma spraying in air (*APS*) method is also used, but the sprayed coatings are more porous and less hard (Fig. 5.5c). Finally, the *CGSM* method starts to be used often, but the coatings is not yet very hard (Fig. 5.5d,e).

Table 5.3 Atomistic techniques and parameters of carbide deposits whose microhardnesses are shown in Fig. 5.3.

		Deposition process				Deposits characteristics			
Symbol	References	Type	Target or gas precursor	Working pressure, hPa	Substrate temperature, K	Carbide	Crystalline phases/ content of C, at. %	Thickness, μm/ substrate	Test method
A	[15]	Magnetron sputtering with *IP*	B_4C target	1.5×10^{-3}	300	B_4C		/Si	Microindentation ($H = 72$ GPa)
B	[16]	Reactive *RF* sputtering	W as target and $Ar + CH_4$ as working gas	2.5×10^{-3}	423	WC	WC, W_2C, and WC_{1-x}/40–82	0.36/Si	Berkovich indenter ($H \leq 22$ GPa)
C	[17]	Magnetron sputtering	SiC as target	8×10^{-3}	—	SiC	Amorphous/40–61	/Si	Nanoindentation ($H \leq 21$ GPa)
D	[18]	*CVD* with hot filament	$H_2 + CH_4$ as precursors	27	1073	TaC	Polycrystalline TaC	8/Ta	Microindentation ($H = 14.6$ GPa)

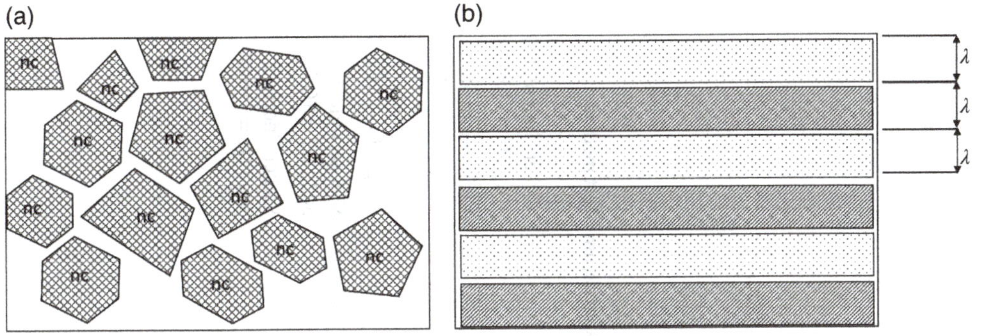

Figure 5.4 Two types of composite thin films: (a) *nanocomposites* with nanocrystalline reinforcement grains in amorphous matrix; and (b) *superlattice* with binary layers deposited with a nanodimensional period λ.

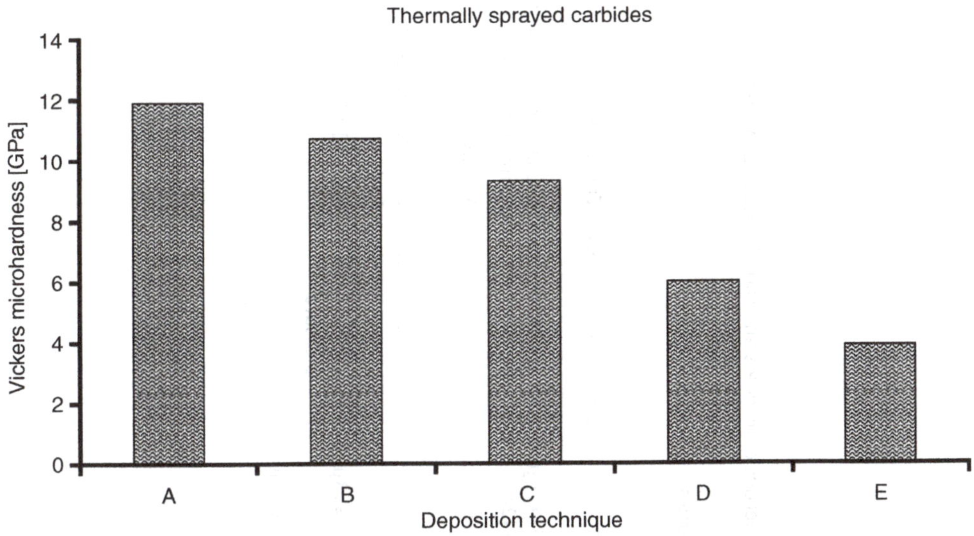

Figure 5.5 Microhardness of carbide deposits obtained by thermal spray techniques. The description of the technology and properties of the deposits is in Table 5.4.

The hardness of WC-Co deposits depends also on the powder manufacturing technology and on the operational spray parameters [28]. The quantity, grain size, and content of crystal phases of carbide are essential. In particular, the deposits including greater quantity of carbide reinforcement are harder. As the carbides may decompose in hot gas, the coating obtained using WC powder may include W_2C. The use of rapid flames such as that in *HVOF* method shortens the time of powders particles interaction with the flame and reduces this decomposition.

Table 5.4 Thermal spraying techniques properties of powders and carbide deposits whose microhardnesses are shown in Fig. 5.5.

		Powder properties			Coating's properties			
Symbol	**References**	**Chemical composition, wt. %**	**Particles size, μm**	**Spray technique**	**Crystalline phases**	**Porosity, %**	**Thickness, δ, μm**	**Test method**
A	[23]	WC + 12Co	16–45 (given in [24])	*HVOF*	WC and traces of W_2C (only carbides tested)	0.3	—	Vickers indenter of 300 g ($H = 11.9$ GPa)
B	[25]	WC + 17Co	<63 μm	*VPS*	Mainly WC	3	300	Vickers indenter of 300 g ($H = 10.7$ GPa)
C	[23]	Cr_2C_3 + 25NiCr	−53 + 11	*APS* (Ar + He)	Cr_2C_3 and traces of Cr_7C_3	3.3	—	Vickers indenter of 300 g ($H = 9.3$ GPa)
D	[26]	WC + 17Ni	$d_{50} \approx 13$ μm	*CGSM* – low pressure	WC and Ni	—	100–250	Vickers indenter of 500 g ($H = 6$ GPa)
E	[27]	B_4C + 8 Ni	$d_{50} \approx 39$ μm	*CGSM* – low pressure	—	—	—	Vickers indenter of 300 g ($H = 3.9$ GPa

Oxide Deposits

The oxides are sprayed mainly by plasma in air. This results from their high melting point and the necessity to melt the sprayed particles at processing. Sometimes the *HVOF* or *D-gun* techniques are also used.

Typical values of microhardness of different oxides' coatings sprayed using powder feedstock are shown in Fig. 5.6. The microhardness of coatings sprayed using such feedstock depends on the powder manufacturing technique. Consequently, dense crystalline $Cr_2 O_3$ powder sprayed by plasma in air results in coatings being are harder than that sprayed using powder produced by sintering or spray drying [32]. The deposits of this oxide are the hardest among the sprayed coatings as seen in Fig. 5.6. Sprayed deposits of alumina with small additions of other oxides have a hardness lower than $H = 10$ GPa. Finally, the coatings of zirconia with yttria and of titania have hardness above $H = 5$ GPa. The hardness of the coatings increases by pre-heating the substrate before spraying. Consequently, the hardness of alumina coating sprayed at $T_d = 300$ K was $H = 9.5$ GPa (see Fig. 5.6b) and increased up to $H = 11.8$ GPa when deposition was made at substrate temperature of 773 K [29].

Deposits of Metals and Alloys

Deposition of pure metals and alloys by thermal spraying is difficult because of possible oxidation of the particles surfaces occurring at high temperatures in contact with air. The application of low-pressure spray techniques, such as *LPPS* or *VPS* or the one realized in inert atmosphere (*IPS*), is a possible way to preserve the chemical composition of the coatings. Another possibility is reducing the oxidation by the use of flame having low temperatures like in the technique *FS* or high-velocity flames, which reduce the time of contact of particle with air, such as that generated in *HVOF* gun. Nowadays, there is a possibility to

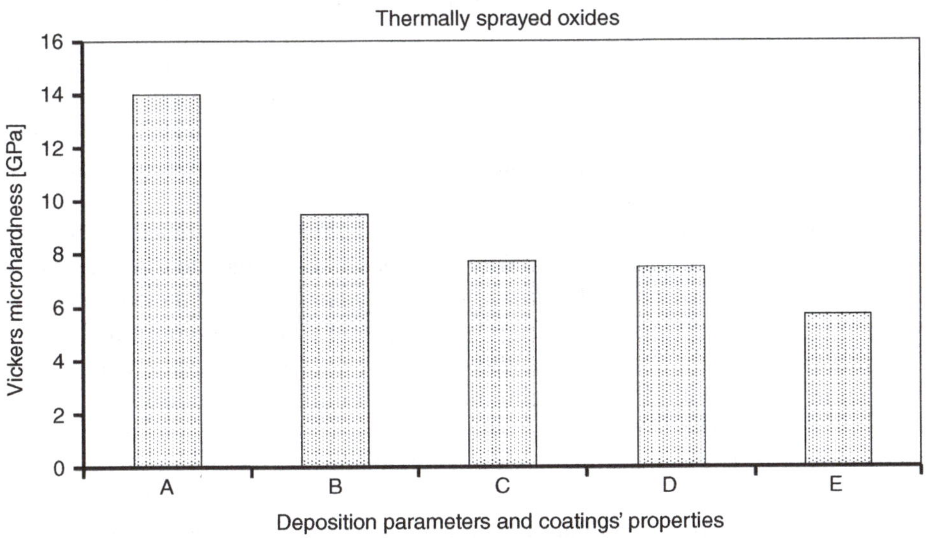

Figure 5.6 Microhardness of oxide coatings obtained by APS. The deposition properties and some coatings properties are shown in Table 5.5.

Table 5.5 Powders and *APS* coatings properties of oxides whose microhardnesses are shown in Fig. 5.6.

		Powder properties		Coatings properties			
Symbol	**References**	**Chemical composition (wt. %)**	**Particles size (μm)**	**Crystalline phases**	**Porosity (%)**	**Thickness (μm)**	**Test method**
A	[23]	Cr_2O_3	22–45	—	2.7	—	Vickers indenter 300 g (H= 14 GPa)
B	[29]	$95Al_2O_3 + 2SiO_2$	10–50	$\gamma - Al_2O_3$[a]	8.7	400	Vickers indenter 300 g (H= 9.5 GPa)
C	[30]	$94.5Al_2O_3 + 2.7TiO_2 + 2.1\ SiO_2$	−50	as above	—	—	Vickers indenter (H = 7.7 GPa)
D	[23]	$ZrO_2 + 8Y_2O_3$	11–45	—	6.0	—	Vickers indenter 300 g (H= 7.5 GPa)
E	[31]	TiO_2	about 50	Brookite		250	Vickers indenter 500 g (H= 5.75 GPa)

[a] The author did not made X-ray tests, but γ -Al_2O_3 is generally obtained when plasma spraying alumina.

spray without flame or discharge by using cold spraying (*CGSM*). Finally, it should be mentioned that an appropriate chemical composition of an alloy could lower its melting temperature. This can be achieved, e.g. by addition of silicium and/or boron to some nickel alloys.

Fig. 5.7 shows the microhardnesses of some alloys and metals sprayed in open atmosphere. The microhardnesses are smaller than $H < 10$ GPa. The greatest values were found in alloys such as medium- and high-entropy ones (Fig. 5.7a,c) and in nickel-based alloy shown in Fig. 5.7b. The cold sprayed metals as Ni and Ti had microhardnesses slightly greater than $H > 2$ GPa (Fig. 5.7d,e).

Thermal spraying, including possible reaction of metal with gases included in air or in other working gas, enables obtaining hard deposits. Consequently, the Mo coating plasma sprayed in air had the Vickers microhardness of $H = 5.2$–6 GPa [36], and the one obtained by arc in air had reportedly hardness of $H = 3$–4 GPa [37]. This effect is confirmed by the microhardness of Ti coatings. The ones sprayed using arc in inert atmosphere of Ar had Vickers microhardnesses of $HV = 3$ GPa. On the other hand, the microhardness increased up to $H = 13$ GPa when the nitrogen atmosphere was applied [38]. This increase resulted from the formation of TiN nitrides and titanium solid solutions with nitrogen α-Ti and β-Ti.

The alloys used against wear such as that shown in Fig. 5.7a–c are relatively hard. This is another, very frequent application of alloys as the bond coats does not require high values of hardness. Consequently, the coatings of ductile alloys such as NiAl, NiCr, or NiCrAlY have typical microhardnesses ranging from $H = 1.8$ to 5.4 GPa [39].

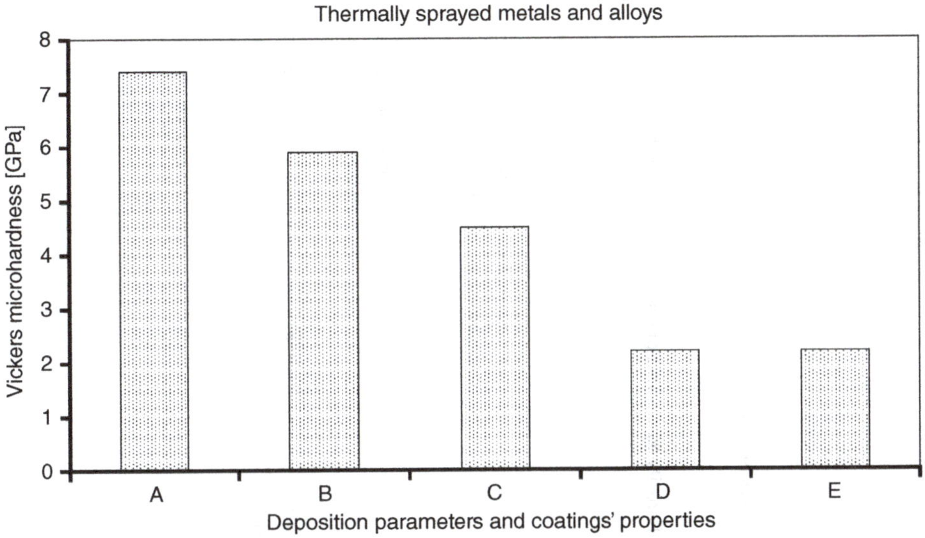

Figure 5.7 Microhardness of metals and alloys coatings obtained by thermal spraying. The deposition properties and some properties of coatings are shown in Table 5.6.

5.2.1.3 Bulk Deposits

The microhardness and hardness of bulk deposits were tested mainly for cermets of carbides or oxides and for alloys. Deposition techniques such as plasma transferred arc as well as laser assisted remelting were applied. The microhardness was tested under loads ranging from 50 to 1000 g using a Vickers indenter on the cross-sections of the deposits.

Ceramic and Composite Deposits

Bulk deposits of ceramic on metallic substrates can be difficult because of an incompatibility of the thermal expansion coefficient (*TEC*), which was associated with a high rate of growth of the deposits, may lead to significant cracking. This is an important reason to use the metal matrix composites (*MMC*) in which the ceramic particles are used as reinforcement or to deposit the ceramic having *TEC* similar to those of the substrate are applied.

The hardnesses of some ceramic and *MMC* coatings obtained using bulk deposition methods are shown in Fig. 5.8. The hardness was generally measured using polished cross-section along a line perpendicular to the surface of coating down to the heat-affected zone. Some data for composite tested using small indenter of 100 g, such as Fig. 5.8a, show hardness of carbide particles and, consequently, are much greater than the others, which correspond mainly to the matrix. This is confirmed by the hardness of cermet NiCrAlY with 13YSZ which has an average value of $H = 3.9$ GPa (see Fig. 5.8e), but the authors did a few measurements showing $H = 14.7$ GPa [44].

The hardnesses of bulk composite deposits with ceramic reinforcement depend mainly on: (i) volume fraction of reinforcement; (ii) dilution; and (iii) hardness of the used matrix. Consequently, this parameter depends also on phases of matrix crystallinity.

Table 5.6 Powders and thermally sprayed metal and alloys coatings properties having microhardnesses shown in Fig. 5.7.

		Powder properties			Coating's properties			
Symbol	**References**	**Chemical composition, wt. %**	**Particles size, μm**	**Spray technique**	**Crystalline phases or structure**	**Porosity, %**	**Thickness, μm**	**Test method**
A	[33]	AlCoCrFeNi[a] (each composant about 20 at. %)	15–53	*HVOF*	*BCC* with two sublattices	—	—	Vickers indenter of 300 g ($H = 7.4$ GPa)
B	[34]	Ni + 23 Co + 20 Cr + 8.5 Al + 4 Ta + 0.6 Y	13–52	*CGSM*	solid solution of Ni, β-NiAl, segregated phase of Ta	<0.5	≈300 (estimated from the cross -section)	Vickers indenter of 100 g ($H = 5.9$ GPa)
C	[33]	CoCrFeNi (each composant about 25 at. %)[b]	15–53	*HVOF*	*FCC*	—	—	Vickers indenter of 300 g ($H = 4.5$ GPa)
D	[35]	Ti	21–40	*CGSM*	Pure metals with oxidation level ≤0.24 wt. %	0.3	≥10 000	Vickers indenter of 1 kg ($H = 2.2$ GPa)
E		Ni				0.4		

[a] Composition corresponds to medium-entropy alloy (*MEA*).
[b] Composition corresponds to high-entropy alloy (*HEA*).

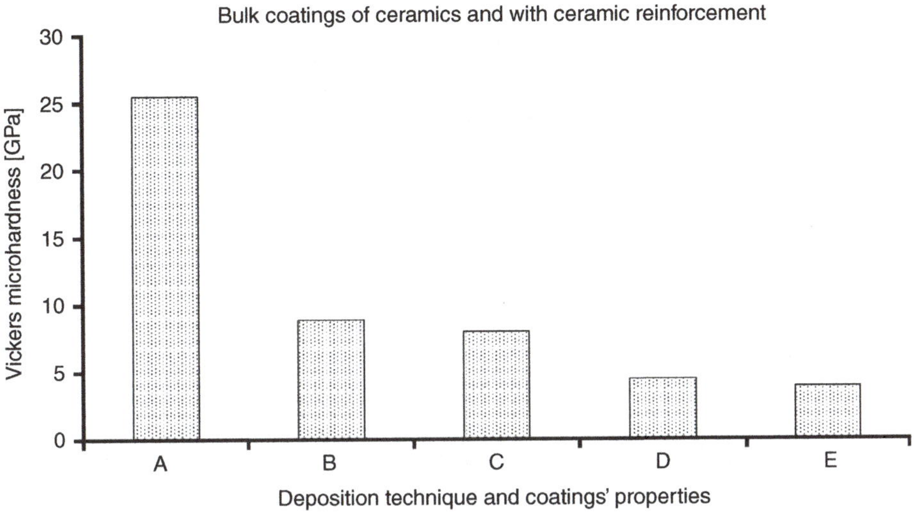

Figure 5.8 Microhardness of ceramic coating and of composite coatings reinforced with ceramics obtained using bulk methods. The deposition technology and some properties of the deposits are shown in Table 5.7.

The reinforcement makes the composite harder, but it is difficult to exceed 50% vol. of its fraction in the deposit. Finally, the hardness of composite deposits with the soft matrix is also very low [45].

Metal and Alloy Deposits

The alloy deposits were obtained using lasers or using plasma transferred arc processes by cladding or alloying techniques. The alloys tested were frequently on the basis of Ni (see Fig. 5.9a) and Co (Figs 5.9b–e). The values of hardnesses are generally lower than 10 GPa. These values depend mainly on:

- crystal structure of the deposit;
- dilution (in the cladding process).

In particular, Yellup [48] tested the influence of dilution on the hardness for Stellite six alloy obtained by laser cladding. The author found that at the dilution of 5%, the hardness was about 5.4 GPa (see Fig. 5.9c) and drops down to about 3 GPa for the dilution of 40%.

5.2.2 Adhesion

The choice of the method with which the adhesion is measured depends strongly on the thickness of the deposit. Consequently, the methods applied to thin films are generally quite different from those used to test thick coatings. Therefore, it is seldom possible to compare the adhesion of the deposits with different thicknesses. Moreover, the factors influencing the adhesion of deposits depend on many more parameters than their material. That is why, the discussion in the following section will focus on these factors.

Table 5.7 Powders and some properties of ceramic and composite coatings having ceramics reinforcement obtained using different laser deposition methods and using plasma transferred arc having microhardnesses shown in Fig. 5.8.

			Powder properties			Coatings' properties		
Symbol	References	Substrate	Chemical composition, wt. %	Particles sizes, μm	Deposition technique	Crystalline phases or structure	Thickness, μm	Test method
A	[40]	Steel AISI 1044	Matrix: Ni + 20Cr + 4.5B + 5Si + 5Fe and reinforcement: WC + W_2C	Matrix and reinforcement: −125 + 45	*PTA*	γ-Ni, WC, W_2C, $M_{23}C_6$ (M = Cr, Fe, W, Ni)	3000	Reinforcement Vickers indenter of 100 g (matrix) (H_{max} = 25.5 GPa), and of 300 g reinforcement (*H* = 8.6 GPa)
B	[41]	Steel AISI 5115	Cr + 29Fe + 6.5C + 1.6Si	—	*PTA*	α-(Fe-Cr), M_2C, M_3C, M_7C_3 and $M_{23}C_6$ (M = Fe, Cr)	1500 (estimated from the cross-section)	Vickers indenter of 100 g (*H* = 8.9 GPa)
C	[42]	Steel 35CD4	Matrix: Ni and reinforcement: WC	−150 + 50	*1SLD*, hard phase dispersion	$M_{23}C_6$, $M_{12}C$	—	(*H* = 8 GPa)
D	[43]	Ti	Ti and C	—	*1SLD*, laser alloying	Ti, TiC, Ti_{1-x} C, TiC_{1-x}	10–40	Vickers indenter of 200 g (*HV* = 4.4 GPa)
E	[44]	Steel AISI 1035	Matrix: Ni + 22Cr + 11 Al + 0.8Y and reinforcement: 13YSZ	Matrix: −125 + 45 and reinforcement: −45 + 22	*PTA*	γ-Ni, β-NiAl and cubic and tetragonal ZrO_2	—	Vickers indenter of 300 g (*HV* = 3.9 GPa)

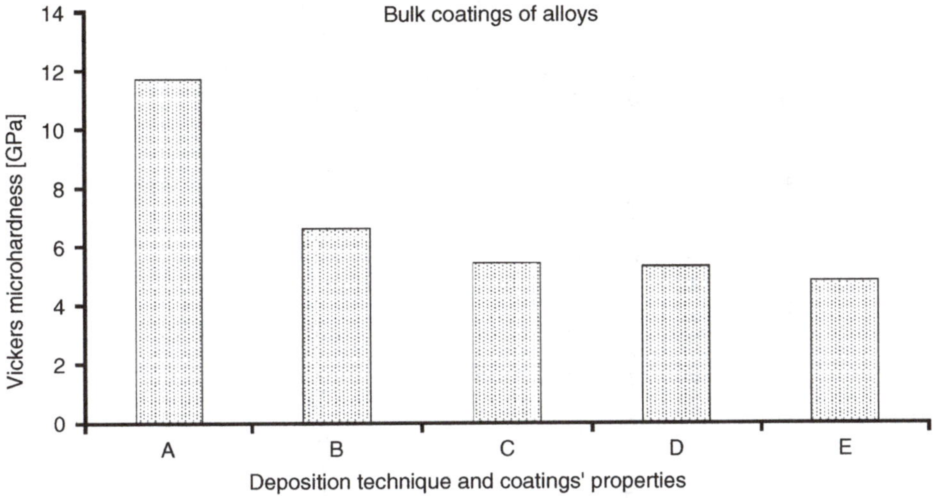

Figure 5.9 Microhardness of coatings of some alloys obtained using *PTA* and laser hardfacing. The description of the technology and some properties of the deposits are shown in Table 5.8.

The scratch test is most often applied to determine the adhesion of ceramic thin films. This test is standardized in Europe[2]. Similarly, the American standard ASTM 633–2021[3] is frequently applied for the adhesion measurements of thermally sprayed granular deposits. However, many other tests are used or about to be developed. An example is an acoustic test described by Devos *et al.* [50]. The tests of adhesion have similar problems, namely:

- distinction between cohesive or adhesive bond strength is made *a posteriori* using a microscope;
- establishing a precise value of the critical load, $F_{cr,}$ at scratch test or the critical strength at tensile adhesion test which correspond to the detachment of deposit's off the substrate. These values can be, e.g. established using acoustic emission techniques. The critical values correspond to the peak of acoustic activity as found Ichimura and Rodrigo [51] by testing adhesion of arc evaporated TiN films. This method using picoseconds impulses enables also to find out the adhesion defects influencing the reflection of acoustic waves [50].

Finally, the adhesion of the bulk deposits is not measured. The interface between such deposit obtained by *PTA* or by laser-assisted methods has metallurgical character which renders the adhesion much greater than in other types of deposits.

---

2 UNI EN 1071-3, European standard, Advanced technical ceramics methods of test for ceramic coatings part 3: determination of adhesion and other mechanical failure modes by a scratch test, October, 2005.
3 ASTM C633-2021, Standard test Method for Ahesion or Cohesion Strangth of Thermal Spray Coatings, last updated November 2, 2021, American Society for Testing Materials (at present ASTM International), West Conshohock, Pennsylvanie (PA), USA.

Table 5.8 Powders and some properties of alloys coatings obtained using different laser deposition methods and using plasma transferred arc, having microhardnesses shown in Fig. 5.9.

			Powder properties			Coatings' properties		
Symbol	References	Substrate	Chemical composition, wt.%	Size, μm	Deposition technique	Crystalline phases	Thickness, μm	Test method
A	[46]	AlSi alloy	Ni + 11Cr + 2Fe + 2.5C + 2Ni + 2Si + 1B	—	*2SLD*, laser alloying	Al, AlNi, Al_3Ni_2, Al_3Ni	500	Vickers indenter of 50 g ($H = 11.7$ GPa)
B	[47]	Steel AISI 304	Co + 30Cr + 13W + 3Fe + + 2.5C + 2Ni + 2Si + 1Mo	—	*1SLD*, laser cladding	Co + Cr + Fe solid solution, Cr_7C_3	—	Vickers indenter of 1000 g ($H = 6.6$ GPa)
C	[48]	Stainless steel 304 L	Stellite 6, Co + 28Cr + 1Si + 1C	—		—	—	As above ($H = 5.4$ GPa)
D	[49]	Stainless steel 4140	Stellite 12, Co + 30Cr + 8.5 W + 2Ni + 2Fe + 2Si + 1.4C	−150 + 53	*2SLD*, CO_2 laser treatment of *PTA* coating	Co + Cr_7C_3 + $Cr_{23}C_6$	2500	As above ($H = 5.3$ GPa)
E					*PTA*			As above ($H = 4.8$ GPa)

5.2.2.1 Atomistic Deposits

The adhesion of atomistic deposits depends mainly on the following factors:

- deposition process and, in particular, on these parameters of the process which influence the microstructure and the residual stresses of the film;
- parameters of the method applied at testing, which should be similar when comparing the adhesion of tested films;
- characteristics of deposited substrate's surface including such factors as roughness and oxidation;
- materials of film and of substrate and, in particular, their physico-chemical affinity.

Table 5.9 enables to compare the adhesion obtained by scratch test for frequently used TiN thin films. The films were obtained using different deposition methods on steel and tungsten carbide substrates and have thickness of 4 μm.

A comparison between the adhesions of the TiN films obtained by the arc evaporation associated with ion plating (arc *IP*) and by *PECVD* shows that better films' adhesion was obtained at arc *IP* processes (lines 1 and 2 of Table 5.9). This can be correlated with much higher kinetic energy of species evaporated at *IP* compared to that at *PECVD*. Therefore, it can be concluded that the more energetic process results in the films adhering better to the substrate. Consequently, it is important to optimize the process parameters to improve adhesion of the film. By taking the example of the magnetron sputtering process, one can see the influence of gas pressures at deposition. Consequently:

- **Total pressure** of gas in deposition chamber. Ward and Datta [52] discussed the influence of pressure at Nb deposition. The variations of the total argon pressure from 0.14 to 1.8 Pa did not influence the adhesion of the 6–8 μm thick films on the 316 L stainless steel substrate. The increase of the pressure up to 3 Pa made the deposits less adherent at scratch test. This effect results from the change in the microstructure of

Table 5.9 Adhesion determined by scratch test of TiN films obtained on steel and WC-Co substrates [51].

			Deposition process			Scratch test	
Number	Substrate	Substrate preparation	Method	Source and precursor (evaporation) or reactant gas (*PECVD*)	Substrate polarization (V)/deposition temperature T_d, K	Indenter's radius, mm	Critical load, F_{cr}, N
1.	HSS steel	Polishing with SiC #600 grit and cleaned with isopropyl alcohol	Arc evaporation associated with ion plating (arc -*IP*)	Ti and N_2	−300/between 600–700	0.2	45
2.	WC-Co						82
3.	HSS steel		*PE CVD*	$H_2 + N_2 + TiCl_4$	−300/823		18
4.	WC-Co						32

the deposit predicted by the *SZM* model discussed in Section 3.2.4. Thus, the fine structure, obtained at low pressure, is more resistant to deformation during the scratch test.

- **Partial pressure** of reactive gas in the chamber. An example was the pressure of nitrogen mixed with argon at Cr deposition investigated by, e.g. Guilbaud-Massereau *et al.* [53]. The authors found that the critical load of the scratch test increased from $F_{cr} = 11$ N to $F_{cr} = 18$ N by increasing nitrogen concentrations in the mixture from 0 at. % to 40%. The reason for this improvement was explained by the reduction of grains size in the nitrogen containing films and to the reduction of tensile stresses in the films

The way of carrying the scratch test, used frequently to characterize the films adhesion, may influence the obtained data. The most important parameters are:

- indenter radius;
- speed of its movement;
- load increase rate.

The round-robin test carried out in five different laboratories for TiN and TiC films obtained by *PVD* and *CVD* on cemented carbide and high-speed steel substrate did show that critical load depends on diamond indenter [54]. The radius of the indenter radius that has a major influence on the critical load found for various film–substrate combinations studied by Randall et al. [55] and shown in Table 5.10 (lines 1 and 6–8). For example, a *DLC* film deposition by *CVD* on silicon substrate is characterized by the critical load of $F_{cr} = 5$ N measured with an indenter of radius $r = 50\,\mu m$ and the value of $F_{cr} = 15$–20 N for an indenter of radius $r = 200\,\mu m$. The analytical expression developed by Ichimura and Rodrigo [51] shows that the critical load is proportional to this radius. Finally, the other test's parameters such as indenter speed and load's increase rate influence much less the measured values of F_{cr}.

The examples of adhesion values obtained using scratch test for some ceramic and metallic films obtained using different deposition methods on typical substrates are collected in Table 5.10.

The values of critical load characterizing adhesion are, generally, lower than 100 N. The adhesion is strongly influenced by the interface between film and substrate. This influence can be described by the following factors:

- **Physical state of surface** characterized by their cleanliness before deposition and by the roughness, which promotes the adhesion of the deposits. Surface cleaning of a substrate by ion bombardment seems to be a particularly effective method in this respect. For example, this type of pre-treatment using Ar ions improved the interfacial toughness between the *DLC* film on aluminum and steel substrates [59]. Similarly, the adhesion of such film to the WC-Co substrate improved by roughening using diamond paste (line 2 of Table 5.10). The ion etching before sputtering was applied prior to Ni deposition on alumina substrate [60].
- **Chemical state of surface** being its chemical such as oxidation. For example, easily detachable oxides formed on the surface of steel or titanium result in lower adhesion of *DLC* films than dense oxides on aluminum [59].

Table 5.10 Examples of adhesion determined by scratch test for atomistic films obtained on different substrates.

					Deposition process			Scratch test	
Number	References	Film/thickness, μm	Substrate	Substrate preparation	Kind	Target/gas	Polarization (V)/T_d, K	r, mm	F_{cr}, N
Carbonic films									
1.	[55]	*DLC*/0.56	Si	—	*CVD*	—	—	0.05–0.2	5–20
2.	[56]	*DLC*/0.78	WC-Co	Diamond paste preparation, ultrasonic bath	*PECVD*	/Ar + CH_4	—	0.2	33
Ceramic films									
3.	[57]	Cr_xN_y[a]/9.3	Ti6Al4V	Surface polished to reach R_a = 0.04 μm	Sputtering	Cr/N_2	−40/	—	30–70
4.	[58]	ZrN/9	Haynes 25 (Co-based alloy)	—	Arc evaporation	—	/less than 623	0.1–0.13	51
5.		TiN/2–3					—		20–30
Metal films									
6.	[55]	W/1.2	Steel 440 C	—	*CVD*	—	—	0.02–0.2	2–25
7.		Al/0.5	Si		Sputtering		-		3–25
8.		Au/0.5							2–20

[a] $0.5 \leq x \leq 1$ and $0 \leq y \leq 1$ depending on N_2 flow.

- **Preheating of the substrate** before deposition can be a good method of preparing the substrate for deposition. The heat treatment cleans up the organic impurities. However, it can happen that this treatment releases gases entrapped at pre-cleaning process [59].
- **Film and substrate affinity** including, e.g. the modulus of elasticity and the coefficient of thermal expansion. The mismatch of these parameters may lead to the generation of residual stresses and to deposits' detachment. The generation of compressive stresses in thin films forms a well-known ***telephone cord buckle shape*** [61]. Finally, the introduction of an interlayer having good affinity to the deposit and to the substrate between these two partners can improve the adhesion of the deposit [62].

5.2.2.2 Granular Deposits

The adhesion of granular deposits sprayed using powder feedstock is typically measured using tensile test standardized as ASTM C633–2021. Most of the authors follow the recommendations of this standard, and there is no need to give the details of the test. Consequently, the adhesion of granular deposits can be categorized by considering the following factors:

- **Deposition process and typical materials** used to spray. In general, metals and alloys are deposited using *CGSM*, *FS*, or *AS*. Ceramic materials such as oxides are deposited using *APS* and carbides being used as composites with metals or alloys are applied using *HVOF* or *DS* processes.
- **Preparation of a substrate before** spraying, which includes its preheating. The surface of a substrate is often activated by sandblasting with silica or corundum, but laser or water jet treatment is also applied. The vacuum plasma spraying process allows the surfaces to be cleaned by the transferred arc, which increases adhesion. The softening of the metallic substrate by vacuum improves adhesion of cold sprayed metals.
- **Coating and substrate physico-chemical affinity,** which, e.g. renders difficult to obtain any coating on substrates with high thermal conductivity (such as copper) or an oxide coating on graphite substrate. Finally, there are methods to improve this affinity by an application of an intermediate coating.

Some examples of granular coatings adhesion are collected in Table 5.11. The choice of the spraying process considerably influences the adhesion of the deposits. For example, composite carbide coatings with a metal matrix adhere better when sprayed using *HVOF* technique (see lines 1–3 of Table 5.11) than with others. The difference results from the process in which the high-velocity flame is able to melt the metal matrix of the composite particles and, simultaneously, accelerate well the powder particles. This promotes the adhesion. On the other hand, the plasma torch jet has lower velocity, and its high temperature is not necessary to melt metal in composite particles. On the contrary, the metal can evaporate in these conditions and create, after condensation, a vapor phase deposit which lowers adhesion at the interface between the sprayed particles and a substrate or previously deposited layer.

The preparation of the substrate surface by blasting with sand, grit, alumina, or iron particles is a standard in granular coatings deposition. There are some other techniques, such as laser texturing (see line 4 of Table 5.11), which can also be used. The hard particles

Table 5.11 Examples of adhesion determined by tensile test for typical granular coatings sprayed using powders obtained on different substrates.

			Substrate			Spray process		
Number	References	Feedstock's chemical composition (wt. %)/size, μm	Type and tradename	Preparation	R_a, μm	Kind	Initial temperature, K	Adhesion strength, MPa
Composites including carbide								
1.	[63]	WC + 12Co/−45 + 5.6	Carbon steel	Corundum blasting	—	*HVOF*	—	>100
2.	[64]	WC + 9Co + 5Cr/−53 + 11	Steel 4340	Grit blasting	—			>89
3.	[65]	Cr_2C_3 + 25 NiCr	Stainless steel	Aceton cleaning followed by alumina blasting	5			82.8[a]
Oxide coatings								
4.	[66]	ZrO_2 + 7Y_2O_3 + 1.7 HfO_2/−100 + 16	Superalloy AM1 monocrystal	Grit blasting/Laser surface texturing	—	*APS*	—	11/14 to 33
5.	[67]	Al_2O_3	Low-carbon steel	Alumina blasting/alumina blasting and NiCrAl undercoating	—			16/26
6.	[68]	Al_2O_3/−45 + 22	Stainless steel 304 L	Sandblasting	12		300/573	15/34
Coatings of metals and alloys								
7.	[69]	Zn/3200[b]	Concrete	Silica blasting	RMS^c = 15/108	*AS*	323	0.87/2.21
8.	[70]	Ni/−75 + 30	Steel	Iron blasting	—	*FS*	—	40
9.	[71]	Ni/d_{50} = 25	Stainless steel AISI 304	Grit blasting/softening by vacuum annealing	R_z = 34	*CGSM*		15/65

[a] Chinese norm GB8642-1988 was used.
[b] Wire diameter.
[c] Roughness was measured as root mean square (*RMS*) of peaks and valleys of the surface.

blasting results in an increase of the substrate's surface roughness which promotes adhesion (for example, a deposit of Zn on concrete shown in line 7 of Table 5.11).

The application of intermediate undercoatings may lead to the reduction of the residual stresses generated in the deposition of coatings with a low physical affinity with respect to the substrate (see line 5 of Table 5.11). This affinity concerns mainly thermal expansion and modulus of elasticity, which are very different in ceramic and metallic materials.

Another possibility is preheating of the substrate prior to coating deposition (see line 6 of Table 5.11) being useful at deposition of oxide coating on metals. Finally, the softening of metallic substrate by, e.g. vacuum annealing was tested to be a useful method to achieve good adherence of cold sprayed Ni coating (see line 9 of Table 5.11).

5.2.3 Elasticity, Strength, and Toughness

The mechanical properties analyzed in this section correspond to the analysis of an elastic and plastic behavior of the deposit. The elastic behavior is described by Young's modulus. The modulus of atomistic deposits is generally found using a nano- or microindentation test made on the section or surface of the deposits by analyzing the area of an indentation described in Section 4.3.4. This test requires the knowledge of Poisson modulus, ν_d, and many authors do not precise this value.

The hardness and elastic modulus of the mechanical properties of thin films have been tested, but other properties are more difficult to test being rarely published in the literature. The test required often the development of new apparatus. An example is the four-point bending set-up made by Ulrich and Zhao [72] to test ultimate strain of Si_3N_4 deposits obtained by *PECVD* on stainless steel. The ultimate strain was found by the authors to be about 0.2%.

The granular deposits are thicker and, therefore, their mechanical properties are easier to test than those of atomistic deposits. Therefore, it is possible to find such properties as tensile strength, fracture toughness, toughness, and the description of plastic deformation at plastic behavior. It is important to emphasize that the mechanical properties of the deposits are generally anisotropic. The atomistic deposits often have columnar grains, while the structure of thermally sprayed granular coatings is lamellar. This is why the direction of measurements is shown in the tables showing the mechanical properties. Finally, the elastic and plastic behavior of bulk deposits is close to the properties of solid materials and is not presented.

5.2.3.1 Atomistic Deposits

The presented values of modulus of elasticity of atomistic deposits depend mainly on three factors:

- **Film deposition process** and the parameters influencing its microstructure, i.e. porosity, grain shape, contacts between grains, imperfections, etc.
- **Method of measuring the modulus**, because the measured value can be influenced more or less by the substrate, the intermediate layer, or the measurement on film's surface or its cross-section. It concerns, in particular, frequently used nanoindentation method.
- **Film's materials** are important in knowing that some materials (e.g. the multilayers composites including *DLC*) are synthesized mainly as the films. Many others can be obtained as a deposit, but exist also as a bulk material. In the last case, the moduli of elasticity of materials obtained in different ways can be compared, and the value can be correlated with the manufacturing method.

Some examples of elasticity modulus for atomistic films are shown in Table 5.12. The deposition process should be adapted to the film material. Some materials have a modulus of film lower than that of the bulk material. The well-known example is diamond having modulus of $E = 1180$ GPa as single crystal [81]. On the other hand, the modulus of polycrystalline diamond films obtained using *PLD* does not exceed $E_d = 550$ GPa and that obtained using *HFCVD* the value of $E_d = 694$ GPa (lines 1 and 3 of Table 5.12). The family of techniques belonging to thermal *CVD* are associated with high deposition temperatures (see Table 2.7) which enables obtaining films having less structural imperfections promoting higher values of E_d. An example confirming this statement is TiC films, which have greater modulus when obtained using thermal *CVD* compared with those synthesized using magnetron sputtering (lines 4 and 5 of Table 5.12). Finally, the deposition process may lead to the generation of residual stresses. This was observed, e.g. at *IBAD* deposition of CN_x films (see line 2 of Table 5.12) enabling to find out that the films with compressive internal stresses have the values of modulus larger by 5–10% with regard to that without stress removed by annealing [73].

The mentioned measurement of TiC films was carried out using different methods, namely 4-point bending and nano-indentation, which could have influenced the obtained values. The anisotropy of the direction of measurements of 7YSZ by nano-indentation also resulted in different values of measured modulus. The anisotropy is one of the consequences of the microstructure of atomistic deposits described by the *SZM* model and shown in Fig. 3.4. This effect was visible for *EBPVD* deposits as shown in line 6 of Table 5.12 having modulus of elasticity different when measured on the film's surface and on its cross-section.

Atomistic deposition techniques make it possible to synthesize composite films with high modulus of elasticity. The composites including nano-crystalline (TiN or W_2N) and amorphous grains of Si_3N_4 obtained using *PECVD* method which enabled to reach the modulus of about $E_d = 500$ GPa (lines 7 and 8 of Table 5.12). The composites may also include multilayers with films obtained using different techniques as sputtering and *PLD* (see line 9 of Table 5.12). The multilayers enable finding out the films' configuration adapted to the substrate and to its protection against friction and wear [5].

5.2.3.2 Granular Deposits

The granular deposits are thicker than the atomistic ones, and the measurements of the mechanical properties are easier to carry out. These properties are comparable with those of sintered materials. The lower values of the moduli of elasticity or of the resistance of the sprayed layers result from their microstructure and in particular from the porosity and the poor contacts between the lamellas. Heat treatment after deposition can then improve these properties. The microstructure, in turn, depends on [39]:

- **Deposition process** and its parameters. The process determines the velocity and degree of melting of particles impacting the substrate.
- **Feedstock used to spray** and, in particular, whether it is initially solid (powder) or liquid (suspension or solution). The feedstock determines morphology and size of liquid particles at impact.
- **Impact of particles** resulting in splats (Fig. 3.8) and includes frequently unmelted particles and pores (Fig. 3.13). Consequently, the measurement of mechanical properties depends the direction of applied stress.

Table 5.12 Modulus of elasticity of typical atomistic films.

				Process			Test		
Number	References	Film[a]/thickness, μm	Substrate	Kind	Target/ reactant gas	Polarization, V/T_d, K	Kind	Direction of test[b]	E_d, GPa
Carbonic films									
1.	[5]	*DLC*/0.5	Steel 440 C	*PLD*	Graphite	—	Nano-indentation	⊥	550
2.	[73]	CN_x/0.06	Si	*IBAD*	Graphite/N^+	−500/		P	120–150
3.	[74]	Diamond[c]		*HFCVD*	/H_2 and CH_4	−30/1123–1223			694
Ceramic deposits									
4.	[75]	TiC/5	Stainless steel SUS304	*CVD*	—	/1298	4-point bending	⊥	513
5.	[76]	TiC/2	Stainless steel 440 C	Magnetron sputtering	/CH_4	−100/	Nano-indentation		420
6.	[77]	7YSZ	NiAl	*EBPVD*	7YSZ/O_2	/873–1373		⊥ and ∥	10–160 for ⊥ 20–120 for ∥
Composite deposits									
7.	[78]	nc-TiN + a-Si_3N_4	Stainless steel	*PECVD*	/N_2 + H_2 + $TiCl_4$ + SiH_4	/823	Nano-indentation	⊥	290–520
8.	[79]	nc-W_2N + a-Si_3N_4	Si		—		Micro-indentation		150–500
9.	[5]	20×[*DLC* -*Ti*] multilayer + TiC + Ti[d]	Steel 440C	Ti by magnetron sputtering, TiC, *DLC* by *PLD*	Ti, TiC, graphite/	/573–673 for *DLC*	Nano-indentation		240
Metallic deposits									
10.	[80]	MoTa/1–2	Si	Magnetron sputtering	Ta and Mo/	—	Nano-indentation	⊥	255

[a] nc – nanocrystalline, a – amorphous.
[b] Direction with regard to the film's surface.
[c] Films described as ultra-nanocrystalline.
[d] Films composition is shown from the surface down to the substrate.

The feedstock used to spray can be molten at processing. Consequently, the carbides being hard and interesting for application as wear-resistant coating have very high melting point and are used generally as composite with metals or alloys. On the other hand, the oxides are more difficult to melt than metals and alloys. The spray techniques used to deposit these materials are different, and the demonstration of the elastic properties follows the materials.

Carbide Deposits

The mechanical properties of sprayed carbides are shown in Table 5.13. The properties of popular tungsten carbide with cobalt can be compared with those of a sintered materials such as:

- WC + 10 wt. % Co, which has Young modulus of $E = 575\,\text{GPa}$, fracture stress of $\sigma_f = 6\text{–}8\,\text{GPa}$, and fracture toughness of $K_{Ic} = 10\,\text{MPam}^{0.5}$ [87];
- WC + 12 wt. % Co having $E = 410\,\text{GPa}$ and fracture stress of $\sigma_f = 3.6\,\text{GPa}$ [82].

It is important to mention that the carbides are sprayed as a composite powder with metals (typically Co, Ni, or Cr) or alloys as NiCr. Consequently, the method of powder manufacturing described in Section 2.3.1, which may be cladding (line 1 of Table 5.13) or agglomeration (line 2 of Table 5.13), influences the modulus of elasticity. The modulus is lower than that of sintered carbides and varies from 50% to 80% of value measured for such material. The specification of spray processes is to metal or alloy without melting the carbide. Consequently, the spray techniques used are frequently *D-gun* or *HVOF*, which is characterized by relatively low temperature of flame rather than *APS* (line 5 of Table 5.13).

The size of carbide in powder feedstock influences considerably the fracture toughness of sprayed coatings. The greatest toughness is reached for nanocrystalline size of carbides (see line 4 of Table 5.13).

Oxide Deposits

The mechanical properties of some thermally sprayed oxides are shown in Table 5.14. Young's modulus of the sprayed oxides is, in general, much lower than that of the sintered oxide. Plasma sprayed alumina coating has the modulus of $E_d = 110\,\text{GPa}$ (line 1 of Table 5.14) compared to that of dense alumina having modulus of $E = 370\,\text{GPa}$ [94]. The main reason for this difference is the porosity occurring in sprayed coatings.

The critical strain energy release rate, G_{Ic}, tested for zirconia coatings (lines 2 and 3 of Table 5.14) using double cantilever beam method was suggested to be a useful method to characterize the coatings' toughness.

The use of fine particles in the feedstock does allow obtaining high elasticity modulus. Consequently, the use of suspension plasma sprayed zirconia stabilized with 8 wt. % Y_2O_3 enabled to obtain coating with modulus $E_d = 87.5\,\text{GPa}$ (line 4 of Table 5.14] compared to the modulus of dense zirconia reported to be $E = 168\,\text{GPa}$ [94].

The comparison of the mechanical properties of coatings plasma sprayed using solid feedstock (powder) and the liquid one (suspension and solution) was carried out for hydroxyapatite coatings (see lines 5–7 of Table 5.14). The modulus of elasticity was greater for coatings sprayed using solid feedstock.

Another important feature influencing mechanical properties of sprayed deposits is their lamellar structure. This type of microstructure may lead to sliding of lamellas submitted to the mechanical stress being parallel to the coatings surface. Berndt [95] analyzed this effect

Table 5.13 Mechanical properties of carbides containing coatings obtained by thermal spraying.

Number	References	Powder		Deposition process	Porosity, %/ density, ρ, kg/m^3	E_d, GPa	σ_f, MPa	K_{Ic}, MPa.m$^{0.5}$	Testing method
		Composition, wt. %	Particles size, μm/size of carbide grain, μm						
1.	[82]	WC + 12Co (cladded)	30–60	*D-gun*	—	210–280[a]		—	Indentation
2.	[83]	WC + 12Co (agglomerated)	—		—	331	820	1.2	4-point bending
3.	[84]	WC + 10Ni	15–45/0.8–2.5	*HVOF*	6–10/	150–200		3.5–4.5	Indentation
4.	[85]	WC + 10Co + 4Cr	15–45/0.2		0.43			7.12	
			15–45/0.35		0.29			4.97	
			15–45/0.86		0.89			6.03	
5.	[86]	Cr_2C_3 + 7NiCr	—	*APS*	5.4/	120[b]		6.1[c]	

[a] The values were depending on the coatings thickness and on residual stresses.
[b] Testing load 2.94 N.
[c] Testing load 294 N.

Table 5.14 Mechanical properties of oxides obtained by thermal spraying.

Number	References	Powder		Deposition process	Mechanical properties			Testing method
		Composition, wt. %	Particles size, μm/precursors		Porosity, %/ density, ρ, kg/m^3	E_d, GPa	G_{Ic}, J/m^2/K_{Ic}, MPa.m$^{0.5}$	
1.	[88]	Al_2O_3	25–100/[a]	*APS*	5.4/	110	—	Indentation
2.	[89]	$ZrO_2 + 5CaO$	10–105/		/5200	—	41/	*DCB*
3.		$ZrO_2 + 8Y_2O_3$			/5100	—	32/	
4.	[90]		$d_{vs} = 4.6$/[b]	*SPS*	—	87.5	—	Indentation
5.	[91]	*HA*	20–45/	*APS*	—	31[c]	/1.14	
6.	[92]		3–10^{22}/	*SPS*	—	28	—	
7.	[93]		/calcium nitrate tetrahydrate and triethyl phosphite	*SPPS*	—	5–22		

[a] Agglomerated nanometric powder with particles size of 200–500 nm.
[b] Suspension was formulated using 20 wt. % of powder, 40 wt. % of water and 40 wt. % of ethanol.
[c] Modulus increased up to $E_d \approx 35$ GPa after heat treatment at $T = 1173$ K.

and called it *pseudo-ductility*. The microstructure can be homogenized and densified at post-spray heat treatment. Such treatment and $T = 1173$ K, applied for plasma sprayed *HA* coating (see line 5 of Table 5.14), enabled to increase Young's modulus from $E_d = 31$ to 35 GPa.

Metal and Alloy Deposits

The mechanical properties of some sprayed metals and alloys are shown in Table 5.15. As the metals are ductile the value of the ultimate stress, σ_u, is also presented (see also Fig. 4.13).

The powders of metals and alloys heated to high temperatures in open atmosphere may oxidize forming oxides of different properties. The oxidation can be reduced by the use of appropriate deposition methods. Consequently the methods: (i) reduce the ambient gas pressure such as *VPS* (see lines 1 and 6 of Table 5.15); (ii) have working gas of low temperature such as *CGSM* technique (lines 3 and 5 of Table 5.15); and (iii) enable to shorten the time of interaction between sprayed particle and working gas by high velocity of gas such as *HVOF* method (lines 4 and 7 of Table 5.15).

The sprayed metals and alloys coating may reach the mechanical properties being close to that of bulk materials produced by metallurgical techniques. For example, the elasticity modulus of bulk tantalum is $E = 184$ GPa [101], and the coating sprayed under vacuum reaches $E_d = 148$ GPa (line 2 of Table 5.15), i.e. the ratio is about 80% of bulk material value. Gouldstone *et al.* [102] report that this ratio for some sprayed metals and alloys is between 58% and 95%. The porosity of sprayed coating seems to be the main microstructural characteristics influencing the mechanical properties (line 2 of Table 5.15)

The mechanical properties of the alloys are adapted to their application. The coatings, which serve as bondcoat in, e.g. *TBC* should have Young's modulus close to that of the substrate and possibly high ductility. The latter helps to absorb residual or service stresses without the generation of cracks. In many other applications, it is important to achieve the best mechanical properties. The example are high-entropy alloys (line 7 of Table 5.15), which can be annealed after spraying to improve these properties [100].

Composite Deposits

The mechanical properties and mainly elastic modulus of some composite coatings obtained by thermal spraying are shown in Table 5.16. Thermally sprayed composites may have following possible functions [39]:

- **Intermediate coating** between metallic substrate and ceramic topcoat applied, e.g. in thermal barriers (lines 1 and 2 of Table 5.16). TBCs are used at high temperatures and should include the neighboring metal and ceramic coatings having thermal expansion and elastic modulus possibly similar. Otherwise, generated thermal stresses would lead to fractures cracking. Consequently, the elastic modulus of intermediate composite coating should be between that of bond and top coating.
- **Improving mechanical properties** of functional coatings. An example is the *HA* coating used for its biocompatibility in the implants. The improvement of mechanical properties by an application of a composite of ceramic *HA* and *8YSZ* with Ti alloy leads to better implant reliability (line 3 of Table 5.16). Similarly, the strengthening of ductile Al coating can be reached by spraying of its composite with graphene nanotubes (line 4 of table 5.16).

Table 5.15 Mechanical properties of metals and alloys obtained by thermal spraying.

		Powder			Mechanical properties			
Number	References	Composition, wt. %	Particles size, μm	Deposition process	Porosity, %/ density, ρ, kg/m^3	E_d, GPa	σ_u, MPa	Testing method
Metals coatings								
1.	[96]	W	5–25	*VPS*[a]	/17400	280–305[b]	—	Ultrasonic resonance
2.	[101]	Ta	5–53		0.6/	148[c]	—	Indentation
3.	[97]	Cu	5–25	*CGSM*[d]	—	—	45	Tensile test in plane
4.			31–88	*HVOF*	—	—	145	
5.	[98]		21–104	*CGSM*	—	108.5	—	Indentation
Alloys coatings								
6.	[99]	Ni + 22Co + 21Cr + 8Al + 4Ta + 0.5Y	−38	*VPS*	–	190	–	Tensile test
7.	[100]	AlCoCrFeNiCu[e]	15–45	*HVOF*	–	162	–	Indentation

[a] Pressure in vacuum chamber was $p = 26.7$ kPa (line 1) and $p = 15$ kPa (line 2).
[b] Depending on direction of measurements (parallel and perpenicular to the surface) and of substrate.
[c] Values for inner layer.
[d] N_2 was used as working gas.
[e] High-entropy alloy (*HEA*) of each metal content was about 16.7 at. %.

Table 5.16 Elastic modulus of composites obtained by thermal spraying.

Number	References	Matrix of composite		Reinforcement of composite		Deposition process	Elastic modulus, E_d, GPa	Testing method
		Chemical composition, wt. %	Size of powder feedstock, μm	Type/chemical composition, wt. %	Fraction, vol. %			
1.	[103]	$ZrO_2 + 24MgO$	—	Powder/Ni + 5Al	0–100	*APS*	65–90[a]	Ultrasonic
2.		*8YSZ*		Powder/Ni + 20Cr			74–97	
3.	[104]	*HA* + *8YSZ*	−20 (*HA*) and −60 + 10 (*8YSZ*)	Powder/Ti + 6Al + 4V	35 (*HA*) + 15 (*8YSZ*) + 50 (Ti6Al4V)		55[b]	Indentation
4.	[105]	Al	−45 + 6.5	Nanotubes/C – graphene	1 wt. %	*CGSM*	90.4 (matrix 60.4)	

[a] The value of $E_d = 90$ GPa was reached for the fraction of reinforcement of 75 vol. %.
[b] The shown value is for as-sprayed sample and Young modulus increased to $E_d = 65$ GPa after heat treatment at $T = 873$ K for 9 h.

Young's modulus is roughly proportional to the volume fraction of the reinforcement. It also remains valid and leads to the reduction of residual stresses enabling to avoid cracking of the brittle ceramic [103].

5.2.4 Friction and Wear

Wear resistance is separated in this book from corrosion resistance, although some authors describe these subjects together, e.g. [106]. The reason is the difference of wear mechanisms that can be characterized by its nature being mechanical or chemical. Friction and wear are part of *tribology*, and the methods of its characterizations are discussed in detail in Section 4.3.6. The basic wear mechanisms of two surfaces in contact include the following phenomena [107]:

- ***abrasive wear*** which can be ***mild*** or ***severe*** includes ploughing and cutting of surface irregularities;
- ***adhesive wear*** mechanism includes shearing of these irregularities and is associated with material transfer of mass between two surfaces in contact;
- ***fatigue*** wear mechanism consists of initiation of cracks and their growth;
- ***tribochemical reactions*** leading to formation of thin microfilms on the surface of one of the surfaces in contact. The microfilm may act as lubricant lowering the friction between the surfaces described by Chen et al. [108] or as an abrasive agent such as oxide formed on the surface of a metal [109].

Although it is difficult to assign a family of deposit to a type of wear, it can be said that thin films obtained using one of atomistic method are most often applied to resist against *friction* and adhesive type of wear. Apart from thickness, the *friction* and wear between deposit and the surface in contact is also depending on [109]: (i) hardness of deposits; (ii) its surface roughness; and (iii) presence or formation of debris in contact. Consequently, the type of material being in contact with the deposit influences considerably the mechanism of wear and *friction* mechanisms. As far as *adhesive wear* is concerned, it is possible to introduce the wear rate, w, being equal to volume of material removed by this wear, V, normalized by the distance, d, traveled during the test and by applied load, F, [11]:

$$w = \frac{V}{Fd} \tag{5.1}$$

Finally, especially, for thin films, the materials of substrate and particularly its hardness should be considered. Consequently, the adhesive wear tests require specifying: (i) type of test; (ii) shape and kind of material in contact; (iii) atmosphere at testing, e.g. wet, dry, type of gas, and vacuum; (iv) distance traveled during the test or test time; and (v) presence of a lubricant between the surfaces in contact.

Wear by *abrasion* or by *erosion* requires deposits with good cohesion and good tenacity, which can tested by specifying, in addition to the factors mentioned above, the shape and kind material of the second and, possibly, third body. As far as *erosion* is concerned, it is also necessary to have specified [110, 111]: (i) size; (ii) impact velocity; and (iii) angle of impact of particles of the erosive material or the conditions of generation of turbulences, which lead to *erosion by cavitation*.

Finally, in the analysis of wear by *fretting*, it is advisable to specify the frequency of oscillations and the amplitude of movement of the surfaces in contact.

5.2.4.1 Atomistic Deposits

Atomistic deposits have been mainly applied against *adhesive* wear. Only a few studies have been devoted to their applications against wear by *abrasion* or *erosion*. Therefore, the wear discussed in this section and in Table 5.17 is exceptionally categorized using types of materials rather than that of wear. Among the materials used as atomistic deposits against wear the various modifications of the carbonic films must be taken into account, including compounds such as nitrides and carbides and composites. Metals and alloys are rarely used for this application.

Carbonic Films

Carbonic films are characterized by a unique combination of tribological properties, i.e. small coefficient of friction, typical for soft materials, associated with a high wear resistance being characteristic for the hard ones. Such materials are called ***solid lubricants*** [117]. Diamond-like carbon (*DLC*) films obtained using different deposition methods and tested using oscillating sphere (alumina or tungsten carbide) tribometer had the coefficient of friction $\mu \leq 0.15$, and the wear rate is in the range $10^{-11} < w < 10^{-5}$ mm^3/(N.m) (see lines 1–5 of Table 5.17). Some studies indicate that diamond is transformed in graphite at high temperatures in presence of H_2O molecules [118]. This corresponds to the replacement of sites sp^3 into sp^2 called graphitization. This transformation may result also from the stresses generated at friction from the sensitivity of *DLC* films to the humidity [119]. The authors observed that presence of graphite in the wear track might lead to lowering wear rate because graphite transformed from the diamond creates a lubricating layer, which improves the tribological behavior of the deposit.

The diamond-like carbon films were synthesized by different techniques such as *PECVD*, *IP*-arc, and *PLD* against fretting wear [120]. Under the same experimental conditions, the most resistant against such wear was the *DLC* films obtained by *IP*-arc. Moreover, these films obtained by any deposition method resisted better against this type of wear than the TiN films obtained by *PVD* being used a reference.

Films of Compounds

The films of the compounds used against different types of wear are shown in Table 5.18, and lines 6 and 7 of Table 5.17 show the details of films used against adhesive wear. The compound being most often applied against all types of wear is TiN. With regard to adhesive wear, this deposit was tested against alumina and the coefficient of friction varied between $\mu = 0.2$ and 0.8. Unlike the *DLC* deposits, this coefficient μ decreases with increasing humidity (line 6 of Table 5.17). Another nitride deposit often tested adhesive and impact wear is CrN tested by, e.g. Zhou et al. [124]. The authors found the coefficient of friction slightly lower than that of TiN. This could have resulted from the formation CrO_3 and Cr_2O_3 thin layers during wear test. These layers lowered this coefficient more than TiO_2 particles formed during wear test of TiN film.

The deposits of MoS_2 compound have tribological properties similar to *DLC* films having small μ and low wear (lines 7 and 10 of Table 5.17). This is why such film is used as a solid lubricant. However, these properties of MoS_2 degrade in humid air, and the element as Ti can be added to the film by, e.g. co-sputtering using two targets to render the films less sensitive to water vapors at high humidity [114].

Table 5.17 Detailed description of some atomistic films applied against adhesive wear.

Number	References	Film		Deposition process	Fiction and wear							
					Test parameters						Test results	
		Chemical composition, at. %	Thickness, δ, µm		Wear type	Test method	Gas/*RH*, %	Counter body	*v*, cm/s/ ν; Hz	Force, *N*, N	µ	*w*, mm^3/(Nm)/reduction of thickness, µm
Carbonic films												
1.	[5]	*DLC*	0.5	*PLD*	Adhesive	Sliding against sphere (see Fig. 4.22)	Air/	Sapphire	—	—	0.08	10^{-9}–10^{-8}/
2.		*DLC* + 25H	1							5	0.07	10^{-6}–10^{-5}/
3.	[112]	*DLC*	3	*PACVD*			Air/65	Alumina	100/	5	0.1	1.1×10^{-7}–2.9×10^{-7}/
4.		*DLC* + 5.7 Si									0.07	0.8×10^{-7}/
5.	[11]	*DLC*	2	*PVD*			Air/	WC-Co	10/	10	0.15	5.2×10^{-11}–1.6×10^{-10}/
Compounds films												
6.	[113]	TiN	4	*IP*	Fretting	Fretting against sphere	Air/ 10; 50; 80	Alumina	/10	1–10	0.8; 0.2; 0.2	/3.1; 2.5; 1.5
7.	[114]	MoS_2	0.8	*PVD* – sputtering	Adhesive	Sliding against sphere	Air/40–50	WC-Co	0.25/0.5	2; 10; 40; 80	.058; 0.055; 0.028; 0.019	2.4×10^{-6}; 2.6×10^{-6}; 5.5×10^{-7}; 3×10^{-7}/
Metallic films												
8.	[115]	Cr	15	*PVD* – sputtering	Adhesive	Sliding against sphere	Air/50	WC	0.1/	10	0.75	/6
Composite films												
9.	[116]	B_4C + 45 Mo	2.0	*PVD* – sputtering	Adhesive	Sliding against sphere	Air	WC-Co	—	15	—	1.8×10^{-4}/
10.	[114]	MoS_2 + (10–20) Ti	0.8				Air/40–50		0.25/0.5	2; 10; 40; 80	0.105; 0.082; 0.059; 0.042	4.21×10^{-7}; 1.76×10^{-6}; 5.95×10^{-8}; 1.01×10^{-8}/

Table 5.18 Atomistic films of compounds against different types of wear.

Film	Deposition process	Wear type	References
Nitrides			
TiN	*Arc PVD, CVD, PACVD, PVD*-sputtering, *IP*	Adhesive, erosion, 2-body abrasion, fretting, impact wear	[113, 121–126]
CrN	*IBAD, PVD* – sputtering	Adhesive, impact wear	[58, 124, 126–128]
TiAlN	*PVD* – sputtering	Adhesive	[126]
SiN_x	*IBAD*		[129]
Carbides			
TiC	*PVD* – sputtering, *CVD*	Erosion, adhesive, impact wear	[76, 122, 123, 127]
TiC_xN_y	*CVD, PVD* – sputtering PACVD	Adhesive, impact wear	[126, 130]
Oxides			
Al_2O_3	*CVD*	2-body abrasion	[131]
Sulfides			
MoS_2	*PVD* – sputtering	Adhesive	[132]

CVD films of TiN, TiC, and Al_2O_3 on the WC-Co substrate having thickness between 2 and 9 μm were tested against erosion caused by alumina sand [122]. The erosion resistance of the TiC film was superior to the others, and the resistance of all deposits was superior than that of the substrate. The erosion rate of the films was strongly dependent on the velocity of eroding particles.

The efficiency of erosion in removing films from their substrates led Shipway and Hutchings [133] to propose the erosion test to estimate the durability of films.

Metallic Films

The deposits of hard metals such as Cr on steel substrate have the advantage of well adhering to the substrate being tough. However, the wear resistance is, generally, not high. For example, lower friction coefficient and much smaller wear volume than that of Cr film described in line 8 of Table 5.17 was obtained when the *PVD* film was doped with 1 wt. % of C. Moreover, it is worth mentioning that compounds such as TiN or TiC having high resistance to adhesive wear are characterized by metallic bond [134].

Composite Deposits

The composite films used in the tribological applications can be roughly divided into two types:

- ***granular inclusions composites***, in which the inclusions (being sometimes reinforcements) can influence the mechanical properties in different ways:
  - small crystals, shown in Fig. 5.4a, are added to increase the films' hardness and toughness e.g. B_4C reinforcement added to Mo matrix and described in line 9 of Table 5.17;

  - soft and rounded islands e.g. soft metals as Sn or Pb acting as solid lubricants in Al matrix [135]. Such composite film is adapted well to the geometry of a contact surface and the debris rich in Sn or in Pb ensured a small coefficient of friction and low wear.
- ***multilayer composites***, which are composed of a few of single layers. The multilayer composites can be divided into three different types [134]:
  - Composites with a limited number of single layers being most popular at present (Fig. 5.10a). This type of deposit is currently the most popular. Thanks to multilayers, the columnar structure of the deposits is interrupted, and two different materials can be combined. The authors analyzed up to seven successive layers of TiN-TiBC-TiN sputtered coatings.
  - Composites with a great number of single layers (Fig. 5.10b). Such composites have a great number of interfaces useful for deviation of cracks generated at different kinds of wear. Deposits with a large number of layers are advantageous because of increase of the deposit's toughness. The number of single layers can reach several hundreds for a coating having thickness of a few micrometers. An example of 100 successive layers of sputtered TiC and of TiN coating is presented.
  - *Superlattices* deposits, already discussed in Section 5.2.1.1 including very thin single layers (having size comparable to the crystal lattice) of two materials having similar chemical bonding and atomic sizes.

Fan *et al.* [136] proposed a concept of a ***mixed composite*** presented in Fig. 5.11. The first, layer adhering to the WC-Co substrate, was a ***granular inclusion composite*** with large diamond monocrystals having size of a few micrometers obtained with help of *PE CVD*. The monocrystals were surrounded subsequently by TiC or TiN films deposited using *PLD*. The second and outer layer was and second polycrystalline diamond, condensed by *PE CVD*.

5.2.4.2 Granular Coatings

The granular coatings are much thicker than the atomistic ones. Their thickness may reach a few millimeters and, consequently, the influence of the substrate on their wear resistance, being an important application of such coatings, is negligible. These deposits have been tested and applied in industrial practice against all types of wear presented in the following section.

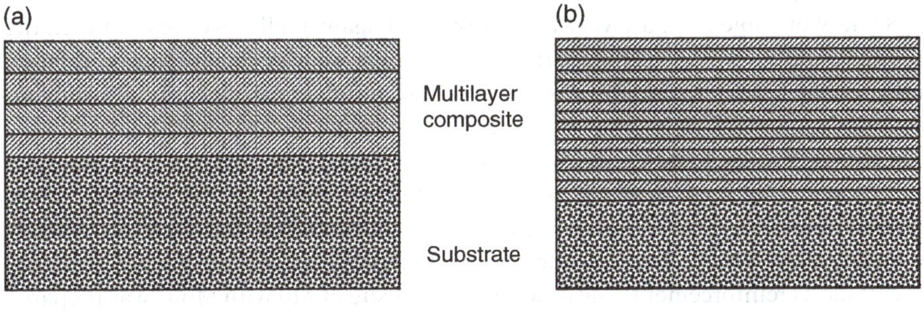

Figure 5.10 Multilayers composites with small number (a) and great number (b) of single layers. Inspired by Holleck and Schier [134].

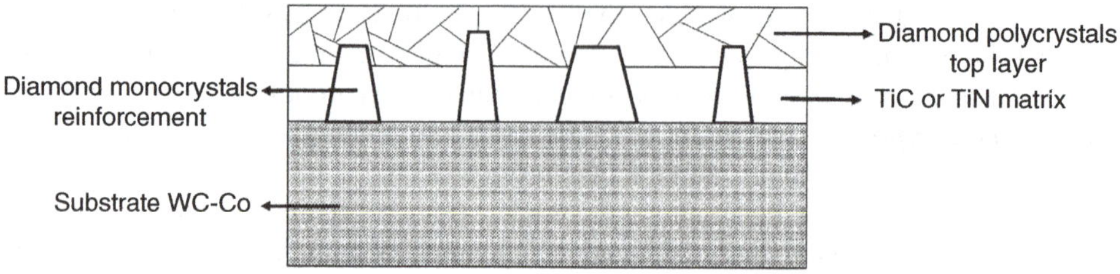

Figure 5.11 *Mixed composite* deposited on WC – Co substrate proposed by Fan *et al.* [136].

Adhesive Wear

Ceramic Coatings The oxide coatings of Cr_2O_3, Al_2O_3, and of TiO_2 have been most frequently applied against adhesive wear (lines 1–3 of Table 5.19). In particular, Cr_2O_3 coatings resist well against adhesive wear. However, it must be mentioned that hexavalent and trivalent chromium, which can be present as vapors at thermal spray processing of chromia, was found to be toxic [147]. The studies of alloys of chromia and titania, shown in line 5 of Table 5.19 and of chromia with some other oxides such as alumina and zirconia, showed higher coefficient of friction and higher wear rate [142]. The oxides are most frequently plasma sprayed in atmosphere using powders as a feedstock. However, recent studies are devoted to the deposition of nano-structural coatings using liquids such as fine powders suspensions or solution (lines 4 and 6 of Table 5.19).

Coatings of Metals and Alloys Deposits of molybdenum, bronze, and of alloys rich in nickel and iron sprayed using flame, arc, and plasma have been traditionally applied to protect the surfaces against adhesive wear [39]. The coatings of Mo agglomerated with MoO_2 powder sprayed using *APS* were tested against steel by Houck and Whisenant [148]. Similarly, Habig *et al.* [149] tested wear resistance of Mo coatings against steel type X155CrVMo121H using the same spray technique. Some alloys, e.g. including tin with copper and antimony, shown in line 7 of Table 5.19, are useful as antifriction coatings in hydrodynamic bearings.

Composite Coatings The composite deposits with tungsten carbide reinforcement, shown in lines 8 and 9 of Table 5.19, are very frequently used against adhesive wear. In fact, the cheapest and simplest way of manufacturing the adhesive-wear-resistant coatings is based on flame spraying of *self-fluxing* alloys as matrixes and carbides or oxides as reinforcement. Such coatings can be also post-spray remelted [39]. An example of such matrix is an alloy NiCrBSiFe tested by Sandt and Krey [150]. Technology of the composite coating by thermal spraying, proposed by Lugscheider [151], consisted of simultaneous use of two torches. The plasma torch was used to spray the refractory reinforcement and the flame gun to spray the *self-fluxing* alloy. Finally, Sampath and Wayne [152] proposed the use of metal as matrix with its carbide as reinforcement. The described composite of Mo with Mo_2C was prepared as agglomerated powder used by plasma spraying.

Table 5.19 Examples of granular coatings applied against adhesive wear.

		Feedstock			Fiction and wear					
					Test parameters				Test results	
Number	References	Chemical composition, wt. %	Particles size, μm	Deposition method	Test method	Counter body	Velocity, cm/s	Force, N	Coefficient of friction	Wear, $mm^3/(Nm)$
Ceramics										
1.	[137]	Cr_2O_3	—	*APS*	Block-on-ring	—	42	400	0.23 – 0.3	$\sim 6 \times 10^{-7}$
2.	[138]	TiO_2	5–45		Ball-on-disc (ASTM G99)	Si_3N_4	5	10.4	—	6.8×10^{-6}
3.	[139]	Al_2O_3	5–25		Pin-on-disc	Al_2O_3	3	1–6	—	$3.5–5.2 \times 10^{-4}$
4.	[140, 141]	Al_2O_3 in isopropanol	0.15	*SPS*	Ball-on-disc		20	5	—	5×10^{-5}
5.	[142]	$Cr_2O_3 + 25\ TiO_2$	15–45	*APS*	Ball-on-disc (ASTM G99)		—		0.75	1.5×10^{-4}
6.	[143]	SiC + 50 YAG in polyacrylic acid	YAG –precursor salts, SiC with $d_{50} \approx 3$	Suspension and solution spraying using *HVOF*			6.3	10	0.9	1.0×10^{-6}
Metals and alloys										
7.	[144]	Sn + 6 Sb + 2 Cu[a]	Wires	*FS* and *AS*	Ball-on-disc (ASTM G99)	Al_2O_3	10	5	~0.5	—
Composites										
8.	[145]	WC + 10 Co	0.3–2	*HVOF*	Ball-on-disc	Si_3N_4	3	9.8	—	0.55×10^{-6}
9.	[146]	WC + 15 FeCrAl	15–45		Ball-on-disc (ASTM G99)	Al_2O_3	10	20	0.4	0.53×10^{-6}

[a] Babbit alloy ASTM B23/2.

Two- and Three-bodies' Abrasion

Two- and three-body abrasion wear resistance will be discussed together. The more recent studies are concentrated mainly on ceramic and cermet coatings and on *HVOF* and *HVAF* spray techniques. The section shows also some studies made on metals and alloys.

Ceramic Coatings Barbezat *et al.* [23] tested air plasma and vacuum plasma sprayed oxide deposits against 2-body and 3-body abrasion. Among the coatings sprayed by plasma in air those of Al_2O_3 and of Cr_2O_3 seem to have highest abrasion resistance comparing with other oxides (lines 1–2 of Table 5.20). An addition of 40 wt. % ZrO_2 to Al_2O_3 tested by the authors results in slight improvement of 2-body abrasion wear resistance. This addition could have resulted in well-known effect of preventing of crack progression by phase transformation of ZrO_2. This effect is known as ***transformation toughening***. Similar effect has an addition of 10 wt. % ZrO_2 to Cr_2O_3 shown in line 3 of Table 5.20. On the contrary, the addition of 10–30 wt. % of TiO_2 to Cr_2O_3 and to Al_2O_3 does not improve the 2-body abrasion resistance of air plasma sprayed coating [23]. The post-spray treatment may influence the abrasion resistance. In particular, the sealing by epoxy resin and by phosphoric acid did improve wear resistance of Al_2O_3 + 13 wt. % TiO_2 plasma sprayed coatings as found by Liscano *et al.* [155] and shown in Fig. 5.12.

Composite Coatings The composites obtained by thermal spraying reinforced by carbides such as WC or Cr_2C_3 are shown in lines 4 and 5 of Table 5.20. Also TiC carbides in Co matrix have been tested against 3-body abrasion. The reinforcement of oxides such as Al_2O_3 and Cr_2O_3 in *self-fluxing* alloy matrix was *FS* deposited and post spray melted and tested against 2-body abrasion [150]. The microscopic effects of 2-body abrasion occurring at the surface of composite coatings were studied by Barbezat *et al.* [156] and are shown in Fig. 5.13. These effects depend on reinforcements' type (particles, whiskers, fibers), size, and volume fraction. Theoretically, the best resistance against 2-body wear abrasion would have a composite with medium size and a large volume fraction of reinforcement (see Fig. 5.13c). Composites with a small volume fraction of the reinforcement would have the matrix worn out quickly upon abrasion (Fig. 5.13a,d). Similarly, a composite with a reinforcement particle size that is too large shown in Fig. 5.13b may have these particles destroyed during abrasion. The experimental study made by Usmani *et al.* [157] of *HVOF* coating of composite WC + 17 wt. % Co enabled to conclude that the best 2-body abrasion resistance has WC grains of the size 5–7 μm. The authors indicated that small carbide reinforcements, having mean size of about 1 and 4 μm, may decompose to W_2C or W more intensively at spraying. Such decomposition decreases their abrasion resistance. Finally, the authors suggested that the abrasion resistance correlates well with coatings' toughness. Other opinion presented Ishikawa *et al.* [158] who suggest rather a correlation of abrasion resistance of *HVOF* sprayed WC + 20 wt. % Cr_3C_2 + 7 wt. % Ni powder with coating's microhardness.

Khan and Clyne [159] tested 3-body abrasion resistance of plasma sprayed WC-Co composite coatings. The authors found that high-volume fraction of WC reinforcement is more advantageous against such wear by comparing coatings 9 wt. % Co with that including 17 wt. % Co.

Table 5.20 Properties of some granular deposits tested against 2-body and 3-body abrasion.

		Powder			Abrasion					
Number	References	Chemical composition, wt. %	Size, μm	Deposition method	Kind of abrasion	Test method	Abrasive body	Velocity, cm/s	Force, N	Wear rate, mm^3/(Nm)/mg/m
Oxides										
1.	[23]	Al_2O_3	—	*APS*	2-body abrasion	See Fig. 4.23a	Al_2O_3, SiO_2 and SiC with 89 μm grains	6	10–50	/0.8–1.5
2.		Cr_2O_3								/0.7–1.3
3.	[142]	$Cr_2O_3 + 10\ ZrO_2$	$-36 + 10$		3-body abrasion	See Fig. 4.23b	SiO_2 size 0.1–0.5 mm	164	22	3×10^{-4}/
Composites										
4.	[153]	WC + 10Co4Cr	$-30 + 5$	*HVAF*	3-body abrasion	See Fig. 4.23b	SiO_2 size 0.1–0.5 mm	164	22	9.2×10^{-6}/
5.		$Cr_3C_2 + 37WC + 18NiCoCr$	$-30 + 10$							1.2×10^{-5}/
Metals										
6.	[154]	Mo	—	*APS*	2-body abrasion	See Fig. 4.23a	Suspension of Al_2O_3	—	230	5.7×10^{-2}/[a]

[a] Measured after 50 rotations of disc.

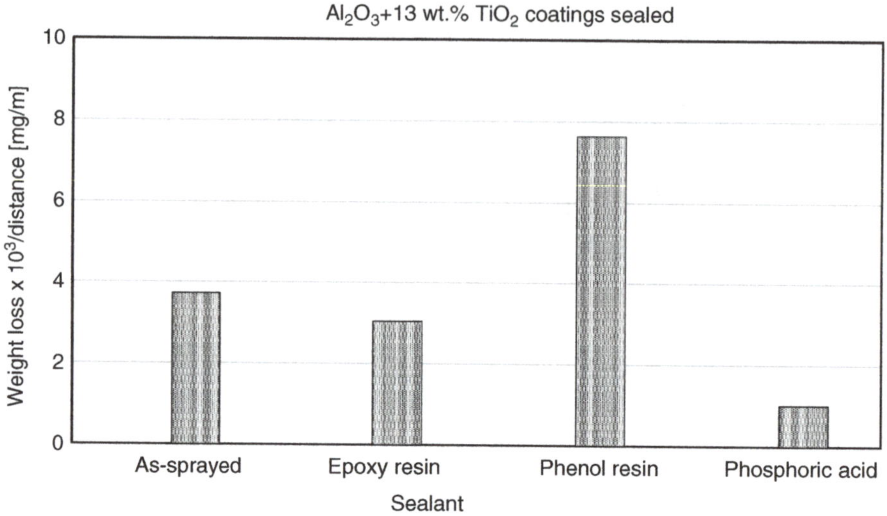

Figure 5.12 3-body abrasion wear test made following ASTM G65-00 norm for air plasma sprayed $Al_2O_3 + 13\,wt.\,\%TiO_2$ coatings treated after deposition using different sealants by Liscano *et al.* [155]. Reproduced by permission.

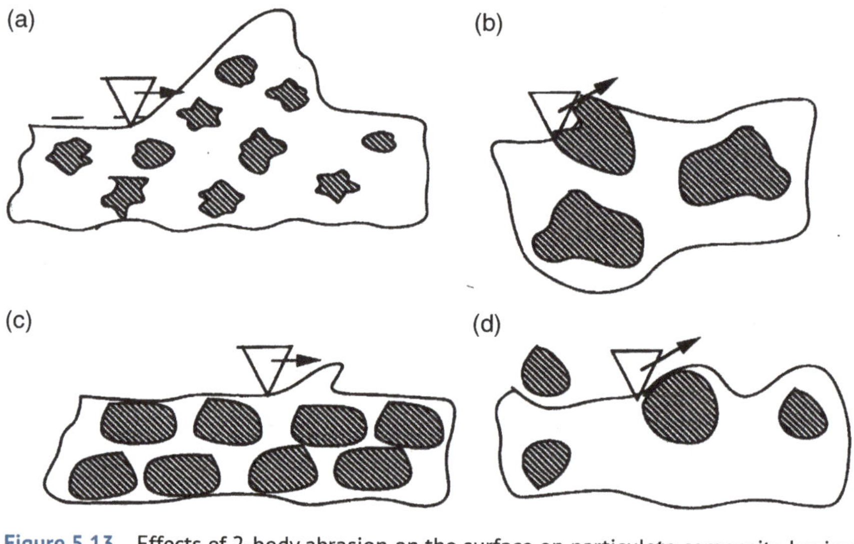

Figure 5.13 Effects of 2-body abrasion on the surface on particulate composite having reinforcement of: (a) small size and large volume fraction; (b) large size and large volume fraction; (c) medium size and large volume fraction; and (d) large size and small volume fraction. After Barbezat *et al.* [156].

Finally, the composite coatings including 20 and 50 wt. % of Cr_3C_2 reinforcement in NiCr matrix were successfully tested against 3-body abrasion [160, 161]

Coatings of Metals and Alloys The molybdenum coatings plasma sprayed using *APS* were tested against wear by 2-body abrasion in the study [154]. The results, summarized in line 6 of Table 5.20, show the wear rate is much greater than that of ceramic and composite coatings. The alloy coatings having composition of Cr + 16 wt. % Mn + 3 wt. % Si + 1 wt. % V obtained by *FS* was successfully tested in study [162] to resist well against 3-body abrasion.

Erosion

Erosion depends on the eroding powder, namely on its: (i) nature, hardness, size, and morphology; (ii) velocity and angle of impact. The medium in which the powder is suspended may play a role in corroding the worn surface. Water is an example of such medium, and ***hydroerosion*** occurs in many hydraulic machines such as slurry pumps [163]. The properties of eroded materials such as hardness, toughness, modulus of elasticity are also important, and some authors, such as Lansdown and Price [164], present a simplified guide, whose materials are recommendable for different velocities and angles of eroding particles. As thermally sprayed coatings having lamellar microstructure are frequently applied against abrasion wear, the selection of materials to spray is related to the mechanism of erosion and includes [165]: (i) porosity and hardness of coatings; (ii) optimal microstructure which corresponds to that of cermets.

Ceramic Coatings The ceramic coatings of Cr_2O_3 and Al_2O_3 and of Cr_2C_3 carbide were tested against erosion resulting from action of ceramic powder as water slurry and of suspension in compressed air. The results are shown in Table 5.21 (lines 1–3). The erosion wear is greater at smaller angle of test and at higher temperatures. The resistance against erosion is correlated with bonding of ceramic lamellas, and Li *et al.* [167] recommend the optimization of the operational spray parameters enhancing this bonding.

An application of Cr_3C_2 carbide sprayed using *HVOF* technique is a rare example of deposition of pure carbide. Low porosity (lower than 2%) and high hardness (*HV* is reportedly between 8 and 9 GPa) result in high abrasion resistance of the coatings [168]. The studies of vacuum plasma sprayed alumina made by Kingswell *et al.* [171] enabled to find the cracking as the mechanism of abrasion at the angle close to 90°. Takeuchi *et al.* [172] found out that this oxide sprayed using *VPS* results in coating having low porosity which promotes the abrasion resistance.

Composite Coatings The wear resistance against slurry of the WC-CoCr composite, shown in line 4 of Table 5.21, was better than that of Cr_2O_3 coating, shown in line 1 of this table and, of another composite WC + 12Co, tested by Karimi *et al.* [166]. The authors explain the possible reasons of this by:

- toughening of matrix by the presence of Cr improving the binding of carbide particles (as in Fig. 5.13b) and improving its corrosion resistance;
- optimal volume fraction of the matrix with regard to the reinforcement (as in Fig. 5.13c) improving the distribution of carbides particles.

Table 5.21 Properties of granular coatings tested against erosion wear.

		Deposition			Erodent			Erosion		
Number	References	Powder composition, wt. %	Particles size, μm	Method	Material/particles' sizes, μm	Angle at impact,°	Velocity at impact, m/s	Method of test	Temperature, K	Wear (different units)
Ceramic coatings										
1.	[166]	Cr_2O_3	−45 + 22	*APS*	Sand/−150 + 20	small	90	Water slurry erosion	—	1.7 mm^3 after 1 h testing
2.	[167]	Al_2O_3	−44 + 10		$Al_2O_3/d_{50} = 330$	90	63	Test in compressed air	*RT*	0.75×10^{-3} g/g (worn mass/mass of erodent)
3.	[168]	Cr_2C_3	—	*HVOF*	SiO_2 and Al_2O_3/740	30 and 90	2.5	Test in compressed air		1 μm and 0 μm of worn thickness
									1023-air	2 and 1 μm worn thickness
Composite coatings										
4.	[166]	WC + 14 CoCr	−45 + 20	*HVOF*	Sand/−150 + 20	small	90	Water slurry erosion	—	1.3 mm^3 after 1 h testing
5.	[168]	Cr_3C_2 + 25NiCr	—		SiO_2 and Al_2O_3/740	30 and 90	2.5	Test in compressed air	*RT*	6 μm and 0 μm (see line 3)
									1023-air	8 and 4 (see line 3)
6.	[169]	WC + 14 Cr_3C_2 + 7 Ni	$d_{50} = 37$		$Al_2O_3/d_{50} = 50$	30	—	ASTM G76 (hot air)	773	0.29×10^{-3} g/g (see line 2)
				APS						0.39×10^{-3} g/g
Alloy coatings										
7.	[168]	Fe + 25Cr + 3B + 2Si	wire	*AS*	SiO_2 and Al_2O_3/740	30 and 90	2.5	Test in compressed air	*RT*	10 and 3
8.	[170]	Ni + 20 Cr	−53 + 22	*HVOF*	$Al_2O_3/d_{50} = 175$	30 and 90	50	ASTM G76 (hot air)	*RT*	1.3×10^{-5} and 1.2×10^{-5} cm^3/g (worn volume/mass of erodent)

In addition to tungsten carbide, chromium carbide in the NiCr matrix sprayed by *HVOF* was tested against wear by erosion (line 5 of Table 5.21). Their resistance was higher for the angle of impact of 90° than that of 30°. This resistance is lower at higher temperatures of testing. The high temperature resulted in the implantation of erodent particles observed using *EDX* technique by the Wang and Lee [168]. The coatings sprayed powder using two carbides, namely 79 wt. % of WC and 14 wt. wt. % of Cr_3C_2 and 7 wt. % of Ni, were less porous and harder when *HVOF* technique was used when compared to the *APS* one [169]. Consequently, the hot air abrasion resistance against alumina erodent was clearly greater for the high velocity oxy-fuel sprayed deposit as seen in line 6 of Table 5.21. The composite coatings having a ductile matrix and hard reinforcement were proved to be useful against hot erosion. Consequently, the powder of alloy composed in wt. % of Co + 34Ni + 22Cr + 7.7Al + 0.7Y was mixed with Al_2O_3 to be vacuum plasma sprayed, and the obtained coatings were tested using ash at $T = 970$ K. It turned out that deposits with a volume fraction of reinforcement 40–60% had the greatest wear resistance [173]. Finally, Wang *et al.* [174] pointed out that the similarity of the moduli of elasticity of the matrix and the reinforcement improve the resistance of the composite to hot erosion.

Alloy Deposits The alloy in wt. % of Fe + 25 Cr + 3B + 2 Si called ***Armacor M*** was arc sprayed and tested against erosion at ambient (line 7 of Table 5.21) and high temperatures. The coatings had lower erosion resistance than that of Cr_2C_3 carbide and of its composite with NiCr and tested against silica with alumina sand. The alloy of Ni + 20 wt. % Cr, deposited using *APS* and *HVOF* (line 8 of Table 5.21), was also tested against erosion using alumina sand. The tests showed that the erosion rate was greater at angle 30° than at 90°. An interesting result was the improvement of wear resistance at higher temperature (up to 1073 K) of more porous coatings obtained by plasma spraying.

Fretting and Impact Wear

The plasma sprayed coatings including oxides as chromia with addition of silica were successfully tested against fretting wear [175]. The addition of boron carbide to chromia and titania improved wear resistance and lowered the coefficient of friction of ceramic coatings (lines 1 and 2 of Table 5.22). The aluminum bronze coatings sprayed using *HVOF* were proved to useful against fretting at temperature of 473 K. The friction coefficient and wear resistance were dependent upon frequency of testing as shown in line 3 of Table 5.22. Another alloy of copper, that with nickel, was optimized by addition of 4 wt. % of boron (line 4 of Table 5.22). This addition reduced the inflight oxidation of droplet at spraying leading to better bonding of splats in coating and, consequently, to the improvement of fretting wear resistance. Finally, the tests of fretting wear made at temperature as high as 1339 K proved the usefulness of composite coating sprayed using *D-gun*. The composite had matrix composed of alloy in wt. % of Co + 25Cr + 10Ta + 7Al + 0.5Y + 2C and the reinforcement of 10 wt. % Al_2O_3 [178].

Table 5.22 Examples of granular coatings applied against fretting wear.

		Deposition			Fretting wear				
Number	References	Powder composition, wt. %	Particles size, μm	Method	Counterbody/ diametermm	Load (N)/ amplitude of movement, mm	Frequency, Hz/ time of testing, s	Temperature of test, K	Wear
Ceramic coatings									
1.	[175]	$Cr_2O_3 + 5SiO_2 + 3TiO_2+$	—	*APS*	Sapphire/10	1/0.1	8/	*RT*	friction coefficient ≈ 0.5 after 10^5 cycles
2.		$B_4C + 42Cr_2O_3 ++28TiO_2$							friction coefficient ≈ 0.2 after 10^5 cycles
Alloys coatings									
3.	[176]	Cu + 10Al + 1Fe	−48 + 17	*HVOF*	Steel GCr15/10	10/0.05	5–20/1250	473	Friction coefficient 0.65 up to 0.93; Wear rate 0.3×10^{-9} up to 24.1×10^{-9} $mm^3/(Nm)$
4.	[177]	Cu + 31Ni + 3In + 3B	−50 + 30	*APS*	Ti + 5Al + 4V/10	100/0.1	10/ 2×10^5	*RT*	Friction coefficient 0.8; worn volume 12.5×10^{-3} mm^3

5.2.4.3 Bulk Coatings

The bulk coatings, tested against different kinds of wear, use powder feedstock similar to that obtained by the granular methods of deposition. The processes used applied to obtain the wear resistant coatings include plasma transferred arc (*PTA*)[4] and laser treatments in one-step and two-steps laser deposition, i.e. *1SLD* and *2SLD*. The coatings are thick and used in mainly metallic alloys and their composites with ceramic reinforcement. The bulk coatings are applied against all types of wear.

Friction and Adhesive Wear

The bulk coatings applied against adhesive wear are alloys and composites. This deposition method is not useful for ceramics on metal substrates because of the significant cracking of their surfaces. Some examples of the tested deposits are shown in Table 5.23.

Cobalt-based alloys such as *Stellite 6* are corrosion resistant and preserve their mechanical properties at elevated temperatures. The *PTA*-deposited coatings of this alloys were tested against AISI 01 steel and against similar coating in temperatures up to 473 K [183]. The authors found that the wear increases with sliding distance, temperature at the test, and with load applied to the bodies in contact. Their relative speed at testing did not have a clear effect on wear. Similar Co-alloy, namely *Stellite 21,* was tested against alumina in *RT,* and the details are shown in line 1 of Table 5.23 and in temperature 873 K. The test at high temperature showed the reduction of wear resulting from the formation of oxide layer on the surface of tested coating. The Ni-based alloy, *Colmonoy 5* coating, deposited by *PTA* was tested at *RT* (line 2 of Table 5.23) and at temperatures up to 823 K. The wear loss did not change a lot with sliding distance but was clearly smaller at higher temperatures. The *PTA* deposited coating of Fe-based alloy *Norem02* shown in line 3 of Table 5.23 had wear rate against WC-Co increasing considerably with the load applied to the counter body.

Finally, the *2SLD* deposition process including CO_2 laser treatment of air plasma sprayed WC-Co composite coating was tested against adhesive wear (see line 4 of Table 5.23). The wear resistance of laser-remelted coatings was favorably compared to that of as-sprayed ones. This improvement can be attributed to the increase of coatings' hardness and to the reduction of their porosity.

Three-body Abrasion The bulk coatings applied against *three-body abrasion* are discussed for alloys and composites coatings. Some examples of tested coatings are shown in Table 5.24. The cobalt and nickel-based alloys are used as the coatings materials (see lines 1 and 2 of Table 5.24) and as the reinforcements for the composite coatings (lines 3 and 4 of this table).

The wear resistance of composite coatings having reinforcement of tungsten carbides is two orders of magnitude greater than that of the nickel alloys used as matrices coatings as tested by Techel *et al.* [184] and shown in Fig. 5.14. Other carbides tested as the reinforcements of bulk coating composites are ZrC, TiC, and Cr_3C_2 (line 5 of Table 5.24). Some composite coatings, shown in Table 5.24 (lines 3–5), were tested also against

---

4 Some authors use the abbreviation *PTAW,* i.e. plasma transferred arc welding.

Table 5.23 Examples of bulk coatings applied against adhesive wear.

		Powder feedstock			Fiction and wear					
					Test parameters				Test results	
Number	References	Chemical composition, wt. %	Particles size, μm	Deposition method	Test method	Counter body	Velocity, cm/s	Load, N	Coefficient of friction	Wear
Alloys coatings										
1.	[179]	Co + 26.5Cr + 5.6Mo + 3.7Ni + 1.2Si + 1.4Fe + 0.2C, ***Stellite 21***	−100 + 50	*PTA* including heat treatment after deposition	Ball-on-disc, (ASTM G99)	Al_2O_3	25	20	0.4 after 1.5 km sliding	Weight loss of 16 mg after 1.5 km sliding at *RT*
2.	[180]	Ni + 12.6Cr + 4.6Fe + 4.0Si + 2.5 + 0.5C, ***Colmonoy 5***	—	*PTA*	Pin-on-disc (ASTM G99-05)	Stainless steel 316 L coated with *Colmonoy 5*[a]	25	60	0.4 after 1.6 km sliding at *RT*	Wear rate of 3.01 g/(Nm) at *RT*
3.	[181]	Fe + 24.8Cr + 3.3Ni + 2.1Mo + 3Si + 1.2C, ***Norem 02***	—		Pin-on-disc (sliding distance 100 m)	WC-Co	5	2, 5, 10	—	Wear rates of 7.8×10^{-6}, 79×10^{-6} and 440×10^{-6} mm^3/m
Composite coatings										
4.	[182]	WC + 17Co	−53 + 11	*2SLD: APS* followed by remelting using CO_2 *cw* laser	Block-on-ring	Steel AISI 1043	26	333	—	Wear volume 1 mm^3 after sliding distance of 6.3 km

[a] Probably the same coatings were on pin and on disc.

Table 5.24 Examples of bulk coatings applied against wear resulting from *three-body abrasion*.

		Powder			Abrasion					
Number	References	Chemical composition, wt. %	Size, μm	Deposition method	Kind of abrasion	Test method	Abrasive body	Velocity, *v*, cm/s	Force, *F*, N	Wear
Alloys coatings										
1.	[184]	Ni + 1.3B + 2.3Si	−125 + 45	*1SLD* using CO_2 *cw* laser	3-body abrasion	Pin-on disc	SiC	—	—	Wear rate $w = 0.140\,mm^3/m$
2.	[185]	Co + 29Cr + 4.5W + 3Ni + 3Fe + 1.5Mo + 1.5Si + 1.2C, *Stellite 6*	−180 + 53	*PTA*	3-body abrasion	Rubber-wheel, ASTM 65	Sand		130	Volume loss $20.5\,mm^3$ after 2000 disc revolutions
Composite coatings										
3.	[186]	ZrC + 10Ni	−310 + 150	*PTA*	3-body abrasion	Steel-wheel, ASTM 65	Silica sand	100	—	Wear rate $w = 0.047\,mm^3/m$ at *RT*
4.	[187]	TiC + 20(66Ni + 33 Mo)	−310 + 150			Cyclic impact abrasion	SiO_2	—	0.8	Volume loss $5\,mm^3$ at *RT* after 7200 impacts
5.		Cr_3C_2 + 20Ni								Volume loss $6.5\,mm^3$ in the same conditions

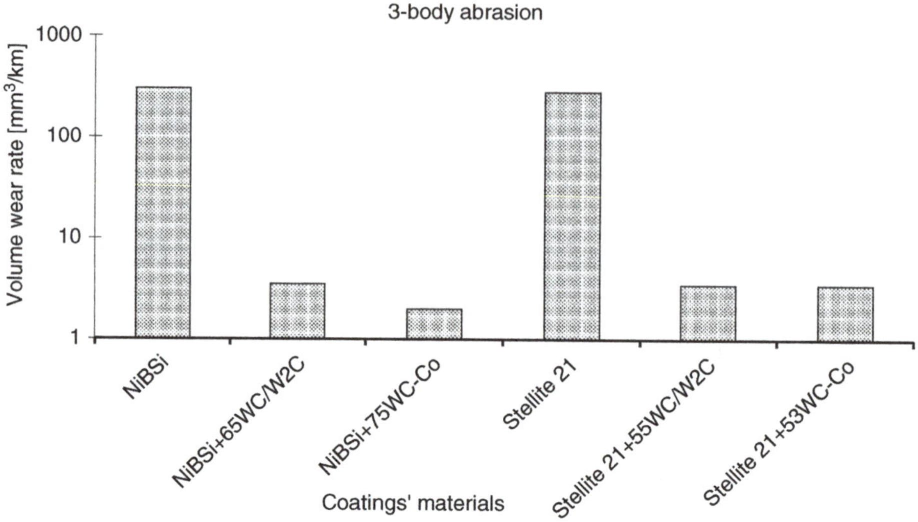

Figure 5.14 *3-body abrasion* wear test of alloys and composites coatings obtained by *1SLD* laser cladding [184]. The composites have a matrix of alloys such as NiBSi or *Stellite 21* alloy and different contents of WC-Co or WC/WC_2 reinforcements.

abrasion at high temperatures, ranging up to 974 K. The wear rate at high temperatures decreases, resulting from the oxidation of matrix and its densification. At high temperatures the reinforcements of TiC and ZrC are more resistant against oxidation than WC/W_2C [186].

Erosion The resistance against erosion wear was tested for bulk coatings of, e.g. (i) alloys Ni-base, Co-base (*Stellite 6*), and of Fe-base (Hadfield steel) and (ii) metal matrix composite of self-fluxing NiCrBSi alloy with WC reinforcement and of tool steel *H13* with TiC reinforcement. The results of the tests are shown in Table 5.25.

The erosion wear for the alloys shown in lines 1–3 of Table 5.25 shows that the wear rate depends slightly on the impact angle at the erosion test between 30° and 90° (perpendicular to the coating's surface). The rate was more depending on the impact angle for the *MMC* composite of *H13* steel with TiC (line 5 of Table 5.25). However, this erosion test was carried out using slurry. Finally, the erosion rate of the composite self-fluxing alloy with WC reinforcement was strongly dependent on the velocity, ranging from $v = 30$ to 70 m/s, of impacting quartz sand particles as shown in line 4 of Table 5.25.

In general, the erosion rate as a function of angle of impact indicates the wear mechanism. Consequently, high erosion rate for:

- low angles of impact indicate mechanism typical or ductile materials, i.e. the detachment of the parts of the coating surface cut by the eroding particles;
- high angles of impact indicate the typical mechanism for brittle materials, i.e. the development of micro-cracks followed by their detachment.

The wear mechanism for the alloy coatings shown in Table 5.25 indicates rather the mechanism typical for ductile materials, i.e. high erosion rate for low-impact angles.

Table 5.25 Examples of bulk coatings applied against wear resulting from erosion.

		Deposition			Erodent			Erosion test		
Number	References	Powder composition, wt. %	Particles size, μm	Method	Material/particles' sizes, μm	Angle[a] at impact,°	Velocity at impact, m/s	Method of test	Temperature, K	Wear rate[b]
Alloys coatings										
1.	[188]	Ni + 1.5Fe + 3Si + 1.4B + 0.1C	—	*PTA*	Al_2O_3/125	30° to 90°	75	Sand blasting	*RT*	Greatest wear rate for 30° = 0.4 mg/g, lowest for 70° = 0.25 mg/g
2.		*Stellite 6*	—	*TIG*						Greatest wear rate for 50° = 0.45 mg/g, lowest for 70° = 0.35 mg/g
3.	[189]	Fe + 12Mn + 1.2C	−100 + 40	*1SLD* using CO_2 *cw* laser	Sand/ d_{50} = 250		100	Bench test		Greatest wear rate for 60° = 0.18 mg/g, lowest for 90° = 0.11 mg/g
Composite coatings										
4.	[190]	NiCrBSi + 35 WC	—	*FS* followed by furnace heating	SiO_2/−180 + 150	30°	30; 50; 70	Rotating erosion machine	*RT*	Wear rate for: 30 m/s = 0.3×10^{-4} g/g; for 50 m/s = 0.6×10^{-4} g/g; for 70 m/s 0.9×10^{-4} g/g
5.	[191]	Matrix *H13* tool steel: (Fe + 5.2Ti + 4.3Cr + 1.5Mo + 0.8Mn + 1.9C + 0.2Si) + 40 vol. % of reinforcement TiC	Steel and TiC: −100 + 50	*1SLD* using Nd: YAG *cw* laser	SiO_2 slurry with water/−850 + 425	30° to 90°	25	*Coriolis* test		Greatest wear rate for 45° = 11 mg/min; lowest for 30° = 5 mg/min

[a] Angle with regard to coating's surface.
[b] Wear defined in mass worn/mass of erodent or mass worn/test time.

Contrarily, the results of erosion testing for metal matrix composites indicated the mechanism typical for brittle materials, i.e. high erosion rate for high-impact angles. It must be underlined that the mechanism of wear may change with the modification of the coating's microstructure at deposition (homogenization, dissolution of oxides in liquid materials, formation of an amorphous structure, etc.).

Fretting Pelletier *et al.* [189] tested Hadfield steel coating obtained by laser hardfacing on low-carbon steel against wear by fretting. The authors observed the evolution of the coefficient of friction against with the number of cycles (see line 1 of Table 5.26) and explained it by formation of debris, hardening of the zone in contact with the testing sphere. The value of friction coefficient indicated the modification of the contact zone from being more ductile to become more brittle. *PTA* method enabled the syntheses of TiC carbide reinforcement in the bath of plain-carbon steel by injection of separate powders including Ti and C (line 2 of Table 5.26). The synthesized carbides solidified in the metallic matrix of Fe and Ni and considerably improved the fretting wear resistance of clad coating comparing to the initial substrate. Another composite including Co-based alloy as matrix and WC as reinforcement with addition of CaF_2 to lower the coefficient of friction is shown in line 3 of Table 5.26. This composite coating, synthesized using fiber laser, was tested against fretting wear in *RT* and in the temperature of 773 K and showed much better resistance against wear and lower coefficient of friction than stainless steel substrate.

5.2.5 Example of Coatings Applications: Anilox Rolls

The mechanical properties of coatings are important in many industrial applications. The ***anilox rolls*** used in printing industry are an example. These anilox rolls are used in the machines working with the flexographic system and should be resistant against adhesive wear generated by contact with doctor blade shown in Fig. 5.15.

The *anilox roll* transports a precisely defined quantity of ink from the container to printing the roll. The ink quantity is controlled by depth and density of the cell engraved using laser on the plasma sprayed coating the roll surface shown in Fig. 5.15. At present, chromium oxide coating by *APS* is mainly applied. The market of ceramic *anilox rolls* includes [39]:

- rolls for new flexographic machines;
- replacements of traditional rolls having metallic coatings obtained using galvanic methods;
- replacements of used ceramic anilox rolls.

The new developments in the technology should be firstly related to the replacement of chromium oxide by more environmental and worker-friendly titania and alumina–titania alloys or tungsten carbides [195, 196]. On the other hand, the possible progress in laser engraving technology is an application of solid-state Nd: YAG laser for engraving of fine cells [197].

Table 5.26 Examples of bulk coatings applied against wear resulting from fretting.

		Deposition			Fretting wear				
Number	References	Powder composition, wt. %	Particles size, μm	Method	Counterbody/ diameter, mm	Load, N/ amplitude of movement, mm	Frequency, Hz/time of testing, s	Temperature of test, K	Wear
Alloy coatings									
1.	[189]	Fe + 12Mn + 1.2C	−100 + 40	*1SLD* using CO_2 *cw* laser	Steel 100 C6	300/0.03	1	*RT*	Friction coefficient≈0.7 after 10^5 cycles
Composite coatings									
2.	[192]	Matrix in at.% (Fe + 36Ni + 2.3 Ti) + 60 vol. % TiC	−150 + 70	*PTA* alloy powder of (wt. %) Ti + 30Ni + 14C injected	Steel GCr15/10	20/2	25/1800	*RT*	Wear mass loss 1.8 mg; substrate carbon steel without coating had mass loss of 35 mg
3.	[193]	Matrix: 85(Co + 28Cr + 5W + 2Ni + 1Fe + 1C + 1Si) + reinforcements: 10WC + $5CaF_2$	Co-alloy and CaF_2:−80 + 50; WC:−90 + 50	*1SLD* using Ytterbium fiber laser	Steel 316/9.5	60/0.04	30/	*RT* and 773K	Wear volume at $RT = 0.3 \times 10^6\ \mu m^3$ and at $773\ K = 0.2 \times 10^6\ \mu m^3$

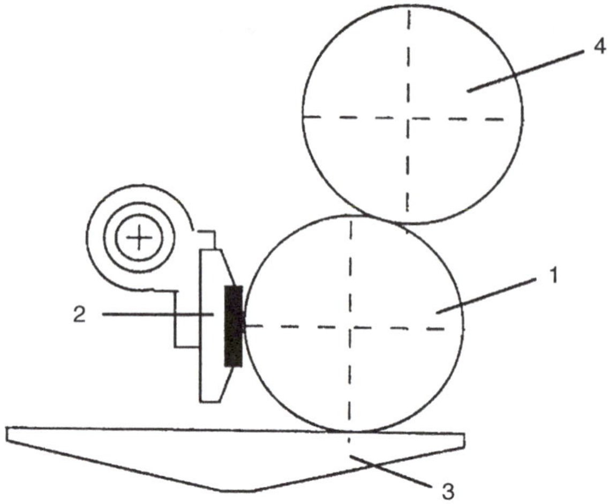

Figure 5.15 Part of a flexographic printing machine [194], 1, anilox roll; 2, doctor blade; 3, ink container; 4, printing roll.

5.3 Electric and Magnetic Properties

The films and coatings obtained using different methods have been applied since 70 years in the electrical and electronic industry, and there are many sources that describe the electrical and magnetic properties of thin films and coatings, e.g. [198–200]. Above all, the development of integrated circuits has been a main driving force in the study of physical deposits. Therefore, the electrical and magnetic properties of atomistic deposits are described in this work in general way. The granular deposits started to be tested for electrical and magnetic applications since the end of the 1960s [201]. In the beginning, the thermal spray methods could have become a valid alternative to *screen-printing*. However, the methods of thermal spraying were probably more expensive and more complicated, and *screen-printing* became the favorite method of thick-film deposition in electronic applications. The electrical and magnetic properties of granular coatings were thoroughly reviewed by Pawłowski [39]. Finally, it should be mentioned that thickness influences the electrical properties less than the mechanical ones. This results from small size of electrons, ions, and holes being the carriers of electric energy. Finally, the technologies of bulk coating are seldom used for the applications related to the electrical or magnetic properties and will not be reviewed in the following chapter.

5.3.1 Conductors and Resistors

Conductors and resistors are characterized by their resistance, R, depending on the resistivity of material, ρ, and on the geometrical dimensions of the elements as seen in Eq. 4.28. Resistance is one of the physical units, which varies on a very wide range of 10^{25} [202]. The resistivity of materials having free electrons can be separated into two parts (see also Eq. 5.2): (i) intrinsic resistivity, ρ_I, depending on the microstructural element such as dislocations, impurities, contact between grains and films thickness for very thin films[5]; and (ii) thermal resistivity, ρ_T, which depends on temperature. This separation of the resistivities is known also as ***Matthiessen's law***:

$$\rho = \rho_I + \rho_T \tag{5.2}$$

The thermal component is equal to $\rho_T = 0\,\mu\Omega\text{cm}$ at the temperature $T = 0\,\text{K}$. At higher temperatures the thermal coefficient of resistance (*TCR*) should be considered. The coefficient varies with temperature [202, 203].

Transport of electric energy is different in:

- metals and alloys in which free electrons transport electric current and at higher temperatures the lattice vibration leads to the reduction of their movement and to increase of the resistivity component ρ_T;
- semiconductors, which have a ***valence band*** and the electrons should have some excitation energy to jump in a ***conduction band*** enabling movement and transport of electric energy. The band gap is narrow (few eV) and excited electrons leave a positively charged hole being also able to transport electric current;
- ions conducting solids in which the ions can move and transport electric energy.

5.3.1.1 Atomistic Deposits

The mean free path of electrons influences the resistance of metals and alloys in atomistic deposits having rather small thickness. The typical values of this path accompanied by the values of ρ and *TCR* for some pure metals are shown in Table 5.27.

The techniques used to obtain the conductive films are mainly thermal evaporation and sputtering. Deposit materials belong to three main groups [202]:

- single metals, such as that shown in Table 5.27;
- binary alloys, such as Cr-Ag, Cr-Au, Mo-Au;
- ternary alloys, such as Ti-Pt-Au, Ti-Pd-Au, Cr-Ag-Au.

Table 5.27 Mean free path of electrons and different temperatures, resistivities, and thermal coefficients of resistivity for some bulk pure metals [204].

		Mean free path, nm			ρ at	*TCR* at
Number	Metal	$T = 73\,\text{K}$	$T = 273\,\text{K}$	$T = 373\,\text{K}$	$T = 300\,\text{K}$, $\mu\,\Omega$.cm	$273 \leqslant T \leqslant 373\,\text{K}$, ppm/K
1.	Cu	296.5	42.1	29.4	1.69	4330
2.	Ag	242.5	57.5	40.5	1.47	4100
3.	Au	153	40.6	29.0	2.44	4000
4.	Ni	—	13.3	8.0	7.24	6750
5.	Pt	72	11.0	7.9	9.83	3902

5 Some authors such as Richardt and Durand [203] add the term of resistivity depending on method of preparation.

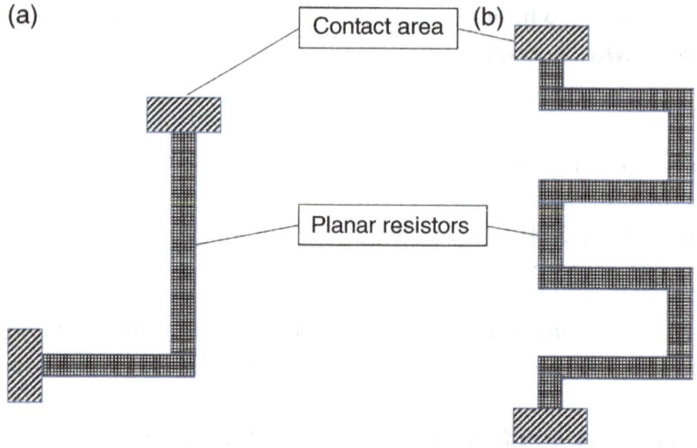

Figure 5.16 Planar resistors with: (a) L-shape and (b) shape of meander.

However, highly conductive films can be also obtained using other techniques and other classes of materials. An example is an IrO_2 deposit having thickness in the range 0.25–0.65 μm obtained by KrF excimer pulsed laser ablation (*PLD*) using an iridium target in the oxygen atmosphere. The obtained film resistivity depended on the temperature of substrate at deposition and was measured to be equal at *RT* to $\rho = 39\,\mu\Omega$.cm [205].

The thin-film resistors are obtained generally using the techniques of thermal evaporation, sputtering, and *PECVD* [202, 206]. The resistance depends not only on the material used, i.e. on the resistivity but also on its geometry (see Eq. 4.26). Consequently, the planar resistors may be L- shaped or have a shape of meander as shown in Fig. 5.16. The shape of planar resistor can be also laser trimmed to reach the desired value of resistance.

The materials used for thin-film resistors can be divided into three groups [202, 203, 206]:

- **Pure metals**, such as e.g. tantalum or titanium. These metals, as bulk materials, have at *RT*, the resistivity of [207]: (i) Ta, $\rho = 13.1\,\mu\Omega$.cm and; (ii) Ti $\rho = 39\,\mu\Omega$.cm. However, the deposition of these metals by reactive sputtering in nitrogen enables to increase the resistivity of tantalum up to $\rho = 250\,\mu\Omega$.cm forming the tantalum nitrides above partial pressure higher than $p = 6.7$ mPa. *TCR* is close to 0 ppm/K at this pressure and becomes negative at higher partial pressures [208]. Similarly, TiN_x films prepared using reactive sputtering had the resistivity in the range $\rho = 80$–$300\,\mu\Omega$.cm at *RT* depending on nitrogen flow [209]. The authors were able to increase the resistivity up to $\rho = 4\times10^4$ μΩ.cm by changing the angle of incidence of species deposited on the substrate. This angle is equal to 0° at the geometry shown in Fig. 2.5 in which the substrate is above the target.
- **Metallic alloys**, such as NiCr. The deposits are obtained by flash evaporation or by sputtering because of the difference in vapor pressure between Ni and Cr. The alloy composition is frequently Ni with 20 wt % of Cr, and operational parameters of the deposition make it possible to obtain the films having resistivity $\rho = 100$–$130\,\mu\Omega$.cm and the *TCR* = +/−100 ppm/K.

- **Cermets**, i.e. composites of metals with ceramics. An example is a Cr-SiO cermet whose resistivity varies from $\rho = 100$ to $4000\,\mu\,\Omega$.cm, and the *CTR* depends on the composition and the heat treatment after deposition[6], being between −600 and +100 ppm/K. Other cermets, e.g. Au-SiO or Au-Ta_2O_5 or Pt-WO_3, were also mentioned to be useful [206].

5.3.1.2 Granular Deposits

The application of granular coatings in microelectric is an interesting perspective, and a lot of studies have been done. The review of the properties is categorized in function of carriers of electric energy.

Electronic Conductors

The conductors are mainly metals. Table 5.28 shows the resistivities of some air plasma sprayed metals tested by Smyth and Anderson [210].

The resistivity of air plasma sprayed copper coating decreases after heat treatment in H_2 at 1070 K from $\rho = 14$ to $7\,\mu\Omega$.cm [211]. Another possibility to avoid the post-spray treatment is the use of vacuum plasma spraying [212]. The copper coatings deposited in this way had the resistivity of $\rho = 2$–$5\,\mu\Omega$.cm and $TCR \approx 2600$–3100 ppm/K being close to that of bulk metal shown in line 1 of Table 5.27. Finally, low-pressure cold spray method can be also used to obtain copper coatings onto polymer substrates [213]. The measured resistivity of Cu coatings was considerably greater and equal to $\rho = 26\,\mu\Omega$.cm. The authors explained this value by high porosity of the coatings.

The borides were also thermal sprayed to obtain conducting coatings. Ananthapagmanabhan *et al.* [214] used *APS* method to obtain TiB_2 deposit with resistivity of $\rho = 10$–$70\,m\Omega$.cm. Knowing that the resistivity of the sintered material is $\rho = 30\,\mu\,\Omega$.cm, the authors explained the

Table 5.28 Resistivity of some air plasma sprayed metals [210].

Number	Metal (alloy)	ρ at $T = 300$ K, $\mu\,\Omega$.cm
1.	Al	25
2.	Cu	20
3.	Cr	180
4.	Ni	40
5.	Stainless steel	80
6.	Ta	200
7.	Sn	70
8.	Ti	170

6 The deposition was made most probably by sputtering.

increase of resistance by the oxidation of the boride particles during spraying. Another boride, ZrB_2, was sprayed by plasma under inert atmosphere (*IPS*), and the obtained coating had resistivity of $\rho = 320\,\mu\Omega$.cm [215]. The application of an inert argon atmosphere at deposition enabled to keep the initial composition of boride powder.

Ionic conductors

Granular deposits were tested to obtain the ionic conductors used as an electrolyte in solid oxide fuel cells (*SOFC*). The ions transported in the cells are oxygen anions (O^{2-}), and the cells operate at high temperatures. Consequently, the resistivities are tested in temperatures up to $T = 1273$ K. The stabilized zirconia coatings are frequently tested. For example, zirconia stabilized with 9–14 mol. % yttria coatings were air plasma sprayed and post spray treated as possible electrolyte. ZrO_2 with 14 mol. % Y_2O_3 coatings had resistivity decreasing with temperature from $\rho \approx 100\,M\Omega$.cm at $T = 473$ K to $\rho \approx 10\,\Omega$.cm at $T = 1273$ K [216]. Another example is zirconia stabilized with 10 ml. % scandia coatings [217, 218]. The resistivity at $T = 1273$ K was found to be $\rho \approx 14\,\Omega$.cm. Finally, the application of solution precursor plasma spraying (*SPPS*) technology to obtain all *SOFC* elements such as anode, electrolyte, and cathode was reviewed by Pawłowski [219].

Semiconducting Coatings

The p- and n- doped silicon-germanium alloy coatings obtained using *APS* were tested for thermoelectric applications in the temperature range $T = 673$–1173 K [220]. Akani *et al.* [221] prepared silicon powder prepared by mixing with phosphorus and boron particles. The powders were air plasma sprayed on single-crystal silicon to obtain p- and n- doped semiconducting coatings. Resistivity of the coatings was measured to be in the range from $\rho = 0.1$ to $100\,\Omega$.cm. Titanium oxide can be also counted among the semiconductors with wide band gap between the valence and conduction band. The band gap depends on the crystal structure of TiO_2 being equal to [222]: (i) 3 eV for rutile; (ii) 3.2 eV for anatase; and (iii) 3.4 eV for brookite.

Thermally sprayed TiO_2 powder enables obtaining different phases including suboxides called ***Mangéli phases*** defined as Ti_nO_{2n-1} ($n \geq 1$). [223]. Suboxides, such as $Ti_2O_3$and Ti_3O_5 were observed by Ohmori *et al.* [224] in the *APS* and *VPS* deposited coatings using rutile. The authors found out resistivity in the range from $\rho = 0.01$ to $1\,\Omega$.cm depending on the hydrogen flow rate in the plasma gas. Finally, the suspension plasma sprayed coatings were tested by Tomaszek *et al.* [225]. The authors used impedance spectroscopy to represent the coatings as highly resistive rutile and anatase phase grains immersed in matrix composed of *Magnéli phases* having much lower resistivity.

The resistivity of coatings including Al_2O_3 with addition of 13 wt. % or 40 wt. % TiO_2 obtained by *APS* changes with increasing temperature [226]. The resistivity of $Al_2O_3 + 13$ wt. % TiO_2 decreases from $\rho \approx 6\,\Omega$.cm at *RT* to $1.5\,\Omega$.cm at $T = 673$ K as in a semiconductor. Above this temperature the resistivity increases up to $\rho \approx 8\,\Omega$.cm at $T = 873$ K as in a metal. This behavior results from the modification of chemical composition of titania part of the coatings. Consequently, the oxygen lost at deposition and leading to the formation of titanium suboxides was probably picked up during measurements at high temperatures. The resistivity of $Al_2O_3 + 40$ wt. % TiO_2 is lower and equal to $\rho \approx 0.3\,\Omega$.cm at *RT* and shows similar behavior at higher temperature reaching $\rho \approx 1.4\,\Omega$.cm at $T = 1073$ K.

Resistor Coatings

The thermally sprayed coatings tested as resistor ones belong to: (i) metals and alloys; (ii) oxides; and (iii) cermets.

Plasma sprayed Ni as well as plasma and high-velocity oxy-fuel sprayed NiCr were tested by Prudenziati and Lassinantti Gualtieri [227]. Ni coatings were sprayed and annealed, and their resistivity at *RT* varied between $\rho \approx 27$ and $35\,\mu\Omega$.cm. The resistivity of NiCr plasma sprayed and annealed was in the range of $\rho = 217$–$463\,\mu\Omega$.cm, and the thermal coefficient of resistance had much greater dispersion varying between $TCR = 1331$ and 184 ppm/K, respectively. The coatings obtained by *HVOF* and annealing had lower resistivity of $\rho = 183\,\mu\Omega$.cm and greater $TCR = 2833$ ppm/K. The variation in ρ and *TCR* values was explained by oxidation during spraying.

The deposits of a mixture of oxides NiO + 45 wt. % Fe_3O_4 plasma sprayed in air had resistivity varying between $\rho = 1$ and 12 k Ω.cm depending on operation processing parameters. Thermal coefficient of resistance is positive and about $TCR = 100$–600 ppm/K [228]. The coatings tested by the same authors showed mixture of oxides $Al_2O_3 + Fe_3O_4$ and CoO with NiO [229]. The resistivity of the coating depended on the fraction of the volume fraction of the oxides used in the mixtures and varied in the range $\rho = 1$–10^8 Ω.cm.

Cermets seem to be an interesting composite to obtain coatings having resistivity controlled by the fraction of metal. Consequently, the powder mixtures of powders V_2O_5 + Ni with and without addition of Al_2O_3 were plasma sprayed to obtain resistors by Smith and Anderson [229]. Taylor *et al.* [230] tested the following cermets for the heaters working at $T = 773$ K: (i) Al_2O_3 with W, Mo, Cu, and Ni; and (ii) 13YSZ with Ni and W.

5.3.2 Dielectric Properties

Dielectric films and coatings are applied industrially in: (i) ***capacitors*** and (ii) ***insulators***:

- *Capacitors* are an important part of integrated circuits (*IC*) and should be prepared to work in different frequencies up to microwaves. Their important parameter is the capacitance determined by *relative dielectric constant,* ε_r, explained in Eq. 4.29, which decreases generally with frequency depending on polarization mechanism described in Section 4.6.2. Knowing that each thin-film capacitor is characterized by the capacitancy, *C*, and resistance, *R*, it is possible to define the loss factor, tan δ, as [202]:

$$\tan\delta = \frac{1}{2\pi\nu RC} \tag{5.3}$$

  The factor is often used in characterizing the planar capacitors (see Fig. 4.30) and denotes the dissipation of energy depending on the frequency of voltage applied at the testing [231].
- *Insulators*, such as substrates of *IC* working in different temperatures preserving their high resistivity, ρ. The *dielectric constant* and the loss factor are less important parameters. The important one is *dielectric strength, E*.

  The *capacitors* deposits should have relatively largeε_r enabling reduction of their surface and low loss factor tan $\delta < 10^{-4}$, while *insulators* should have high resistivity, ρ, and dielectric strength $E > 500$ kV/cm. The materials being used in these applications are mainly oxides and nitrides.

5.3.2.1 Atomistic Deposits

The films being tested for their dielectric properties have been deposited mainly using oxides, including ferroelectric ones, and nitrides obtained by reactive evaporation, sputtering, or *PECVD* [232]. More recent examples of the atomistic films having characterized their dielectric properties and shown in Table 5.29 were deposited also using *PLD* and *IP*. The as-deposited films are generally amorphous and should be annealed to restore the crystallinity (lines 3, 4, 5 of Table 5.29).

The dielectric films applied as capacitors should be thin to reach greater capacitance (see Eq. 4.29). In practice, the typical thickness is lower than 1 μm. Thinner deposits may have structural defects such as the contacts between the columnar grains and the pores between these columns (see Fig. 3.3) influencing their resistivity and loss factor as well as the *dielectric strength*. Consequently, by decreasing the thickness, the *loss factor* of capacitor can become too important.

Oxides such as Al_2O_3, Ta_2O_5, or TiO_2 (lines 1 to 4 of Table 5.29) can be used as dielectric films in capacitors of small capacitance. The *HA* ceramic and be applied in the future to manufacture electronic elements having contact with bone, i.e. being used in dental and orthopedic implants (line 5 of Table 5.29).

The ferroelectric oxides have a dielectric constant, which can be greater than $\varepsilon_r > 1000$ and even the amorphous $BaTiO_3$ film deposited without annealing have this constant useful to be applied as capacitor (see line 6 of Table 5.29). Another ferroelectric oxide, $SrTiO_3$, was tested because of its chemical stability and attractive opto-electronic properties (see line 7 of Table 5.29). Consequently, these dielectrics are useful if small surface of a capacitor is required. Ferroelectricity is an effect occurring in the materials with orientation type of polarization. This type of polarization decreases with the increasing vibration of the atoms, i.e. with increasing temperature. For example, for $BaTiO_3$, the orientation polarization ceases above the ***ferroelectric Curie temperature***, which is equal to 393 K [242].

Nitrides films can also be used as dielectrics in the integrated circuits capacitors. Consequently, Si_3N_4 is a possible candidate to replace SiO_2 as the dielectrics because of high dielectric strength and BN has high dielectric strength being useful in some circuits (lines 8 and 9 of Table 5.29).

5.3.2.2 Granular Deposits

Thermal spraying was used to obtain *insulators* and ferroelectric oxides having high *dielectric constant*. Examples of such deposits are presented in Table 5.30.

The well-known *insulator* is alumina coating. The *undercooling* effect, discussed for plasma spraying method in Section 3.3.3, results in formation of coating composed of very hygroscopic γ-Al_2O_3 phase while spraying any Al_2O_3 powder. The presence of water in coating results in very high *dielectric constant* of as-sprayed coating [243]. Consequently, the post-spray annealing and sealing at temperature, leading to evaporation of water, are necessary in order to reduce the constant (see lines 1 and 2 of Table 5.30). The *dielectric strength* of plasma sprayed alumina ($E < 0.20$ MV/cm) is lower than that of sputtered alumina films ($1.6 < E < 3.1$ MV/cm). This low *breakdown strength* can be explained by the greater thickness of the sprayed deposits and of possible accumulation of structural defects such as cracks and pores.

Table 5.29 Examples of dielectric properties of some atomistic films[a].

Number	References	Film	Thickness, nm	Deposition method	ε_r	tan δ	E, MV/cm	Remarks
Oxides								
1.	[233]	Al_2O_3	150	*RF magnetron sputtering*	8.2 – 8.7	2×10^{-3} -4×10^{-3}	—	Tests at $\nu = 0.2–100$ kHz
2.	[234]		210–640	*dc magnetron sputtering*	—	—	1.64–3.10	—
3.	[235]	Ta_2O_5	60–250	*dc magnetron sputtering* and annealing at 1173 K	8.8 to 18.7	—	0.54	ε_r tested at $\nu = 500$ Hz decreased with frequency, reaches $\varepsilon_r = 0$ at $\nu = 6$ MHz
4.	[236]	TiO_2	20–220	*IP* and annealing at 423 K	12–13	0.1	—	ε_r tested at $\nu = 1$ kHz decreased with frequency
5.	[237]	*HA*	1000	*PLD* using ArF pulsed laser and annealed at 823 K	5.7	4×10^{-2}	0.01	ε_r and *tan* δ values for $\nu = 1$ kHz
Ferroelectric oxides[b]								
6.	[238]	$BaTiO_3$	2700	*RF magnetron sputtering*	18.5	4×10^{-3}	—	amorphous film, ε_r tested at $\nu = 100$ kHz decreases with frequency
7.	[239]	$SrTiO_3$	<300		30 to 66	—	—	ε_r tested at $\nu = 100$ kHz
Nitrides								
8.	[240]	Si_3N_4	275	*RF magnetron sputtering*	15–31	—	—	ε_r tested at $\nu = 1$ kHz
9.	[241]	BN	80		5.9	—	15	amorphous film, ε_r tested at $\nu = 10$ MHz

[a] Data mainly at $T = 300$ K and, in the not precised cases, at frequency of $f = 1$ kHz, with exception of E tested using *dc*.
[b] Ferroelectric oxides having generally structure of perovskites are electrically polarized without external electric field [242].

Table 5.30 Examples of dielectric properties of some granular coatings[a].

Number	References	Coating	Thickness, μm	Spray method	ε_r	tan δ	E, MV/cm	Remarks
Oxides								
1.	[243]	Al_2O_3	500–1500	*APS* and post-spray polymer impregnation	5–8	0.01–0.05	0.10–0.18	ε_r and *tan* δ tested at $\nu = 1$ kHz
2.	[244]	Al_2O_3 + 13 wt. % TiO_2	100–1000		—	—	0.16–0.23	Penetration depth of sealant reached 300 μm
3.	[245]	Y_2O_3	300	*WSP* of powder with two injectors	10.0–10.2	0.0008–0.0013	—	ε_r and *tan* δ tested at $\nu = 1$ MHz and $T = 323$ K
4.			130	*SPS* using *WSP* plasma torch	10.6	0.0008–0.0016	—	—
5.	[246]	$CaZrO_3$	2000	*WSP*	10–18	0.01–0.09	—	ε_r and *tan* δ tested at $\nu = 1$ MHz and $T = 293$ K
Ferroelectric oxides								
6.	[247]	$BaTiO_3$	900–1000	*APS*	400	0.45	—	ε_r and *tan* δ tested at $\nu = 1$ kHz and $T = 293$ K
7.	[248]	$PbTiO_3$	50–100	*APS* supersonic	180	—	—	ε_r tested at $\nu = 1$ kHz and $T = 293$ K
8.	[249]	*PZT*	200	*APS* annealed at 773–1273 K	200–840	0.03–0.05	—	ε_r tested at $T = 293$ K

[a] Data mainly at $T = 300$ K and, in the not precised cases, at frequency of $\nu = 1$ kHz, with exception of E tested using *dc*.

Y_2O_3 oxide coating can be useful in the production of integrated circuits because of the resistance of this oxide against erosion occurring at interaction with plasma used at manufacturing. The dielectric properties of this oxides coating are shown in lines 3 and 4 of Table 5.30. $CaZrO_3$ coating (see line 5 of Table 5.30) can be useful for production of capacitors because of high dielectric constant.

Ferroelectric oxides coatings (lines 6–8 in Table 5.30) are useful in many applications because of their high dielectric constant ($BaTiO_3$) and piezoelectric properties as $PbTiO_3$ or *PZT*. These properties are interesting for such applications of deposits as capacitors or as the pressure sensors. Thermal spraying of these oxides having complex crystalline structure requires: (i) optimization of the operational process parameters; and (ii) post-spray treatment. This optimization and the treatment, being generally annealing, restore the crystalline microstructure and reduce the effects of formation of metastable phases and selective evaporation of some oxides, such as PbO at $PbTiO_3$ spraying [248].

5.3.3 Magnetic Properties

Atomistic deposits of magnetic materials have been studied and applied since 1950s [250]. The films and coatings were deposited mainly using *ferromagnetic* and *ferrimagnetic* materials. The authors describe the effect of magnetization using vector $\vec{M}$. The magnetization describes, in microscopic scale, the organization of magnetic domains in external magnetic field $\vec{H}$ (see Section 4.6.3). The magnetization reaches the maximum value of saturation M_S, which corresponds to H_{max} in Fig. 4.32. The tests were made using external magnetic field parallel to the direction of magnetized domains (in plane) or in other directions (out of plane).

5.3.3.1 Atomistic Deposits

The films having recently tested their magnetic properties are shown in Table 5.31 by separating the types of materials as follows: (i) metals and alloys; and (ii) ceramics. The deposition technique was frequently *sputtering*, and the tested films were *ferromagnetic* and *ferrimagnetic*.

The *ferromagnetic* film of Fe was obtained by evaporation in presence of high external magnetic field, which influenced the films' microstructure and resulted in augmentation of magnetization of saturation (line 1 of Table 5.31). A MnBi alloy deposited by sputtering can be a candidate for the nanomagnets.

The magnetic properties of these films and the direction of axis of easy magnetization depend on the atomic ratio of Mn to Bi (see line 2 of Table 5.31). Similar deposition technique of *sputtering* enabled obtaining *soft magnetic* films of CoFeB and FeSi multi-films with Cr interlayers shown in lines 3 and 4 of Table 5.31. The presented films have similar magnetization of saturation, but the coercive field is smaller in multi-films.

Ceramic films such as *YIG* and CeO_2 were deposited using *RF magnetron sputtering* followed by annealing (lines 5 and 6 of Table 5.31). The post-deposition heat treatment restored the crystallinity and enabled obtaining crack-free *YIG* films. In addition, the magnetic properties of these ceramics were considerably improved rendering possible the application of these films in integrated circuits. The Fe_4N films obtained by *dc sputtering* was characterized very high magnetization M_S (line 7 of Table 5.31). This parameter was

Table 5.31 Magnetic properties of some atomistic deposits[a].

Number	References	Film	Thickness, nm	Deposition method	Temperature of test, K	Coercive field, H_c, A/m	Magnetization of saturation, M_s, A/m	Remarks
Metals and alloys								
1.	[251]	Fe	170	*Evaporation* in presence of external magnetic field	—	0.47×10^3–1.26×10^3	1.2×10^6–1.53×10^6	*Ferromagnetic* nanocrystalline films
2.	[252]	MnBi with atomic ratio Mn/Bi of 0.4–1.32	170	*dc magnetron sputtering* followed by annealing at 653 K	*RT*	4×10^5–1.3×10^6	1.78×10^4–1.52×10^5	*Hard magnetic* films for permanent magnet
3.	[253]	CoFeB	26	*RF magnetron sputtering* followed by annealing up to 773 K	—	2.4×10^3–3.3×10^3	up to 1.28×10^6	*Soft magnetic* films
4.	[254]	100 nm FeSi films separated by 10 nm Cr interlayers	Total 2000	*dc magnetron sputtering*	—	0.25×10^3	1.03×10^6	
Ceramics								
5.	[255]	$Y_3Fe_5O_{12}$ (*YIG*)	180–220	*RF magnetron sputtering* followed by annealing up to 1223 K	—	2.55×10^3	8×10^4	*Soft magnetic* films
6.	[256]	CeO_2	295–327	*RF magnetron sputtering* followed by annealing up to 1073 K	*RT*	4×10^3–6×10^3	3.6×10^3–27.3×10^3	
7.	[257]	Fe_4N	10–120	*dc magnetron sputtering*		2.4×10^3–4.8×10^3	8.4×10^5–13.2×10^5	
8.	[258]	$SiC_xFe_yN_z$	2500	*PE CVD*	1.77–300	—	6×10^4 at $T = 1.77$ K 10^4 at $T = 300$ K	*Ferromagnetic* films having composition by *XPS* (at. %): Si + 62% C + 2% N + 2% Fe + 25% O

[a] Magnetic properties measured probably at *RT*.

dependent on the film's thickness reaching the maximum at $\delta = 30$ nm. Finally, the ceramic film of $SiC_xFe_yN_z$ obtained using *PE CVD* had many crystal phases including magnetic ones such as FeSi and Fe_3C rendering the films ferromagnetic with Curie temperature above *RT* as shown in line 8 of Table 5.31.

5.3.3.2 Granular Deposits

The studies of thermal spraying magnetic materials started in the year 1970. NiZn-ferrites and intermetallic compounds $SmCo_5$ were studied in University of Wales in UK and Charles Stark Draper Laboratory in MA in USA [259, 260]. The initial works concerning *soft* and *hard magnetic* materials were reviewed in [39].

The recent works about thermal spraying of magnetic materials are presented in Table 5.32. The studies are separated into: (i) alloys; (ii) ferrites including metal or alloy and iron oxide, and (iii) composites. The coatings were developed frequently for the applications in microelectrons.

The soft magnetic alloy FeSi was *HVOF* sprayed and obtained coatings' porosity was influenced by spray distance. The porosity influenced the tested magnetic properties, i.e. coercive field, H_c, and magnetization of saturation, M_s (line 1 of Table 5.31). The plasma spray process operational parameters and with dry-ice blasting at deposition influenced the content of amorphous phase content and the magnetic properties od FeSiB alloy shown in line 2 of Table 5.32. Magnetic shielding capacity was tested for the $Fe_{73.5}Cu_1Nb_3Si_{13.5}B_9$ alloy coatings having amorphous and nanocrystalline crystal grains. The coatings were obtained by *APS* as shown in line 3 in Table 5.32. Similar microstructure of $Co_{58}Ni_{10}Fe_5B_{16}Si_{11}$ plasma sprayed alloy was obtained in plasma sprayed coating having magnetic properties shown in line 4 of Table 5.32.

The air plasma sprayed MnZn (*MZF*) and cobalt ferrites were processed using powders manufactured by mixing of oxides and optimizing deposition parameters as shown in lines 5 and 6 in Table 5.32, respectively. The authors analyzed the influence of chemical composition modification during deposition process, such as Zn evaporation in plasma jet, as well as of the shape of lamellae on coating's magnetic properties.

The use of cold spray method to obtain the deposits using hard particles of FeSiAl magnetic alloy required the cladding of powder particles with ductile Ni 5–14 μm thin film. The cladding, explained in Section 2.3.1 and shown in Fig. 2.24, enabled obtaining composite coating with magnetic properties shown in line 7 of Table 5.32. The cold spray process was modified by the use of powder preheating up to the temperature of 773 K.

5.3.4 Superconductors

The industrial applications of superconducting materials are limited by the necessity of cooling down to low temperatures. Many superconductors require cooling with liquid helium having temperature about 4 K. As helium is a rare and expensive gas, the progress in this field was the discovery of high T_{cr} superconductors by Bednorz and Müller [268]. These superconductors can be cooled by liquid nitrogen at 77 K. Phillips [269] reviewed the physics of high T_{cr} superconductor, and Pawłowski [270] described in detail the most frequently tested superconducting oxide, namely *YBCO*.

Table 5.32 Magnetic properties of some granular deposits[a].

Number	References	Film	Thickness, μm	Deposition method	Temperature of test, K	Coercive field, A/m	Magnetization of saturation[b], $A.m^2/kg$ or A/m	Remarks
Alloys								
1.	[261]	FeSi	200	*HVOF*	—	1193	1170 $A.m^2/kg$	*Soft magnetic*
2.	[262]	FeSiB	250–270	*APS* with dry-ice blasting	*RT*	1200–1512	—	
3.	[263]	$Fe_{73.5}Cu_1Nb_3Si_{13.5}B_9$	200–300	*APS*	—	144	about 2×10^5 A/m	*Soft magnetic*, amorphous and nanocrystalline coating
4.	[264]	$Co_{58}Ni_{10}Fe_5B_{16}Si_{11}$	2000			24	about 48×10^3 A/m	*Soft magnetic* $\mu=20\times10^3$, $T_c=640$ K
Metal-oxide ferrites								
5.	[265]	$Mn_{1-x}Zn_xFe_2O_4$ (*MZF*)	100–600	*APS*	—	about 10^4	40–60 $A.m^2/kg$	*Soft magnetic* the properties were correlated with spray parameters using *DoE*
6.	[266]	$CoFe_2O_4$	—		*RT*	8.4×10^3–21.9×10^3	44–72 $A.m^2/kg$	*Soft magnetic* the properties were correlated with plasma gun working power
Composite								
7.	[267]	Ni + FeSiAl	150–375	*CGSM*	*RT*	4.6×10^3–5.3×10^3	58–67 $A.m^2/kg$	*Soft magnetic* powder of FeSiAl particles coated with Ni was preheated before injection to nozzle at *RT*, 573 K, 673 K and 773 K

[a] Magnetic properties measured probably at *RT*.
[b] The unit used by some authors was [emu/g], which correspond to SI unit: 1 emu/g = 1 $A.m^2/kg$.

The discovery of high T_{cr} superconductors stimulated the development of laser ablation technique (*PLD*). This technique, as well as the sputtering, enables congruent evaporation, i.e. obtaining the multicomponent films preserving the chemical composition of laser-ablated target (see Section 2.2.4). On the other hand, thermal spraying techniques were used very frequently in the period of initial studies concerning high T_{cr} superconductors. Unfortunately, it turned out that the critical current transported by the sprayed coating is limited by intrinsic structural defects, and the number of studies decreased considerably.

5.3.4.1 Atomistic Deposits

The examples of superconducting deposits obtained using atomistic techniques are shown in Table 5.33.

The sputtered Nb films were developed recently to be used in superconducting microelectronic devices (line 1 in Table 5.33). Nb_3Sn has greater critical temperature and was developed to be used in superconducting cavities. The cavities work at radio frequency and are used in particles accelerators. The films having properties shown in line 2 of Table 5.33 may possibly replace Nb ones.

The *YBCO* as well as $SmBa_2CuO_{7-x}$ oxide films (see line 3–5 of Table 5.33) have been studied since the discovery of high T_{cr} superconductivity as possible material to transport energy or for high-intensity magnets [274]. The films were deposited on $SrTiO_3$ single crystal substrates. The titanate is often used in microelectronics because of low dielectric loss and low thermal expansion in low temperatures [276]. The use of single crystal enables the epitaxial growth of films during the post deposition heat treatment.

5.3.4.2 Granular Deposits

Thermal spraying techniques have been studied to obtain high T_{cr} superconductor$_s$ since the end of the 1980s. The used techniques were mainly atmospheric or vacuum plasma spraying. The heterogeneous microstructure of sprayed coatings limited the critical current density to $j_{cr} = 4.6\text{–}6.9 \times 10^6$ A/m^2 [270, 277, 278]. The critical current was considerably lower than that obtained in thin films shown in lines 3 and 4 of Table 5.33. It could have been one of the reasons for decrease of the number of studies about thermally sprayed coatings of high T_{cr} superconductors. However, Kumar *et al.* [279] recently published an interesting application of cold spraying of niobium coating for particles accelerators. The obtained coating had critical temperature of $T_{cr} = 8.6$ K, and electrical properties were comparable to those of films obtained by atomistic methods.

5.3.5 Electron Emitters

The emission of electrons from solid to vacuum occurs when sufficient energy is delivered to the electrons close to the body's surface. Type of emission depends on the energy source. The typical emission and their sources are as follows [280]:

- ***thermionic emission*** takes place when the material is heated up;
- ***Schottky emission*** is similar but part of energy is delivered by external electric field;
- ***photoemission*** occurs when energy is delivered by the photons;

Table 5.33 Examples of superconducting films obtained using atomistic methods.

				Deposition process			Superconducting properties		
Number	References	Film	Thickness, μm	Technique	Substrate material/ temperatureK	Heat treatment	Critical temperature, K	Critical current density, A/m^2	Remarks
Metal & alloys superconductors									
1.	[271]	Nb	100 and 200	*Sputtering*	Si covered with SiO_2 film/	Some films were treated with pulsed laser	8.7–8.9	about 10^{11} depending on testing temperature	Some films were treated with pulsed laser
2.	[272]	Nb_3Sn	—		Sapphire/773	up to 1223 K for 3 h	16.94–17.61	–	Annealing temperature increased T_{cr}
Oxides superconductors									
3.	[273]	*YBCO*	500	*PLD* using XeCl excimer laser, $\lambda = 308$ nm	$SrTiO_3$ single crystals/848 up to 1193	—	87 up to 90	10^7 up 10^{10}	Optimal temperature with greatest T_{cr}, j_{cr} was 998 K
4.	[274]		300	*PLD*, $\lambda = 248$ nm	$SrTiO_3$ single crystals/1053 under O_2, N_2O and N_2, $p = 30–40$ Pa	723 K for 340 min under O_2, $p = 50$ Pa	up to 91	up to 6.75×10^{10}	Heat treatment increased j_{cr}
5.	[275]	$SmBa_2CuO_{7-x}$	100	*PLD* using Nd-YAG laser, $\lambda = 532$ nm	$SrTiO_3$ single crystals/1013	723 K under O_2, $p = 0.1$ MPa	87	—	Deposition geometry off-axis

- ***emission by tunneling*** [7] of electrons occurring in presence of electric field. The field can be enhanced locally by the shape of emitter; and:

$$E_{\text{loc}} = \beta E \tag{5.4}$$

Consequently, the density of electrons emitted, *j*, by a cathode is given the ***Fowler–Nordheim*** relationship, which in the simplified form, can be written as follows:

$$j = c_1 \frac{E_{\text{loc}}^2}{\phi} \exp\left(c_2 \frac{\phi^{\frac{3}{2}}}{E_{\text{loc}}} \right) \tag{5.5}$$

in which c_1 and c_2 are the constants and Φ is work function being the minimum energy required to extract an electron from the surface of solid. For example, $\Phi = 4.55\,\text{eV}$ for polycrystalline tungsten [207]. It can be concluded that chemistry of materials as well as morphology of its surface are the main factors determining *emission by tunneling*.

In general, the emission is influenced by many factors and in industrial practice, many of the shown emission's types may occur simultaneously.

5.3.5.1 Atomistic Films

Some examples of atomistic films tested as electron emitters are collected in Table 5.34. The types of emission varied from thermionic emission from rare-earth metal borides deposited on tungsten substrates to be used as electron guns (see line 1 of Table 5.34). The double mechanism of electric-field-induced *photoemission* under *uv* radiation in sputtered semiconducting *ITO* films is presented in line 2 of Table 5.34. The nitride compounds of AlN, BN, and NbN films on Si substrates have been studied for numerous applications in microelectronics and optoelectrics in many studies (see, e.g. lines 3–5 of Table 5.34). Finally, the carbonic films such as *DLC* were frequently tested for application in field emission displays [280]. Some properties of very thin carbon films obtained by *CVD* are presented in line 6 of Table 5.34.

5.3.5.2 Granular Coatings

A research team around Znamirowski and Pawłowski has intensively studied the electron emission of granular coatings. These authors tested plasma sprayed and laser engraved Cr_2O_3 coatings [287]. The emission was enhanced by the asperities and tips created by laser engraving (see Fig. 5.17) and occurred from thin Ti film obtained by *PVD* on the coatings' surface. The plasma sprayed and laser engraved coatings of TiO_2 as well as of $Al_2O_3 + 13$ wt. % TiO_2 and $Al_2O_3 + 40$ wt. % TiO_2 alloys were also tested [288]. The greatest emission was measured for plasma sprayed TiO_2 coating, laser engraved with the density of 60 cells/cm. The emitting sites on the coatings' surface were attributed to electrically conducting Magnéli phases. Finally, suspension plasma spraying was used to obtain TiO_2 and bio-glasses coatings [289, 290]. The titania coatings were used directly in the emission tests, and the Magnéli phases were acting as emitting inclusions inside the conducting rutile and anatase. The emission current reached 400 μA making the coatings applicable for

7 This type of emission is called cold electrons emission by many authors.

Table 5.34 Examples of films tested for electrons' emission obtained using atomistic methods.

				Deposition process		Electron emission			
Number	References	Film	Thickness, nm	Method	Substrate	Type	Enhancement factor, β	Emission barriers	Remarks
					Compounds				
1.	[281]	LaB_6, CeB_6,SmB_6	3000	*Sputtering*	Mo, W	Thermionic	—	—	Working temperature limited to 1673 K
2.	[282]	*ITO*	10–200		glass	Photo- and tunneling emission	—	—	Optimal $\delta = 10$ nm
3.	[283]	AlN	20–150		Si wafer	Tunneling	2.1–11	Surface of AlN film	Optimal $\delta = 44$ nm
4.	[284]	BN, BCN	8–10	*PECVD*	Si wafer and SiO_2		—	Surface of films	—
5.	[285]	Nb-Si-N	160	*Sputtering*	Si wafer		—	Nanogaps between the phases	tested for flat displays
					Carbonic materials				
6.	[286]	Carbon	1	*CVD*	Si wafer	Tunneling	92–285	Carbonic islands	Emission from sp^2 sites

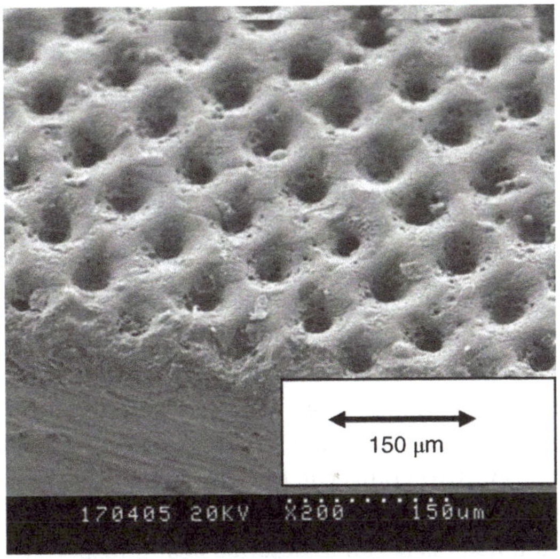

Figure 5.17 The surface of air plasma sprayed Cr_2O_3 coating polished after deposition and CO_2 pulsed laser engraved to obtain the array of cells [287]. The cells presented are aligned and have the density of 140 cells/cm and the depth of 34 μm.

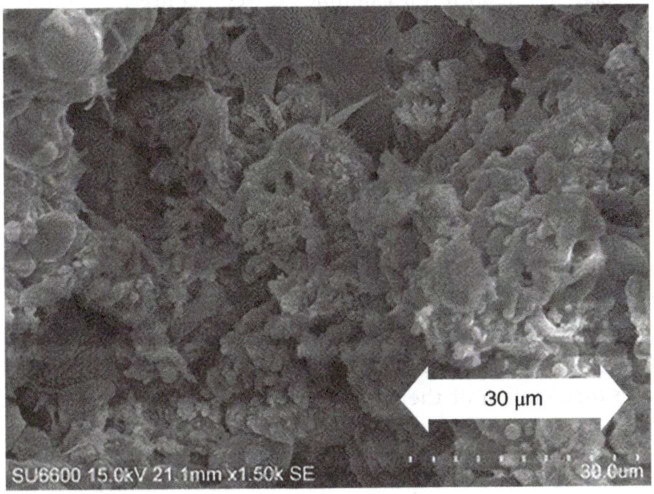

Figure 5.18 SEM micrograph (secondary electrons) of the surface of suspension plasma coating of bioglass and covered with thin Zr film emitting electrons with the enhancement factor of $\beta = 280$ [290].

electronic devices. The bioglasses were covered with thin Zr film to render the surfaces electrically conducting. The morphology of the *SPS* coatings included fine droplets enhancing the electron emission and shown in Fig. 5.18 has the factor value of $\beta = 280$ being comparable to the values reached in the nanotubes.

5.4 Thermophysical Properties

Thermophysical properties include: (i) thermal conductivity; (ii) thermal diffusivity; (iii) specific heat; (iv) thermal dilatation; and (v) emissivity. Thermal conductivity characterizes transport of heat and depends on thermal diffusivity, density, and specific heat (see *Eq.* 4.19).

Thermal diffusivity describes evolution of temperature[8]. Thermal conductivity concerns energy transport by free electrons in metals and alloys and by phonons in ceramic materials. The transport by electrons and phonons depends on the microstructure of the deposits and in particular on: (i) crystal phases; (ii) crystal grains geometry with regard to the direction of energy transport; (iii) contacts between the grains; (iv) porosity; and (v) pore's shape.

Emissivity describes the transport by photons in the voids and pores and on the surface of materials at high temperatures. This transport can take place outside the deposit from its surface or inside it. The latter case occurs when crystal phase of deposit material is transparent to an electromagnetic wave and when porosity of a deposit is considerable.

Specific heat and density influence thermal conductivity energy as shown in *Eq.* 4.19. These parameters may be modified with regards to initial material by the deposition process by, e.g. oxidation by the nucleation conditions resulting in different shapes of crystal grains.

Finally, the data shown in this section will concern properties related to the transport of energy. Byron Bird *et al.* [291] discuss the physical phenomena of such transport in detail. In some cases, it is useful to mention the thermal contact resistance, R_T, defined in *Eq.* 3.8. This parameter is useful to describe heat transport: (i) perpendicular to the surface of a deposit, for example between deposit and substrate or a neighboring deposit; and; (ii) parallel to the surface of a deposit, for example between columnar grains in atomistic deposits shown in Fig. 3.3.

Finally, the bibliographical research enabled to find thermophysical properties of atomistic and granular deposits.

5.4.1 Thermal Conductivity and Diffusivity

Thermal conductivity can be determined using static and dynamic methods described in the Section 4.4.3. Thermal diffusivity values are added as the property being most frequently directly measured.

5.4.1.1 Atomistic Deposits

The mechanism of thermal conductivity transport of atomistic deposits is influenced by three major factors [292]:

- thickness of the deposits resulting from the dependence of thermal conductivity on mean free path of heat carriers. This factor concerns mainly transport by free electrons in thin metals and alloys films, and the calculations for Al and Au deposits showed that this effect is significant for thicknesses $\delta < 100\,\text{nm}$.
- columnar grain microstructure, shown in Fig. 3.3, resulting from the atomistic films' nucleation. Thermal contact resistance can characterize the contacts between the columns. The resistance, being in the range of $10^{-10} \leq R_T \leq 10^{-8}$ m^2 W/K, can affect considerably the lateral heat transport.

---

8 The German language makes it possible to understand this difference well because thermal conductivity is *die Wärmeleitfähigkeit* which literally translates into *the capacity for transporting heat* while thermal diffusivity is *die Temperaturleitfäfigkeit* which translates word for word as *the capacity for transporting temperature.*

- defects and impurities and interface between substrate and film, which affect the vertical heat transport. This factor depends strongly on the deposition operational parameters.

The thermophysical properties of some atomistic deposits are shown in Table 5.35. The results of measurements confirm that thermal conductivity of atomistic deposits is much lower than that of the bulk material (see lines 1 and 2 of Table 5.35 for metals). This difference is also considerable for ceramics as shown in line 4 of this table and even greater for carbonic film shown in line 9.

Thermal conductivity increases with the thickness of deposit of metals (see lines 1 and 2 of Table 5.35). Finally, the substrates being used for the tested films are frequently silicon (see lines 3, 9, and 10 of Table 5.35). This results for the growing number applications of these films in microelectronics.

5.4.1.2 Granular Deposits

The thermophysical properties of thermal sprayed coatings are frequently tested because of some important industrial applications such as [39, 301]:

- thermal barrier coatings (*TBC*) having often duplex configuration of thermally insulating outer coating being frequently yttria zirconia-stabilized deposited onto adherent coating of nickel-based alloy.
- dielectric coatings, such as alumina, used as electric insulators evacuating heat generated by an electrical circuit.
- wear-resistant coatings dissipating heat generated during friction without cooling. For this application, mainly chromium or molybdenum oxide is used.

The insulating coatings used for *TBC* and dielectric ones used as electric insulators are often oxides. That is why, the oxide coatings are presented in Table 5.36 separately from metals and alloys coatings shown in Table 5.37. Fig. 5.19 shows the values of thermal conductivity of oxides in function of temperature.

Thermal barriers are currently made using plasma sprayed powders of ZrO_2 coatings stabilized with Y_2O_3 (lines 1–3 of Table 5.36). The deposition technology of suspension plasma spraying of *YSZ* enables obtaining columnar microstructure, shown in Fig. 3.19, resulting in high thermal shock resistance, and this has been also tested (lines 2 and 5 of Table 5.36). Compounds including zirconia such as $Gd_2Zr_2O_7$ and $La_2Zr_2O_7$ are also possible candidates for thermally insulating coatings for the *TBC*s (lines 4 and 6 of Table 5.36). Finally, some silicates such as Mg_2SiO_4 and titanates such as $CaTiO_3$ have low thermal conductivity and can be considered as thermally insulating coatings (see lines 8 and 9 of Table 5.36). In addition to the chemical composition, the porosity of thermally sprayed plays an important role in thermal energy transport. In particular, the gas included in the pores of a deposit contributes to the heat conduction [316].

Thermal conductivity of another oxide coating, namely alumina being frequently used as dielectric coatings, has been tested to establish evacuation of heat generated by Joule effect (see line 7 of Table 5.36 and Fig. 5.20).

Table 5.37 shows the thermal conductivities of molybdenum and of some nickel base alloys obtained by thermal spraying and some details of their fabrications and some other properties.

Table 5.35 Examples of thermophysical properties of some atomistic deposits at room temperature.

				Deposition process		Thermophysical properties			
Number	**References**	**Film**	**Thickness nm**	**Method**	**Substrate**	**λ_d, W/(mK)**	**a_d, m^2/s**	**R_T (film – substrate), $m^2.K/W$**	**Remarks**
Metals and alloys									
1.	[292]	Au	200	*EBPVD*	Glass BK7	25.2	—	—	Photothermal method was used, bulk conductivity, $\lambda = 311\,W/(mK)$
			500			31.5	—	—	
2.		W	94	*RF Sputtering*	Fused SiO_2	11.5	—	—	Photothermal method was used, bulk conductivity, $\lambda = 173\,W/(mK)$
3.	[293]	Cu + 50 at. % Ta	2500	*Magnetron Sputtering*	Si wafer and Mo foil	60	3.5×10^{-5}	—	Laser flash method used to measure diffusivity
Ceramics									
4.	[294]	Al_2O_3	100 000	*Sputtering*	—	2.5	—	—	Bulk conductivity, $\lambda = 30\,W/(mK)$
5.	[295]	*7YSZ*	100 000	*EBPVD*	Inconel 738	1	—	—	Laser flash method used to measure diffusivity
6.	[296]	*YBCO*	95	*PLD* using KrF laser, $\lambda = 248\,nm$	$SrTiO_3$	0.82	0.25×10^{-6}	—	Photothermal methods were used
7.	[297]	GaN	2500	*MOCVD*	Sapphire	128	—	—	
8.	[298]	TiN	2700	*Ion plating*	Steel Z160CDV12	—	5.6×10^{-6}	5×10^{-8}	
Carbonic materials									
9.	[299]	*DLC*	3 000	*PLD* using Nd:YAG laser, $\lambda = 1.06\,\mu m$	Si	6.06	3.7×10^{-4}	—	Measured using holographic interferometry, sample surface 25 × 10 mm, bulk conductivity $\lambda = 1800\,W/(mK)$
10.	[300]	Hydrogenated amorphous carbon (*a-C:H*)	286–873	*PECVD*		0.14–0.35	—	—	Photothermal method was used, thermal conductivity increases with film thickness

Table 5.36 Thermophysical properties of some thermally sprayed oxide coatings. Thermal conductivities in function of temperature are shown in Fig. 5.19.

		Deposition process				Thermophysical properties		
Number	References	Method/feedstock	Coating, wt. %	Crystal phases	Porosity,%/ density, ρ, kg/m^3	λ_d, W/(mK	a_d, m^2/s	Remarks
1.	[302]	*APS*/powder	*8YSZ*	Mainly tetragonal	/5200	see Fig. 5.19	4×10^{-7} at 600 K	Tested as-sprayed using laser flash (a_d), scanning calorimeter (c_p) and dilatometer (*TEC*)
2.	[303]	*SPS*/suspension	*14YSZ*	Cubic	15.2/		—	
3.	[302]	*APS*/powder	*20YSZ*		20/4700		10^{-7} at 600 K	
4.	[304]		$La_2Zr_2O_7$	Pyrochlore phase	16/		4.5×10^{-7} at 373 K	
5.	[305]	SPS/suspension	ZrO_2 + 24 CeO_2 + 2.5Y_2O_3	Mainly tetragonal	18.7/		5×10^{-7} at *RT*	Tested as-sprayed using laser flash (a_d) other data obtained from literature for measured porosity
6.	[306]	*APS*/powder	$Gd_2Zr_2O_7$	Cubic	4.7/		4.7×10^{-7} at *RT*	Measured as a part of TBC including bond coating and substrate
7.	[307]		Al_2O_3	γ- Al_2O_3	—		—	Measured using *ir* camera, tested sample is shown in Fig. 5.20
8.	[308]		*Forsterite* (Mg_2SiO_4)	Mg_2SiO_4 with small quantity of $MgSiO_3$	—		—	Tested as-sprayed using laser flash (a_d) other data obtained from literature
9.	[309]	*WSP*/powder	$CaTiO_3$	—	6.95/3684		6×10^{-7} at *RT*	Tested as-sprayed using laser flash (a_d), scanning calorimeter (c_p) and dilatometer (*TEC*)

Table 5.37 Thermophysical properties of metal and alloys coatings thermally sprayed using powders.

		Deposition method				Thermophysical properties		
Number	**References**	**Method**	**Coating, wt. %**	**Crystal phases**	**Porosity, %/ density, ρ, kg/m^3**	**λ_d, W/(mK)**	**a_d, m^2/s**	**Remarks**
Metal								
1.	[310]	*APS*	Mo	—	/7710	see Fig. 5.21	—	Tested as-sprayed using laser flash (a_d), Bunsen ice calorimeter (c_p), and dilatometer (*TEC*)
Alloys								
2.	[311, 312]	*APS*	Ni + 5Al	γ – Ni	/7200		4.25×10^{-6} at 600 K	Tested as-sprayed using laser flash (a_d), scanning calorimeter (c_p), and dilatometer (*TEC*)
3.		*VPS*			/7800		7×10^{-6} at 700 K	
4.	[313]	*HVOF*	Ni + 20Cr		-	see Fig. 5.21	3.5×10^{-6} at *RT*	—
5.	[314]	*APS*	NiCoCrAlY	—	7.5/		1.2×10^{-6} at *RT*	Tested as-sprayed using laser flash (a_d), scanning calorimeter (c_p), and dilatometer (*TEC*)
6.	[315]	*APS*	$Al_{44}Ni_{45}Ru_{11}$	B2	/5360		5.88×10^{-6} at *RT*	

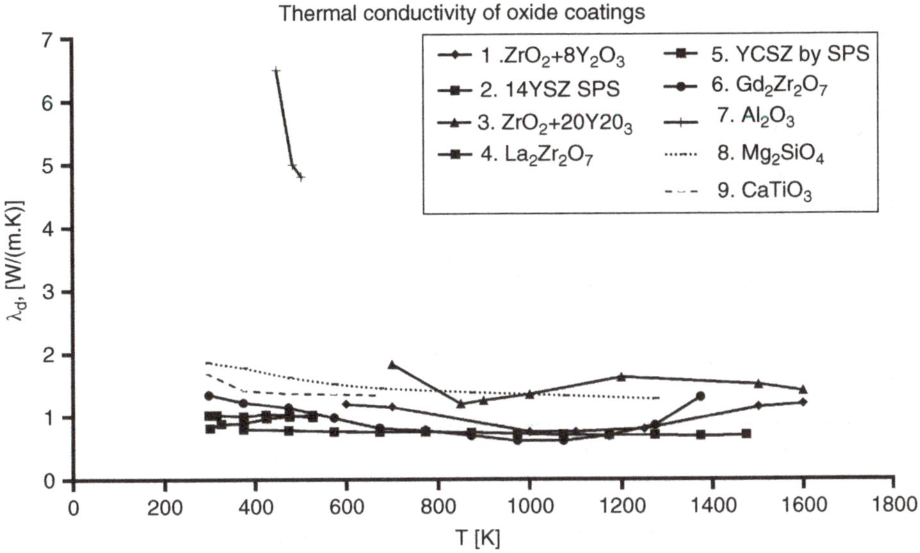

Figure 5.19 Thermal conductivity of oxide coatings obtained by plasma spraying of powders and suspensions in function of temperature. The properties of coatings are shown in Table 5.36 and the numbers of coatings correspond to that of the table lines.

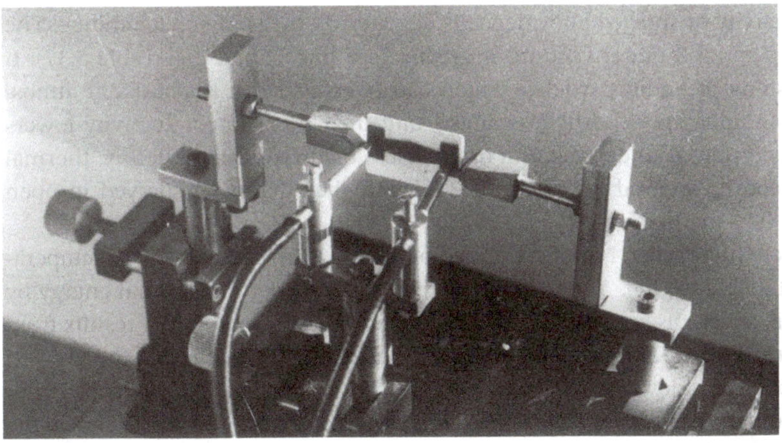

Figure 5.20 Alumina sample with deposited planar resistor of NiO+ Fe_3O_4 and Cu contact areas heated by electric current used in *ir* method of thermal conductivity measurements [307].

Thermal conductivity of the molybdenum coatings is around 20 W/(mK) as seen in Fig. 5.21. This value is relatively high which makes it possible to evacuate easily the heat during service of this deposit against wear. Wilkes and Lagedrost [310] observed that the measurement of λ_d of molybdenum coatings made under helium atmosphere shows greater values of conductivity than those made under vacuum. This confirms the conduction of

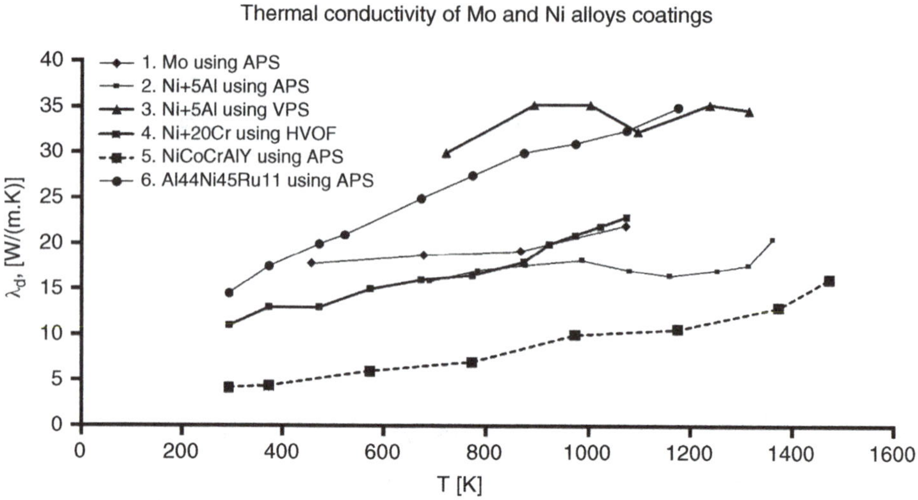

Figure 5.21 Thermal conductivity of Mo and of some Ni alloys coatings obtained by thermal spraying using powders in function of temperature. The properties of coatings are shown in Table 5.37 and the numbers of coatings correspond to lines of the table.

heat by convection in the gas inside the coating's pores. Another observation is the increase of conductivity after heat treatment of the samples resulting in their sintering and in the reduction of porosity.

Thermal conductivity of sprayed alloy deposits depends on the spraying technique. The density of coatings sprayed under vacuum is greater (see lines 2 and 3 of Table 5.37). It results from splashing of liquid particles, which can be oxidized in the ambient atmosphere. The oxides render the deposits less dense, and their thermal conductivity lowers down. The coatings sprayed under vacuum are not/less oxidized. Consequently, thermal conductivity of deposits sprayed by *VPS* is 2–3 times higher than that sprayed in open atmosphere using *APS* method.

Thermal conductivity of NiCr, NiCoCrAlY, and of $Al_{44}Ni_{45}Ru_{11}$ increases with temperature up to about 1200 K (see Fig. 5.21). This may result in transport of thermal energy by free electrons. This effect, discussed in Section 4.4.3 and in the review [301], results from ***law of Wiedemann–Franz–Lorenz*** concerning mean free path of electrons.

The bondcoats of nickel alloy with elements such as Cr, Co, Al, or Y and an outer coating of zirconia are used as the parts of thermal barriers. The temperature drop on a barrier is in the range of $\Delta T = 100$–200 K [317]. This drop also depends on thermal contact resistance between neighboring coatings. This resistance was determined for example by Gitzohofer et al. [318] for the contact between ZrO_2 and Mo deposits to be about $R_T = (2.0–3.5) \times 10^{-4}$ m^2.K/W and to decrease with temperature.

5.4.2 Emissivity

The solid materials may contain atoms and molecules emitting the photons. The photons may collide with other atoms and molecules and can be absorbed inside the material. On the other hand, the photons close to the solid's surface may leave the solid in a way similar to that of energetic electrons described in the Section 5.3.5. Emission of photons is consequently mainly a surface phenomenon and concerns the zone being in distance of about 1 μm from the solid surface [319]. The transport of radiative energy from the surface is function of three main factors: (i) wavelength of photons, λ; (ii) temperature of emitting surface, T; and (iii) directional distribution of emission or emission angle θ ($\theta = 0°$ for emission perpendicular to the surface).

The discussion about emission is facilitated by the concept of ***blackbody***, which emits radiation only in function of temperature, T, and wavelength, λ, and having the spectral emittance given by ***distribution of Planck*** given by following equation [320]:

$$\frac{dI(\lambda,T)}{d\lambda} = \frac{2\pi hc^2}{\lambda^5\left(e^{\frac{hc}{\lambda kT}} - 1\right)} \tag{5.6}$$

in which h is Planck constant, c is light speed, and k is Boltzmann constant. An example is sun, which emits as the *blackbody* having temperature of $T = 5800$ K and an important part of emitted spectrum is in the visible region with the wavelengths in the range of $0.38\,\mu\text{m} \le \lambda \le 0.75\,\mu\text{m}$.

The emissivity of *blackbody* is constant for all wavelengths and equal to $\varepsilon(\lambda) = 1$. It can be added that there also is a concept of ***graybody*** whose spectral emissivity is constant for all wavelengths, $\varepsilon(\lambda) = \text{const}$. Consequently, the emissivity of any surface can be defined as the ratio of radiation from this surface at given wavelength and temperature to the radiation emitted by *blackbody* at the same λ and T. As real surfaces may emit differently in function of angle θ and the emissivity is integrated over the emission hemisphere and $\varepsilon(\lambda, T)$ is called ***hemispherical emissivity***. However, a reasonable approximation is such that this emissivity has the value close to that at the angle $\theta = 0°$ being called ***normal emissivity***. Finally, ***total emissivity,*** $\varepsilon(T)$, is defined as *normal emissivity* averaged over all wavelengths.

Total emissivity of dielectrics varies with temperature in a different way than metals [319, 320]. Namely, the value of $\varepsilon(T)$ of dielectrics decreases with increasing temperature and increases with it for metals. However, it must be underlined that the surface condition and, in particular, the roughness and the level of oxidation can considerably influence the real values for different deposits:

- atomistic deposits generally have a smooth surface and their emissivity is similar to that, which characterizes bulk materials. For certain wavelengths and for thin and very thin layers, the substrate contributes to the *total emissivity* value. This situation may occur happen when the substrate has high value of $\varepsilon(\lambda, T)$ and, at the same time, the deposit is transparent for the emitted spectrum.
- granular deposits have much rougher surfaces. Even their polishing does not render them perfectly smooth because of the presence of open porosity. Therefore, the values of $\varepsilon(\lambda, T)$ depend not only on coating material but also on the microstructure and, consequently, on entire deposition process.

The emissivity of some metallic and ceramic deposits obtained by different methods including the details of deposition processes and the deposits properties influencing emissivity are presented in Table 5.38[9].

9 The experimental points of emissivity at different temperatures made in Ref. 307 and shown in lines 4–8 of Table 5.38 were approximated by linear regression equation, and the coefficients of correlation were in the range of 0.752–0.963.

Table 5.38 Emissivity of some atomistic and granular films and coatings of metals, alloys, and ceramics.

		Deposition method			Emissivity		
Number	**References**	**Method**	**Deposit's material, wt. %/ substrate, wt. %**	**Deposit's thickness, μm/ surface roughness, μm**	**Measurement**	**Values**	**Remarks**
Metals and alloys							
1.	[321]	*DC sputtering*	Au/Ni alloy K424 with Co, Cr and Al,Ti, Mo and Fe	1000/	*ir* emissometer	0.18 to 0.28 increasing after heat treatment	Tested at wavelength, $\lambda = 3–14\,\mu m$
2.	[322]	*Magnetron sputtering*	Ag/stainless steel 316	0.04 to 0.19/	—	0.02 to 0.05 increased with films thickness	Calculated for $\lambda = 2–25\,\mu m$
3.	[323]	*APS*	Mo	$/R_t = 15$	Photoelectric detector	See Fig. 5.22	$\theta = 0°$
4.	[307]		Ni + 20Cr/stainless steel	100/	*ir* camera		Tested at wavelength, $\lambda = 2–5.6\,\mu m$ for $\theta = 0°$
Ceramics							
5.	[307]	*APS*	Al_2O_3/stainless steel	100/	*ir* camera	See Fig. 5.23	Tested at wavelength, $\lambda = 2–5.6\,\mu m$ for $\theta = 0°$
6.			*30CSZ*/ stainless steel				
7.			Cr_2O_3/ stainless steel				
8.			NiO + 45 Fe_3O_4/stainless steel				
9.	[324]		*8YSZ*/stainless steel 316 L	135/2.1 and 10.5	Reflectance tested using spectrometer	0.43 and 0.54 depending on roughness	Tested at the wavelength $\lambda = 3–5\,\mu m$, coating porosity was about 8 %
10.	[325]		*LMA*/free- standing	200–300/	Tested using spectrometer	See Fig. 5.23	
11.	[326]		$MoSi_2$/stainless steel 316 L	$30–40/R_a = 6.7$	*ir* thermograph		Tested at the wavelength $\lambda = 0.78–2.5\,\mu m$
12.	[327]	*APS* and *IPS* (Ar atmosphere)	ZnO + 3 and 22 Al_2O_3/steel	—	Spectrophotometer	See Fig. 5.24	Tested at the wavelength $\lambda = 1.6–15\,\mu m$ at *RT*

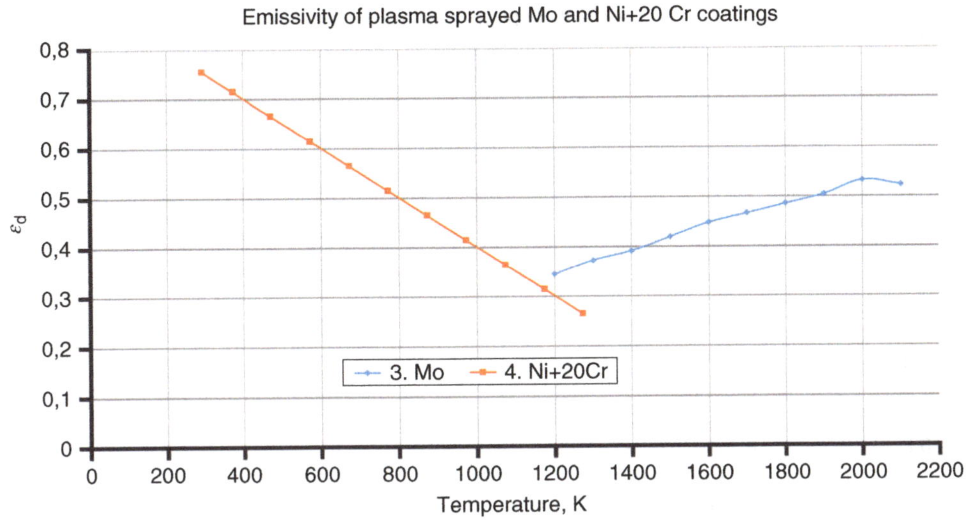

Figure 5.22 Emissivity of Mo and of NiCr alloy coatings obtained by plasma spraying in function of temperature. The properties of coatings are shown in Table 5.38 and the numbers of coatings corresponds to the lines of this table.

Low emissivity thin films of Au and Ag were obtained by sputtering (see lines 1 and 2 of Table 5.38, respectively). The values of emissivity were in the range of 0.05 (Ag film) to 0.18 (Au film) and depended on heat treatment after deposition and on film thickness. Fig. 5.22 shows emissivity of plasma sprayed Mo and NiCr plasma sprayed coatings. The as-sprayed molybdenum deposit has an emissivity, which increases with temperature. The surface of coating just after spraying is rough which could have influenced the measured emissivity. The emissivity of plasma sprayed NiCr alloy decreases with temperature. This is contrary to expected behavior for metal and alloys. This effect could have resulted from oxidation of particles at spraying as well as from the oxidation during emissivity measurement.

The emissivity of plasma sprayed ceramic coatings is greater than that of metals and alloys films and coatings in the temperature ranging from *RT* to 1700 K (see lines 5–11 of Table 5.38). The values of emissivity decreased with temperature. Hexaaluminate, such as $LaMgAl_{11}O_{19}$ (*LMA*) having magnetoplumbite structure, is a material tested for high-temperature application such as *TBC* (see line 10 of Table 5.38). The emissivity of this oxide, shown in Fig. 5.23, can increase by addition of multiple rare – earth elements [325]. Similarly, the emissivity of $MoSi_2$ coatings in high temperature can be increased by plasma spraying of mixture of $MoSi_2$ + 20 wt. % of SiC from about $\varepsilon_d = 0.84$ at $T = 1473$ K shown in Fig. 5.23 up to $\varepsilon_d = 0.90$ [326]. Finally, the plasma sprayed coatings using mixed powders of zinc oxide and aluminum oxide have the emission different in medium *ir* ($\lambda > 3\,\mu m$) than in solar emission range i.e. $3\,\mu m > \lambda > 0.3\,\mu m$. This spectral selectivity is useful in the application such as ***solar energy receivers***. The emissivity of the coating sprayed using APS and IPS techniques in function of radiation wavelength is presented in Fig. 5.24

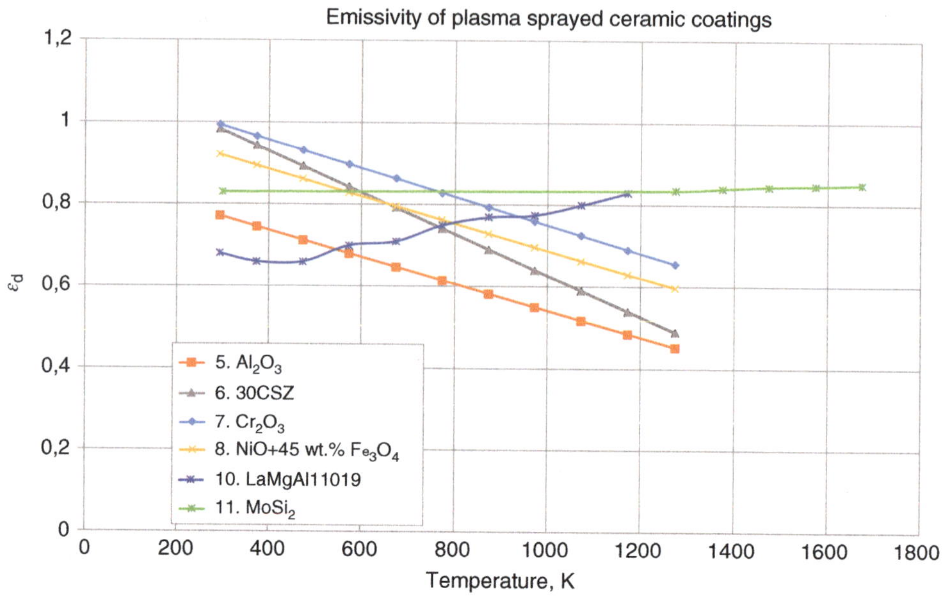

Figure 5.23 Emissivity of ceramic coatings obtained by plasma in function of temperature. The properties of coatings are shown in Table 5.38 and the numbers of coatings corresponds to the lines of this table.

5.5 Corrosion Resistance

Corrosion can be defined as a chemical attack of a deposit by an aggressive environment, such as an aqueous solution, molten salts, molten metals, or corrosive gases [106]. Films and coatings are mainly applied against two types of chemical corrosion, discussed already in Section 4.5, namely: (i) aqueous corrosion; and (ii) hot corrosion caused by hot gases.

Each of these types requires different deposits, and the corrosion mechanisms are not the same. Moreover, it should be added that in the practical applications of deposits, corrosion is accompanied by wear. For example, seawater corrodes structural elements of a boat. These elements can be simultaneously submitted to cavitation erosion by the sand. Another example is the hot corrosion of the blades of aeronautical turbines being oxidized by a gas and associated by erosion caused by the small solid particles present in the air sucked into a turbine.

5.5.1 Aqueous Corrosion

Aqueous corrosion, sometimes called galvanic corrosion, occurs when metals or alloys are in contact with a liquid. Protection can be another metal with a more positive reduction potential or more noble as shown in Table 4.5. The corrosion tests are frequently made using *polarization cell* using an electrolyte and a reference electrode, shown in Fig. 4.26. The tested sample attached to the working electrode is anode, and the reference electrode being a cathode is frequently calomel being mineral of Hg_2Cl_2. The working electrode is

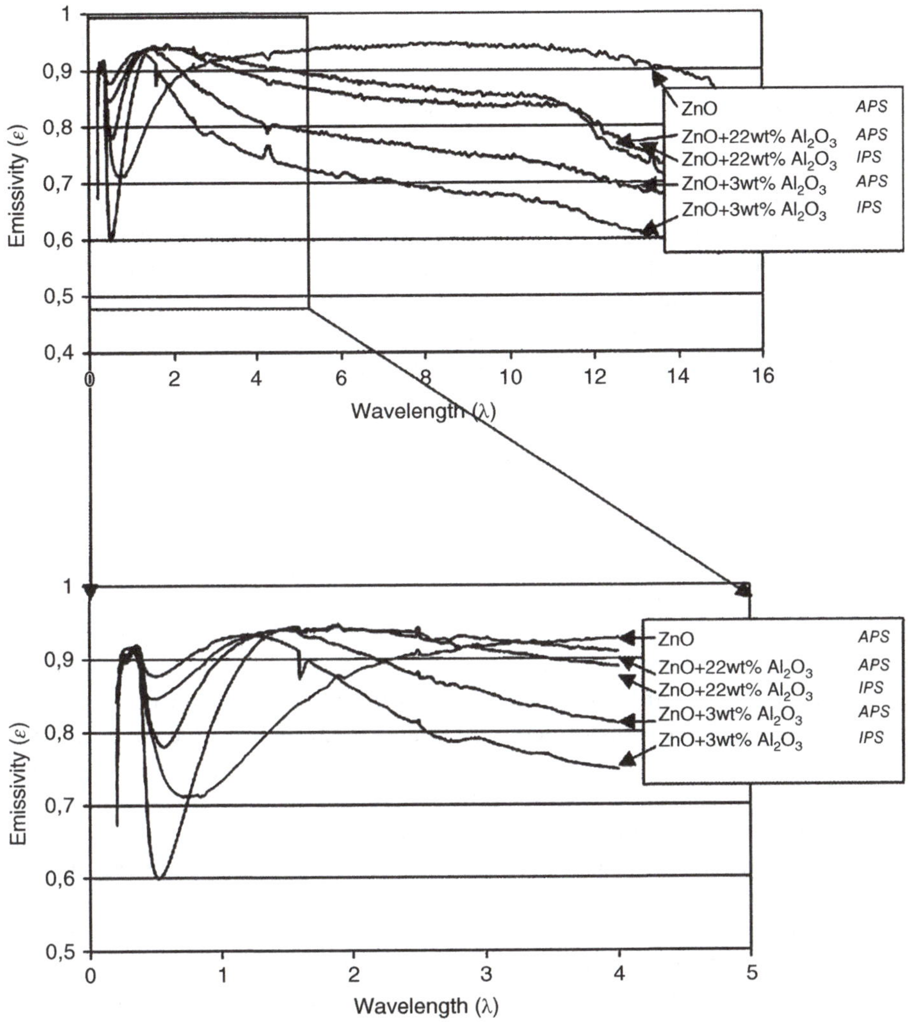

Figure 5.24 Emissivity at *RT* at different wavelengths of coatings plasma sprayed using powders ZnO, ZnO + 3 wt.% Al_2O_3 and ZnO + 22 wt.% Al_2O_3 in air (*APS*) and in inert gas (*IPS*) [327].

tested by application of the potential scanned from positive to negative values, and a typical curve of potential vs. current density is shown in Fig. 4.27. The materials with high aqueous corrosion resistance would have this curve with small passivation current and extended passive zone [328].

The corrosion can be also categorized by taking into account the microstructural effects in the materials as [329]:

- ***galvanic corrosion*** of two metals with different oxidation potential;
- ***intergranular corrosion*** resulting in preferential dissolution of crystal grain boundaries;

- ***pitting corrosion*** which starts with the anode potential between passive and transpassive zone and corresponds to the fracturation of the dense film created in the passive zone;
- ***crevice corrosion*** which develops under the deposits.

5.5.1.1 Atomistic Films

Corrosion-resistant atomistic deposits must ensure the hermetic protection of a substrate. Even in the case of less noble (sacrificial) anodic metallic films, there is no interest of getting them dissolved it too quickly. This is why the deposits must be dense and the columnar structure, such as in zone 1 of the structure zone model (*SZM*) discussed in Section 3.2.4, is not recommended. It is preferable to have a grain structure similar to characterizing zone T of the *SZM*. Some examples of atomistic deposits useful against aqueous corrosion and tested using *polarization cell* are shown in Table 5.39.

Anodic (sacrificial) films of zinc or aluminum have been on steel substrates by evaporation or ion plating for several decades. The industrial installations allow obtaining 15 μm thick films on steel strips moving with a speed of 30 m/min and having temperature up to 573 K by evaporating 45 kg/h of zinc [106]. Another metallic film of chromium described in line 1 of Table 5.39 protects against corrosion by forming dense structure and by avoiding the formation of columnar microstructure being permeable to corrosive liquid. The dense structure was achieved by ion bombardment and by optimizing the angle of sample orientation at Cr film deposition. Consequently, the film microstructure could be transformed from that characteristic for the zone 1 to that of zone T of *SZM* shown in Fig. 3.4b. Similar effort of the microstructure transformation from columnar to dense one was made for Ta_2O_5 film obtained by *LCVD* method (see line 2 of Table 5.39). The microstructure was modified by increase of the temperature of substrate at deposition and by decreases of laser beam density of energy and of laser pulses repetition rate.

The sputtered films of biocompatible Nb_2O_5 and of TiC were tested to protect against corrosion caused by the liquids present in human body acting on medical-grade stainless steel and on Ti6Al4V alloy used in, e.g. orthopedic devices (see lines 3 and 4 of Table 5.39, respectively).

Hard and wear-resistant films of TiN or TiAlN obtained by sputtering have good corrosion resistance in HCl and NaCl aqueous solutions [334]. The tested films were deposited by sputtering, and the parameters were optimized to reach possibly dense microstructure. Finally, another titanium including compound TiB_2 was sputtered of steel substrate improving only slightly its corrosion resistance against sulfuric acid solution [335]. This resistance increased when multilayer with Ti layer on the bottom, followed by TiB_2 and, by an outer layer of Al_2O_3, was applied (see line 6 of Table 5.39).

5.5.1.2 Granular Coatings

The resistance to aqueous corrosion of granular deposits depends mainly on the kind of sprayed material and on the coating's porosity. On the other hand, although the sprayed metals and alloys coatings have low porosity, the corrosion resistance of bulk oxides is superior. Consequently, the deposition of oxide coatings should be made using energetic

Table 5.39 Examples of atomistic films applied against aqueous corrosion tested using *polarization cell*.

			Deposition method				Corrosion test				
Number	**References**	**Substrate**	**Method**	**Film material/ thickness, μm**	**Deposition temperature, K**	**Microstructure**	**Electrolyte**	**Reference electrode**	**Corrosion current density[a], A/cm²**	**Corrosion potential, V**	**Remarks**
1.	[330]	Iron AISI 1018	*Magnetron sputtering* (plasma enhanced)	Cr /6.8	758	Zone T of *SZM* Fig. 3.4b	NaCl water solution 0.5 mol/L	Calomel	$4–5 \times 10^{-5}$	−0.35	Corrosion potential at calomel electrode
2.	[331]	Iron	*LCVD* using KrF laser	Ta_2O_5/3	673	Dense			Passivation current of 10^{-10} density	−0.5	Corrosion potential on sample electrode (cathode)
3.	[332]	Stainless steel AISI 316 L	*Magnetron sputtering*	Nb_2O_5/0.15	—	Amorphous	NaCl 8.9 % water solution		about 10^{-5}	−0.15 to 0	Pitting corrosion observed for positive potential
4.	[333]	Ti6Al4V	*RF sputtering*	TiC/	353	Dense	*SBF*	Pt	10^{-4}	−0.25	Microstructure depended on RF power
5.	[334]	Stainless steel AISI 316	*Magnetron sputtering*	TiN/2.5	573	Fine grained	HCl 3.4 % water solution	Pt and calomel	8.3×10^{-7}	−0.3 (open circuit)	Ti film of 0.7 μm used as intermediate film
6.	[335]	Steel ASP 23		TiB_2/6.3	473 −523	Dense, hexagonal	H_2SO_4 water solution 0.4 mol/L	Calomel	10^{-2}	−0.419	Multilayer of Ti, TiB_2, and Al_2O_3 lower corrosion current density to 10^{-6} A/cm²

[a] Corrosion current was supposed corresponding to the cathode potential equal to 0 V for published curves.

spray techniques such as *D-gun* or *APS*, which enable obtaining low-porosity deposits. Moreover, the post-spray treatment such as impregnation using polymer sealers can be used to decrease the open porosity.

Some examples of granular deposits tested against aqueous corrosion are shown in Table 5.40. The coatings of metals such as Al or Zn were deposited using cold spray method (*CGSM*) as shown in lines 1 and 2 of Table 5.40. This technique enabled obtaining low porosity and well-adhering coatings to protect magnesium alloy and steel substrates against corrosion, respectively. The coating of iron alloyed with chromium and other metals was sprayed using *APS* and *HVOF* techniques and tested against HCl aqueous solutions (lines 3 and 4 of Table 5.40). The *HVOF* method enabled obtaining denser coating having better corrosion resistance. An important aspect of this resistance was the nitrogen being used as working gases at spraying. This gas dissolved close to the surface of sprayed coatings enabled formation of nitrides improving their corrosion resistance.

The plasma sprayed oxides, in particular, alumina are frequently tested against aqueous corrosion (see line 5 of Table 5.40). Porosity is an important factor, which determines the corrosion resistance. To limit this influence, the following technological means can be used:

- sealing of ceramic to fill coatings' open porosity;
- deposition of thicker deposits to avoid the formation the channels connecting their surface with the substrate [339];
- spraying of a dense bond-coat of metal or alloy being more corrosion resistant than the substrate;
- reduction of the porosity of the oxide coatings by mixing and, eventually, alloying with another oxide, such as titanium oxide added to alumina (see line 6 of Table 5.40).

The composite coatings including metal or alloy matrices are applied against adhesive wear and abrasion (see Section 5.2.4.2). As wear and abrasion are frequently accompanied at coatings' service by corrosion by, e.g. seawater, the composite coatings sprayed using *HVOF* were tested. The results of long-term corrosion test of as-sprayed composite coatings are shown in line 7 of Table 5.40. The authors of this study tested also the post-spray treatment of the coatings by sealing using silicone resin applied under vacuum. Such treatment improved their corrosion resistance by hindering the penetration of salt spray into coating [341].

5.5.1.3 Bulk Deposits

Bulk deposits tested against aqueous corrosion, shown in this section, were obtained using the techniques of plasma (*PTA*) and laser melting (*1SLD*). Some studies presented also laser remelting (*2SLD*). The substrates were typically steels and aluminum alloys, and the coatings were the alloys having the chemical composition, which improved by corrosion resistance of deposits. The bulk deposits obtained using laser remelting had the microstructure comparable to the alloys obtained by rapid solidification [342]. The remelting modified the initial coating by:

- increase of the volume of heat affected zone (*HAZ*) shown in Fig 2.52;
- formation of new crystalline phases;
- reduction of porosity and homogenization of the coating's structure.

Table 5.40 Examples of thermal sprayed coating tested against aqueous corrosion.

			Thermal spray process			Corrosion test				
Number	**References**	**Substrate, wt. %**	**Method**	**Coating material, wt. %/thickness, μm**	**Porosity, %/ microstructure**	**Electrolyte**	**Reference electrode**	**Corrosion current, A/cm^2/ corrosion rate, $mg/(cm^2 . day)$**	**Corrosion potential, V**	**Remarks**
Metals										
1.	[336]	Mg + 3Al + 1Zn	*CGSM*	Al/200-250	very low/	Fog of NaCl solution	—	/0.45	—	Test ASTM B117
2.	[337]	Carbon steel		Zn/108	0.8/HCP – Zn	solution of Na_2SO_4	Hg_2SO_4	1.2×10^{-5}	-1.574	Tested using *polarization cell* to optimize spray parameters
Alloys										
3.	[338]	Steel St 37	*APS*	Fe + 23Cr + 6Ni + 4Mo + 2Mn	1.5/lamellar	solution of 0.1 N HCl	Calomel	2×10^{-4}/	−0.4	Tested using *polarization cell* corrosion current at cathode potential of 0 V
4.			*HVOF*		0.2/dense no lamellar			4×10^{-4}/	−0.3	
Oxides										
5.	[339]	Steel Q235 steel, pre-coated with 200 μm Ni	*APS*	95 Al_2O_3 /300–400	/γ-Al_2O_3	5% HCl boiling	—	/48	—	Corrosion rate was given in [$mg/(cm^2.h)$]
6. 7	[340]	Steel Q235 steel, pre-coated NiAl		Al_2O_3 + 13TiO_2 /	1.75 (connected porosity)/dense lamellar	HCl 1N boiling	—	/27	—	Test duration up to formation of bulges i.e. 23 h
Composites										
7. 9	[341]	Steel Q345	*HVOF*	WC + 20Cr_2C_3 + 7 Ni /40 (estimated)	1.5 as-sprayed/ dense	NaCl 5 wt. % solution	'—	3.4×10^{-7}/	−0.13	Tested after 70 d

Some examples of bulk deposits tested against aqueous corrosion are shown in Table 5.41. The alloy of Co, Cr, Cu, Fe, and Ni shown in line 1 of Table 5.41 was prepared by mixing of powders and tested against corrosion after *PTA* preparation with and without the addition of Nb powder. The NiCrSiB alloy, used frequently against aqueous corrosion, was obtained by optimized *PTA* technique as shown in line 2 of Table 5.41. Other discussed coatings were deposited by laser alloying.

CO_2 *cw* laser was used to obtain Cr alloys with Fe included in steel substrate in one step (line 3 of Table 5.41). Another alloy, included also Ni and Mo to have chemical composition similar to that of stainless steel AISI 316, is shown in line 5 of this table.

The coating obtained by laser remelting in two stages (*2SLD*) was obtained using Al substrate and ytterbium fiber *cw* laser included Al alloy with W. The coating had better corrosion resistance than the substrate (see line 4 of Table 5.41). Finally, the *2SLD* processing using Al and Al alloy as substrate to obtain alloys with Cr using not precised laser by injection of powder of Al with Cr followed by laser remelting is shown in line 6 of Table 5.41. The authors observed the formation of intermetallic phases and of passive film including CrOOH and/or Cr_2O_3 increasing the pitting potential at aqueous corrosion tests.

5.5.2 Hot Corrosion

Hot corrosion results from the chemical reactions resulting in contact of films or coatings surfaces being in contact with hot gases or hot liquids. As the most frequent reaction in practical use of deposits is oxidation of metals and alloys, the definitions concerning the chemical reaction at high temperatures are focused, for simplification, on contact with hot oxygen.

The kinetics of oxidation can be described by the increase of mass on unit area, w, as shown in the following equation:

$$w = \frac{\Delta m}{A} \tag{5.7}$$

The increase of mass Δm can be determined by the thermogravimetric measurements carried out in the oxidizing atmosphere under controlled temperatures for chosen time duration. The volume of oxide formed on the oxidized surface, V_O, divided by the volume of oxidized metal at the same temperature, V_m, is defined as a coverage factor Δ [329, 348]:

$$\Delta = \frac{V_O}{V_m} \tag{5.8}$$

The coverage factor enables predicting the morphology of oxide layer growing with time. Consequently, if the value of factor is:

- $\boldsymbol{\Delta < 1}$, the oxide layer grows with time in a linear way and is not continuous and does not protect the metal as shown in left side of Fig. 5.25. As the volume of created oxide is lower than that of the metal coating, tensile stresses are generated in the oxide layer leading to its cracking[10]. Such kinetics of oxidation regime is characteristic for light alkali metals such as Li, Na, or K.

10 The difference in volumes corresponds also to the difference of thermal expansion coefficients enabling calculation of thermal stresses as shown in *Eq.* 3.4.

Table 5.41 Examples of coatings obtained using powders feedstock using bulk methods including plasma and laser melting and tested against aqueous corrosion.

			Coating deposition					Corrosion test				
Number	References	Substrate	Method	Powder composition, wt. %/size, μm	Power, kW/power density, kW/cm^2	Thickness, μm	Microstructure	Electrolyte	Reference electrode	Corrosion current, A/cm^2	Corrosion potential, V	Remarks
1.	[343]	Steel Q235	*PTA*	Co + Cr + Cu + Fe + Ni equiatomic/ 25–110	4.8	<2000 of preplaced deposit	*FCC* phase	HCl 6 mol/L water solution	Calomel	11×10^{-5}	−0.403	Initial powder was preplaced and glued to the substrate
2.	[344]	Steel 15HM		Ni + 16Cr + 4Si + 3B + 1C/d_{50} = 57	Arc current = 70 A	—	Cr-rich dendrites in Ni-rich matrix	HCl 3.5 % water solution	Ag/AgCl	12.8×10^{-6}	−0.468	Arc current tested up to 100 A
3.	[345]	Steel with Cr and Mo	*1SLD*, alloying by injection of Cr powder	Cr/	3	—	Finely grained	H_2SO_4 1 mol/L water solution	Calomel	0.487×10^{-3} up to 0.669×10^{-3}	−0.5094 up to −0.5019	CO_2 *cw* laser was used
4.	[346]	Al 1100	*2SLD*, alloying with W including air sprayed slurry	W/1–5	0.8 to 1.2/	220 prior to laser melting	Al, W and Al_4W phases	NaCl 3.5 wt. % water solution		0.17×10^{-6} up to 0.47×10^{-6}	−0.736 up to −0.733	Ytterbium fiber *cw* laser was used
5.	[347]	Steel ST 37	*1SLD*, alloying	Cr + 38 Ni + 8Mo/d_{max} = 56;90; 45 for Cr, Ni, and Mo	/30 to 150	1500 (typical)	Alloys martensitic, austenitic	NaCl 3% water solution		10^{-5} up to 10^{-4}	−0.05 up to 0.05	CO_2 *cw* laser used, laser scan speed was tested
6.	[342]	Al and Al alloy ANSI 7175	*2SLD*, Cr + Al powders injection and remelting	Al + 25 Cr/ d_{max} = 150;56 for Al and Cr	/110 to 260	600 to 700	$Al_{17}Cr$, $Al_{11}Cr_2$ and Al_7Cr			$<6 \times 10^{-4}$	−0.6 up to −0.3	Content of Cr powder varied from 5 wt % to 12 wt %

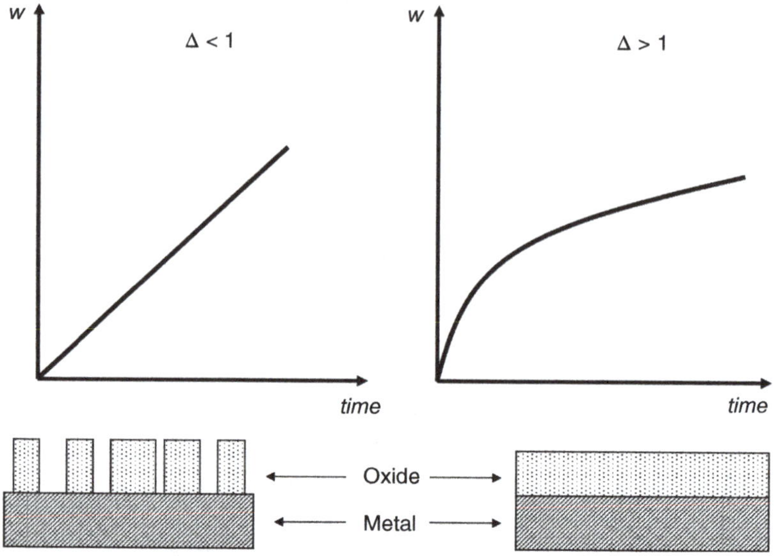

Figure 5.25 Increase of mass of metal surface per unit area, *w*, in function of oxidation time for different coverage factors (inspired by [329]).

- **Δ > 1**, the kinetics of oxide growth decreases with time and the formed layer is continuous, protecting the metals surface against hot gas. The layer is submitted to compressive stresses being easier to withstand. The curve of mass vs. time can have the following shapes:
  - parabolic, shown on the right side of Fig. 5.25, being characteristic for metals as Fe, Ni, or Cu;
  - logarithmic, observed at oxidation of Zn;
  - logarithmic, with a horizontal asymptote corresponding to formation of impermeable passivation layer typical for Al or Cr;
  - cubic, being characteristic for the metals as Ni, Cu, Zr at medium temperatures.

The degradation mechanism of a deposit subjected to hot corrosion depends on three main factors [348]:

- **material** of film or coating, i.e.:
  - metal or alloy;
  - ceramic;
  - duplex of a metal or alloy and ceramic;
- **environment** in contact with deposit's surface:
  - hot gas;
  - suspension of hot gas with solid particles (leading to corrosion with erosion);
  - suspension of hot gas and liquid;
  - suspension of hot gas and solid;
- **temperature** in contact of environment and surface:
  - isothermic contact with constant temperature;
  - cyclic variation of temperature.

Typical materials applied for hot corrossion resistant deposits are shown in Table 5.42. Considering only the resistance to oxidation, the alloys MCrAlY (M = Ni, Co, Fe, or the intermetallics of these metals) seem to be very often used at present.

Table 5.42 Examples of materials used as films or coatings against different types of hot corrosion.

Deposits' materials	Hot corrosion origins: Oxidation	Sulfidation	Liquid metals	Liquid salts	References
Metals	Cr	Nb, Ta	—	—	[348, 349]
Alloys	Intermetallic alloys, such as: NiAl, NiCr, FeCr, MCrAlY (M = Ni, Co, Fe)	MCrAlY (M = Ni, Co, Fe)	—	Intermetallic alloys, such as: NiAl, NiCr or FeCr, MCrAlY (M = Ni, Co, Fe)	[348, 350, 351]
Oxides	SiO_2, Al_2O_3, Y_2O_3 (alloyed with ZrO_2), ZrO_2 used often as duplex with MCrAlY	ZrO_2	Al_2O_3, ZrO_2, HfO_2	Al_2O_3, Y_2O_3 (alloyed with ZrO_2), ZrO_2 often used as duplex with MCrAlY	[348, 351, 352]
Nitrides	—	TiN, CrTiN	Si_3N_4, BN	—	[352, 353]
Silicides	$MoSi_2$, $TiSi_2$, WSi_2 (up to 2100K)	$MoSi_2$	—	—	[352]
Carbides	SiC	—	—	—	[352, 354]
Borides	HfB_2 (up to 1020K)	—	TiB_2 (tested against liquid Al)	—	[352]

The environment in contact with the deposit's service is specified in its service conditions. An example of the severe service conditions is the gases in contact with the coated turbines blades inside aeronautical engines. The gases are the product of the combustion of fuel (having such impurities as sulfur of vanadium) mixed with the gases entering to the turbine from the outside (such as oxygen). These gases may lead to the reactions of oxidation, carburization, or sulfidation of a deposit. During airplane flights over the sea, this gas entering from the outside may contain molecules of NaCl, which may result in the reactions leading to the formation of salt Na_2SO_4. This salt causes the degradation of the deposit [130, 351]. Finally, when airplane flies over a desert, the gas entering a turbine may contain small solid particles, and hot corrosion is then accompanied by erosion.

The temperature of environment being in contact with the deposit is rarely constant. Isothermic conditions are applied mainly in physico-chemical tests of coatings properties. On the other hand, the cyclic tests are closer to the service conditions and simulate, e.g. start and stop of an aeronautical engine turbine, cyclic contact with hot gas of a turbine in an energy conversion station, etc. This type of test is typically performed with a *burner rig* (see also Section 4.4.3.4). In the cyclic conditions, the rapid change of temperature associated with of formation of reaction products may lead to the rapid degradation of deposit called ***scale spalling***. Consequently, the evolution of deposit's mass with time, $w = f(t)$, is more difficult to predict than on isothermal condition, illustrated in Fig. 5.25 and the final mass can be lower than the initial one.

All techniques of physical methods of films and coatings deposition can be applied to obtain dry corrosion-resistant deposits. Generally, the deposits must have sufficiently low porosity to avoid the transport of the hot gas or liquid toward the protected substrate. Otherwise, the deposits must be sufficiently thick. Many applications require that hot corrosion resistance be accompanied by thermal insulation. The use of multilayers of different materials is a possible solution. An example is frequently used *TBC* having duplex configuration. Such multilayer is composed of hot corrosion-resistant 100 μm thick bond coating of MCrAlY obtained by *VPS* on the substrate. The thermal insulation is reached by 380 μm thick *8YSZ* external coating deposited by *APS* or by *EB PVD*.

5.5.2.1 Atomistic Deposits

To protect the substrate against hot corrosion a deposit should isolate it from the ambient atmosphere. Knowing that the thickness of atomistic films is generally limited to a few tens of micrometers, their microstructure should be impermeable to hot gas. This can be achieved by optimizing the deposition temperature to obtain the microstructure corresponding to the *zone 3* of the *SZM* model shown in Fig. 3.4 in entire film or in its part as shown by Buchanan and Little [354].

The atomistic deposition techniques make possible the synthesis of all types of materials used against hot corrosion. In particular, the techniques of:

- ***PVD*** by sputtering can be applied to obtain deposits of MCrAlY alloys (see line 1 of Table 5.43) oxides as Al_2O_3 or nitrides as TiAlN (lines 2 and 3 of Table 5.43);
- ***CVD, PE CVD* or *LCVD*** can be used to obtain deposits of intermetallic alloys, oxides, silicides, and borides [348].

The films of nitrides, being used frequently against wear and for decorative applications, do not resist against oxidation at high temperature. However, such films can be applied to protect against sulfidation as shown in line 3 of Table 5.43.

5.5.2.2 Granular Deposits

The most important application of thermally sprayed coatings is in thermal barriers (*TBC*) having duplex configuration. The MCrAlY alloys are used as a bondcoats, which resists against oxidation. *YSZ* as topcoat reduces the temperature of the substrate and is in direct contact with the products of fuel combustion such as liquid salts. The mechanism of *TBC* degradation includes:

- **thermal fatigue** resulting from the stresses generated by the action of heat flux on the materials included in the thermal barrier and having different coefficients of thermal expansion.
- **oxidation** of MCrAlY alloy and formation of oxides as, e.g. α-Al_2O_3, Y_2O_3, or $Ni(Cr,Al)_2O_4$ which increase their mass in parabolic way with time under isothermal conditions (lines 1 and 2 of Table 5.44).
- **peeling off** the ceramic outer coating in contact with molten salts observed, e.g. in 7*YSZ* (line 3 of the Table 5.44).

Table 5.43 Examples of atomistic deposits tested against hot corrosion.

			Deposition method				Corrosion test				
Number	References	Substrate	Method	Film material, wt. %	Thickness, μm	Microstructure	Temperature evolution	Temperature, K	Environment	Mass increase, w, mg/cm^2/ after time, t, h	Remarks
1.	[355]	Ni – base (Ni + Cr + W + Al + Ti)	*Magnetron sputtering*	Ni + 30 Cr + 8Al + 0.5Y	40	Columnar grains	Isothermic	1273–1373	$75Na_2SO_4 + 25K_2SO_4$ and NaCl	0.8/300	Water solution of salts composition in wt. %, corrosion product α-Al_2O_3
2.	[356]	Ni-base (Ni + Cr + Co + Al + W + Mo + Ta + Nb)		Al_2O_3	6–8	Amorphous		1123	$75Na_2SO_4 + 25K_2SO_4$	0.45/100	Sulfates were molten, their composition in wt. %
3.	[357]	Waspaloy (Ni + Cr + Co + Mo + Ti + Al + Fe)	*PVD arc heating*	Ti + 28Al + 23N	3–4	FCC cubic		973	$75Na_2SO_4 + 25NaCl$	1/25	Salts composition in wt. %

Table 5.44 Examples of thermally sprayed coatings applied against hot corrosion.

			Deposition method				Corrosion test				
Number	References	Substrate	Method	Coating's material, wt. %	Thickness, μm	Microstructure	Temperature evolution	Temperature, K	Environment	Mass increase, w, mg/cm^2/ after time, t, h	Remarks
1.	[358]	Rene' N5 (Ni + Co + Cr + Ta + Al + W + Re + Mo)	*VPS*	Ni + 22Cr + 10Al + 1Y	150	Crystals of Ni_3Al, NiAl	Isothermic	1423	Dry air	1/100	Coatings' surfaces were polished after spraying
2.	[359]	Steel	*HVOF*	NiCoCrAlYRe	—	Crystals of Ni_3Al, NiAl, σ -phase of Co + Cr		1323	He + 10 vol. % O_2	0.63/1500	Corrosion product was mainly α-Al_2O_3
3.	[360]	Inconel 718	*APS*	*7YSZ*	300	—		1273	Molten salt of $45Na_2SO_4 + 55V_2O_5$ as in norm ASTM G 41	Mass loss 1 g after 50 h	Bond coating was 100 μm of NiCrAlY, doping with Gd_2O_3 improved corrosion resistance

The MCrAlY alloys are sprayed frequently using technique *VPS* or *HVOF*. The direct comparison of oxidation resistance of coatings sprayed using these two techniques indicated better resistance of the *HVOF* ones [358].

The resistance against degradation caused by molten salts of ceramic topcoat in *TBC* can be improved by addition Gd_2O_3 to *YSZ* [360]. Otherwise, some new oxides, e.g. $La_2Ce_2O_7$, tested recently by Kandasamy *et al.* [361], can also be applied.

5.5.2.3 Bulk Deposits

The coatings obtained using bulk deposition methods are useful against hot corrosion. The examples of such coatings are shown in Table 5.45. Laser glazing of *APS* predeposits was applied to improve the oxidation resistance of *TBC* having duplex configuration (see line 1 of Table 5.45). The spallation of the laser-glazed specimens tested using cyclic oxidation tests occurred after 4 times more thermal-cycles than of the plasma sprayed ones (Fig. 5.26). Zaplatynsky [365] found out similar improvement of the resistance to oxidation of *TBC* resulting from laser glazing.

Laser glazing plasma sprayed Ni and Cr predeposited coating improved their oxidation resistance at 1073 K (line 2 of Table 5.45). The mass of oxides formed after 100 hour of testing was fivefold smaller.

The coatings of Ni-alloy obtained using plasma transferred arc (*PTA*) were useful to protect cast iron against oxidation by air at 1073 and 1173 K (line 3 of Table 5.45). The oxidation follows parabolic oxidation mass evolution in function of time at testing at 1073 K which results from the formation of Si-O oxide layer. The oxidation at 1173 K has mass evolution including also linear increase of oxide mass resulting from the formation of other oxides.

5.6 Optical Properties

The optical properties of the atomistic deposits fabricated, in particular, by *PVD*, have been traditionally tested [366, 367]. This application does not require deposits having important thickness but, on the other hand, require their high quality, i.e. control of: (i) impurities and of chemical composition; (ii) microstructure; and (iii) surface roughness. The fact that atomistic deposits are made frequently under vacuum and that their constituent species are atoms or molecules renders their quality better suited for the optical applications. On the other hand, an important advantage of granular methods of coatings deposition is their capacity to coat rapidly large surfaces. This was the frequent motivation for the studies on the optical properties of thermally sprayed coatings. Among the optical applications of films and coatings the most popular are probably the decorative and functional ones. Consequently, the presentation of optical properties is focused on these two applications.

5.6.1 Decorative Deposits

Decorative deposits are characterized by their color. As the visibility of colors results from the sensibility of human eye, it is not possible to express them by the physical constants discussed in Section 4.7. On the other hand, it is important to quantify them in a

Table 5.45 Examples of bulk deposits tested against hot corrosion.

			Deposition method						Corrosion test				
Number	**References**	**Substrate**	**Method**	**Laser power, kW**	**Scan speed, cm/s**	**Coating material, wt. %**	**Thickness, μm**	**Microstructure**	**Temperature evolution**	**Temperature, K**	**Environment**	**Mass increase, *w*, mg/cm^2/ after time, *t*, h**	**Remarks**
1.	[362]	Hastelloy X	2*SLD* with pre-deposits by *APS*	0.45	3.7	Ni + 22Cr + 10Al + 1Y	60	—	Cyclic	1373 for 1 h followed by cooling down to T_{amb} for 10 min	Air	see Fig. 5.26	Duplex TBC, glazed with pulsed CO_2 laser of $\nu = 800$ Hz
						12*YSZ*	220	Cubic after glazing					
2.	[363]	Mild steel		3	0.83–1.25	Ni	75	Smooth and poresfree after glazing	Isothermic	1073		5/100	Predeposits are Cr as bondcoat and Ni as outer one, glazing made by CO_2 *cw* laser
						Cr	75						
3.	[364]	Cast iron	*PTA*	—	—	Ni + 2.6SI + 2.5Cr + 1.3Al + 1.1Fe	—	Ni-Fe solid solution		1173		0.4/2	Arc current was 100 A

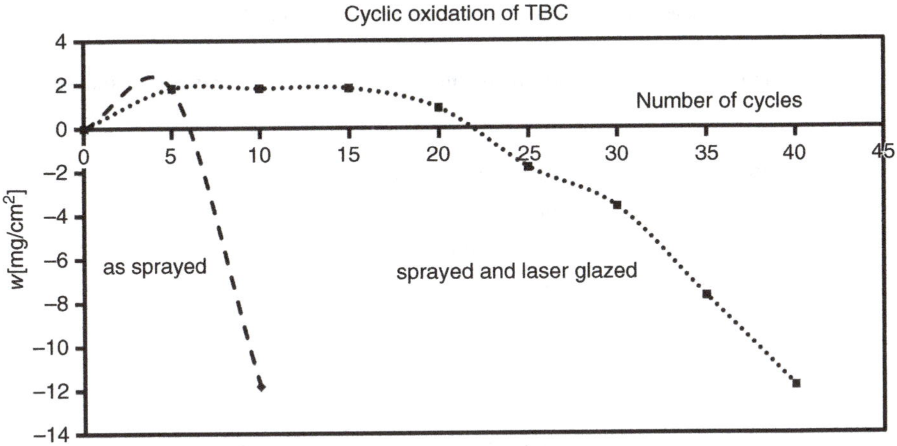

Figure 5.26 Kinetics of cyclic oxidation of a thermal barrier of the duplex type including 12*YSZ* ceramic and NiCrAlY alloy as a bond coating substrate obtained by plasma spraying on the Hastelloy-X and glazed by laser after deposition as reported in line 1 Table 5.45.

Table 5.46 Axes of the sphere of color space $L^*a^*b^*$ [368, 369].

Axis	Values	Remarks
L^*	Brightness ranging from $L^* = 0$ (black) to $L^* = 100$ (white)	Vertical axis placed at $a^* = 0$ and $b^* = 0$
a^*	Chromatic axis green-red ranging from negative values (green) to positive values (red)	Horizontal axes[a]
b^*	Chromatic axis blue-yellow ranging from negative values (blue) to positive values (yellow)	

[a] The values of the chromatic axes are, generally, comprised between −120 and +120.

way enabling the measurements. The popular method of the quantification, recommended by ***International Commission on Illumination***, corresponds to $L^*a^*\ b^*$ colors space[11]. The space corresponds to the three-dimensional sphere with the axes shown in Table 5.46. The colors can be measured using colorimeters, spectrophotometers, and other methods [369].

The colors of films and coatings vary with their electronic structure resulting in selective absorption of incident light. The electronic structure of thin films is influenced by their stoichiometry or/and microstructure. For example, the color of TiN is determined rather by the film's stoichiometry. Another nitride film, namely ZrN, has the color dominated by the microstructure [370]. The practical applications of thin films specify frequently some additional properties, e.g. wear resistance correlated with the hardness. Some examples of thin-films colors together with their hardness are presented in Table 5.47.

11 Some sources use also abbreviation *CIE* $L^*a^*\ b^*$ colors space.

Table 5.47 Colors and hardness of thin films synthesized using *PVD* [370].

Number	Film	Color variations	Vickers microhardness, GPa
Nitrides			
1.	TiN_x	Brown → brown yellow	24
2.	ZrN	Yellow → green	26
3.	Cr_2N, CrN	Metallic → brown	25
4.	TaN	Blue → gray	—
5.	(Ti,Al)N	Gold → dark blue	24–29
6.	(Ti, Zr)N	Golden	—
Carbides			
7.	TiC	Bright gray	15–29
8.	TiC/WC	Dark gray	14
9.	TaC_x	Yellow → brown	18
10.	SiC	Black	—

The color of thin film depends also on its thickness of and on the substrate roughness. For example, the brightness[12] of TiN film obtained by sputtering was influenced by the roughness of the steel substrate [369]:

- $L^* \approx 42$ for polished substrate having roughness of $R_a = 0.02\,\mu m$;
- $L^* \approx 81$ for corundum sandblasted substrate with $R_a = 0.78\,\mu m$.

Similarly, the operational deposition parameters have an influence on the films' color. For example, the variation of nitrogen flow rate applied at reactive sputtering modifies color of TiN_x film from brown to yellowed brown as shown in line 1 of Table 5.47. The increase of the flow rate resulted in the colors of films resulting in the following values of the horizontal axes in color space measured by Beck *et al.* [371]:

- $a^* \approx 2$ and $b^* \approx 30$ for N_2 flow rate of $3.5\,cm^3/min$;
- $a^* \approx 8$ and $b^* \approx 7$ for N_2 flow rate of $30\,cm^3/min$.

Thermal sprayed coatings was also tested to be used as color coatings. The flame sprayed (*FS*) coatings of copper on glass substrate were characterized by Tikkanen *et al.* [372]. Copper powder oxidized at deposition to Cu_2O gave red to the coating or to CuO, rendering the deposits black. The authors sprayed also different glasses composed of SiO_2, Na_2O, BaO, CaO, K_2O, B_2O_3, and other oxides and found out that the temperature of the substrate enabled obtaining non-porous and uniform glass coatings. The air plasma sprayed (*APS*) coatings of SnO_2 doped with the pigments as ***casserite*** and ***malayite*** on ceramic and metallic substrates and tested by Dosta *et al.* [373]. The colors of obtained coatings depended on pigment content and on the spray parameters such as: (i) arc current and (ii) working gases composition. The greatest modifications were observed in the brightness, which varied from $L^* \approx 56$ to $L^* \approx 72$ and in chromatic green-red axis ranging from $a^* \approx 3$ to $a^* \approx 19$.

12 Measured using goniospectroscopy under angle of 10°.

5.6.2 Functional Deposits

The functional optical deposits are realized mainly using atomistic methods and can be used in a multitude of applications. The applications results from the values of the refraction index as well as of the transmittivity and reflectivity in function of the wavelength. Their physical meaning is related to the interaction of light with deposit's presented in Section 4.7. In this section, the examples of following functional films are presented in Table 5.48:

- antireflecting;
- transparent;
- transparent and electrically conducting;
- nanocomposites.

Antireflecting films, being useful for optoelectronic elements, can be deposited in the direction of semiconducting substrates. An example is ZnO sputtered on Si substrate studied by Khoshman *et al.* [380]. The authors found the refraction coefficient in the range $n = 1.65–1.73$ decreases with the wavelength rising from $\lambda = 0.8$ to $2.5\,\mu m$. Another example of antireflecting film is amorphous *DLC* shown in line 1 of Table 5.48. The reflectivity for visible light is smaller than 20% as shown in Fig. 5.27. The film VO_2 having antireflecting properties was obtained using *CVD* technology under atmospheric pressure (see line 2 of Table 5.48).

The transparent deposits are presented in lines 3 and 4 of Table 5.48. CdS films were deposited by laser evaporation and were transparent for visible light. Slightly better transmissivity, for similar wavelengths, had Nb_2O_5 deposited by ion-assisted evaporation (*IBAD*) technique. The films had also index of refraction greater than $n > 2$ in visible light spectrum useful in many optoelectronic applications. Similar properties have nanocrystalline TiO_2 films obtained by magnetron sputtering assisted by ion bombardment and tested by Sarma *et al.* [381]. The films are transparent to visible and near *ir* light and can be used to filter out *uv* spectrum. Indium tin oxide (*ITO*) film is also frequently applied in optoelectronics. The film obtained by reactive sputtering and shown in line 5 of Table 5.48 can have low electrical resistivity being useful in some applications. Thin films of HfO_2 and Y_2O_3, being hard and having refraction index $n \approx 2$ at the wavelength of $\lambda = 0.5\,\mu m$ and obtained using laser beam evaporation, were tested by Thomsen-Schmidt *et al.* [382]. The films are useful in the applications requiring also high mechanical and chemical stability.

Several optical deposits are made using multilayers. An example of film having small reflectivity is a nanocomposite including very thin SiO_2 layers being 2.5 nm thick, separating 85 nm thick layers of Mo. The deposition method and reflectivity of such multilayers deposited onto Si substrate is shown in line 6 of Table 5.48.

5.7 Biomaterials

Biomaterials have been used intensively for prosthetic corrections of knee, shoulder, hip, and other parts of human body since World War II [383]. Initially, the first generation of materials implanted in the human body were bioinert metals, and polymers were studied starting from 1950s. Then, the second and third generations of biomaterials concerned mainly ceramics being bioactive and/or biodegradable initiated in 1980s. The last generation of biomaterials, which

Table 5.48 Examples of functional optical deposits obtained by atomistic techniques.

		Deposition process					Spectral optical properties				
Number	References	Method	Film material	Thickness, μm	Microstructure	Substrate	Wavelength, μm	Index of refraction	Transmittivity, %	Reflectivity, %	Remarks
Antireflecting films											
1.	[374]	*PE CVD*	*DLC*	up to 0.08	Amorphous	GaAs	0.4–0.8	2.4	—	see Fig. 5.27	a:*DLC* film on solar cells, *n* was constant in the wavelengths
2.	[375]	*CVD*	VO_2	0.32–0.35	Crystals in amorphous matrix	Glass	0.6–1.8	1.5	up to 80	–	Film deposition was at atmospheric pressure
Transparent films											
3.	[376]	*Laser evaporation*	CdS	0.5–1.2	Hexagonal and cubic CdS	Different substrates	0.5–0.8	—	70–80 for $\lambda > 0.56\,\mu m$	—	Nd:YAG *cw* laser was used in deposition, structure changed with deposition temperature
4.	[377]	*IBAD*	Nb_2O_5	0.3	Columnar structure	Quartz, silicon	0.4–1.2	2.42–2.14 for $0.4\,\mu m \leq \lambda \leq 1.2\,\mu m$	60–90 for $0.4\,\mu m \leq \lambda \leq 0.8\,\mu m$		Refractive index increased with ions current
Transparent and electrically conducting films											
5.	[378]	*Reactive sputtering*	*ITO*	0.05–0.1	Metallic to amorphous	Glass or quartz	0.55	—	up to 80 depending on annealing temperature	—	In_2O_3 observed after annealing at 623 K, lowest resistivity $\rho = 366\,\mu\,\Omega.cm$
Nanocomposite films											
6.	[379]	*Magnetron sputtering*	Mo/ SiO_2	(0.085/ 0.0025) × 5	Columnar structure	Si	0.25–1	—	—	15 to 0 for $0.83\,\mu m \geq \lambda \geq 0.25\,\mu m$	Periodic multilayer composite

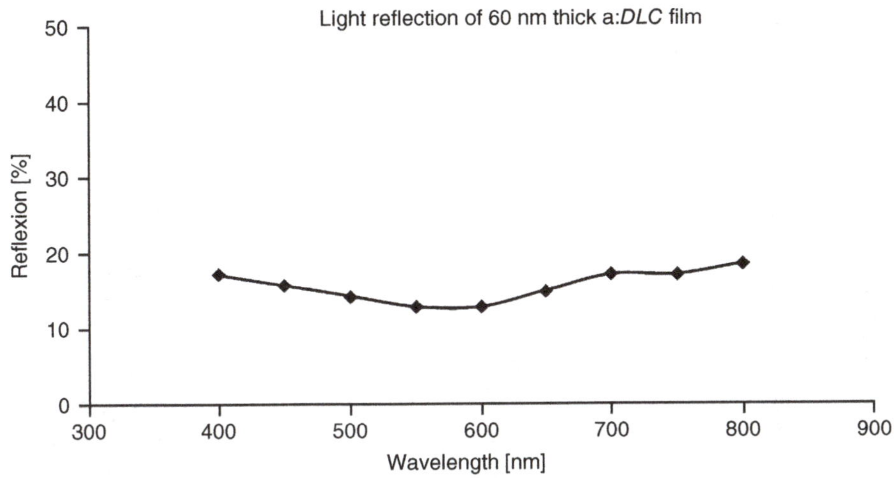

Figure 5.27 Reflection coefficient vs. light wavelength of an amorphous diamond deposit obtained by *PE CVD*. The description of the technology and deposit properties are shown in Table 5.48.

started to be studied in 2000s, are biomimetic ones [384]. The properties of biomaterials and implanted films and coatings required to reach a successful bioimplantation include [383–385]:

- **compatibility**, e.g. biocompatibility (deposits should be robust anti-bacteria and free of biotoxicity), positive human tissue reaction, chemical stability with time, controlled modification of properties in contact with the bone and, in short, material should mimic with regards to the natural bone;
- **mechanical properties** such as high adhesion to the substrate, modulus of elasticity and density, which should be equivalent to that of the human bone, porosity adequate enabling cell seeding, low friction coefficient, and small wear;
- **corrosion resistance** resulting from composition of material and its microstructure which should be excellent in the physiologic milieu (tested using *SBF*).

5.7.1 Atomistic Deposits

The films obtained using atomistic methods being actually tested to be used as surface modification in biomedical applications enable to achieve the mentioned requirements. The films belong to the following groups [385]: (i) hard films; (ii) thin metallic glasses; (iii) diamond-like carbon coatings; and (iv) bioceramics. Some examples of the studies including these films are shown in Table 5.49.

The chemical composition of films results from their specification. Thin-films technologies enable deposition of many materials having many useful properties such as high hardness and modulus of elasticity and corrosion resistance. The addition of some metals improves the antibacterial properties of the film. An example, presented in line 1 of Table 5.49 is frequently deposited TiN film with addition of Ag and Ta. The deposition method is magnetron sputtering of nitride associated with ion implantation of metals. The hardness of TiN films is shown in Fig. 5.2 and its corrosion resistance in Table 5.39. The metals were chosen to be added to improve the antibacterial properties (Ag) and corrosion resistance (Ta) of the deposit.

Table 5.49 Examples of atomistic films of biomaterials.

		Deposition process					Biomedical characterization					
Number	References	Method	Film material	Substrate	Thickness, μm	Microstructure	Category of biomaterial	*In vitro* test	Results	*In vivo* test	Results	Remarks
Hard films												
1.	[386]	*Magnetron sputtering* followed by *IP*	TiN with implanted Ag and Ta ions	Ti6Al4V	—	—	*Bioinert*	Test with *Escherichia coli* bacteria	96 % reduction of bacteria for Ag/Ta(4:1)	Test with rats	Positive results for all tested	Ratio of Ag/Ta from 1:1 to 4:1
Thin metallic glasses												
2.	[387]	*RF magnetron sputtering*	NbTaTiVZr (about 20 at. % each metal) and 6 at. % O	Si	2	Amorphous/ nanocrystalline	*Bioinert*	Cells culture test for up to 72 h	High cell viability	—	—	Composition belongs to *HEA*
Diamond like carbon films												
3.	[388]	*PLD* using ArF excimer laser	DLC with SiO dopant (0–30%)	Fused silica	0.03–0.11	—	*Bioinert*	*SBF* soaking for 48 weeks at $T = 310$ K	No delamination observed	—	—	Useful as antifogging film in laparoscope lens
Bioceramics												
4.	[389]	*PLD* using KrF excimer laser	*HA* with $CaSiO_3$ (10:1 in mass ratio)	Ti6Al4V	—	HA, $CaSiO_3$ crystalline, droplet grains	*Bioactive*	*SBF* soaking up to 28 days at $T = 309.5$ K	Weight increase due to formation of apatite layer	Test with rabbit sine bone	Good osseointegration	Observed apatite film on the surface of deposited film at *SBF* soaking

High-entropy alloys (*HEA*) having amorphous microstructure are intensively tested because of interesting mechanical properties, bio-corrosion, and *in vitro* biocompatibility. An example of such alloy film composed of NbTaTiVZr is presented in line 2 of Table 5.49 having high cells' viability inspiring further studies.

DLC films are frequently used for their excellent mechanical properties such as hardness (shown in Fig. 5.1) and modulus of elasticity (see Table 5.12). The addition of small quantity of Ag to DLC films obtained by PECVD technique improved the antibacterial properties [390]. The SiO doping of *DLC* deposited using *PLD* added the antifogging properties by reducing the contact angle of water droplets. This property is important in laparoscopes applied at surgery and the technology of films and the correct results of *in vitro* test are shown in line 3 of Table 5.49.

The bioactive hydroxyapatite (*HA*) is used as bone substitute because of the similarity of their chemical composition. *HA* promotes tissue growth which renders its application useful in orthopedic and dental applications [389]. The addition of $CaSiO_3$ (***wollastonite***) accelerates the dissolution rate of *HA* in physiological liquids. The films of this composition, obtained using PLD technique, were favorably tested *in vivo* and *in vitro* as shown in line 4 of Table 5.49.

5.7.2 Granular Deposits

The deposition using thermal spraying has an important practical advantage of coating large surface resulting from deposition in open atmosphere. Moreover, the microstructure of the coating is different as the coating grows up from the molten splats splashed on the substrate and on the sprayed coating surface. The melting process of the feedstock in open atmosphere results in some oxidation of metal and alloys, which is reduced in the processes carried out under vacuum and inert atmosphere as *VPS* or *IPS*. In addition, the open porosity of coatings is greater than that in the atomistic deposits, which increases the surface of contact with cells after implantation. The feedstock used to spray is most frequently powder used in deposition techniques as *APS* or *FS*. The powder manufacturing and properties are described in Section 2.3.1. More recently, the liquid feedstock such as suspension and solution started to be used in spray processes as *SPS* and *SPPS* presented in Section 2.3.3. The main groups of biomaterials tested as thermally sprayed coatings are: (i) bioactive ceramics as *HA* used, e.g. to coat a ***femoral stem*** as shown in Fig. 5.28 and being doped frequently with metallic ions; (ii) bioactive bioglasses; and (iii) metals having antibacterial properties.

Some examples of recently tested and described in the literature thermally sprayed coatings are presented in Table 5.50. Bioactive ceramics such as hydroxyapatite (*HA*) are used to coat metallic implants, but some of its mechanical and biological properties, e.g. brittleness

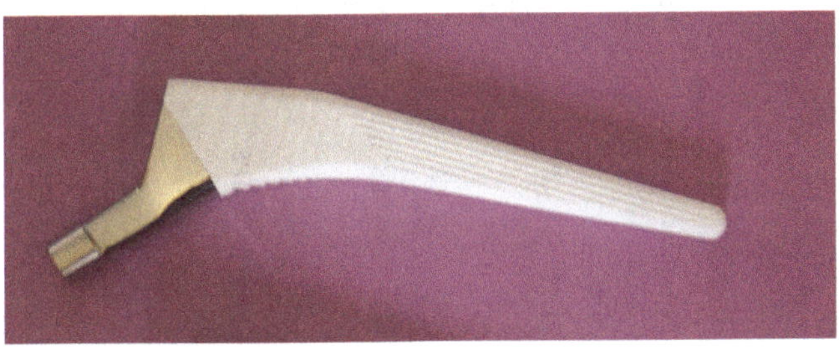

Figure 5.28 Hip prosthesis with a *femoral stem* coated by plasma spraying with bioactive *HA*.

Table 5.50 Examples of coatings of biomaterials obtained by thermal spraying on Ti6Al4V substrates.

		Thermal spray process					Biomedical characterization				
Number	**References**	**Processing method**	**Feedstock**	**Coating material**	**Thickness, μm/ porosity, %**	**Microstructure**	***In vitro* test**	**Results**	**Other tests**	**Results**	**Remarks**
Bioactive ceramics											
1.	[391]	*SPPS*	Solution of Cu-doped (3–10 mol. %) calcium phosphate	Cu doped *HA*	30–50/	*HA*, *TCP*, CuO	*SBF* immersion up to 21 d	Modification of structure (see Fig. 5.29)	Cytotoxicity test	Best viability of coating with 3 mol. % Cu	Formation apatite at *in vitro* test deduced from atomic emission spectroscopy
2.	[392]	*FS*	Powder, size 74–177 μm	*HA* and *HA* with ZnO	/26	Crystals of *HA* and $Ca_3(PO_4)_2$, ZnO, $CaZn_2(PO_4)_2$	Test with *Escherichia coli* bacteria	High inhibition rate after 3h	Cytotoxicity test	Good biocompatibility	Porosity was greater than in commercial HA coatings
3.	[393]	*APS*	Powder agglomerated, size 50–100 μm	*HA* and *HA* + 10 or 20 wt. % $CaSiO_3$	50–175/16–19	Crystals of *HA*, $CaSiO_3$ and of decomposition phase	Test in ringer solution (NaCl + $CaCl_2$+ KCl + $NaHCO_3$) for 24 h	Increase of oxygen content	Corrosion tests	Corrosion potential lowest for HA	Better corrosion resistance, greater microhardness observed in coating with $CaSiO_3$
Bioglass											
4.	[394]	*APS*	Powder of glass 62 W composed of [wt.%] 46 CaO, 40 SiO_2, 11 P_2O_5 and 3 MgO, size 63–100 μm	Bioglass of similar to powder composition	105/	Amorphous	*SBF* immersion up to 7 d	Formation of hydroxyl carbonate apatite	Biodegradation in ***buffer solution*** up to 10 weeks	Increase up to 5 weeks	The research will continue
Metals											
5.	[395]	*VPS* followed –(or not) by NaOH soaking and heat treatment	Powder of size 5–120 μm	Ta with 3 layers: inner; middle; outer	350–380 in total/1 inner layer – 8 upper layer	Crystals of Ta and its oxides, compound of Ta, O and Na as $Na_5Ta_{11}O_{30}$	*SBF* immersion up to 2 weeks	Formation of apatite after 3 d in bioactivated and heat treated sample	Cell culture test	Better biocompatibility than Ti6Al4V	NaOH activation: (i) improves bioactivity; (ii) not influence biocompatibility

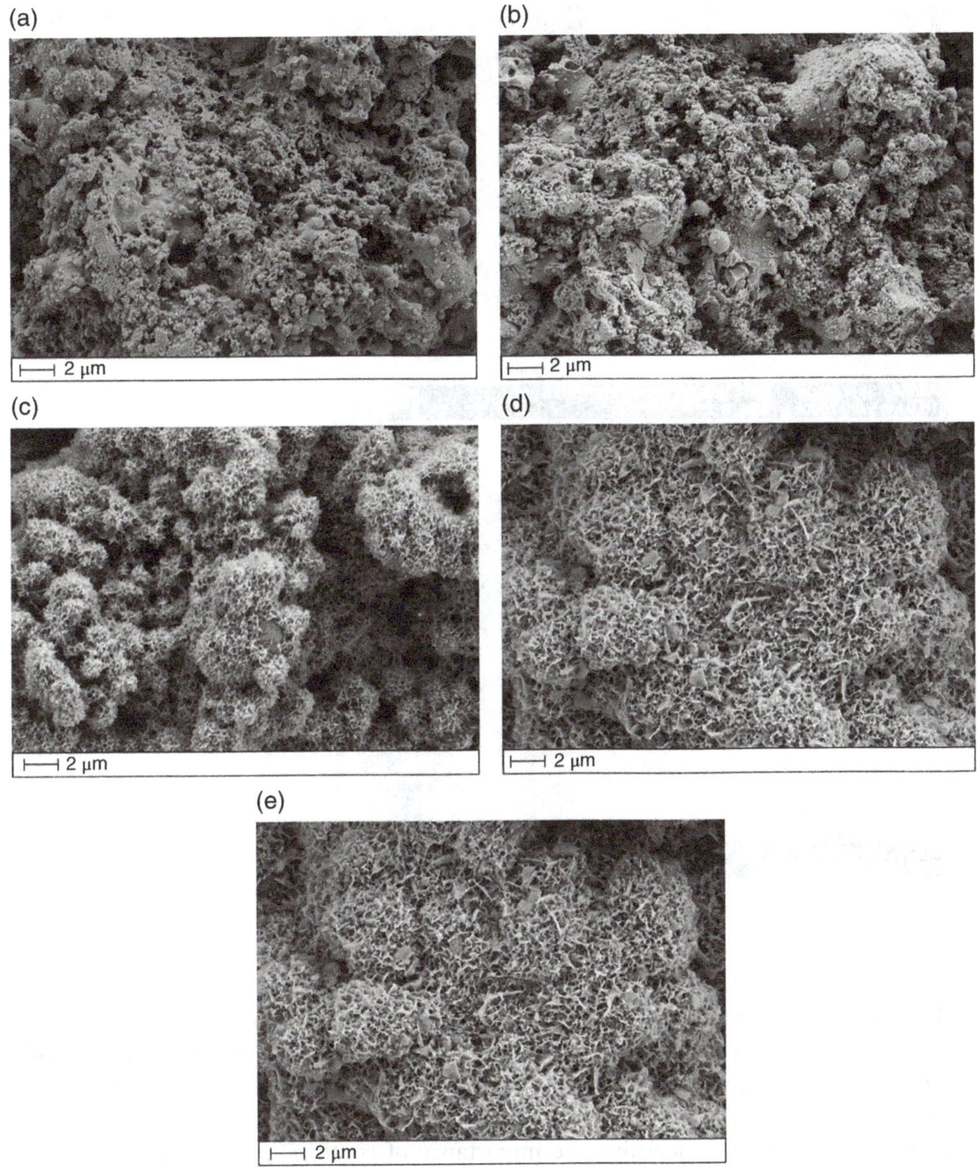

Figure 5.29 SEM (secondary electrons) images of the 3% copper-doped hydroxyapatite coatings obtained by *SPPS* submitted to immersion in *SBF* for [391]: (a) 0 day, (b) 3 days, (c) 7 days, (d) 14 days, and (e) 21 days.

as well as lack of antimicrobial effects, require addition of the dopants [396]. The dopants integrate the structure as singles ions, e.g. Cu^{2+}, shown in line 1 of Table 5.50 or Zn^{2+} shown in line 2 of this table and were also studied by Sergi *et al.* [397]. The doping can be also made using binary ions, e.g. Ag^{2+} and Sr^{2+} [398]. The corrosion resistance and greater hardness of plasma sprayed *HA* coating can be also achieved by addition of calcium silicate as shown in line 3 of Table 5.50.

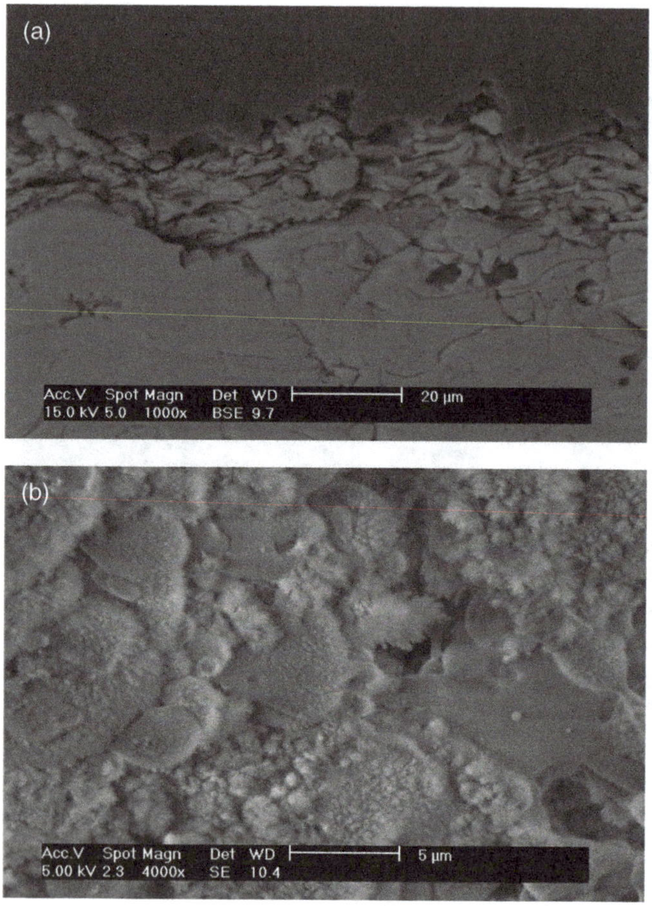

Figure 5.30 SEM images (secondary electrons) of bioglass film obtained by *SPS* on the top of *HA* coating obtained by *APS* [399]: (a) cross-section, (b) surface.

The bioglasses were studied as entire coating obtained by *APS* (line 4 of Table 5.50) or as a duplex composed of a thin topcoat obtained by *SPS* onto thick *HA* plasma sprayed deposit (see Fig. 5.30). The outer bioglass film was porous and has rough surface promoting bioactivity of the duplex.

Similar consideration concerning the importance of coatings' porosity and roughness was observed in the study of Ti coatings obtained by flame spraying on stainless steel substrates [400]. The authors compared these coatings with the denser ones, obtained using atmospheric plasma spraying, and found out that porosity was the factor promoting the incorporation of biomolecules at *in vitro* testing using *SBF*. Another metallic deposit, obtained using vacuum plasma spraying of Ta coating on Ti6Al4V substrates, is shown in line 5 of Table 5.50. The authors modified the operational spray parameters to obtain coating having greater porosity near to the surface. The post-spray soaking with NaOH solution followed by heat treatment at 573 K enabled to form apatite during in vitro test proving acceptable bioactivity of deposits.

References

1 Lazic, Z.R. (2004). *Design of Experiments in Chemical Engineering: A Practical Guide.* Weinheim, Germany: Wiley – VCH.

2 Pierlot, C., Pawłowski, L., Bigan, M., and Chagnon, P. (2008). Design of experiments in thermal spraying: a review. *Surf. Coat. Technol.* 202 (18): 4483.

3 Al-Asadi, M.M. and Al-Tameemi, H.A. (2022). A review of tribological properties and deposition methods for selected hard protective coatings. *Tribol. Int.* 176: 107919.

4 Savvides, N. and Bell, T.J. (1993). Hardness and elastic modulus of diamond and diamond-like carbon films. *Thin Solid Films* 228: 289.

5 Voevodin, A.A., Donley, M.S., and Zabinsky, J.S. (1997). Pulsed laser deposition of diamond-like carbon wear protective coatings: a review. *Surf. Coat. Technol.* 92: 42.

6 Stallard, J., Mercs, D., Jarratt, M. et al. (2004). A study of the tribological behaviour of three carbon-based coatings, tested in air, water and oil environments at high loads. *Surf. Coat. Technol.* 177–178 (3): 545. (cited after [3]).

7 Kabir, M.S., Zhou, Z., Xie, Z., and Munroe, P. (2023). Scratch adhesion evaluation of diamond like carbon coatings with alternate hard and soft multilayers. *Wear* 518–519: 204647.

8 Raveh, A., Martinu, L., Hawthorne, H.M., and Wertheimer, M.R. (1993). Mechanical and tribological properties of dual plasma-deposited diamond-like carbon. *Surf. Coat. Technol.* 58: 45.

9 Neuville, S. (1999). Dépôts physiques à base de carbone amorphe. *Galvano-Organo Traitements de Surface* 699: 897. (in French).

10 McGuire, G.E. (1982). Characterization of thin films. In: *Deposition Technologies for Films and Coatings* (ed. R.F. Bunshah), 548–569. Park Ridge, New Jersey: Noyes Publications.

11 Łępicka, M., Grądzka-Dahlke, M., Pieniak, D. et al. (2017). Effect of mechanical properties of substrate and coating on wear performance of TiN- or DLC-coated 316 LVM stainless steel. *Wear* 382–383: 62.

12 Tian, S., Xu, F., Ye, P. et al. (2018). Deposition of cubic boron nitride films by anode layer linear ion source assisted radio frequency magnetron sputtering. *Thin Solid Films* 653: 13.

13 Attar, F. (1996). Hardness evaluation of thin ceramic coatings on tool steel. *Surf. Coat. Technol.* 78: 78.

14 Ehrhardt, H. (1995). New development in the field of superhard coatings. *Surf. Coat. Technol.* 74–75: 29.

15 Ulrich, S., Theel, T., Schwan, J., and Ehrhardt, H. (1997). Magnetron-sputtered superhard materials. *Surf. Coat. Technol.* 97: 45.

16 Abdelouahdi, K., Sant, C., Legrand-Buscema, C. et al. (2006). Microstructural and mechanical investigations of tungsten carbide films deposited by reactive RF sputtering. *Surf. Coat. Technol.* 200: 6469.

17 Wang, L., Wang, L., Li, H. et al. (2023). Microstructure, mechanical and high-temperature tribological properties of amorphous SiC films by pulsed magnetron sputtering. *J. Non-Cryst. Solids* 599: 121949.

18 Dua, A.K. and George, V.C. (1994). TaC coatings prepared by hot filament chemical vapours deposition: characterization and properties. *Thin Solid Films* 247: 34.

19 Vepřek, S. (1997). Conventional and new approaches towards the design of novel superhard materials. *Surf. Coat. Technol.* 97: 15.

20 Deng, Y., Yin, S., Hong, Y. et al. (2020). Microstructures and properties of novel nanocomposite WNx coatings deposited by reactive magnetron sputtering. *Appl. Surf. Sci.* 512: 145508.

21 Sproul, W.D. (1996). Reactive sputter deposition of polycrystalline nitride and oxide superlattice coating. *Surf. Coat. Technol.* 86–87: 170.

22 Kim, Y.J., Byun, T.J., and Han, J.G. (2008). Tribological and mechanical behaviors of CrN/a-CNx superlattice thin films. *Surf. Coat. Technol.* 203: 790.

23 Barbezat, G., Nicoll, A.R., and Sickinger, A. (1993). Abrasion, erosion and scuffing resistance of carbide and oxide ceramic thermal sprayed coating for different applications. *Wear* 162–164: 529.

24 de Villiers Lovelock, H.L., Kinds, J., and Young, P.M. (1998). Characterization of WC-12Co thermal spray powders and HPHVOF wear resistant coatings. *Powder Metall.* 41 (4): 292.

25 Liao, H., Normand, B., and Coddet, C. (2000). Influence of coating microstructure on the abrasive wear resistance of WC/Co cermet coatings. *Surf. Coat. Technol.* 124: 235.

26 Sai Jagadeeswar, A., Kumara, S., Venkataraman, B. et al. (2020). Effect of thermal energy on the deposition behavior, wear and corrosion resistance of cold sprayed Ni-WC cermet coatings. *Surf. Coat. Technol.* 399: 126138.

27 Lee, Y.T.R., Ashrafizadeh, H., Fisher, G., and McDonald, A. (2017). Effect of type of reinforcing particles on the deposition efficiency and wear resistance of low-pressure cold-sprayed metal matrix composite coatings. *Surf. Coat. Technol.* 324: 190.

28 Vinyao, M.E., Kassabji, F., Guyonnet, J., and Fauchais, P. (1985). Plasma sprayed WC-Co coatings: influence of spray conditions (atmospheric and low pressure plasma spraying) on crystal structure, porosity and hardness. *J. Vac. Sci. Technol.* A3: 2483.

29 Sarikaya, O. (2005). Effect of some parameters on microstructure and hardness of alumina coatings prepared by the air plasma spraying process. *Surf. Coat. Technol.* 190: 388.

30 Sahab, A.R.M., Saad, N.H., Kasolang, S., and Saedon, J. (2012). Impact of plasma spray variables parameters on mechanical and wear behaviour of plasma sprayed Al_2O_3 3% wt TiO_2 coating in abrasion and erosion application. *Procedia Engineering* 41: 1689.

31 Dinesh Babu, P., Prasannakumar, B., Marimuthu, P. et al. (2019). Microstructure, wear and mechanical properties of plasma sprayed TiO_2 coating on Al–SiC metal matrix composite. *Arch. Civ. Mech. Eng.* 19: 756.

32 L. Pawłowski., R. Zacchino., R. Dal Maschio, V.M. Sglavo, J. Andresen, and F.J. Driller, (1993). Structure-properties relationship in plasma sprayed chromium oxide coatings, [in:] *Thermische Spritzkonferenz, TS93*, Band 152, DVS Verlag, Düsseldorf, Germany, 1993, pp. 132–138.

33 Meghwal, A., Singh, S., Sridar, S. et al. (2023). Development of composite high entropy-medium entropy alloy coating. *Scr. Mater.* 222: 115044.

34 Huang, J., Sun, W., Chu, X. et al. (2023). Effect of high temperature exposure on microstructure, mechanical and tribological properties of cold sprayed NiCoCrAlTaY coatings. *Surf. Coat. Technol.* 462: 129466.

35 Seiner, H., Cizek, J., Sedlák, P. et al. (2016). Elastic moduli and elastic anisotropy of cold sprayed metallic coatings. *Surf. Coat. Technol.* 291: 342.

36 Adam, P. (1977). Merkmale thermischer Spritzverfahren und ihr Einfluss auf die Eigenschaften der Schichten – Verfahren Gesichtspunkten. *VDI Ber.* 333: 97. (in German).

37 W. Milewski and P. Milewski. (1980). Some properties of molybdenum coatings produced by electric arc process, [in:] *9th ITSC*, 19–23 May 1980, Hague, The Netherlands, Nederlands Instituut voor Lastechniek, Hague, The Netherlands, pp. 290–298.

38 K. N. Muller (1973). Structure and properties of arc-sprayed titanium coatings, [in:] *7th ITSC*, 10–14 September 1973, London, UK, The Welding Institute, Abington Hall, UK, pp. 165–172.

39 Pawłowski, L. (2008). *The Science and Engineering of Thermal Spray Coatings*, 2nd ed. Chichester, UK: Wiley.

40 Zhou, Y., Liu, R., Liu, H. et al. (2023). Influence of tungsten carbide raw materials to microstructure and wear performance on PTA hard-facing materials with its micro-mechanism analysis. *Surf. Coat. Technol.* 454: 129200.

41 Ulutan, M., Kilicay, K., Celik, O.N., and Er, Ü. (2016). Microstructure and wear behaviour of plasma transferred arc (PTA)-deposited FeCrC composite coatings on AISI 5115 steel. *J. Mater. Process. Technol.* 236: 26.

42 Song G.Q., Hlawka F., Cornet A., and Marchione T. (1995). Modification des propriétés de surface de l'acier 35CD4 par refusion laser avec ou sans injection de carbures, [in:] *10e Colloque Université-Industrie*, Besançon, France, 15 juin 1995, CEE, Besançon, 1995, pp. C8-1-C8-6 (in French).

43 Fouilland-Paille, L., Ettaqi, S., Benayoun, S., and Hantzpergue, J.J. (1996). Structural and mechanical characterization of Ti/TiC cermet coatings synthesized by laser melting. *Surf. Coat. Technol.* 88: 204.

44 Demian, C., Denoirjean, A., Pawłowski, L. et al. (2016). Microstructural investigations of NiCrAlY + Y_2O_3 stabilized ZrO_2 cermet coatings deposited by plasma transferred arc (PTA). *Surf. Coat. Technol.* 300: 104.

45 D. Fischer, W. Löschau, and W. Reitzenstein (1996). Mit Laser erzeugte grobkörnige Hartstoff-Dispersionsschichten auf Aluminium–Legierungen zum Verschleißschutz, [in:] *ECLAT'96*, 16–18 September 1996, Stuttgart, Germany, F. Dausinger, H.W. Bergmann, and, J. Sigel (eds), AWT, Stuttgart, 1996, pp. 455–460 (in German).

46 Liang, G.Y. and Wong, T.T. (1997). Microstructure and character of laser remelting of plasma sprayed coating (Ni-Cr-B-Si) on Al-Si alloy. *Surf. Coat. Technol.* 89: 121.

47 Bozzi, A.C., Ramos, F.D., and Vargas, D.B.O. (2023). Microabrasive wear behavior of different stellites obtained by laser cladding and casting processes. *Wear* 524–525: 204857.

48 Yellup, J.M. (1995). Laser cladding using the powder blowing technique. *Surf. Coat. Technol.* 71: 121.

49 Ali Dilawari, S.A., Motallebzadeh, A., Paksoy, A.H. et al. (2017). Influence of laser surface melting on the characteristics of Stellite 12 plasma transferred arc hardfacing deposit. *Surf. Coat. Technol.* 317: 110.

50 Devos, A., Vital-Juarez, A., Chargui, A., and Cordill, M.J. (2021). Thin-film adhesion: a comparative study between colored picosecond acoustics and spontaneous buckles analysis. *Surf. Coat. Technol.* 421: 127485.

51 Ichimura, H. and Rodrigo, A. (2000). The correlation of scratch adhesion with composite hardness for TiN coatings. *Surf. Coat. Technol.* 126: 152.

52 Ward, L.P. and Datta, P.K. (1995). Scratch test adhesion and hardness evaluation of sputtered Nb coatings as a function of the argon gas pressure. *Thin Solid Films* 271: 101.

53 Guilbaud-Massereau, V., Celerier, A., and Machet, J. (1995). Study and improvement of the adhesion of chromium thin films deposited by magnetron sputtering. *Thin Solid Films* 258: 185.

54 Perry, A.J., Valli, J., and Steinmann, P.A. (1988). Adhesion scratch testing: a round-robin experiment. *Surf. Coat. Technol.* 36: 559.

55 Randall, N.X., Favaro, G., and Frankel, C.H. (2001). The effect of intrinsic parameters on the critical load as measured with the scratch test method. *Surf. Coat. Technol.* 137: 146.

56 Park, H.S. and Kwon, D. (1997). An energy approach to quantification of adhesion strength from critical loads in scratch tests. *Thin Solid Films* 307: 156.

57 Herr, W., Matthes, B., Broszeit, E. et al. (1993). Influence of substrate material and deposition parameters on the structure, residual stresses, hardness and adhesion of sputtered Cr_xN_y hard coatings. *Surf. Coat. Technol.* 60: 428.

58 Cockeram, B.V. and Wilson, W.L. (2001). The hardness, adhesion and wear resistance of coatings developed for cobalt-base alloys. *Surf. Coat. Technol.* 139: 161.

59 Peng, X.L. and Clyne, T.W. (1998). Mechanical stability of *DLC* films on metallic substrates: part II- interfacial toughness, debonding and blistering. *Thin Solid Films* 312: 219.

60 Venkataraman, S.K., Kohlstedt, D.L., and Gerberich, W.W. (1993). Metal-ceramic interfacial fracture resistance using continous microscratch technique. *Thin Solid Films* 223: 269.

61 Lassnig, A., Putza, B., Hirn, S. et al. (2021). Adhesion evaluation of thin films to dielectrics in multilayer stacks: a comparison of four-point bending and stressed overlayer technique. *Mater. Des.* 200: 109451.

62 Kusano, E., Kitagawa, M., Kuroda, Y. et al. (1998). Adhesion and hardness of compositionally gradient TiO_2/Ti/TiN, ZrO_2/Zr/ZrN and TiO2/Ti/Zr/ZrN coatings. *Thin Solid Films* 334: 151.

63 Kreye, H., Fandrich, D., Müller, H.-H., and Reiners, G. (1986). Microstructure and bond strength of WC-Co coatings deposited by hypersonic flame spraying (Jet Kote process). In: *Advances in Thermal Spraying*, 121–128. New York, NY: Pergamon Press.

64 Agüero, A., Camon, F., Garcıa de Blas, J. et al. (2011). HVOF- deposited WCCoCr as replacement for hard Cr in landing gear actuators. *J. Thermal Spray Technol.* 20 (6): 1292.

65 Xie, M., Zhanga, S., and Li, M. (2013). Comparative investigation on HVOF sprayed carbide-based coatings. *Appl. Surf. Sci.* 273: 799.

66 Kromer, R., Costil, S., Verdy, C. et al. (2018). Laser surface texturing to enhance adhesion bond strength of spray coatings – cold spraying, wire-arc spraying, and atmospheric plasma spraying. *Surf. Coat. Technol.* 352: 642.

67 Yang, Y., Liu, Z., Luo, C., and Chuang, Y. (1997). Measurements of residual stresses and bond strength of plasma sprayed laminated coatings. *Surf. Coat. Technol.* 89: 97.

68 Bianchi, L., Grimaud, A., Blein, F. et al. (1995). Comparison of plasma-sprayed alumina coatings by RF and DC plasma spraying. *J. Thermal Spray Technol.* 4 (1): 59.

69 Legoux, J.-G. and Dallaire, S. (1995). Adhesion mechanisms of arc-sprayed zinc on concrete. *J. Thermal Spray Technol.* 4 (4): 395.

70 Wielage, B., Milewski, W., and Drozak, J. (1990). Einfluβ der Spritzverfahren auf die Eigenschaften von Nickelschichten. In: *Thermisches Spritzen TS90*, Band 130, 206–209. Düsseldorf: DVS Verlag.

71 Bruera, A., Puddu, P., Theimer, S. et al. (2023). Adhesion of cold sprayed soft coatings: effect of substrate roughness and hardness. *Surf. Coat. Technol.* 466: 129651.

72 Ulrich, R.K. and Zhao, G. (1993). A new apparatus for measuring the ultimate strain of thin films. *Thin Solid Films* 224: 63.

73 Bai, M., Kato, K., Umehara, N., and Miyake, Y. (2000). Nanoindentation and FEM study of the effect of internal stress on micro/nano mechanical property of thin CN_x films. *Thin Solid Films* 377–378: 138.

74 Feng, J., Xie, Y.-N., Li, Z. et al. (2015). Microscopic mechanical characteristics analysis of ultranano-crystalline diamond films. *Trans. Nonferrous Metals Soc. China* 25: 3291.

75 Asada, H., Kishi, Y., and Hirose, Y. (1993). Measurements of Young's moduli of TiC coated film by the X-ray method. *Thin Solid Films* 236: 247.

76 Feng, L., Tang, J., and Zabinsky, J.S. (1998). Tribological properties of magnetron-sputtered TiC coatings. *Mater. Sci. Engn A* 257: 240.

77 Lugscheider, E., Bobzin, K., Bärwulf, S., and Etzkorn, A. (2001). Mechanical properties of EB-PVD-thermal barrier coatings by nanoindentation. *Surf. Coat. Technol.* 138: 9.

78 Veprek, S. and Reiprich, S. (1995). A concept for the design of novel superhard coatings. *Thin Solid Films* 268: 64.

79 Veprek, S. (1998). New development in superhard coatings: the superhard nanocrystalline-amorphous composites. *Thin Solid Films* 317: 449.

80 Zak, S., Trost, C.O.W., Kreiml, P., and Cordill, M.J. (2022). Accurate measurement of thin film mechanical properties using nanoindentation. *J. Mater. Res.* 37 (7): 1373.

81 Klein, C.A. and Cardinale, G.F. (1993). Young's modulus and Poisson's ratio for CVD diamond. *Diam. Relat. Mater.* 2: 918.

82 Wang, T.-G., Zhao, S.-S., Hua, W.-G. et al. (2010). Estimation of residual stress and its effects on the mechanical properties of detonation gun sprayed WC–Co coatings. *Mater. Sci. Engn A* 527: 454.

83 Kim, M.C., Kim, S.B., and Hong, J.W. (1997). Effect of powder type on mechanical properties of D-gun coatings. In: *Thermal Spray: A United Forum for Scientific and Technological Advances* (ed. C.C. Berndt), 791–795. Materials Park, OH: ASM International.

84 Yao, H.-L., Yang, C., Yi, D.-L. et al. (2020). Microstructure and mechanical property of high velocity oxy-fuel sprayed WC-Cr_3C_2-Ni coatings. *Surf. Coat. Technol.* 397: 126010.

85 Wang, D., Zhang, B., Jia, C. et al. (2017). Influence of carbide grain size and crystal characteristics on the microstructure and mechanical properties of HVOF-sprayed WC-CoCr coatings. *Int. J. Refract. Met. Hard Mater.* 69: 138.

86 Li, J. and Ding, C. (2000). Improvement in the properties of plasma-sprayed chromium carbide coatings using nickel-clad powders. *Surf. Coat. Technol.* 130: 15.

87 Kelly, A. and Macmillan, N.H. (1986). *Strong Solids*, 3e. Oxford, U.K.: Clarendon Press.

88 Ben Ghorbal, G., Tricoteaux, A., Thuault, A. et al. (2020). Mechanical properties of thermally sprayed porous alumina coating by Vickers and Knoop indentation. *Ceram. Int.* 46: 19843.

89 Guo, D.Z. and Wang, L.J. (1992). Measurement of the critical strain energy release rate of plasma sprayed coatings. *Surf. Coat. Technol.* 56: 19.

90 Łatka, L., Chicot, D., Cattini, A. et al. (2013). Modeling of elastic modulus and hardness determination by indentation of porous yttria stabilized zirconia coatings. *Surf. Coat. Technol.* 220: 131.

91 Kweh, S.W.K., Khor, K.A., and Cheang, P. (2000). Plasma-sprayed hydroxyapatite (HA) coatings with flame-spheroidized feedstock: microstructure and mechanical properties. *Biomaterials* 21: 1223.

92 Łatka, L., Pawłowski, L., Chicot, D. et al. (2010). Mechanical properties of suspension plasma sprayed hydroxyapatite coatings submitted to simulated body fluid. *Surf. Coat. Technol.* 205: 954.

93 Mejias, A., Candidato, R.T. Jr., Pawłowski, L., and Chicot, D. (2016). Mechanical properties by instrumented indentation of solution precursor plasma sprayed hydroxyapatite coatings: analysis of microstructural effect. *Surf. Coat. Technol.* 298: 93.

94 Samsonov, G.V. (ed.) (1978). *Physical and Chemical Properties of Oxides*. Moskva, Russia: Metallurgija (in Russian).

95 Berndt, C.C. (1986). Determination of material properties of ceramic coatings. In: *Advances in Thermal Spraying*, 149–158. New York, NY: Pergamon Press.

96 Neiser, R.A., Smolik, G.R., Hollis, K.J., and Watson, R.D. (1993). Evaluation of plasma sprayed tungsten for fusion reactors. *J. Thermal Spray Technol.* 2: 393.

97 Gärtner, F., Stoltenhoff, T., Voyer, J. et al. (2006). Mechanical properties of cold-sprayed and thermally sprayed copper coatings. *Surf. Coat. Technol.* 200: 6770.

98 Shikalov, V.S., Vidyuk, T.M., Filippov, A.A., and Kuchumova, I.D. (2022). Microstructure, mechanical and tribological properties of cold sprayed Cu-W coatings. *Int. J. Refractory Metals and Hard Mater.* 106: 105866.

99 Veys, J.M., Riviere, A., and Mevrel, R. (1988). Mechanical properties of LPPS NiCoCrAlTa alloy. In: *1st Plasma Technik Symposium*, vol. 2 (ed. H. Eschnauer, P. Huber, A.R. Nicoll, and S. Sandmeier), 115–123. Wohlen, Switzerland: Plasma Technik AG.

100 Zhou, Y.-K., Kang, J.-J., Zhang, J. et al. (2022). Effect of vacuum heat treatment on microstructure and mechanical properties of HVOF sprayed AlCoCrFeNiCu high-entropy alloy coating. *Mater. Lett.* 323: 132551.

101 Kuo, T.-Y., Chin, W.-H., Chien, C.-S., and Hsieh, Y.-H. (2019). Mechanical and biological properties of graded porous tantalum coatings deposited on titanium alloy implants by vacuum plasma spraying. *Surf. Coat. Technol.* 372: 399.

102 Gouldstone, A., Choi, W.B., Chi, W. et al. (2007). Mechanical, thermal and electrical properties of cold sprayed coatings. In: *The Cold Sprayed Materials Deposition Process* (ed. V.K. Champagne), 245–263. Cambridge: Woodhead.

103 Gault, C., Boilevin, S., and Desplanches, G. (1988). Elastic properties of ceramic and ceramic metal composite TBC. *Sci. Ceram.* 14: 389.

104 Khor, K.A., Gu, Y.W., Pan, D., and Cheang, P. (2004). Microstructure and mechanical properties of plasma sprayed HA/YSZ/Ti–6Al–4V composite coatings. *Biomaterials* 25: 4009.

105 Kang, K., Bae, G., Won, J., and Lee, C. (2012). Mechanical property enhancement of kinetic sprayed Al coatings reinforced by multi-walled carbon nanotubes. *Acta Mater.* 60: 5031.

106 Bunshah, R.F. (1994). Metallurgical applications. In: *Handbook of Deposition Technologies for Films and Coatings* (ed. R.F. Bunshah), 740–762. Park Ridge, N.J.: Noyes Publications.

107 Föhl, J., Weissenberg, T., and Wiedermayer, J. (1989). General aspects for tribological applications of hard particle coatings. *Wear* 130: 275.

108 Chen, Q., Li, A., Wu, G., and Wang, Y. (2023). Tribochemistry of TiN films sliding against ceramic counterparts in vacuum condition and N2 atmosphere. *Ceram. Int.* 49: 8907.

109 Holmberg, K. and Matthews, A. (1994). Coatings tribology: a concept, critical aspects and future directions. *Thin Solid Films* 253: 173.

110 Taylor, M.I., Murphy, J.G., and King, H.W. (1987). Brittle/ductile erosion of plasma sprayed ceramic and metallic coatings. In: *Thermal Spray: Advances in Coatings Technology* (ed. D.L. Houck), 143–149. Materials Park, OH: ASM International.

111 Tiwari, R., Sampath, S., Herman, H., and Anekawa, Y. (1993). Cavitation-erosion of plasma sprayed nickel aluminides. In: *Thermal Spray Coatings: Research, Design and Applications* (ed. C.C. Berndt and T.F. Bernecki), 423–428. Materials Park, OH: ASM International.

112 Gilmore, R. and Hauert, R. (2000). Comparative study of the tribological moisture sensitivity of Si-free and Si-containing DLC films. *Surf. Coat. Technol.* 133–134: 437.

113 Wu, P.-Q., Chen, H., van Stappen, M. et al. (2000). Comparison of fretting wear of uncoated and PVD TiN coated high speed steels under different testing conditions. *Surf. Coat. Technol.* 127: 114.

114 Renevier, N.M., Vox, V.C., Teer, D.G., and Hampshire, J. (2000). Coatings characteristics and tribological properties of sputter deposited MoS_2/metal composite coatings deposited by closed field unbalanced sputter ion plating. *Surf. Coat. Technol.* 127: 24.

115 Darbeïda, A., von Stebut, J., Bartile, M. et al. (1994). Comparative tribological study of chromium coatings with different specific hardness. *Surf. Coat. Technol.* 68/69: 582.

116 Kustas, F., Mishra, B., and Zhou, J. (2001). Wear behavior of B_4C-Mo co-sputtered wear coatings. *Surf. Coat. Technol.* 141: 48.

117 Szeri, A.Z. (2001). Tribology. In: *Encyclopedia of Physical Sciences and Technology* (ed. R.A. Meyers), 127–152. Cambridge, MA: Elsevier, Academic Press.

118 O'Bannon, E., Xia, G., Shi, F. et al. (2020). The transformation of diamond to graphite: experiments reveal the presence of an intermediate linear carbon phase. *Diam. Relat. Mater.* 108: 107876.

119 Voevodin, A.A., Phelps, A.W., Zabinski, J.S., and Donley, M.S. (1996). Friction induced phase transformation of pulsed laser deposited diamond-like carbon. *Diam. Relat. Mater.* 5: 1264.

120 Blanpain, B., Celis, J.P., Roos, J.R. et al. (1993). A comparative study of the fretting wear of hard carbon coatings. *Thin Solid Films* 223: 65.

121 Klaffke, D., Skopp, A., and Lenke, I. (1995). Tribological characterization of coatings by oscillating sliding testing. *Surf. Coat. Technol.* 74–75: 977.

122 Shanov, V., Tabakoff, W., and Metwally, M. (1992). Erosive wear of CVD ceramic coatings exposed to particulate flow. *Surf. Coat. Technol.* 54–55: 25.

123 Tabakoff, W., Hamed, A., and Shanov, V. (1998). Blade deterioration in a gas turbine engine. *Int. J. Rotating Machinery* 4 (4): 233.

124 Zhou, Y.M., Asaki, R., Higashi, K. et al. (2000). Sliding wear behavior of polycristalline TiN/CrN multilayers against alumina ball. *Surf. Coat. Technol.* 130: 9.

125 Wei, J., Fouvry, S., Kapsa, P., and Vincent, L. (1997). Fretting behaviour of TiN coating. In: *Surf. Modif. Technol. X* (ed. T.S. Sudarshan, K.A. Thor, and M. Jeandin), 25–36. London, book no. 668: Institute of Metals.

126 Bienk, E.J., Reitz, H., and Mikkelsen, N.J. (1995). Wear and friction properties of hard PVD coatings. *Surf. Coat. Technol.* 76–77: 475.

127 Bantle, R. and Matthews, A. (1995). Investigation into the impact wear behaviour of ceramic coatings. *Surf. Coat. Technol.* 74–75: 857.

128 Bin, T., Xiadong, Z., Naisai, H., and Jiawen, H. (2000). Study on the structure and tribological properties of CrN coatings by IBAD. *Surf. Coat. Technol.* 131: 391.

129 Cerny, F., Suchanek, J., and Hnatowicz, V. (1998). IBAD SiNx coating: preparation and tribological properties. *Thin Solid Films* 317: 490.

130 Knotek, O., Lugscheider, E., Löffler, F. et al. (1994). Behaviour of CVD and PVD coatings under impact wear. *Surf. Coat. Technol.* 68–69: 253.

131 Kato, K. (1995). Microwear mechanisms of coatings. *Surf. Coat. Technol.* 76–77: 469.

132 Simmonds, M.C., Savan, A., Pflüger, E., and van Swygenhoven, H. (2000). Mechanical and tribological performance of MoS_2 co-sputtered composites. *Surf. Coat. Technol.* 126: 15.

133 Shipway, P.H. and Hutchings, I.M. (1995). Measurement of coating durability by solid particle erosion. *Surf. Coat. Technol.* 71 (1): 1.

134 Holleck, H. and Schier, V. (1995). Multilayer PVD coatings for wear protection. *Surf. Coat. Technol.* 76–77: 328.

135 Bangert, H., Eisenmenger-Sittner, C., and Bergauer, A. (1996). Deposition and structural properties of two-component metal coatings for tribological applications. *Surf. Coat. Technol.* 80: 162.

136 Fan, W.D., Chen, X., Jagannadham, K., and Narayan, J. (1994). Diamond-ceramic composite tool coatings. *J. Mater. Res.* 9: 2850.

137 Li, J., Zhang, Y., Huang, J., and Ding, C. (1998). Properties of plasma-sprayed Cr_3C_2-NiCr, WC-Co, and Cr_20_3 coatings. *J. Therm. Spray Technol.* 7: 242.

138 Leblanc, L. (2003). Abrasion and sliding wear of nanostructured ceramic coatings. In: *Advancing the Science & Applying the Technology* (ed. C. Moreau and B. Marple), 291–299. Materials Park, OH: ASM International.

139 Bull, S.J., Kingswell, R., and Scott, K.T. (1996). The sliding wear of plasma sprayed alumina. *Surf. Coat. Technol.* 82: 218.

140 Hu, H., Mao, L., Yin, S. et al. (2022). Wear-resistant ceramic coatings deposited by liquid thermal spraying. *Ceram. Int.* 48: 33245.

141 Bolelli, G., Rauch, J., Cannillo, V. et al. (2009). Microstructural and tribological investigation of high-velocity suspension flame sprayed (HVSFS) Al_2O_3 coatings. *J. Therm. Spray Technol.* 18 (1): 35.

142 Bolelli, G., Steduto, D., Kiilakoski, J. et al. (2021). Tribological properties of plasma sprayed Cr_2O_3, Cr_2O_3–TiO_2, Cr_2O_3–Al_2O_3 and Cr_2O_3–ZrO_2 coatings. *Wear* 480–481: 203931.

143 Memon, H., Romero, A.R., Derelizade, K. et al. (2022). A new hybrid suspension and solution precursor thermal spray for wear-resistant silicon carbide composite coatings. *Mater. Design* 224: 111382.

144 Alcover, P.R.C. Jr. and Pukasiewicz, A.G.M. (2019). Evaluation of microstructure, mechanical and tribological properties of a Babbitt alloy deposited by arc and flame spray processes. *Tribol. Int.* 131: 148.

145 Qiao, Y., Liu, Y.R., and Fischer, T.E. (2001). Sliding and abrasive wear resistance of thermal-sprayed WC-CO coatings. *J. Therm. Spray Technol.* 10 (1): 118.

146 Picas, J.A., Menargues, S., Martin, E., and Baile, M.T. (2023). Cobalt free metallic binders for HVOF thermal sprayed wear resistant coatings. *Surf. Coat. Technol.* 456: 129243.

147 Hossini, H., Shafie, B., Niri, A.D. et al. (2022). A comprehensive review on human health effects of chromium: insights on induced toxicity. *Environ. Sci. Pollut. Res. Int.* 29 (47): 70686.

148 Houck, D.L. and Whisenant, W. (1987). Wear resistance of five molybdenum containing plasma sprayed coatings. In: *Thermal Spray: Advances in Coatings Technology* (ed. D.L. Houck), 55–61. Materials Park, OH: ASM International.

149 K.-H. Habig, D. Liedtke, W.-D. Münz, and G. Wahl (1985). Eigenschaften von Verschleiβ-Schutzschichten auf Stählen, [in:] *41 Härterei-Kolloquium*, Wiesbaden, Allemagne, 9–11 october 1985, pp. 1–10. (in German).

150 A. Sandt. and J. Krey (1984). NiCrBSi-Schichten mit Hartstoff-Beimischungen, Teil I. Herstellung und Charakterisierung der Schichten, *Metall*, 39, 1171 (in German).

151 Lugscheider, E. (1987). Plasma spraying for wear application. In: *Thermal Spray: Advances in Coatings Technology* (ed. D.L. Houck), 105–122. Materials Park, OH: ASM International.

152 Sampath, S. and Wayne, S.F. (1994). Microstructure and properties of plasma-sprayed Mo-Mo_2C composite. *J. Therm. Spray Technol.* 3: 282.

153 Matikainen, V., Koivuluoto, H., and Vuoristo, P. (2020). A study of Cr_3C_2-based HVOF- and HVAF-sprayed coatings: abrasion, dry particle erosion and cavitation erosion resistance. *Wear* 446–447: 203188.

154 Cheney, R.F., Port, D.J., and Lafferty, W.D. (1978). Plasma sprayed coatings for high temperature wear resistance. *Engn. Mater. Design* 22 (3): 25.

155 Liscano, S., Gi, L., and Staia, M.H. (2005). Correlation between microstructural characteristics and the abrasion wear resistance of sealed thermal-sprayed coatings. *Surf. Coat. Technol.* 200: 1310.

156 Barbezat, G., Hochstrasser, J., and Nicoll, A.R. (1991). *Caractérisation et contrôle des revêtements dêpôses par projection thermique, study unpublished*. Wohlen, Suisse (in French): Plasma Technik AG.

157 Usmani, S., Sampath, S., Houck, D.L., and Lee, D. (1997). Effect of carbide grain size on sliding and abrasive wear behavior of thermally sprayed WC-Co coatings. *Tribol. Trans.* 40 (3): 470.

158 Ishikawa, Y., Kawakita, J., Osawa, S. et al. (2005). Evaluation of corrosion and wear resistance of hard cermet coatings sprayed using and improved HVOF process. *J. Therm. Spray Technol.* 14 (3): 384.

159 Khan, M.S.A. and Clyne, T.W. (1996). Microstructure and abrasion resistance of plasma sprayed cermet coatings. In: *Thermal Spray: Practical Solutions for Engineering Problems* (ed. C.C. Berndt), 113–122. Materials Park, OH: ASM International.

160 Reardon, J.D., Mignona, R., and Longo, F.N. (1981). Plasma- and vacuum plasma sprayed Cr_3C_2 composite coatings. *Thin Solid Films* 83: 345.

161 Eronen V., Ahmaniemi S., Niemi K., and Vuoristo P. (2005). Microwear in thermally sprayed hard coatings by different abrasives [in:] *ITSC'05*, E. Lugscheider (ed.), CD-Rom, ISBN 3-87155-793-5, DVS-Verlag, Düsseldorf, Germany.

162 Kretschmar, E. (1986). Protection against wear by powder flame spraying. In: *Advances in Thermal Spraying*, 367–375. New York, NY: Pergamon Press.

163 Haque, U., Singh, G., Gill, H.S. et al. (2022). Erosion wear analysis of thermally sprayed WC + TiO_2 micron layers on SS-404. *Mater. Today: Proc.* 48: 1246.

164 Lansdown, A.R. and Price, A.L. (1986). *Materials to resist wear: a guide to their selection and use*. Oxford, U.K.: Pergamon Press.

165 Kulu, P. and Pihl, T. (2002). Selection criteria for wear resistant powder coatings under extreme erosive wear conditions. *J. Therm. Spray Technol.* 11 (4): 517.
166 Karimi, A., Verdon, C., Martin, J.L., and Schmid, R.K. (1995). Slurry erosion behaviour of thermally sprayed WC-M coatings. *Wear* 186–187: 480.
167 Li, C.J., Yang, G.J., and Ohmori, A. (2006). Relationship between particle erosion and lamellar microstructure for plasma-sprayed alumina coatings. *Wear* 260: 1166.
168 Wang, B.Q. and Lee, S.W. (1997). Elevated temperature erosion of several thermal-sprayed coatings under simulated erosion conditions of in-bed tubes in a fluidized bed combustor. *Wear* 203–204: 580.
169 Bhosale, D.G., Pabhu, T.R., and Rathod, W.S. (2020). Sliding and erosion wear behaviour of thermal sprayed WC-Cr_3C_2-Ni coatings. *Surf. Coat. Technol.* 400: 126192.
170 Vicenzi, J., Marques, C.M., and Bergmann, C.P. (2008). Hot and cold erosive wear of thermal sprayed NiCr-based coatings: influence of porosity and oxidation. *Surf. Coat. Technol.* 202: 3688.
171 Kingwell, R., Rickerby, D.S., Scott, K.T., and Bull, S.J. (1991). Comparison of the erosive wear behaviour of the vacuum plasma sprayed and bulk alumina. In: *Thermal Spray Research and Applications* (ed. T.F. Bernecki), 179–185. Materials Park, OH: ASM International.
172 Takeuchi, J., Nakahira, H., and Nagai, J. (1991). Physical properties of some oxide coatings by low pressure plasma spraying. In: *2nd Plasma Technik Symposium*, vol. 2 (ed. S. Blum-Sandmeier, H. Eschnauer, P. Huber, and A.R. Nicoll), 141–151. Wohlen, Switzerland: Plasma-Technik AG.
173 Gudmundsson, B., Jacobson, B.E., Berglin, L. et al. (1988). Microstructure and erosion resistance of vacuum plasma sprayed CoNiCrAlY/Al_2O_3 composite. In: *1st Plasma Technik Symposium*, vol. 2 (ed. H. Eschnauer, P. Huber, A.R. Nicoll, and S. Sandmeier), 105–114. Switzerland: Plasma-Technik AG.
174 Wang, B.Q., Geng, G.Q., Levy, A.V., and Buchanan, E.R. (1992). Elevated temperature erosion of carbide-metal composite coatings. In: *Thermal Spray: International Advances in Coatings Technology* (ed. C.C. Berndt), 735–742. Materials Park, OH: ASM International.
175 de Bonte M., Economou S.N., Celis J.P. et al. (1993). Wear behaviour of plasma sprayed ceramic coatings: laboratory ball-on-disk testing, [in:] *Thermische Spritzkonferenz, TS93*, Band 152, DVS Verlag, Düsseldorf, Germany, pp. 447–449.
176 Shi, Z., Xu, L., Deng, C. et al. (2023). Effects of frequency on the fretting wear behavior of aluminum bronze coatings. *Surf. Coat. Technol.* 457: 129306.
177 Zhu, Y.-S., Luo, X.-T., Sun, Y.-Q. et al. (2023). Atmospheric plasma-sprayed CuNiInB coatings of high fretting wear performance enabled by oxide-free metallic droplet deposition. *Surf. Coat. Technol.* 464: 129537.
178 Wolfla, T.A. and Tucker, R.C. Jr. (1978). High temperature wear-resistant coatings. *Thin Solid Films* 53 (3): 353.
179 Rajeev, G.P., Kamaraj, M., and Bakshi, S.R. (2019). Comparison of microstructure, dilution and wear behavior of Stellite 21 hardfacing on H13 steel using cold metal transfer and plasma transferred arc welding processes. *Surf. Coat. Technol.* 375: 383.
180 Kesavan, D. and Kamaraj, M. (2010). The microstructure and high temperature wear performance of a nickel base hardfaced coating. *Surf. Coat. Technol.* 204: 4034.
181 Ardigo-Besnard, M.R., Tellier, A., Besnard, A., and Chateau-Cornu, J.-P. (2021). Effect of the microstructure on the tribological properties of HIPed and PTA-welded Fe-based hardfacing alloy. *Surf. Coat. Technol.* 425: 127691.

182 Mateos, J., Cuetos, J.M., Fernández, E., and Cadenas, M. (1997). Effect of laser treatment on tungsten carbide coatings. In: *Transaction on Engineering Sciences*, vol. 17 (ed. M.H. Aliabadi and C.A. Brebbia), 239–246. Southampton: WIT Press.

183 Yang, L.J. and Loh, N.L. (1995). The wear properties of plasma transferred arc cladded stellite specimens. *Surf. Coat. Technol.* 71: 196.

184 Techel, A., Luft, A., Müller, A., and Nowotny, S. (1995). Production of hard metal-like wear protection coatings by CO_2 laser cladding. *Opt. Quant. Electron.* 27: 1313.

185 Swamy, G.M., Anil Kumar, B.S., Kumarswamy, H.S. et al. (2023). Microstructural analysis and mechanical properties of a dissimilar material of AISI 4130 steel substrate coated with Stellite 6 manufactured by transferred ARC plasma welding (PTA). *Materials Today: Proceedings* in press[1], https://doi.org/10.1016/j.matpr.2023.07.280.

186 Yung, D., Zikin, A., Hussainova, I. et al. (2017). Tribological performances of ZrC-Ni and TiC-Ni cermet reinforced PTA hardfacings at elevated temperatures. *Surf. Coat. Technol.* 39: 497.

187 Zikin, A., Antonov, M., Hussainova, I. et al. (2013). High temperature wear of cermet particle reinforced NiCrBSi hardfacings. *Tribol. Int.* 68: 45.

188 Hejwowski, T. (2009). Erosive and abrasive wear resistance of overlay coatings. *Vacuum* 83: 166.

189 Pelletier, J.M., Sauger, E., Gachon, Y., and Vannes, A.B. (1999). Mechanical and tribological properties of Hadfield steel coatings manufactured by laser processing. *J. Mater. Sci.* 34: 2955.

190 Guo, D.Z., Li, F.L., Wang, J.Y., and Sun, J.S. (1995). Effects of post-coating processing on structure and erosive wear characteristics of flame and plasma spray coatings. *Surf. Coat. Technol.* 73: 73.

191 Jiang, W.H. and Kovacevic, R. (2007). Laser deposited TiC/H13 tool steel composite coatings and their erosion resistance. *J. Mater. Process. Technol.* 186: 331.

192 Chen, D., Liu, D., Liu, Y. et al. (2014). Microstructure and fretting wear resistance of γ/TiC composite coating in situ fabricated by plasma transferred arc cladding. *Surf. Coat. Technol.* 219: 28.

193 Ding, H., Cao, Y., Hua, K. et al. (2023). Fretting wear resistance at ambient and elevated temperatures of 316 stainless steel improved by laser cladding with Co-based alloy/WC/CaF_2 composite coating. *Opt. Laser Technol.* 163: 109428.

194 Pawłowski, L. (1996). Technology of thermally sprayed anilox rolls: state of art, problems, and perspectives. *J. Thermal. Spray Technol.* 5 (3): 317.

195 Tomaszek, R., Pawłowski, L., Zdanowski, J. et al. (2004). Microstructure transformations of TiO_2, Al_2O_3+13 TiO_2 and Al_2O_3+40 TiO_2 at plasma spraying and laser engraving. *Surf. Coat. Technol.* 185 (2–3): 137.

196 Langer G., Wewel M., and Wand R. (2005). Carbide coatings on fibre-reinforced anilox rollers using TeroStar™ process abrasives, [in:] *ITSC'05*, E. Lugscheider (ed.), CD-Rom, ISBN 3-87155-793-5, DVS-Verlag, Düsseldorf, Germany.

197 Meiners E. (1996). Mikrostrukturierung von Makrooberflächen-Laserawendungen in der Druckformenherstellung, [in:] *ECLAT'96,* 16–18 September 1996, Stuttgart, Germany, F. Dausinger, H.W. Bergmann, and, J. Sigel (eds), AWT, Stuttgart, 1996, pp. 633–674 (in German).

1 The papers was withdrawn by the authors.

198 Bunshah, R.F. (1982). Deposition technologies: an overview. In: *Deposition Technologies for Films and Coatings* (ed. R.F. Bunshah), 1–18. Park Ridge, New Jersey: Noyes Publications.

199 Berry, R.W., Hall, P.M., and Harris, M.T. (1968). *Thin Film Technology*. Princeton, N.J.: D. van Nostrand.

200 Meiksin, Z.H. (1976). *Thin and Thick Films for Hybrid Microelectronics*. Lexington, Mass: Lexington Books.

201 Harris, D.H. and Janowiecki, R.J. (1970). Arc plasma deposits may yield some big microwave dividends. *Electronics* 2: 108.

202 Ghate, P.B. (1982). Deposition techniques and microelectronic applications. In: *Deposition Technologies for Films and Coatings* (ed. R.F. Bunshah), 514–547. Park Ridge, N.J.: Noyes Publications.

203 Richardt, A. and Durand, A.-M. (1994). *Le Vide: les Couches Minces – les Couches Dures*. Paris, France (in French): In Fine.

204 Maissel, L.I. (1970). Electrical properties of metallic thin films. In: *Handbook of Thin Film Technology* (ed. L.I. Maissel and R. Glang), 305–345. New York, 1970, Russian translation, 1977: McGraw Hill (in Russian).

205 El Khakani, M.A. and Chaker, M. (1998). Reactive pulsed laser deposition of iridium oxide thin films. *Thin Solid Films* 335: 6.

206 Maissel, L.I. (1970). Thin film resistors. In: *Handbook of Thin Film Technology* (ed. L.I. Maissel and R. Glang), 578–618. New York, 1970, Russian translation, 1977: McGraw Hill (in Russian).

207 Lide, D.L. and Frederikse, H.P.R. (ed.) (1997). *CRC Handbook of Chemistry and Physics*. Boca Raton, FL: CRC Press.

208 Petrović, R., Nenadović, T., Kraljević, N., and Dimitrijević, T. (1979). Electrical and structural properties of tantalum nitride thin films deposited by sputtering. *Thin Solid Films* 57: 333.

209 Lopes, C., Rodrigues, M.S., Ferreira, A. et al. (2023). The influence of the nanostructure design on the optical, electrical and thermal properties of TiN_x thin films prepared by reactive magnetron sputtering. *Mater. Chem. Phys.* 306: 127981.

210 R.T. Smyth and J.C. Anderson (1976). Electronic circuit production by arc plasma spraying, [in:] *Proceedings of 8th International Thermal Spraying Conference*, American Welding Society, Miami Beach, FL, pp. 456–463.

211 M. Braguier, J. Bejat, R. Tueta, M. Verna, G. Aubin, and C. Naturel (1973). Improvements of plasma spraying process for hybrid microelectronics, [in:] *Proceedings of Conference of Hybrid Electronics and Microelectronics*, 25–27 September, Cantebury, UK, pp. 15–37.

212 L. Pawłowski (1980). The vacuum plasma sprayed copper and tantalum coatings, [in:] Proceedings of *9th International Thermal Spray Conference,* 19–23 mai 1980, Hague, The Netherlands, Nederlands Instituut voor Lastechniek, pp. 299–305.

213 Małachowska, A., Winnicki, M., Konat, Ł. et al. (2017). Possibility of spraying of copper coatings on polyamide 6 with low pressure cold spray method. *Surf. Coat. Technol.* 318: 82.

214 Ananthapagmanabhan, P.V., Sreekumar, K.P., Ravidran, P.V., and Venkatramani, N. (1993). Effect of oxygen pick-up on the properties of plasma sprayed diboride coatings. *Thin Solid Films* 224: 148.

215 Tului, M., Ruffini, F., Arezzo, F. et al. (2002). Some properties of atmospheric air and inert gas high-pressure plasma sprayed ZrB_2 coatings. *Surf. Coat. Technol.* 151–152: 483.

216 Scagliotti, M., Parmigiani, F., Chiodelli, G. et al. (1988). Plasma sprayed zirconia electrolytes. *Solid State Ionics* 28–30: 1766.

217 Zhang, S.-L., Li, C.-X., Li, C.-J. et al. (2013). Scandia-stabilized zirconia electrolyte with improved interlamellar bonding by high-velocity plasma spraying for high performance solid oxide fuel cells. *J. Power Sources* 232: 123.

218 Li, C.-X., Li, C.-J., Long, H.-G. et al. (2006). Characterization of atmospheric plasma-sprayed Sc_2O_3–ZrO_2 electrolyte coating. *Solid State Ionics* 177: 2149.

219 Pawłowski, L. (2015). Application of solution-precursor spray techniques to obtain ceramic films and coatings. In: *Future Development of Thermal Spray Coatings* (ed. N. Espallargas), 123–141. Amsterdam, The Netherlands: Elsevier.

220 R.J. Janowiecki, M.C. Wilson, and D.H. Harris (1968). Plasma sprayed thermoelectric materials research. Technical report AFAPL-TR-68-29, Monsanto Research Corp., Dayton, OH, (March).

221 Akani, M., Suryanarayanan, R., and Brun, G. (1987). Resistivity and photoconductivity of plasma-sprayed polycrystalline silicon. *Thin Solid Films* 151 (3): 343.

222 Blanchart, P. (2018). Extraction, properties and applications of titania. In: *Industrial Chemistry of Oxides for Emerging Applications* (ed. L. Pawłowski and P. Blanchart), 255–309. Chichester, UK: Wiley.

223 Kumar, A., Barbhuiya, N.H., and Singh, S.P. (2022). Magnéli phase titanium sub-oxides synthesis, fabrication and its application for environmental remediation: current status and prospect. *Chemosphere* 307: 1135878.

224 Ohmori, A., Park, K.C., Inuzuka, M. et al. (1991). Electrical conductivity of plasma sprayed titanium oxide (rutile) coatings. *Thin Solid Films* 201: 1.

225 Tomaszek, R., Nitsch, K., Pawłowski, L. et al. (2006). Impedance spectroscopy of suspension plasma sprayed titania coatings. *Surf. Coat. Technol.* 201 (5): 1930.

226 Ramachandran, K., Selvarajan, V., Ananthapadmanabhan, P.V., and Sreekumar, K.P. (1998). Microstructure, adhesion, microhardness, abrasive wear resistance and electrical resistivity of the plasma sprayed alumina and alumina-titania coating. *Thin Solid Films* 315: 144.

227 Prudenziati, M. and Lassinantti Gualtieri, M. (2008). Electrical properties of thermally sprayed Ni- and Ni20Cr-based resistors. *J. Thermal. Spray. Technol.* 17 (3): 385.

228 Smyth, R.T. and Anderson, J.C. (1975). Production of resistors by arc plasma spraying. *Electrocompon. Sci. Technol.* 2: 135.

229 Smith, D.P.H. and Anderson, J.C. (1980). Plasma sprayed resistors. *Act. Passive Electron. Compon.* 7: 125.

230 Taylor A.J., Parker N.J., and Wellhofer F. (1989). Plasma sprayed cermets, [in:] *12th International Thermal Spray Conference*, June, 4–9, 1989, London, U.K. The Welding Institute, Abington Hall, U.K., 1989, vol. 2, paper 29.

231 Fantozzi, G., La Gallet, S., and Nièpce, J.C. (2011). *Science & Technologies Céramiques*. Les Ulis, France (in French): EDP Sciences.

232 Harrop, P.J. and Campbell, D.S. (1977). Dielectric properties of thin films. In: *Handbook of Thin Film Technology* (ed. L.I. Maissel and R. Glang), 433–477. New York, NY, 1970, translated in Russian by M.I. Elinson and G.G. Smolko, Sovetskoe Radio, Moskva, Russia, 1977: McGraw Hill (in Russian).

233 Segda, B.G., Jacquet, M., and Besse, J.P. (2001). Elaboration, characterization and dielectric properties study of amorphous alumina thin films deposited by r.f. magnetron sputtering. *Vacuum* 62: 27.

234 Hanby, B.V.T., Stuart, B.W., Grant, C. et al. (2018). Dielectric breakdown of alumina thin films produced by pulsed direct current magnetron sputtering. *Thin Solid Films* 662: 145.

235 Wei, A.X., Ge, Z.X., Zhao, X.H. et al. (2011). Electrical and optical properties of tantalum oxide thin films prepared by reactive magnetron sputtering. *J. Alloy. Compound.* 509: 9758.

236 Babuji, B., Balasubramanian, C., and Radhakrishnan, M. (1983). Dielectric properties of ion plated titanium oxide thin films, 1983. *J. Non-CrystaL Sol.* 55: 405.

237 Hontsu, S., Matsumoto, T., Ishi, J. et al. (1997). Electrical properties of hydroxyapatite thin films grown by pulsed laser deposition. *Thin Solid Films* 295: 214.

238 El Kamel, F., Gonon, P., and Jomni, F. (2006). Electrical properties of low temperature deposited amorphous barium titanate thin films as dielectrics for integrated capacitors. *Thin Solid Films* 504: 201.

239 Goldenberg, E., Bayrak, T., Ozgit-Akgun, C. et al. (2015). Effect of O_2/Ar flow ratio and post-deposition annealing on the structural, optical and electrical characteristics of $SrTiO_3$ thin films deposited by RF sputtering at room temperature. *Thin Solid Films* 590: 193.

240 Chou, C.-Y., Lin, C.-H., Chen, W.-H. et al. (2020). High-dielectric-constant silicon nitride thin films fabricated by radio frequency sputtering in Ar and Ar/N_2 gas mixture. *Thin Solid Films* 709: 138198.

241 Abbas, Q., Liang, H., Shi, J. et al. (2018). Growth and characterization of amorphous boron nitride dielectric films on Si via RF sputtering at room temperature. *Mater. Lett.* 227: 284.

242 Callister, W.D. Jr. and Rethwisch, D.G. (2008). *Fundamentals of Materials Science and Technology*, 3e. Hoboken, NJ: Wiley.

243 Pawłowski, L. (1988). The relationship between structure and dielectric properties in plasma sprayed alumina coatings. *Surf. Coat. Technol.* 35: 285.

244 Kim, H.J., Odoul, S., Lee, C.-H., and Kweon, Y.G. (2001). The electrical insulation behavior and sealing effects of plasma-sprayed alumina-titania coatings. *Surf. Coat. Technol.* 140: 293.

245 Ctibor, P., Sedáčcek, J., Musalek, R. et al. (2022). Structure and electrical properties of yttrium oxide sprayed by plasma torches from powders and suspensions. *Ceram. Int.* 4: 7464.

246 Ctibor, P., Petrášek, J., Sedláček, J., and Neufuss, K. (2020). Dielectric properties of $CaZrO_3$ coatings made by plasma spraying and bulks fired by spark plasma sintering. *Mater. Res. Bull.* 124: 110775.

247 Ctibor, P., Sedláček, J., and Pala, Z. (2015). Structure and properties of plasma sprayed $BaTiO_3$ coatings after thermal post-treatment. *Ceram. Int.* 41: 7453.

248 Xing, Z., Wang, H., Xu, B. et al. (2014). Structural integrity and ferroelectric–piezoelectric properties of $PbTiO_3$ coating prepared via supersonic plasma spraying. *Mater. Des.* 62: 57.

249 Haessler, W., Thielsch, R., and Mattern, N. (1995). Structure and electrical properties of PZT thick films produced by plasma spraying. *Mater. Lett.* 24: 387.

250 Chen, M.S. (1977). Ferromagnetic properties of thin films. In: *Handbook of Thin Film Technology* (ed. L.I. Maissel and R. Glang), 477–578. New York, NY, 1970, translated in Russian by M.I. Elinson and G.G. Smolko, Sovetskoe Radio, Moskva, Russia, 1977: McGraw Hill. (in Russian).

251 Du, J., Li, G., Wang, Q. et al. (2015). Microstructural evolution and magnetic properties of nanocrystalline Fe films prepared in a high magnetic field. *Vacuum* 121: 88.

252 Qin, X., Di, L., Sui, C. et al. (2020). Effects of the Mn/Bi ratio on the magnetic properties of MnBi thin films grown by magnetron co-sputtering. *J. Alloys Compd.* 842: 155694.

253 Xu, Z. and Qun, L. (2021). Effects of sputtering parameters and annealing temperatures on magnetic properties of CoFeB films. *J. Magn. Magn. Mater.* 538: 168302.

254 He, J., Zhang, Z., Bao, Z. et al. (2023). Effect of Cr interlayers on texture and magnetic properties of FeSi films with micrometer thickness. *Appl. Surf. Sci.* 610: 155334.

255 Lian, J., Chen, Y., Liu, Z. et al. (2017). Annealing effects on the microstructure and magnetic properties of $Y_3Fe_5O_{12}$ films deposited on Si/SiO_2 substrates by RF magnetron sputtering. *Ceram. Int.* 43: 7477.

256 Phokha, S., Limwichean, S., Horprathum, M. et al. (2020). Effect of annealing temperature on the structural and magnetic properties of CeO_2 thin films. *Thin Solid Films* 704: 138001.

257 Seema, A., Tayal, P., Gupta, S.C., and Gupta, M. (2022). Thickness-dependent structural and magnetic properties of Fe_4N thin films. *J. Magn. Magn. Mater.* 563: 169999.

258 Plekhanov, A.G., Maksimovskiy, E.A., Kriventsov, V.V. et al. (2022). Composition, structure, and magnetic properties of $SiC_xN_yFe_z$ films produced by the PECVD method from a gas mixture of 1,1,1,3,3,3-hexamethyldisilazane, ferrocene, and helium. *Colloids Surf. A Physicochem. Eng. Asp.* 655: 130333.

259 Andrews, C.W.D., Fuller, B.A., and Preece, I. (1974). Observation of arc-plasma sprayed ferrite films on alumina substrates. *J. Mater. Sci. Lett.* 9: 856.

260 Kumar, K. and Das, D. (1978). Equilibrium and metastable samarium-cobalt deposits produced by arc plasma spraying. *Thin Solid Films* 54: 263.

261 Cherigui, M., Guessasma, S., Fenineche, N. et al. (2005). Studies of magnetic properties of iron-based coatings produced by a high-velocity oxy-fuel process. *Mater. Chem. Phys.* 92: 419.

262 Dong, S., Song, B., Zhang, X. et al. (2014). Fabrication of FeSiB magnetic coatings with improved saturation magnetization by plasma spray and dry-ice blasting. *J. Alloys Compd.* 584: 254.

263 Ni, X., Bo, X., Wu, J. et al. (2011). Microstructure and magnetic properties of plasma-sprayed $Fe_{73.5}Cu_1Nb_3Si_{13.5}B_9$ coatings. *Thin Solid Films* 519: 8287.

264 Denisova, E., Kuzovnikova, L., Iskhakov, R. et al. (2015). Bulk CoNiFe-SiB Amorphous and nanostructured alloys produced by plasma spray deposition and dynamic compaction: formation of soft magnetic properties. *Phys. Procedia* 75: 1238.

265 Shinoda, K., Liang, S., Sampath, S., and Gambino, R.J. (2011). Processing effects on in-flight particle state and functional coating properties of plasma-sprayed manganese zinc ferrite. *Mater. Sci. Eng. B* 176: 22.

266 Miao, Y., Zhu, H., Gao, P., and Li, L. (2020). The effects of spraying power on microstructure, magnetic and dielectric properties of plasma sprayed cobalt ferrite coatings. *J. Mater. Res. Technol.* 9 (6): 14237.

267 Chen, C., Xie, X., Xie, Y. et al. (2018). Cold spraying of thermally softened Ni-coated FeSiAl composite powder: Microstructure characterization, tribological performance and magnetic property. *Mater. Des.* 160: 270.

268 Bednorz, J.G. and Müller, K.A. (1986). Possible high *Tc* superconductivity in the Ba-La-Cu-O system. *Zeitschrift fur Physik B, Condensed Matter* 64 (2): 189.

269 Phillips, J.C. (1989). *Physics of High Tc Superconductors* Academic Press. San Diego, CA.

270 Pawłowski, L. (2018). Synthesis, properties and applications of $YBa_2Cu_3O_{7-x}$. In: *Industrial Chemistry of Oxides for Emerging Applications* (ed. L. Pawłowski and P. Blanchart), 211–253. Chichester, UK: Wiley.

271 Badía-Majós, A., Martínez, E., Angurel, L.A. et al. (2024). Laser nanostructured metasurfaces in Nb superconducting thin films. *Appl. Surf. Sci.* 649: 159164.

272 Sayeed, M.N., Pudasaini, U., Eremeev, G.V., and Elsayed-Ali, H.E. (2023). Fabrication of superconducting Nb_3Sn film by co-sputtering. *Vacuum* 212: 112019.

273 Misra, D.S., Padalia, B.D., Pai, S.P. et al. (1994). Microstructure, flux pinning and critical current density in $YBa_2Cu_3O_{7-\delta}$ films grown by laser ablation. *Thin Solid Films* 245: 186.

274 Guo, M., Li, J., and Jia, J. (2023). Growth of $YBa_2Cu3O_{7-\delta}$ superconducting films under different atmospheres by pulsed laser deposition. *Phys. C: Supercond. Appl.* 605: 1354206.

275 Morone, A., Di Trolio, A., Parisi, G.P., and Gambardella, U. (2000). Some results on the SmBaCuO thin films formation by off axis pulsed laser deposition. *Physica C: Superconductivity* 341–348: 2351.

276 Tsunekawa, S., Watanabe, H.F.J., and Take, H. (1984). Linear thermal expansion of $SrTiO_3$. *Phys. Status Solidi* 83: 467.

277 Pawłowski, L., Hill, A., McPherson, R. et al. (1991). Properties of plasma sprayed $YBa_2Cu_3O_x$ high temperature superconductors. In: *Thermal Spray Research and Applications* (ed. T.F. Bernecki), 641–646. Materials Park, OH: ASM International.

278 Tachikawa, K., Watanabe, I., Kosuge, S. et al. (1988). High T_c superconducting films of Y-Ba-Cu oxides prepared by low-pressure plasma spraying. *Appl. Phys. Lett.* 52: 1011.

279 Kumar, S., Dhavale, A.S., Chavan, N.M., and Acharya, S. (2022). Superconducting niobium coating deposited using cold spray. *Mater. Lett.* 312: 137715.

280 Silva, S.R.P., Carey, J.D., Guo, X. et al. (2005). Electron field emission from carbon-based materials. *Thin Solid Films* 482: 79.

281 Waldhauser, W., Mitterer, C., Laimer, J., and Stori, H. (1998). Sputtered thermionic hexaboride coatings. *Surf. Coat. Technol.* 98: 1315.

282 Olesik, J. and Małachowski, M. (2002). Field- and optically induced electron emission from tin oxide films. *Thin Solid Films* 422: 193.

283 Wang, Y.X., Li, Y.A., Feng, W. et al. (2005). Influence of thickness on field emission characteristics of AlN thin films. *Appl. Surf. Sci.* 243: 394.

284 Sugino, T., Kimura, C., and Aoki, H. (2008). Surface properties and field emission of boron nitride. *Diam. Relat. Mater.* 17: 1764.

285 Wu, H., Wang, J., Songa, Z. et al. (2012). Microstructural and electron-emission characteristics of Nb-Si-N films in surface-conduction electron-emitter display. *Phys. Procedia* 32: 139.

286 Gabdullin, P., Zhurkin, A., Osipov, V. et al. (2020). Thin carbon films: correlation between morphology and field-emission capability. *Diam. Relat. Mater.* 105: 107805.

287 Znamirowski, Z., Czarczynski, W., Le Maguer, A., and Pawłowski, L. (2003). Plasma sprayed and laser engraved field electron emitters. *Surf. Coat. Technol.* 165: 211.

288 Znamirowski, Z., Pawłowski, L., Cichy, T., and Czarczynski, W. Low macroscopic field electron emission from surface of plasma sprayed and laser engraved TiO2, Al_2O_3+13TiO_2 and Al_2O_3+40TiO_2 coatings. *Surf. Coat. Technol.* 187: 37.

289 Tomaszek, R., Znamirowski, Z., Pawłowski, L., and Zdanowski, J. (2007). Effect of conditioning on field electron emission of suspension plasma sprayed TiO_2 coatings. *Vacuum* 81: 1278.

290 Znamirowski, Z., Posadowski, W., Pawłowski, L. et al. (2015). Electron emission from the zirconium coated suspension plasma sprayed bioglass. *Surf. Coat. Technol.* 268: 63.

291 Byron Bird, K., Stewart, W.E., and Lightfood, E.N. (2007). *Transport Phenomena*, 2e. New York, NY: Wiley.

292 Wu, Z.L., Kuo, P.K., Wei, L. et al. (1993). Photothermal characterization of optical thin films. *Thin Solid Films* 236: 191.

293 Qin, W., Fu, L., Zhu, J. et al. (2018). Electrical and thermal properties of Cu-Ta films prepared by magnetron sputtering. *Appl. Surf. Sci.* 443: 97.

294 Guenther, A.H. and Mciver, J.K. (1988). The role of thermal conductivity in the pulsed laser damage sensitivity of optical thin film. *Thin Solid Films* 163: 203.

295 Janga, B.-K., Sakka, Y., Yamaguchi, N. et al. (2011). Thermal conductivity of EB-PVD ZrO_2–4 mol. % Y_2O_3 films using the laser flash method. *J. Alloys Compd.* 509: 1045.

296 Alwi, H.A. and Abd-Shukor, R. (2011). Measurement of thermophysical properties of $YBa_2Cu_3O_{7-x}$ thin film using picosecond thermoreflectance technique. *Physica C* 471: 26.

297 Souissia, M., Ghrib, T., Al-Otaibi, A. et al. (2019). Structural, optical and thermal properties of V-doped GaN thin films grown by *MOCVD* technique. *Thermochim. Acta* 682: 178428.

298 Martinson, C. and Heuret, M. (1998). Recent progress in the measurement of the thermal properties of hard coatings. *Thin Solid Films* 317: 455.

299 Leung, K.M., Cheung, A.C., Liu, B.C. et al. (1999). Measuring thermal conductivity of CVD diamond and diamond-like films on silicon substrates by holographic interferometry. *Diam. Relat. Mater.* 8: 1607.

300 Alwi, H.A., Kim, Y.Y., Awang, R. et al. (2013). Measurement of thermophysical properties of hydrogenated amorphous carbon thin films using picosecond thermoreflectance technique. *Int. J. Heat Mass Transf.* 63: 199.

301 Pawłowski, L. and Fauchais, P. (1992). Thermal transport properties of thermally sprayed coatings. *Int. Mater. Rev.* 37 (6): 271.

302 Pawłowski, L., Lombard, D., and Fauchais, P. (1985). Structure-thermal properties relationship in plasma sprayed zirconia coatings. *J. Vac. Sci. Technol. A* 3 (6): 2494.

303 Sokołowski, P., Kozerski, S., Pawłowski, L., and Ambroziak, A. (2014). The key process parameters influencing formation of columnar microstructure in suspension plasma sprayed zirconia coatings. *Surf. Coat. Technol.* 260: 97.

304 Sivakumar, S., Praveena, K., Shanmugavelayuthama, G. et al. (2017). Thermo-physical behavior of atmospheric plasma sprayed high porosity lanthanum zirconate coatings. *Surf. Coat. Technol.* 326: 173.

305 Sokołowski, P., Łatka, L., Pawłowski, L. et al. (2015). Characterization of microstructure and thermal properties of YCSZ coatings obtained by suspension plasma spraying. *Surf. Coat. Technol.* 268: 147.

306 Moskal, G., Swadźba, L., Hetmańczyk, M. et al. (2012). Characterization of microstructure and thermal properties of $Gd_2Zr_2O_7$-type thermal barrier coating. *J. Eur. Ceram. Soc.* 32: 2025.

307 L. Pawłowski, C. Martin, and P. Fauchais (1983). The application of infrared thermography in testing the coatings and optimizing the plasma spraying process [in:] *10th ITSC*, 2–6 may 1983, Essen, Germany, DVS Verlag, Düsseldorf, Germany, 1983, pp. 31–36.

308 Samadi, H. and García, E. (2014). Thermal conductivity of plasma sprayed forsterite/mullite coatings. *Ceram. Int.* 40: 13995.

309 Neufuss, K. and Rudajevova, A. (2002). Thermal properties of the plasma-sprayed $MgTiO_3$–$CaTiO_3$ and $CaTiO_3$. *Ceram. Int.* 28: 93.

310 K.E. Wilkes and J.F. Lagedrost (1973). Thermophysical properties of plasma sprayed coatings, *NASA Rapport*, no. CR-121144.

311 Brandt, R., Pawłowski, L., Neuer, G., and Fauchais, P. (1986). Specific heat and thermal conductivity of plasma sprayed yttria stabilized zirconia and NiAl, NiCr, NiCrAl, NiCrAlY, NiCoCrAlY coatings. *High Temper. High Press.* 18: 65.

312 Pawłowski, L., Lombard, D., Tourenne, F. et al. (1985). Thermal diffusivity of plasma sprayed NiAl, NiCr, NiCrAl, NiCrAlY, and NiCoCrAlY coatings. *High Temper. High Press.* 17: 611.

313 Abu-warda, N., Boissonnet, G., López, A.J. et al. (2021). Analysis of thermo-physical properties of NiCr HVOF coatings on T24, T92, VM12 and AISI 304 steels. *Surf. Coat. Technol.* 416: 127163.

314 Khor, K.A. and Gu, Y.W. (2000). Thermal properties of plasma-sprayed functionally graded thermal barrier coatings. *Thin Solid Films* 372: 104.

315 Vjunitsky, I., Bandyopadhyay, P.P., St. Siegmann, M.D. et al. (2005). Thermophysical properties and deposition of B2 structure based Al–Ni–Ru–M alloys. *Surf. Coat. Technol.* 192: 131.

316 McPherson, R. (1984). A model for thermal conductivity of plasma sprayed ceramic coatings. *Thin Solid Films* 112: 89.

317 Xu, H., Gong, S., and Deng, L. (1998). Preparation of thermal barrier for gas turbine blades by EB-PVD. *Thin Solid Films* 334: 98.

318 Gitzhofer, F., Pawłowski, L., Lombard, D. et al. (1985). The apparent thermal diffusivity and thermal contact resistance in plasma sprayed multicoatings. *High Temper. High Press* 18: 563.

319 Incropera, F.P., Dewitt, D.P., Bergman, T.L., and Lavine, A.S. (2007). *Fundamentals of Heat and Mass Transfer*, 724–809. Hoboken, NJ: Wiley.

320 Gaussorgues, G. (1981). *La Thermographie Infrarouge*, 21–59. Paris, France: Technique et Documentation (in French).

321 Huang, Z., Zhu, D., Lou, F., and Zhou, W. (2008). An application of Au thin-film emissivity barrier on Ni alloy. *Appl. Surf. Sci.* 255: 2619.

322 Xu, K., Hao, L., Du, M. et al. (2020). Thermal emittance of Ag films deposited by magnetron sputtering. *Vacuum* 174: 109200.

323 Brandt, R., Neuer, G., and Wörner, B. (1979). Thermophysical properties of plasma sprayed materials. *Rev. Int. Hautes Tempér. Réfract.* 16: 309.

324 Yin, J., Zhang, L., Ma, W. et al. (2023). Research on reducing infrared emissivity of 8YSZ coating by regulating microstructure. *Infrared Phys. Technol.* 130: 104587.

325 Xu, M., Yuan, J., Lu, X. et al. (2022). Infrared radiation and thermal cyclic performance of a high-entropy rare-earth hexaaluminate coating prepared by atmospheric plasma spraying. *Ceram. Int.* 48: 26003.

326 Mao, J., Ding, S., Lia, Y. et al. (2019). Preparation and investigation of $MoSi_2$/SiC coating with high infrared emissivity at high temperature. *Surf. Coat. Technol.* 358: 873.

327 Tului, M., Arezzo, F., and Pawłowski, L. (2004). Optical properties of plasma sprayed $ZnO+Al_2O_3$ coatings. *Surf. Coat. Technol.* 179: 47.

328 Ailor, W.H. (1971). *Handbook on Corrosion Testing and Evaluation*. New York, NY: John Wiley.

329 Perrin, R. and Scharff, J.-P. (1999). *Chimie Industrielle*. Paris, France (in French): Dunod.

330 Kenneth Chiang, K.-T. and Wei, R. (2011). Growth morphology and corrosion resistance of magnetron sputtered Cr films. *Surf. Coat. Technol.* 206: 1660.

331 Mukaida, M., Osato, K., Watanabe, A. et al. (1993). Densification of Ta_2O_5 film prepared by KrF excimer laser CVD. *Thin Solid Films* 232: 180.

332 Rojas, P.N. and Rodil, S.E. (2012). Corrosion behaviour of amorphous niobium oxide coatings. *Int. J. Electrochem. Sci.* 7: 1443.

333 Abegunde, O., Akilabi, E., and Oladijo, P. (2021). Examination of electrochemical corrosion properties of titanium carbide thin film grown by RF magnetron sputtering. *Mater. Today: Proc.* 44: 994.

334 Cunha, L., Andritschky, M., Rebouta, L., and Silva, R. (1998). Corrosion resistance of TiN, (TiAl) N and CrN hard coatings produced by magnetron sputtering. *Thin Solid Films* 317: 351.

335 Matthes, B., Broszeit, E., and Kloos, K.H. (1993). Tribological behavior and corrosion performance of Ti-B-N hard coatings under plastic manufacturing conditions. *Surf. Coat. Technol.* 57: 97.

336 Diab, M., Pang, X., and Jahed, H. (2017). The effect of pure aluminum cold spray coating on corrosion and corrosion fatigue of magnesium (3% Al-1% Zn) extrusion. *Surf. Coat. Technol.* 309: 423.

337 Lapushkina, E., Yuan, S., Mary, N. et al. (2020). Contribution in optimization of Zn cold-sprayed coating dedicated to corrosion applications. *Surf. Coat. Technol.* 400: 126193.

338 Zhao, L., Lugscheider, E., Fischer, A., and Reimann, A. (2001). Thermal spraying of a high nitrogen duplex austenitic-ferritic steel. *Surf. Coat. Technol.* 141: 208.

339 Yan, D., He, J., Wu, J. et al. (1997). The corrosion behavior of a plasma spraying Al_2O_3 ceramic coating in dilute HCl solution. *Surf. Coat. Technol.* 89: 191.

340 Yan, D., He, J., Li, X. et al. (2001). An investigation of the corrosion behavior of Al_2O_3 – based ceramic composite coatings in dilute HCl solution. *Surf. Coat. Technol.* 141: 1.

341 Wei, Z., Qin, J., Wang, K., and Hong, S. (2023). Long-term salt spray corrosion behaviors of HVOF-sprayed WC-Cr_3C_2-Ni coatings before and after vacuum sealing with silicone resin. *Int. J. Refract. Met. Hard Mater.* 117: 106388.

342 Almeida, A., Anjos, M., Vilar, R. et al. (1995). Laser alloying of aluminum alloys with chromium. *Surf. Coat. Technol.* 70: 221.

343 Cheng, J.B., Liang, X.B., and Xu, B.S. (2014). Effect of Nb addition on the structure and mechanical behaviors of CoCrCuFeNi high-entropy alloy coatings. *Surf. Coat. Technol.* 240: 184.

344 Sekyi Appiah, A.N., Wyględacz, B., Matus, K. et al. (2024). Microstructure and performance of NiCrBSi coatings prepared by modulated arc currents using powder plasma transferred arc welding technology. *Appl. Surf. Sci.* 648: 159065.

345 Khanna, A.S. and Sridhar, K. (1998). Corrosion and oxidation behavior of laser treated surfaces. In: *Lasers in Surface Engineering* (ed. N.B. Dahotre), 395–431. Materials Park, OH: ASTM International.

346 Rajamure, R.S., Vora, H.D., Srinivasan, S.G., and Dahotre, N.B. (2015). Laser alloyed Al-W coatings on aluminum for enhanced corrosion resistance. *Appl. Surf. Sci.* 328: 205.

347 Anjos, M.A., Vilar, R., Li, R. et al. (1995). Fe-Cr-Ni-Mo-C alloys produced by laser surface alloying. *Surf. Coat. Technol.* 70: 235.

348 Hocking, M.G., Vasantasree, V., and Sidky, P.S. (1989). *Metallic and Ceramic Coatings*. Burnt Mill, Essex, England: Longman Scientific & Technical.

349 Du, H.L., Datta, P.K., Burnell-Gray, J.S. et al. (1995). Sulfidation behavior of Nb and Nb+Ta double layer plasma sprayed coatings. In: *Advances in Inorganic Films and Coatings* (ed. P. Vincenzini), 599–607. Faenza, Italy: Techna.

350 Hodge, P.E., Stecura, S., Gedwill, M.E. et al. (1980). Thermal barrier coatings: burner rig hot corrosion test results. *J. Mater. Energy Systems* 1: 47.

351 Birks, N., Meier, G.H., and Pettit, F.S. (1996). Degradation of coatings by high temperature atmospheric corrosion and molten salt deposits. In: *Metallurgical and Ceramic Protective Coatings* (ed. K.H. Stern), 290–305. London: Chapman & Hall.

352 Pierson, H.O. (1999). *Handbook of Chemical Vapor Deposition*, 2e. Norwich, NY: Noyes Publications.

353 Datta, P.K., Du, H.L., Pischow, K.A., and Ristolainen, E.O. (1995). Degradation of TiN and CrTiN coated AISI 304 stainless steel in an environment of low oxygen and high sulfur potentials at 750°C. In: *Advances in Inorganic Films and Coatings* (ed. P. Vincenzini), 591–598. Faenza, Faenza, Italy: Techna.

354 Buchanan, F.J. and Little, J.A. (1991). The growth and morphology of chemically vapor deposited carbide coatings on carbon-carbon composites. *Surf. Coat. Technol.* 46: 217.

355 Ren, X. and Wang, F. (2006). High-temperature oxidation and hot-corrosion behavior of a sputtered NiCrAlY coating with and without aluminizing. *Surf. Coat. Technol.* 201: 30.

356 Zhu, S., Wang, F., Lou, H., and Wu, W. (1995). Reactive sputter deposition of alumina films on superalloys and their high-temperature corrosion resistance. *Surf. Coat. Technol.* 71: 9.

357 Biava, G., Siqueira, I.B.A.F., Vaz, R.F. et al. (2022). Evaluation of high temperature corrosion resistance of CrN, AlCrN, and TiAlN arc evaporation PVD coatings deposited on Waspaloy. *Surf. Coat. Technol.* 438: 128398.

358 Haynes, J.A., Rigney, E.D., Ferber, M.K., and Porter, W.D. (1996). Oxidation and degradation of plasma sprayed thermal barrier coating system. *Surf. Coat. Technol.* 86–87: 102.

359 Brandl, W., Thoma, D., Krüger, J. et al. (1997). The oxidation behavior of HVOF thermal sprayed MCrAlY coatings. *Surf. Coat. Technol.* 94–95: 21.

360 Kamalan Kirubakaran, A.M., Anderson, A., Rajasekaramoorthy, M. et al. (2020). Hot corrosion behavior of plasma sprayed YSZ and gadolinium oxide doped YSZ thermal barrier coatings. *Mater. Today. Proc.* 33: 954.

361 Kandasamy, P., Kandala, B., Lee, M.W., and Sivakumar, G. (2024). Phase stability and initial phase high-temperature corrosion behavior of non-stoichiometric lanthanum cerium oxide thermal barrier coatings. *Ceram. Int.* 50: 14458.

362 Tsai, H.L. and Tsai, P.C. (1995). Performance of laser-glazed plasma sprayed (ZrO_2-12 wt. % Y_2O_3)/(Ni-22 wt. % Cr-10 wt. % Al-1 wt. % Y) thermal barrier coatings in cyclic oxidation test. *Surf. Coat. Technol.* 71: 53.

363 Khanna, A.S., Pattnaik, A.K., and Wissenbach, K. (1995). Oxidation and corrosion behavior of mild steel plasma coated followed by plasma treatment. In: *Thermal Spraying Current Status and Future Trends* (ed. A. Ohmori), 993–999. Osaka, Japan: High Temperature Society of Japan.

364 Fernandes, F., Cavaleiro, A., and Loureiro, A. (2012). Oxidation behavior of Ni-based coatings deposited by PTA on gray cast iron. *Surf. Coat. Technol.* 207: 196.

365 Zaplatynsky, I. (1982). Performance of laser-glazed zirconia thermal barrier coatings in cyclic oxidation and corrosion burner rig test. *Thin Solid Films* 95: 275.

366 Piegari, A. and Flory, F. (ed.) (2018). *Optical Thin Films and Coatings*, 2e. Oxford, UK: Woodhead.

367 Martinu, L. and Poitras, D. (2000). Plasma deposition of optical films and coatings: a review. *J. Vac. Sci. Technol. A* 18 (6): 2619.

368 Ly, B.C.K., Dyer, E.B., Feig, J.L. et al. (2020). Research techniques made simple: cutaneous colorimetry: a reliable technique for objective skin color measurement. *J. Invest. Dermatol.* 140: 3.

369 Beck, U., Reiners, G., and Witt, K. (1993). Decorative hard coatings: advances in optical characterization techniques. *Thin Solid Films* 236: 184.

370 Reiners, G., Beck, U., and Jehn, H.A. (1994). Decorative optical coatings. *Thin Solid Films* 253: 33.

371 Beck, U., Reiners, G., Kopacz, U., and Jehn, H.A. (1993). Decorative hard coatings: interdependence of optical, stoichiometric and structural properties. *Surf. Coat. Technol.* 60: 389.

372 Tikkanen, J., Eerola, M., and Rajala, M. (1994). Coating glass by flame spraying. *J. Non-Cryst. Solids* 178: 220.

373 Dosta, S., Torrell, M., Cano, I.G., and Guilemany, J.M. (2012). Functional colored ceramic coatings obtained by thermal spray for decorative applications. *J. Eur. Ceram. Soc.* 32: 3685.

374 Alone-Alaluf, M., Appelbaum, J., and Croituru, N. (1998). Properties of GaAs solar cells coated with diamondlike carbon films. *Thin Solid Films* 320: 159.

375 Bodurova, G., Ivanova, T., Abrashev, M. et al. (2013). Thin film optical coatings of Vanadium Oxide and mixed Tungsten/Vanadium Oxide deposited by APCVD employing precursors of Vanadyl Acetylacetonate and a mixture with tungsten hexacarbonyl. *Phys. Procedia* 46: 127.

376 Trujillo, O., Moss, R., Vuong, K.D. et al. (1996). CdS thin film deposition by CW Nd: YAG laser. *Thin Solid Films* 290–291: 13.

377 Ai, W. and Xiong, S. (2022). Structural and optical properties of Nb_2O_5 films prepared by dual ion assisted deposition. *Opt. Laser Technol.* 150: 107850.

378 Davis, L. (1993). Properties of transparent conducting oxides deposited at room temperature. *Thin Solid Films* 236: 1.

379 Vávra, I., Križanová, Z., Dérer, J., and Humlíček, J. (2012). Mo/SiO_2 nanocomposite films for optical coatings prepared by vacuum magnetron sputtering. *Vacuum* 86: 742.

380 Khoshman, J.M., Manda, A.A., and Kordesch, M.E. (2015). Near-infrared optical constants and optical polarization properties of ZnO thin films. *Thin Solid Films* 578: 139.

381 Sarma, B.K., Pal, A.R., Bailung, H., and Chutia, J. (2012). Role of ion energy on growth and optical dispersion of nanocrystalline TiO_2 films prepared by magnetron sputtering with ion assistance at the substrate. *Appl. Surf. Sci.* 258: 5659.

382 Thomsen-Schmidt, P., Schäfer, D., Steiger, B. et al. (1994). Laser- assisted deposition of optical coatings. *Thin Solid Films* 253: 28.

383 Nandi, S.K. and Biswas, S. (2013). *In vivo* characterization of biomaterials. In: *Characterization of Biomaterials* (ed. A. Bandyopadhyay and S. Bose), 255–297. Waltham, MA: Elsevier.

384 Wu, X., Han, H., Jiang, Y. et al. (2024). Opportunities and challenges of the nitride coatings for artificial implants: a review. *Surf. Coat. Technol.* 480: 130587.

385 Thanka Rajan, S., Subramanian, B., and Arockiarajan, A. (2022). A comprehensive review on biocompatible thin films for biomedical application. *Ceram. Int.* 48: 4377.

386 Song, X., Zhao, M., and Li, D. (2022). Controllable Ag/Ta ratios of co-implanted TiN films on titanium alloys for osteogenic enhancement and antibacterial responses. *Surf. Coat. Technol.* 436: 128294.

387 Cemin, F., Artico, L.L., Piroli, V. et al. (2022). Superior *in vitro* biocompatibility in NbTaTiVZr(O) high-entropy metallic glass coatings for biomedical applications. *Appl. Surf. Sci.* 596: 153615.

388 Leonard, R.L., Bull, A.B., Xue, F. et al. (2023). Biocompatibility of antifogging SiO-doped Diamond-Like carbon laparoscope coatings. *Appl. Surf. Sci.* 634: 157606.

389 Chen, Z., Zhai, J., Wang, D., and Chen, C. (2018). Bioactivity of hydroxyapatite/wollastonite composite films deposited by pulsed laser. *Ceram. Int.* 44: 10204.

390 Marciano, F.R., Bonetti, L.F., Santos, L.V. et al. (2009). Antibacterial activity of DLC and Ag–DLC films produced by PECVD technique. *Diam. Relat. Mater.* 18: 1010.

391 Unabia, R.B., Candidato, R.T. Jr., Pawłowski, L. et al. (2020). In vitro studies of solution precursor plasma-sprayed copper-doped hydroxyapatite coatings with increasing copper content. *J. Biomed. Mater. Res.* 108B: 2579.

392 Yang, Y.-C., Chen, C.-C., Wang, J.-B. et al. (2017). Flame sprayed zinc doped hydroxyapatite coating with antibacterial and biocompatible properties. *J. Biomed. Mater. Res.* 43: S829.

393 Singh, J., Singh Chatha, S., and Singh, H. (2021). Characterization and corrosion behavior of plasma sprayed calcium silicate reinforced hydroxyapatite composite coatings for medical implant applications. *Ceram. Int.* 47: 782.

394 Garrido, B., Sara Rodrigo-Vázquez, C., Cano, I.G. et al. (2024). Plasma-sprayed coatings of 62W bioactive glass: mechanical properties and in vitro behavior. *Surf. Coat. Technol.* 476: 130269.

395 T.-Y. Kuo, W.-H. Chin, C.-S. Chien, and Y.–H. Hsieh, 2019, Mechanical and biological properties of graded porous tantalum coatings deposited on titanium alloy implants by vacuum plasma spraying, *Surf. Coat. Technol.*, 372, 399.

396 Ratha, I., Datta, P., Balla, V.K. et al. (2021). Effect of doping in hydroxyapatite as coating material on biomedical implants by plasma spraying method: a review. *Ceram. Int.* 47: 4426.

397 Sergi, R., Bellucci, D., Candidato, R.T. Jr. et al. (2018). Bioactive Zn-doped hydroxyapatite coatings and their antibacterial efficacy against Escherichia coli and Staphylococcus aureus. *Surf. Coat. Technol.* 352: 84.

398 Fielding, G.A., Roy, M., Bandyopadhyay, A., and Bose, S. (2012). Antibacterial and biological characteristics of silver containing and strontium doped plasma sprayed hydroxyapatite coatings. *Acta Biomater.* 8: 3144.

399 Cattini, A., Bellucci, D., Sola, A. et al. (2014). Functional bioactive glass topcoats on hydroxyapatite coatings: analysis of microstructure and in-vitro bioactivity. *Surf. Coat. Technol.* 240: 110.

400 Gkomoza, P., Lampropoulos, G.S., Vardavoulias, M. et al. (2019). Microstructural investigation of porous titanium coatings, produced by thermal spraying techniques, using plasma atomization and hydridedehydride powders, for orthopedic implants. *Surf. Coat. Technol.* 357: 947.

Index

Physical Deposition Methods for Films and Coatings, First Edition. Lech Pawłowski.

f

m

u

v

Printed in the USA/Agawam, MA
September 17, 2026

910661.168